STRATEGIC AND TACTICAL CONSIDERATIONS ON THE FIREGROUND

Fourth Edition

James P. Smith

Deputy Chief, Philadelphia Fire Department (Ret.)
Adjunct Instructor, National Fire Academy

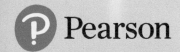

330 Hudson Street, NY, NY 10013

Publisher: Julie Levin Alexander
Publisher's Assistant: Sarah Henrich
Senior Acquisitions Editor: Sladjana Repic Bruno
Editorial Assistant: Lisa Narine
Development Editor: Susan Beach
Director, Publishing Operations: Paul DeLuca
Team Lead, Program Management: Melissa Bashe
Team Lead, Project Management: Cynthia Zonneveld
Manufacturing Buyer: Maura Zaldivar-Garcia
Vice President of Sales & Marketing: David Gesell
Vice President, Director of Marketing: Margaret Waples
Senior Field Marketing Manager: Brian Hoehl
Senior Producer: Amy Peltier
Media Producer and Project Manager: Lisa Rinaldi
Full-Service Project Manager: Monica Moosang, iEnergizer Aptara®, Ltd.
Composition: iEnergizer Aptara®, Ltd.
Printer/Binder: LSC Communications
Cover Printer: LSC Communications
Cover Image: Photograph by Greg Masi, used by permission of the City of Philadelphia
Text Font: Sabon LT Std, 10/12

Credits and acknowledgments borrowed from other sources and reproduced, with permission, in this textbook appear on appropriate pages within the text. Unless otherwise stated, all photos have been provided by the authors.

Library of Congress Cataloging-in-Publication Data
Names: Smith, James P. (James Patrick), 1946- author.
Title: Strategic and tactical considerations on the fireground/James P.
    Smith, Deputy Chief, Philadelphia Fire Department(Ret.), Adjunct
    Instructor, National Fire Academy.
Description: Fourth edition. | New York, NY : Pearson, [2017] | Includes
    bibliographical references and index.
Identifiers: LCCN 2016027411
Subjects: LCSH: Command and control at fires. | Fire extinction.
Classification: LCC TH9310.8 .S65 2017 | DDC 363.37—dc23 LC record available at
    https://lccn.loc.gov/2016027411

2011014765

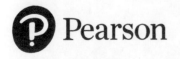 **Pearson**

ISBN 10:   0-13-444264-4
ISBN 13: 978-0-13-444264-8

20 2022

DEDICATION

This book would never have been started without the love and support of my wife and best friend, Pat; my son, Jim, chief of the Ocean City, New Jersey, Fire Department; my lovely daughter, Colleen; and my grandchildren, Ashley, Tyler, Justin, Brendan, and Sarah. I dedicate this book to my parents, James P. and Caroline M. Smith. Mom and Dad taught me through their words and actions: "Love and family above everything else."

CONTENTS

Chapter 1 Preparation 1

Chapter 2 Management Tools 36

Chapter 3 Decision Making 82

Chapter 8 Health Care and High-Risk Populations 319

Chapter 9 Commercial and Industrial 394

Chapter 10 Technical Operations 457

Chapter 11 After the Incident 526

I am honored to have been asked by my good friend and mentor to write the foreword for the newest edition of *Strategic and Tactical Considerations on the Fireground*. There is no doubt that the previous three editions have prepared countless current fire service leaders for the operational challenges faced daily in the United States by the fire service. Chief Smith and I have known each other for more than 25 years. As members of the Philadelphia Fire Department, we have supported each other through numerous challenges in administrative and operational roles. Chief Smith admirably continues to present the knowledge necessary to change critical behaviors commensurate with varying and complicated risks.

The first step in the emergency management process is identifying the risks we face, whether natural, manmade, or technological in type. The road to *avoidance* is *awareness*. The threats to both the public and first responders at the emergency scene continue to evolve. In the fourth edition, Chief Smith expands on an already comprehensive body of knowledge to include information that is changing operational approaches that have been accepted practices for decades in the fire service. An example of this is information in Chapter 3, where UL and NIST testing on ventilation and flow paths have led to the development of the *transitional offensive attack*.

Strategic and Tactical Considerations on the Fireground continues its tradition of providing solid, tested, and rational information for the emergency scene manager, whether he or she is an experienced commander or a newly frocked chief. This authoritative text takes the reader from preincident preparation through the incident management systems and decision making processes, all the while exposing the reader to almost every tactical scenario that could possibly be encountered. The information offered is the result of exhaustive research by the author and his own hard-won experience garnered from operating in one of the largest fire departments in the United States for more than 40 years.

To the fire service member who is about to begin reading this volume, I can confidently state that you will be safer with each chapter that you absorb. The knowledge contained within can only help you make better, more informed, and safer critical decisions. I thank Chief Smith for collecting and sharing this knowledge that we all may be safer as we carry out our mission and duties.

Chief John McGrath
Raleigh Fire Department Raleigh,
North Carolina

This book is written for firefighters, fire officers, and chief officers who hope to improve their firefighting skills. It can be utilized in college fire science courses, or at a fire training academy. It can be beneficial for a candidate preparing for promotion to company or chief officer.

What makes this book better than the other firefighting books? It has been written for firefighters. It is written by one person, not a group of individuals who discuss ideas to produce a book without a consistent theme. It discusses, in easily understood language, what needs to be accomplished for successful fireground operations. It contains systems that prepare a firefighter for handling minor and major fires and emergencies. Other textbooks discuss tactics, but go no further. *Strategic and Tactical Considerations on the Fireground, Fourth Edition,* reviews strategic decisions as well as tactical operations.

This book is a compilation of 41 years of experience in a large metropolitan fire department by a firefighter and fire officer who achieved every civil service rank and held the highest rank of deputy chief for more than 20 years. It is not based upon theory found on library bookshelves. It is a lifetime of experiences, written so that other firefighters can benefit from those experiences. It is further bolstered by more than 30 years of teaching in the classroom to students throughout the United States. This interaction has allowed Chief Smith not only to share his information but to expand his knowledge on firefighting strategies and situations from other experts.

New to this Edition

The success of the previous three editions, as well as the need to include new information since the printing of the third edition in 2011, prompted the publication of this fourth edition. Changes to this edition include the following:

- Chapter 1 now includes Safe Operation of Fire Department Apparatus, which has been moved from the appendices.
  - Shares new research findings and information developed by the Underwriters Laboratories (UL) and the National Institute of Standards and Technology (NIST) on the effects of ventilation on a fire.
- Chapter 2 now contains Emergency Operating Centers (EOCs), which has been moved from the appendices.
- Chapter 3 discusses an Agency Administrator's briefing;
  - Discusses the UL and NIST tests on how flow paths occur and their impact on a fire, as well as the use of an offensive-exterior fire attack in dwelling fires; and
  - Discusses ICS Form 208 Safety Message/Plan.
- Chapter 4 discusses the needed change from Vent, Enter, and Search (VES) to Vent, Enter, Isolate, and Search (VEIS).
- Chapter 6 offers an expanded discussion on the Safety Officer's role;
  - Near miss information;
  - Electrical hazards and firefighting safety; and
  - Additional discussion on accountability and Maydays.
- Chapter 7 discusses new basement fire information;
  - Updated information on wildland urban interface fires; and
  - Numerous USFA statistics gleaned from the National Fire Incident Reporting System on residential fires, vacant building fires, and civilian fatalities in residential fires.
- Chapter 8 expands from acknowledgment and discussion of active shooter incidents in schools to other active shooter events, including churches, movie theaters, government and military installations, and so on.
- Chapter 10 discusses changes in the handling of hazardous materials since the United States adopted the Globally Harmonized System (GHS). It looks at the impact of

GHS on responding firefighters and clarifies the difference between the previously used Material Safety Data Sheets (MSDS) and the current Safety Data Sheets (SDS);
- ■ Looks at the impact of some recent hurricanes and superstorms; and
- ■ Now includes information on derechos under Natural Disasters.
- ■ Chapter 11 introduces a format that can be used for a formal critique.

The reader will also find more On Scene stories. These encapsulated real-life scenes allow readers to share the author's experiences as well as significant incidents that have occurred in other locales. They are not for storytelling purposes or entertainment. Rather, these real-life occurrences permit readers to increase their cue-based decision making skills, skills normally accumulated through training or during actual situations at an incident scene.

The fourth edition also includes more current summaries of many NIOSH investigative reports of firefighter fatalities. Listed is the report number, which can be easily accessed online in Appendix B to read the entire report. Realize that the inclusion of these reports is not intended to point fingers on the reasons for these firefighter deaths, but to learn from what occurred on those incidents and ensure that during our training and fireground operations we operate in a safe manner.

Finally, online student resources can be found at www.pearsonhighered.com/bradyresources, where students can access additional skills and information for more practice and review.

Using this Book

The purpose of writing this book was to share the many things I have been taught or learned the hard way during my career. We as firefighters work in an uncontrolled environment so very different than other occupations. Through training, we are taught the prerequisites needed to perform our jobs. Yet once the alarm is received, there are many factors that can be learned only by experience. Since there are fewer fires today, we must continually train to keep our skills sharp. I hope the information in this book will minimize the number of surprises that you will find on the fireground.

The first chapter of the book covers preincident factors: training, preincident planning, fire behavior, and the fire officer. Chapters 2, 3, and 4 go over management, decision making, and engine and truck company operations. Chapters 5 and 6 take in building construction, collapse, and incident scene safety. Chapters 7, 8, 9, and 10 discuss specific types of occupancies and operations. Chapter 11 looks at the actions involved after the incident: critiques and critical incident stress.

There are repetitive statements and thoughts throughout this book. This was done intentionally. I hope this will not be a distraction and will instead be used to reinforce those important ideas. Though many firefighters who purchase a book read it from cover to cover, some will also use it as reference material. Chapters are designed to stand alone, so firefighters can look up a specific subject without having to constantly reference numerous chapters.

There are terms that can be considered regional or parochial. Some areas of the country use the term *incident command system* or the acronym ICS; other areas use *incident management system*, or IMS. Sectoring can be alphabetical or numerical.

I realize that implementation of a command system and strategy and tactics will be highly dependent upon the number of resources available. I have tried to be generic so that the basic information will be helpful regardless of the terminology. I have included some information on National Fire Protection Association Standard 1710 on Staffing for Career Fire Departments. With all those differences, there is one basic fact. We as firefighters will be called to control and extinguish a fire or mitigate an emergency. The greater the amount of information we have and preparation we have done, the better and safer the overall operation will be performed.

The safety of civilians and firefighters alike is an awesome responsibility. It is extremely important for anyone who is assuming that responsibility at an emergency scene to have a plan. This book will assist in preparing a firefighter for that command role.

James P. Smith

Ancillary Materials (Online Appendices)

Student Resources can now be freely accessed at www.pearsonhighered.com/bradyresources. Here, students will find additional skills and information for more practice and review. The material is organized into several appendices:

Appendices

A. Helpful Websites
B. NIOSH Fire Fighter Fatality Reports
C. Performance Standard 2: *Stang Gun Operation*
D. Quick Action Preplan (QAP)
E. Vital Building Information Sheet (VBI)
F. Incident Command System Organization Chart
G. Scenarios
   Private Dwelling
   Garden Apartment
   Strip Mall
   Commercial Building
   Lumberyard
   Church
   High-Rise Building
H. Case Studies
   Gulf Oil Refinery Fire
   One Meridian Plaza
I. Sample Damage Assessment Forms
J. NIMS Incident Command System Forms Booklet
K. Core Concepts at a Glance

ACKNOWLEDGMENTS

Lieutenant Bill Emery, for constant technical support.

Joe Hoffman, for so many great photos.

Deputy Chief Aide Charlie Armstrong, for the many great years of support and friendship.

Deputy Chief Jim Smith, Ocean City Fire Department, New Jersey, for his technical assistance and making me keep everything in perspective.

Deputy Chief Bill Shouldis, Philadelphia Fire Department (ret), for his excellent technical support and friendship.

Just like on the fireground, you can't do it alone; it takes a team effort.

Reviewers

Fourth Edition:

Charlie Butterfield, M.ED,
    Assistant Fire Chief
*Sun Valley Fire Department, Idaho
    State University, Hailey, ID*

Timothy Flannery,
    M.S., Adjunct Professor
*John Hay College of Criminal Justice,
    New York, NY*

Third Edition:

Kevin S. Bersche, Deputy Chief
*Farmington Hills Fire Rescue
Farmington Hills, MI*

Dane Carley, Captain
*Fargo Fire Department
Fargo, ND*

Lieutenant Anthony Gianantonio/A.S.
    in Fire Science
*Palm Bay Fire Rescue
Palm Bay, FL*

J. Robert Griffin
*Asheville Buncombe Community
    College
Asheville, NC*

Richard W. Hally, Adjunct Professor,
    Fire Science
*University of New Haven
West Haven, CT*

Chuck Hathaway
*Chief/Lead Instructor/Professor, Lone
    Star College—Cy-fair
Fire Chief, Tri-County Fire
    Department
Hockley, TX*

Mark Martin, MPA
*Perry Township Fire Department—
    Chief
Massillon, OH*

Paul Reynolds, BS Education, AAS
    Fire Science
*SWOCC
Coos Bay, OR*

B/C Mark Rosen MPA
*Indian River State College
Ft. Pierce, FL*

James P. Ryan, MS Fire Excutive
    Leadership
*New York State Academy of Fire
    Science
Montour Falls, NY*

Rick Tappan, CSHM, FF/NREMT-P
*Training Coordinator, Office of
    Homeland Security
George Washington University
Lt (ret) WLHFD, Prince Georges
    County, MD
Montgomery County Dept. of Fire
    and Rescue Services (active)
HHS, NDMS-WMD-DMORT*

Greg Ward, Battalion Chief
*Loveland Fire and Rescue
    Department/Aims Community
    College
Loveland, CO*

ABOUT THE AUTHOR

James P. Smith was appointed to the Philadelphia Fire Department on June 29, 1966. He was promoted to lieutenant on December 18, 1972; to captain on December 30, 1974; to battalion chief on August 3, 1981; and to deputy chief on June 27, 1987. Deputy chief is the highest fire department civil service position. Chief Smith reported to the deputy commissioner. He retired on July 7, 2007, and continues to lecture and write on firefighting.

Photo courtesy of Jim Smith, Jr.

Chief Smith has worked on both engine and ladder companies and in every section of the city. He has served as director of the Philadelphia Fire Academy. In this role, he was the departmental Safety Officer and responded on multiple alarm fires performing the Safety Officer's function. Additional areas of responsibility included the research and planning unit. In his role as a field deputy chief, Chief Smith had responsibility for operations for one half of the city. Currently he is a member of the Atlantic County New Jersey All-Hazard Incident Management Team, serving as a Safety Officer.

Chief Smith has developed and taught many programs. He has been associated with the National Fire Academy in Emmitsburg, Maryland, since 1982. As an adjunct instructor since 1984, he has taught numerous operational courses. He is a graduate of the prestigious Executive Fire Officer's Program. He also teaches courses for the Emergency Management Institute.

Chief Smith has authored the Fire Studies column in *Firehouse* magazine since 1987. He has served as a technical consultant for numerous textbooks, and has lectured throughout the United States on incident management, officer development, safety, church fires, building construction, building collapse, strategy and tactics, incident command systems, tank farm fires, and high-rise firefighting.

Chief Smith has also served as a subject matter expert and testified as a technical expert on firefighting strategy and tactics, operations in vacant buildings, dangers of vacant buildings, and the training needs of fire departments.

Chief Smith has also been involved in independent fire investigations involving fire fatalities and firefighter injuries, to determine whether the strategy and tactics that were employed at those fires met national standards.

Chief Smith can be contacted at JPSMITHPFD@aol.com.

INTRODUCTION

I started writing articles for various journals and felt that it was a great way of sharing my experiences in the fire service. I had the opportunity to write the bimonthly Fire Studies column for *Firehouse* magazine, to become an adjunct instructor for the National Fire Academy, and to travel nationwide presenting seminars. Being involved in writing and teaching allowed me to interact with firefighters throughout the United States. This association with many firefighters allowed me to recognize that although we did the same job of emergency response, there were many correct ways to perform our jobs.

I found that firefighter staffing varied from jurisdiction to jurisdiction depending upon budgetary constraints. Large metropolitan areas had fully staffed career departments, while smaller locales had reduced staffing. Combination fire departments consisted of both career and volunteer firefighters, while most fire departments consisted of volunteer members. It did not matter whether it was the east or west coast, I found that firefighting is virtually the same.

I discovered that the methods of learning and the education of firefighters have dramatically advanced. The fire service is changing and evolving continuously, and we need to ensure that we are changing to meet the new demands placed upon us and our fire departments. The Internet has permitted these changes in the learning process by increasing the tools available for firefighters. The use of e-books and online college courses makes the learning process available to everyone. These venues provide information at one's fingertips and provide current information to make the decisions that affect our fire departments.

The National Fallen Firefighters Foundation, in association with the US Fire Administration, has taken the lead in advancing firefighter safety by hosting a National Firefighter Life Safety Summit. This summit produced 16 major initiatives that give the fire service a blueprint for making changes to reduce firefighter deaths. In my many years of firefighting, I have made more than my share of mistakes. I have tried through my writing to share these mistakes along with many positive experiences that I have learned over the years.

I consider myself fortunate to have been in the fire service during the tumultuous 1960s and 1970s, both as a firefighter and a company officer. The reason I say this is that I had the ability to respond to many fires. There were many nights when we went from fire to fire. Many times we would be restoring hose-line and equipment on our apparatus at a fireground and hear another fire being dispatched. We would make ourselves available for service and immediately respond to another working fire. The ability to respond to so many fires permitted us to reach a level of teamwork where companies functioned as well-oiled machines. Another plus was that if we made a mistake, we didn't have too long to dwell upon it before we would respond to another fire and rectify the previous error. After a 14-hour night tour we would be dog-tired, but we had enjoyed every minute of it. My friends who were not firefighters could never understand why I did not transfer from these busy inner-city areas to a quieter station closer to home. If you have never been in the fire service, you too would have a difficult time understanding. There was the feeling of accomplishment, of being challenged, and yes, on a few occasions of being scared, but the final outcome of controlling a fire and hopefully being successful in making rescues made that feeling of tiredness go away.

The hardest things to deal with were the seemingly endless number of children dying in sparse surroundings. Many of them never had a chance. Fires were started by portable heaters being used to keep them from freezing to death or by candles being used because the electricity had long ago been turned off. Along with careless smoking, many fires were started by children playing with matches, or by the cowardice of Molotov cocktails being thrown into a structure, stealing the lives of those sleeping within. Seeing rugged firefighters crying as they carried these lifeless children's bodies from the burned-out structures is embedded in my brain.

Firefighters

The other area that made just as indelible a mark is the loss of life of fellow firefighters. Some were close personal friends, such as Lieutenant Jim Pouliot of Engine 20. Jim was a Minnesota farm boy who, after serving in the US Navy at the Philadelphia Navy Yard, married and settled in Philadelphia and joined the fire department. I had the privilege of working as a firefighter with Jim for close to seven years at Engine 30. We studied hard together for the lieutenant's test and we both were successful. Jim gave his life rescuing a fellow firefighter from certain death at the deadly Gulf Oil Refinery fire on August 17, 1975.

That same deadly fire took a total of eight firefighters' lives, including firefighter John Andrews of Engine 49. I was transferred as the captain of Engine 49 on that deadly date. What a first day it was! The fire started around 6:30 am and went to 11 alarms, destroying tank after tank of flammable chemicals. It engulfed the firefighters in flames around 4:30 pm. When I arrived on the fireground with E-49's "C" platoon around 6:15 pm to relieve the "D" platoon, I was stunned to learn that at that time more than 30 firefighters were reported missing, including three members of Engine 49. In the ensuing hours, I received the sad news that John Andrews of Engine 49 had made the supreme sacrifice. I had the responsibility of informing John's wife Gertrude, his daughter Patty, and his son John of his untimely death. There is no easy way of telling a family that their husband and dad would no longer be coming home. It is an act that I hope never to have to perform again.

That same fire caused career-ending injuries to firefighter George Schrufer of Engine 49, due to burns sustained. The third missing firefighter from Engine 49 was located (see Case Studies: Gulf Oil Refinery www.pearsonhighered.com/bradyresources).

Another was the loss of Captain John Taylor. John was a friend and an excellent firefighter. He lost his life trying to rescue a member of his company while fighting a basement fire in a row house.

Change in the Fire Service

Philadelphia is a city that has experienced problems similar to those of other large metropolitan cities in the northeast section of the United States, which are often referred to as "Rust Belt Cities." The population of Philadelphia has decreased from more than 2.2 million to around 1.5 million citizens. This declining population and subsequent tax loss has impacted the fire department. Philadelphia has seen a reduction in the number of firefighters from more than 3,200 to today's 2,140.

While I have seen many changes in the fire service, many things remain unchanged. We no longer call ravaged inner-city neighborhoods ghettos. Today they are referred to as economically depressed areas or empowerment zones. This change is not because of any improvements; in fact, many of these areas are incredibly worse due to the proliferation of drugs. The terminology is only a case of semantics. Save for the politicians seen in these areas around election time looking for votes and making promises that are rarely kept, only the residents, police, and firefighters can be found.

One of the most positive areas of change is the smoke alarm. I feel that it has had the greatest effect in reducing civilian deaths and injuries. The biggest problem to overcome is the indifference of most citizens. When asked their biggest fear, most people are concerned with crime; few fear fire. Today most fire deaths still occur in residential buildings. My experience is that most of these could have been prevented if a smoke alarm had been installed and maintained. The sad part is the great number of deaths that occur with a smoke alarm present but nonoperable, due to either dead or missing batteries. This indifference has to stop. We must impress upon everyone the importance of maintaining smoke alarms.

An even better method of protection of life is the installation and maintenance of residential sprinkler systems. As a national fire service, we need to impress upon our legislators the need and importance of these systems. They will save not only the lives of residents, but invariably those of firefighters.

Fire and Emergency Services Higher Education (FESHE) Grid

The following grid outlines the course requirements of the Strategy and Tactics course developed as part of the FESHE Model Curriculum. For your convenience, we have indicated specific chapters where these requirements are located in the text.

Course Requirements	1	2	3	4	5	6	7	8	9	10	11	Student Resource Site
Demonstrate (verbal and written) knowledge of fire behavior and the chemistry of fire.	X		X	X		X						Scenarios, and Case Studies
Articulate the main components of prefire planning and identify steps during a prefire plan review.	X	X	X									Quick Action Preplan, and Vital Building Information Sheet
Recall the basics of building construction and how they interrelate to prefire planning.					X	X						
Recall major steps taken during size-up and identify the order in which they will take place at an incident.			X									Scenarios
Recognize and articulate the importance of fireground communications.		X	X			X						
Identify and define the main functions within the ICS system and how they interrelate during an incident.		X	X			X						Scenarios, Incident Command System Forms, and Sample Incident Action Plan
Given different scenarios, the student will set up an ICS, call for appropriate resources, and bring the scenario to a mitigated or controlled conclusion.		X	X	X			X	X	X	X		Scenarios, Incident Command System Forms, and Sample Incident Action Plan
Identify and analyze the major causes involved in line-of-duty firefighter deaths related to health, wellness, fitness, and vehicle operations.	X					X						Case Studies

1
Preparation

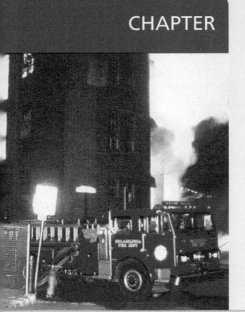

" Firefighting is an art and not a science. You can't always predict the exact outcome of an event even if you have good information. Yet there is a need to make predictions based upon one's experience and training. Since the incident scene is dynamic and constantly changing, we must adjust, or change as it changes. This will ensure that our actions allow us to prevail in accomplishing our goals to protect life and property. "

—James P. Smith

Firefighters must anticipate changing conditions and ensure that apparatus is operating at safe locations. *Used with permission of Joseph Hoffman.*

KEY TERMS

OBJECTIVES

Upon completion of this chapter, the reader should be able to:

- Discuss the behavior of fire.
- Understand the benefits of training for the firefighter, company officer, and the fire department.
- Discuss the benefits of preplanning.
- Understand how to calculate needed fire flow.
- Recognize the duties of both company and chief officers.
- Identify the traits of a person with command presence.
- Identify and discuss the 16 Firefighter Life Safety Initiatives.
- Understand the safe operation of fire department apparatus

Firefighters who respond to emergencies must be prepared to handle whatever they encounter. Their success is dependent on the training they receive. This training must encompass basic and advanced areas. Chapter 1 first examines the **behavior of fire**; it reviews the basic concepts of training and preincident planning, or how prior knowledge of a building can assist in accomplishing an overall safe operation.

Calculating the needed fire flow will assist fire officers to determine the personnel needed to handle hose-lines and to implement tactical operations at an incident scene.

Chapter 1 further discusses company and chief officers, the importance of these positions, and what these men and women must accomplish to serve the fire department and

behavior of fire ■ Fire is a chemical process where fuel, oxygen, and heat come together in an uninhibited chain reaction. It involves the rapid oxidation of a combustible material producing heat and flame.

the firefighters that they lead. It also reviews the 16 life safety initiatives that, if followed, will keep firefighters safe.

Behavior of Fire

Fire is a chemical process in which fuel, oxygen, and heat come together in an uninhibited chain reaction. It involves the rapid oxidation of a combustible material, which produces heat and flame. In order to have fire in common materials, these three elements (fuel, oxygen, and heat) are required and are often represented as a fire triangle. (See Figure 1-1.)

A fire triangle will cease to exist if any one of the three sides is removed. (Fire can exist in atmospheres containing oxidizers other than oxygen, such as chlorine, but for this discussion we will focus only on oxygen.)

- Fuel may be eliminated by removing it.
- Oxygen may be eliminated by excluding air.
- Heat may be eliminated by cooling.

In most cases, the removal of fuel from a fire is impractical. However, flammable liquid storage tanks are sometimes arranged so that if one should catch fire, its contents may be pumped to an isolated empty tank. Thus, the fire is extinguished by removing the fuel. Fires in other flammable liquids that are flowing from a pipe and burning can be extinguished if the pipe contains a valve and the flow of fuel can be shut off. This, in effect, removes the fuel supplying the fire.

The exclusion of oxygen from a fire may be achieved by covering the fire, usually with dirt, foam, or a wet blanket, so that air cannot reach it. Placing a lid over a burning pan of oil on a stove will extinguish the fire by excluding air. Oxygen may also be reduced by blanketing the fire with another gas heavier than air. This displaces oxygen from the fire area, which can occur when applying carbon dioxide onto a fire. The application of the above extinguishing agents excludes air from the fire and in doing so removes the oxygen necessary for combustion.

Reduction in temperature, or cooling, can be accomplished by the application of a substance that absorbs heat. Water is most commonly used for this purpose. It absorbs heat first by being raised to its boiling point and second by being turned from boiling water into steam. The heat that is absorbed is taken from the fire and reduces its temperature accordingly.

FIRE TETRAHEDRON

Research has found that in addition to removing one of the three elements of the fire triangle to extinguish a fire, there is actually a fourth method. This method is referred to as the fire tetrahedron or the fourth side of a triangle. The fire tetrahedron is an uninhibited *chemical chain reaction* that occurs when fuel is broken down by heat. The fire is extinguished as the chemical chain reaction is interrupted by an extinguishing agent. (See Figure 1-2.)

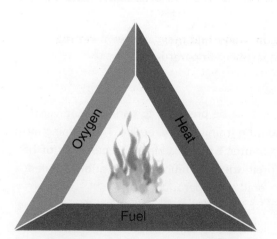

FIGURE 1-1 Fire Triangle. *Used with permission of Pearson Education.*

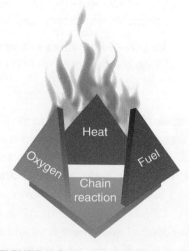

FIGURE 1-2 Fire Tetrahedron. *Used with permission of Pearson Education.*

CLASSES OF FIRES

Different systems for classifying fires are in use worldwide. The United States places fires into one of five classes, for purposes of extinguishment and to identify which fire extinguisher to use. Those classes are A, B, C, D, and K. There are many types of fire-extinguishing methods; some of the more common methods are listed below.

Class A Fires

Class A fires involve ordinary combustible materials: wood, paper, textile, etc.

Class A fires are commonly extinguished with water, or a method that uses water and an additive, or with fire extinguishers.

The additives used with water are wet water, Class A foam, and high expansion foam. They are used to increase water's effectiveness. For additional information, see the following National Fire Protection Association standards: "NFPA 18: Standard on Wetting Agents," "NFPA 18A: Standard on Water Additives for Fire Control and Vapor Mitigation," and "NFPA 1145: Guide for the Use of Class A Foam in Manual Structural Fire Fighting."

- Wet water reduces the surface tension of water, making it much more easily absorbed by a burning material. This can help extinguish a fire that is deep-seated, such as in bales, overstuffed furniture, mattresses, and so on.
- Class A foam is a liquid foam solution that is made by introducing air into a mixture of water and concentrate. Bubbles from the foam blanket adhere to fuels and gradually release the water they contain to continue to wet fuels for a longer period of time than water alone. Due to the addition of the air bubbles and the expansion of the foam, Class A foam provides a larger contact area with the burning surface to increase heat absorption. By adhering to surfaces, it can also be used to protect unburned materials. Surfactants contained in the solution reduce surface tension and allow water to penetrate fuels to reach deep-seated fire.
- High expansion foam can be supplied from expansion rates of 200 to 1 and up to 1000 to 1. It is delivered through large sleeves pushed by high-powered fans. It works by producing a cooling blanket of foam and excluding air from the burning material. It can be used for basement fires as well as holds on ships. For it to be effective, hose-lines that deliver water to fight the fire must be shut down in the intended area to prevent the water from breaking down the foam. There must be an opening for the foam to be introduced and another opening past the fire to allow the foam to flow and fill the area. If a ventilation opening is not provided past the fire area, such as in a confined space with only one opening, the foam will not flow properly into the fire area and will prove to be ineffective. Likewise, any obstructions, such as closed doors or doors blocked by debris, will restrict flow and reduce the foam's effectiveness.

Class A fire extinguishers consist of pressurized water extinguishers, multipurpose dry chemical extinguishers, and so on.

- The pressurized water extinguisher contains water and may contain a wetting agent. The extinguisher is pressurized by an air compressor. It is effective on small fires and can produce a 40- to 50-foot stream of water.
- The multipurpose dry chemical fire extinguisher contains ammonium phosphate and is rated for Class A fires. It is a powder-based agent that extinguishes by smothering, cooling, and radiation shielding, but studies suggest that a chain-breaking reaction in the flame is the principal cause of extinguishment. The rapidity of extinguishment is due to the interference of the dry chemical particles with the propagation of the combustion chain reaction, which reduces the concentration of free radicals present in the flames. To accomplish this, the dry chemical must become thermally decomposed. The discharge of dry chemical into the flames prevents reactive particles from coming together and continuing the combustion chain reaction. It is referred to as the chain-breaking mechanism of extinguishment.

Fire extinguishers for Class A fires are marked with a green triangle containing the letter A. (See Figure 1-3.)

Ordinary combustibles

FIGURE 1-3 Class A Fire Symbol. *Used with permission of Pearson Education.*

FIGURE 1-4 Class B Fire Symbol.
Used with permission of Pearson Education.

FIGURE 1-5 Class C Fire Symbol.
Used with permission of Pearson Education.

FIGURE 1-6 Class D Fire Symbol.
Used with permission of Pearson Education.

FIGURE 1-7 Class K Fire Symbol.
Used with permission of Pearson Education.

Class B Fires

Class B fires involve flammable liquids, combustible liquids, petroleum greases, tars, oils, solvents, lacquers, alcohols, and flammable gases.

These fires are extinguished by using foam and fire extinguishers.

Class B foam is used to fight flammable and combustible fires. It is noted as Class B foam to distinguish it from Class A foam. Some Class B foams can be thick and viscous, forming tough heat-resistant blankets over burning liquid surfaces and vertical surfaces, while other foams can be thinner and spread more rapidly. Some foams are capable of forming a vapor film of surface active water solution on a liquid surface, and some are meant to be used as large volumes of wet gas cells for inundating surfaces and filling cavities.

Dry chemical fire extinguishers can be used to control and extinguish flammable liquid fires.

Fire extinguishers for Class B fires are marked with a red square containing the letter B. (See Figure 1-4.)

Class C Fires

Class C fires involve energized electrical equipment.

These fires are extinguished by using extinguishers that will not conduct electricity.

Dry chemical and carbon dioxide (CO_2) fire extinguishers are used for extinguishing fires involving energized electrical equipment.

Carbon dioxide is a compressed gas that prevents combustion by displacing the oxygen in the air surrounding a fire and by cooling the fuel. An excellent benefit of using carbon dioxide fire extinguishers is that they leave no residue.

Fire extinguishers for Class C fires are marked with a blue circle containing the letter C. (See Figure 1-5.)

Class D Fires

Class D fires involve combustible metals: aluminum, magnesium, titanium, sodium, and potassium.

These fires are commonly fought with special types of fire extinguishers for fighting metal fires. As a general rule, water should not be used on metal fires. The reaction when water is applied to a burning combustible metal can range from minor to explosive. The exception is that in some instances water can be successful if applied in large quantities to a small combustible metal fire, but a severe reaction that could be explosive should be anticipated.

Class D fires can be found in industrial establishments and in automobile fires containing aluminum engines.

Fire extinguishers for Class D fires are marked with a five-pointed yellow star containing the letter D. (See Figure 1-6.) Ensure that the correct Class D extinguisher is used for the burning material.

Class K Fires

Class K fires involve vegetable oils, animal oils, or fats in cooking appliance fires.

Changes in commercial cooking operations have presented challenges to dry chemical fire extinguishers and extinguishing systems. Currently, the most effective extinguishing system for commercial cooking fires is a wet chemical commercial hood suppression system. Fire extinguishers classified as multipurpose ABC, used in combination with the wet chemical suppression system, will threaten the foamy layer and cooling ability of the wet chemical agent, which could result in damage to cooking appliances. Class K fire extinguishers are specifically intended for these types of fires, can be used simultaneously with the hood suppression system, and are generally found in commercial kitchens.

Note that only the Class K fire extinguisher is compatible with wet chemical agents. Class K fires should not be confused with Purple K fire extinguishers, which are designed for Class C fires.

Fire extinguishers for Class K fires are marked with the letter K, or a black hexagon containing the letter K may be indicated on the extinguisher. (See Figure 1-7.)

SPREAD OF HEAT

Heat is generated when a combustible material comes in contact with a heat source. If there is sufficient oxygen, the combustible material will burn. The heat generated by the fire can spread to other materials in the room where the fire started, to adjacent rooms, and to other areas within the original fire building. Depending upon many factors, if the fire is left to burn unabated, it can spread to nearby structures. This spread of fire is dependent upon:

- Amount of heat generated by the fire
- Distance or spacing between buildings
- Duration of the fire
- Height of the buildings (since taller buildings that are exposed to lower buildings can have multiple floors at risk)
- Type of building construction (frame buildings would be more susceptible than a fire-resistive building)
- Other factors

METHODS OF HEAT TRANSFER

In addition to direct contact with the flames of a fire, there are three ways that heat travels and causes a fire to extend to other areas or other buildings. They are through conduction, convection, and radiation.

Conduction

Heat is transferred by conduction through contact of materials. When different materials touch, the material with the greater temperature will transfer heat to the material with the lower temperature until the temperatures of both materials are identical. An example of conduction of fire is when floor joists are set in the same wall socket in adjoining buildings and are in contact with each other. Heavy fire involving a joist in one structure can spread undetected to a joist in the adjacent building. The ease of the spread of fire via conduction depends upon the materials involved. Metal is a good conductor and once heated it can ignite combustibles that are in contact with it, even at a distance from the original fire.

Convection

Convection is heat that is conducted by a gas or liquid. Air currents allow heat to rise through a structure if unimpeded by doors, walls, or ceilings. Once these air currents reach the top of a fire area, the smoke and heat will mushroom and start to spread laterally. As the air current of smoke and heat fills the upper parts of the fire area, it will start to bank downward. This movement of heated air currents can spread the fire to uninvolved areas. Staircases are a ready path for convective air currents to spread fire from lower to upper floors. This is especially true in multistory single-family residential structures where staircases are usually unenclosed. Likewise, the proper use of ventilation in conjunction with fire attack by firefighters can channel the heat and smoke generated by a fire and vent it to the exterior to reduce its spread.

Radiation

Heat from a fire is radiated in all directions, including back toward the fire, which helps accelerate the chain reaction process. The rays travel in straight lines and in all directions from the fire and continue to travel until their heat is dissipated or they meet an object. If the object is combustible, the radiated heat can cause it to ignite. If the object is noncombustible, like a masonry wall, the rays will heat the object and the object absorbs the heat. Depending upon the material, it can dissipate the heat through conduction, or ignite and start to burn. Exterior fires that require protection of exposed buildings can often be protected by water streams being applied to the exposed surfaces to keep them cool and prevent ignition.

THE STAGES OF FIRE

Fire will go through various stages, from inception to extinguishment. There are five stages of fire:

1. Ignition
2. Growth

3. Flashover
4. Fully developed
5. Decay

Ignition Stage

This is the initial stage, as the fuel, oxygen, heat, and the uninhibited chemical chain reaction come together. Material is heated from a source, and the fire normally starts small and is referred to as being in an incipient stage.

Growth Stage

The next stage is the growth stage, in which the fire starts to develop depending upon the amount of fuel and oxygen that is available and the absence of built-in fire-extinguishing systems. An abundance of fuel and oxygen will allow a fire to develop rapidly. The contents of the room or area in which the fire originated will become heated as the temperature of the fire increases. This increase in temperature will generate additional heat, smoke particulates, and toxic products of combustion, primarily in a gaseous form. Heat and smoke will rise throughout the fire area (the fire area can be a room, a number of rooms, an entire floor, or an entire building, depending upon the presence or absence of fire stops) until it reaches the top level and starts to bank down, filling it with superheated smoke and gases. As the temperature builds to over 1000 degrees Fahrenheit (F), superheated gases will increase within the fire area, and if a sufficient amount of oxygen is available, it will near flashover stage. Unlike in the ignition or incipient stage, when a fire can be quickly controlled by a minimum of firefighters, the larger fire will require a greater fire department response. Multiple growth stages may occur, depending on available fuel and oxygen.

Flashover Stage

The flashover stage of a fire does not always occur. There must be sufficient fuel and oxygen for the fire to reach flashover. The flashover stage is the transition between the growth stage and a fully developed stage. What occurs is the temperature in the fire area rises rapidly as the gaseous products of combustion reach their ignition temperature and are ignited, which increases the intensity of the fire. The temperature of flashover is estimated between 900 and 1100 degrees F.

Flashover is a simultaneous ignition of the surface area of the combustible materials in the fire space. The fire's intensity will be dependent upon sufficient oxygen to sustain the burning. Ventilation openings can supply a source of air to the fire as firefighters enter the fire area to extinguish the fire. Ceiling temperatures can easily exceed 1300 degrees F.

Fully Developed Stage

The fully developed stage occurs when all combustible materials in the fire area are involved in fire. The heat released by the burning material will be at its maximum. As in the flashover stage, continued burning will be dependent on a sufficient supply of fuel and oxygen. Temperatures can exceed 2000 degrees F.

Decay Stage

The decay stage occurs when the oxygen or fuel starts to diminish. When the burning materials are consumed, the fire will start to decay. This process will produce large volumes of smoke that contain carbon monoxide. The fire may also start to decay due to the lack of oxygen. What occurs is that the smoke reduces the available oxygen and slows down the burning process so that temperatures decline. This lack of oxygen can, however, create a smoldering state that could set the stage for a backdraft or what is referred to as a *smoke explosion*. (See Figure 1-8.)

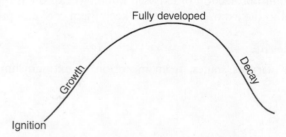

FIGURE 1-8 Traditional Fire Behavior.

ON SCENE

Carbon monoxide (CO) is an odorless, colorless gas commonly encountered by firefighters at every fire. In addition to fires, its presence can be caused by a variety of household defects, including a defective heater or a blocked chimney. Because CO cannot be detected, occupants are often sickened by its fumes. At low levels of concentration, CO can cause headaches and dizziness before incapacitation. The installation of CO detectors in buildings is a positive step toward its early detection.

At one medical response a patient was displaying symptoms (headaches and dizziness) similar to her recent ailment, which had required hospitalization. The medic unit requested an engine company to assist in her removal so she could be transported to the hospital. As the members were preparing her, a male occupant in the same room passed out. As one of the paramedics went to assist him, she too passed out, and the other paramedic started exhibiting signs of lightheadedness. The lieutenant of the engine company had the two paramedics immediately taken to the exterior. He called for additional assistance and members donned self-contained breathing apparatus, reentered the structure, and removed the two occupants who were transported to the hospital along with the two paramedics and three firefighters. Though the engine company had initiated ventilation, on arrival of the battalion chief, the testing of the structure for CO showed extremely high concentrations, especially in the basement where it was found that the source was a defective heater.

This incident was the impetus for installing clip-on CO monitors to first responder medical bags that are carried on medic units and first responder engine and ladder companies. The monitor is always active. It enters a structure with the first person and will sound an alarm when encountering high levels of CO. This relatively inexpensive tool has proven invaluable by alerting firefighters and paramedics on many responses.

MULTIPLE GROWTH AND DECAY STAGES

A fire burning outside in the open air may have sufficient oxygen to allow a material to burn freely. A fire burning within a structure will depend upon the amount of available material (fuel) and the amount of oxygen present to keep it burning. The *growth stage* of a fire begins as air (oxygen) feeds the fire from all directions. A fire in a structure during this stage is called *fuel-limited/controlled*, not because fuel is absent but because it is not involved in the fire yet, and the smoke and hot gases are spreading along the ceiling to all of the open rooms in the building.

As the burning increases, the oxygen within the structure is consumed and the layer of smoke reaches the burning materials. This is still the growth stage, but the fire becomes *ventilation-limited/controlled*. This reduction in air limits the ability for free burning and the oxygen level is reduced to less than 21 percent, slowing the burning rate. As the fire burns, it will most likely spread from the original material and involve other furnishings within the fire room or fire area.

Once the oxygen level gets below 16 percent, the fire begins its *initial decay stage*. Temperatures will remain high in the fire room, but throughout the rest of the structure temperatures will decrease as the heat release rate decreases. A decaying fire must draw in more oxygen, which it will attempt to get from cracks or voids in the structure. This may appear from the exterior as a pulsing action of the smoke. If the fire is unable to pull in more air, it will self-extinguish. (See Figure 1-9.)

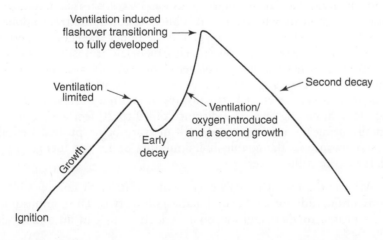

FIGURE 1-9 Modern Content Fire Behavior.

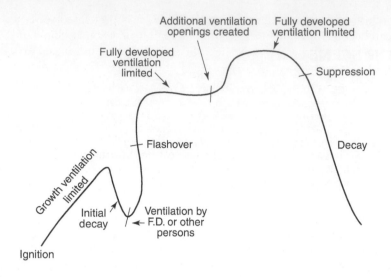

Should air enter the burning room through a window broken by the fire, or a door opened by a firefighter or someone else, the fire will begin a *second growth stage*. If water is not applied immediately to slow the fire, this second growth stage could transition to *flashover*.

After flashover, the fire grows (heat release rate increases) to the point where there is more burning than can be supported by the air coming in through the newly created opening. Fuel-rich smoke and hot gases flow out of the opening, meet the oxygen outside of the house, and burn outside the structure. At this stage, the fire is ventilation-limited and temperatures in the structure will remain high. The fire is not vented, but it is venting, and if no additional windows fail or doors are opened or holes are cut in the roof, the fire will enter the *fully developed stage*. This fire will continue to burn at the same heat release rate, unless additional oxygen is made available to the fire, or fuel is consumed to the point that the fire pulls back into the house (and becomes fuel-limited), or water is applied to the fire.

If fire suppression has not begun and additional ventilation openings are created by the fire department, the newly introduced oxygen can transition the fire into a third growth stage. The heat release rate increases as additional smoke and hot gases are ventilated from these newly created openings. If there is still fuel remaining, the fire will now be showing through these openings. The fire will again become ventilation-limited and enter a second fully developed stage. Now the fire will remain at this stage unless additional oxygen is made available to the fire through additional ventilation openings, or fuel is consumed to the point that the fire pulls back into the house and becomes fuel-limited, or the fire department initiates suppression activities and applies water to the fire to move into the decay stage. (See Figure 1-10.)

ROLLOVER, FLASHOVER, BACKDRAFT

Rollover The term *rollover* is used to describe the fire or flame front that often is observed rolling along in front of burning materials. A combustible gas is produced and liberated from the material that is ablaze. This gas mixes with air (oxygen) in order to burn. Because the combustible material that is burning consumes tremendous amounts of air, there may be a limited amount of oxygen in the upper levels of the room to support combustion of all the fuel being produced. This fuel-rich atmosphere will be pushed in front of the fire by the thermal column of heat from the fire and may not come within its flammable limits until it is several feet away from the main body of the fire. This is especially true in confined areas such as hallways. It is often observed that the fire seems to be rolling along at ceiling level at a distance of 10 to 20 feet ahead of the main fire. What is actually being witnessed is a fuel-rich mixture being pushed well ahead of the fire. When it comes into its flammable limits (mixture of air and fuel gas), it burns. This is described as the fire rolling over.

Flashover A basic definition of *flashover* is the ignition of combustibles in an area heated by convection, radiation, or a combination of the two. The combustible substances in a room are heated to their ignition point, which results in an almost simultaneous combustion of all the materials. Because the entire area is preheated to its ignition temperature, it can become fully involved in fire in a matter of seconds. Some of the warning signs of imminent flashover are intense heat, free-burning fire, unburned articles starting to smoke, and fog streams turning to steam a short distance from the nozzle. To reduce the chance of flashover, temperatures need to be lowered quickly by coordinated ventilation and water application.

Backdraft As a fire develops, the combustion process creates an atmosphere that is deficient in oxygen and can lead to the possibility of a *backdraft* or smoke explosion occurring. The difference between flashover and backdraft is the amount of oxygen present. In flashover, there is adequate oxygen available for combustion and the fire is free-burning prior to flashover. In a backdraft, there is insufficient oxygen for active burning, and the fire is smoldering.

Sufficient oxygen is present during most fires, so the conditions leading to backdraft are rare. However, when oxygen is depleted and the fire begins to smolder, an oxygen-deficient atmosphere is created in the fire area. When a condition like this develops, it produces gases, such as carbon monoxide and carbonaceous-particle smoke that are capable of reacting with oxygen. This poses an explosion threat if oxygen is allowed to enter the structure because the accumulated gases will ignite readily, spreading fire or causing a violent explosion. Due to the high temperatures in the room, the fuel evolves into ignitable vapors, at or above their ignition temperature. Oxygen is all that is needed to complete the fire triangle.

When backdraft conditions are present and oxygen is introduced before the inside pressure is relieved, an explosion can occur. This is especially true when the oxygen is introduced from a lower area. The potential for backdraft exists in buildings, rooms, attics, or any other confined space.

The action required when a backdraft situation is recognized is to provide adequate ventilation above the fire. Ventilation is the first priority and must precede fire attack under backdraft conditions. This will relieve the pressure, venting the heat and smoke to the exterior. After proper ventilation is performed, rapid fire involvement must be anticipated as fresh air is introduced into the previously unvented area.

training ■ The backbone of every fire department; encompasses basic and advanced areas.

Training

Training is the backbone of a fire department. It produces a well-prepared force that, through repetition, can increase the speed of an operation and enhance proper execution while reducing injuries.

Firefighters who arrive at an emergency unprepared can be faced with life-and-death situations and will find themselves under extreme stress to perform the necessary duties.

Training benefits everyone: the firefighter, the company officer, the fire department, and the community. Let's review how training benefits each group.

FIREFIGHTER BENEFIT

All fire department members need training. New members need to learn basic skills. Senior members need the training as a refresher and to keep their skills sharp. (See Figure 1-11.)

As firefighters improve their skills, they experience less fumbling and fewer errors. They gain confidence in themselves, since they can perform their job at a high level. They develop pride in themselves and in their department. Training allows for continuous growth in their ability and prepares them to assume more responsibility while grooming them for promotions. As firefighters are promoted through the officer ranks, they should constantly receive training to allow them to better understand and accept the new roles they will be assuming and the responsibility they will be undertaking. The promotion of a firefighter allows that individual to

FIGURE 1-11 All fire department members need training. New members need to learn basic skills. Senior members need the training as a refresher and to keep their skills sharp.

continue in the role of fighting fires, which he or she should be quite familiar with. Yet there are many other aspects of training, supervision, leadership, and communications that need to be learned.

COMPANY OFFICER BENEFIT

The company officer reaps many benefits from a highly trained crew. There is better control over operations. The training frees the officer from interruption by unnecessary workers' questions, allowing more time for the officer to assume greater responsibility. It improves the firefighters' overall ability and the officer has less fear of emergencies. The relationship between the officer and the firefighters becomes more pleasant and fewer troubles exist. This results in the officer having job satisfaction instead of job headaches.

DEPARTMENT BENEFIT

The department as a whole is a beneficiary, since training allows for constantly improved operations. The efficiency of the fire department is recognized by the citizens it protects, and can be directly linked to good public relations. This leads to public support as well as from politicians and community leaders. It will assist in the passage of bills that could be vital for a fire department's success in providing the necessary services to the public, which will continue to benefit the fire department. Training keeps morale at a high level, which, though intangible, facilitates every function of a department. The firefighter will operate in a pleasant environment and will look forward to participation in the various departmental functions.

PREPARATION IS REQUIRED

Training exercises, though, must be challenging. Reading from a text is boring and counterproductive. To conduct an interesting training exercise, the officer has to be knowledgeable and prepared. This involves prior reading and research to ensure that the goals of the training exercise will be attained. In addition to textbooks in the firehouse and the training bureau, outside sources should be explored.

The National Fire Academy (NFA) is an excellent resource. At no cost, its Learning Resource Center (LRC) offers a wealth of information and research, which can be used for fire department research projects or personal projects such as college work. Using the data provided will also enhance any training exercise. (The Learning Resource Center can be reached at 1-800-638-1821.)

Additionally, there is an abundance of information online that can be downloaded for use. For example, the US Fire Administration (USFA) has developed a program called Coffee Break Training. It hosts a multitude of topics, usually on one page. Coffee Break Training can be used as a stand-alone training tool or easily combined with other data to expound on a subject. Some fire departments use it as a bulletin board informational sheet. [To find an index of the Coffee Break bulletins on a range of topics, go to the USFA website (www.usfa.fema.gov), and search for "coffee break training fire protection series."]

An officer can maintain interest in a subject by asking questions and seeking input from all participants. A tool to add interest in a training exercise is to utilize local buildings or potential hazards and seek input from the members. Drawing from the experience of each member allows for a well-rounded exercise. It also permits the officer conducting the training to reap the benefits of the firefighters' experiences while letting the officer recognize each member's level of expertise.

Training permits mistakes to be made and corrected in a nonemergency setting. The **fire officer** can take the time to stop a training exercise and point out correct procedures. He or she can explain what problems can arise when firefighters fail to use the correct method, including difficulties that can occur if shortcuts are taken.

Training fosters teamwork and cooperation. Training can be accomplished formally through drills and practical evolutions or informally by explaining policies and procedures. Members can set goals and discuss their individual progress with their commanding officers.

fire officer ■ Usually holds the rank of lieutenant or captain and has the responsibility for leading an engine or ladder/truck company.

PERFORMANCE STANDARDS

A department that establishes performance standards or timed evolutions for engine and truck/ladder companies and then trains utilizing that criteria will be better prepared to handle the varied problems that occur at an incident scene. (Go to Appendix C: Performance Standard 2 in www.pearsonhighered.com/bradyresources to view a sample performance standard.)

Development of these evolutions can start by stretching an attack line into the first floor of a structure while hooking up to a hydrant or obtaining a water supply from a water tender. This basic evolution can then be changed to placing a portable ladder and stretching a hose-line up the ladder and through a window.

Each evolution can become more complex by including additional functions. The agenda can be expanded to placing master stream devices in operation. Ladder companies or tower ladders can place their apparatus into operation for simulated fires requiring elevated streams, rescues from upper floors, or rescues utilizing a wire basket from elevated or below-grade locations. The object is to achieve a standardized operation that emphasizes safety. Standardization ensures that members assigned to different units work together seamlessly.

The entire evolution must be specific and documented. There should be a maximum amount of time to complete an evolution. Using time frames simulates the stress found at the incident scene. It also demands teamwork on the part of all members to ensure that the time frames will be met.

An excellent method to keep training interesting is to foster a competitive spirit among the various units. This can be accomplished by recording the time needed to complete each evolution and posting the individual times. Realize that speed alone should not be the determining factor. Safe operations and adhering to the entire performance standard must be judged. There should be methods to penalize units for minor mistakes or omissions.

Videotaping evolutions lets the officer note a unit's strengths and discover areas where improvement is needed. An excellent tool for the training division is to maintain tapes of the units performing the best times. This permits recognition of these outstanding accomplishments while allowing other units reviewing the tapes to take advantage of their experience.

In addition to the benefits gained by company members training on timed evolutions, a fire officer should make note of the amount of time required to perform these evolutions. This can help the officer when commanding a fire scene in assigning tactical operations. By knowing how long it takes to place a master stream device into operation under ideal conditions, the fire officer can utilize these established time frames while recognizing potential limitations.

CROSS-TRAINING

Fire departments should regularly schedule training involving multiple units. This should include the cross-training of members normally assigned to an engine on the operations of a main ladder or tower ladder, while ladder company members get the opportunity to operate the pumps on the engines. Cross-training should include hazardous materials teams and members of specialized units, such as a heavy rescue company. (See Figure 1-12.) This hands-on training helps give members a better understanding of how the various units function, allowing an emergency scene to operate smoothly. Assigned members of a rapid intervention crew will then be able to utilize apparatus on the incident scene to assist in removing trapped firefighters.

Departments should routinely train with mutual-aid departments. These exercises enable members to form friendships and share experiences that will benefit each department when called upon to operate together on future incidents.

FIGURE 1-12 Training should include hazardous materials teams. This exercise was conducted at the Philadelphia Veterans Administration Hospital and permitted Philadelphia firefighters assigned to the hazardous materials team to interact with hospital personnel. *Used with permission of Philadelphia VA Medical Center.*

Preincident Planning

Knowledge is a tool. The more tools that we have at our disposal at an emergency scene, the better the odds of a successful outcome. A well-informed Incident Commander can handle problems effectively, with fewer resources and less chance for error.

Similar to the way that life insurance companies invest their insured dollars and bet that their insured are not going to die, we in the fire service plan for a large fire or incident that we hope will never occur.

preincident planning ■ Method of gathering facts about a building or process prior to an emergency.

Preincident planning provides information. It is a method of gathering facts about a building or a process within a building. It lets a fire department evaluate conditions and situations in its area of responsibility prior to an emergency. Through evaluation, we can compare what we may be called upon to do with what we can do. (Fire department staffing in some areas may not be capable of controlling a major fire due to limited personnel or equipment.) Fire departments that utilize preplanning find that it can mean the difference between success and failure at an emergency incident. Preincident planning allows us to anticipate potential problems and analyze possible solutions to those problems.

PREPLANNING RESPONSIBILITY

The responsibility for preplanning starts with the fire chief. Once a policy is established, all members of the fire department must carry it out.

A chief officer should be assigned as a coordinator to oversee the preincident planning programs. This person should have sufficient authority to ensure compliance by company and chief officers. By placing responsibility with one person, that individual can decide what is best for the program. This ensures standardization and continuity of all preplans. Should any questions arise, they can be directed to that individual, which results in consistency of forms, inspection methods, and record keeping. The chief officers who will be implementing the plan during an emergency must be involved in selecting which target hazards to preplan.

Target hazards for which preincident plans should be prepared include buildings or processes:

- That pose a high threat to life safety of those who work in the building or facility and to those who live nearby
- That would create safety problems for firefighters and other emergency responders (hazardous materials, buildings with lightweight building components, etc.)

- That could create a conflagration hazard
- That would present unusual and demanding situations for responding firefighters
- That have a high frequency of fires
- That would have a large economic impact on the community

Historical data enable us to select the most critical properties or specific problems in our community. The next step is gathering information to analyze the overall situation. The data should include both national and local statistics—what types of fires are common versus the types of buildings and occupancies found in our community. This analysis assists in determining potential problems. A plan of action can be developed based upon what may occur.

TYPICAL TARGET HAZARDS

Nursing Homes	Public Assemblies	Bulk Storage or Tank Farm
College Dormitories	Libraries and Museums	Facilities
Penal Institutions	Courthouses	Large Buildings or Building
Grain Elevators	Enclosed Malls	Complexes
Schools	Major Transportation	Chemical Plants
Hospitals	Carriers	High-Rise Buildings

DOCUMENTATION

Preplanning is a tool that sets forth a framework for interfacing all fire protection components before an emergency occurs. It is a method of gathering facts and collating information. The preplanning process begins with an on-site survey.

The responsibility for gathering information for the preplan is usually assigned to the first-due engine or truck company. The crew should contact the building owner or responsible party and schedule a time to tour the facility. A thorough inspection can reveal locations where a problem may occur. It can identify the immediate life threat and actions that the fire department can initiate to mitigate the problem. This meeting can enable both parties to discuss concerns about problems that firefighters may encounter and how the fire department and facility can solve those problems together.

There are numerous ways of gathering and saving this valuable information. Yet there must be a method of easily recalling the stored information during an emergency, or the preplanning process will be useless.

The stored data:

- Can be very basic and kept on large index cards for easy reference. This will limit the amount of information, but the data will be easily accessible.
- Can be in a booklet consisting of multiple pages. This contains more comprehensive information, though it becomes more difficult to access due to the greater amount of data.
- Can be in a database that stores and retrieves information, including safety information that can be flagged and become immediately available by being printed out at the time of the alarm. The information can be sent automatically to the scene via onboard printers or facsimile machines. One drawback can be the cost of these systems.
- Can be in a palm-sized or handheld computer for quick and easy retrieval
- Can be a combination of an index card system that is backed up by a booklet

Initial concerns can be placed on the index cards to be easily accessed by the initial units. Comprehensive information can be gleaned from the booklet once the initial concerns have been addressed and more specific information is needed.

Quality is the important concern. An overwhelming amount of data can be counterproductive. It can take too long to sift through useless data before locating the needed information.

The booklet method can contain a section on fire department concerns. This can cover areas of general concern and specific areas that would pose a threat to firefighters. It may include:

- Building renovations that might not show on the original building plot plans
- The presence of pressure-reducing valves on standpipes
- Standpipe and sprinkler connections
- Any special extinguishing systems utilized within the building, such as dry chemical or CO_2
- The presence and location of hazardous materials in the building, such as asbestos, radioactive material, and PCBs
- Flammable or explosive processes
- The location of open shaftways or chases
- Special needs of occupants (disabled, infirm, etc.)
- Means of access, egress, and floor or plot plans

Phone numbers can change. Emergency contact personnel can change. Responders should review and update the information during site visits. Scheduling of multiple dates for site visitations can ensure that all members in career and volunteer departments will have the opportunity to visit the site for first-hand knowledge. The data should be incorporated into company drills to permit a constant refresher.

CONSIDERATIONS

We must thoroughly analyze each situation. How will we be able to protect building occupants and those in threatened exposures? What evacuation plans have been formulated? Are there other means by which occupants can be protected, such as lateral evacuation or protecting in place? Are there protective systems in the building?

Our on-site survey should consider how construction features and protective systems assist or impede the fire department once a fire occurs.

These include:

- Is the entire building equipped with automatic sprinklers? Is an engine company assigned to pressurize the system?
- Standpipes will aid firefighters in placing hose streams onto the fire. Is it a wet or dry system? Which fire department unit will pressurize the system?
- Compartmentation will assist in containment of a fire, permitting the fire department an opportunity to control and extinguish it. Do fire doors close automatically? Are they propped open? Are there any obstructions that would prevent their closing?
- Wire-glass windows provide buildings with exterior exposure protection. They may crack from radiant heat but will remain in place. Are any exterior exposure protection systems in place? This could include an exterior deluge sprinkler system that provides protection from nearby buildings for windows and other exposed openings.
- Smoke-proof doors and smoke-removal systems assist in reducing the spread of smoke and can help in minimizing evacuation problems in some facilities. Are these systems in place?
- How are lightweight building components protected?

Once the target hazards are identified, potential problems can be predicted and solutions to those problems can be formulated. These solutions can be addressed by developing standardized procedures for these recurring situations. The implementation of set procedures permits them to be practiced under nonemergency conditions. Firefighters' skills can be honed so that time will not be wasted at an emergency.

Another benefit derived from preplanning is that command decisions can be decided under nonemergency conditions. This reduces the stress on the Incident Commander during an emergency.

RESOURCE UTILIZATION

The preplan should consider the availability of resources. Many combination and volunteer fire departments have a difficult time staffing apparatus at daytime hours during the week. Since there may be a delay in the response of personnel, the number of personnel that will be needed to perform specific functions must be considered in the preplan. Alternative strategies must be formulated to anticipate reduced staffing. In career fire departments, there may be minimum staffing where the number of personnel working remains the same, or a fluctuation of personnel numbers on any given day. Some career departments reduce staffing when personnel call in sick or are off injured. In other locales, companies may be browned out or placed out of service to minimize overtime. If the number of personnel in the career fire departments can fluctuate, only preparation will ensure that the assignments can be accomplished with the staffing that respond.

Resource utilization includes securing the services of outside agencies, such as the police, Red Cross, private security, public works, public health, utilities, or federal, state, or local agencies. Their inclusion in the preplan should be specific about the duties expected of them. Police, for instance, can be used to evacuate an area if they are not placed in jeopardy. This assignment means one less headache for the fire department. A command police officer can be used when relocation of displaced people associated with an incident occurs. Though seemingly a minor point, delegation to another agency will save firefighting personnel for needed assignments.

The use of the National Incident Management System for all-hazard incidents is mandated in the National Response Plan (NRP) in Homeland Security Presidential Directive/HSPD-5. The National Response Framework, which has replaced the NRP, presents guiding principles that enable all response partners to prepare for and provide a unified national response to disasters and emergencies, from the smallest incident to the largest catastrophe.

An incident management system must be used by all agencies. The system will address the multitude of problems associated with large-scale incidents. Anticipating specific problems allows prior thought and research on the part of the individuals who may assume specific roles in the command system.

 ON SCENE

A training exercise at a refinery involved a scenario in which a small aircraft from the nearby airport was simulated to strike a storage tank containing crude oil. The ensuing simulated fire went to two alarms and included training with the city fire department and the plant fire brigade. Surprisingly, within two weeks, a major fire occurred at that refinery. The knowledge learned from the drill was quickly placed into action and the fire brought under control in a timely manner. In addition to the information that was gained on the facility, a major benefit was the mutual respect that developed between the fire brigade and the fire department members.

DISASTER PLANNING

There are certain facilities where, should an emergency occur, there could be a direct impact on the immediate surrounding community. These include chemical plants, refineries, large water purification plants, testing laboratories, and so on. These sites will require large-scale plans with multi-agency planning and community input. This type of plan must have the support of every agency involved for it to succeed. A member of each supporting agency should be conferred with when drawing up the plan. This liaison should have the authority to authorize specific commitments of resources in the event of an emergency. The inclusion of participating agencies in the planning process allows them to buy into the plan. Success or failure is dependent upon them as well.

Community support should be sought. The immediate population around the site should be informed of evacuation plans and any prefire drills. This can be accomplished through independent community groups (church or senior citizen groups) or groups established to identify problems associated with a specific site. Some communities have established strong relationships with industrial plants and work diligently for the betterment of both the community and the plant.

The preplan can stipulate special equipment that can be used effectively at the scene. Mutual-aid response should be reviewed to allow proper deployment. A greater initial response may be indicated by the hazards presented, necessitating changing existing policies.

PLAN REVIEW

Review of the preplan with the personnel of the affected facility is a positive step. It may open their eyes to the distinct possibility of destruction of their facility. It could initiate major changes to prevent a disaster. This may include separating processes that would react unfavorably during an incident, installing sprinkler systems, or increasing the available water supply on the premises. During an emergency, these same individuals can be utilized as technical specialists who can assist the Incident Commander by providing important information.

TESTING THE PLAN

Implementation of a preplan during a simulated exercise assists in adjusting the plan as needed.

- What worked well?
- What needs to be adjusted?
- What problems did the fire department encounter?
- If the community was involved in the exercise, did it have any problems?
- Did the plant or facility find any discrepancies in the plan?

The addition of contingency plans for foreseeable problems and their incorporation into the exercises should be encouraged. The plan should be reviewed annually to see if any modifications are needed.

 **ON SCENE**

Successful Operation Due to Preplanning

There may be no better argument for a preplan than a major fire that occurred in Center City, Philadelphia. A six-story commercial building that encompassed a full city block was undergoing renovations. Missing stairways and open shafts when elevators were removed created vertical openings. Sprinklers and standpipes had been removed before being replaced.

The local engine company and battalion chief were concerned that many basic safety measures were not being complied with. They had responded to a few minor fires in the structure during the renovations, leading to a prefire plan being written and disseminated for the building during the renovation.

Within a month's time, the preplan was placed into operation. A fire occurred in the afternoon hours that went to seven alarms within eight minutes, and eventually nine alarms (that included more than 50 apparatus) with full involvement of the six-story building within ten minutes. Located in an area of high building density, the prefire plan recognized existing building conditions. It designated positions of first-alarm companies and warned of the potential of a large-scale fire. This prior knowledge allowed the Incident Commander to calculate his situation and anticipate problems that could arise. It allowed for immediate commitment of master stream devices for a concentrated attack.

An independent study in reviewing this incident found that the preplan was a major factor in the fire department's ability to control this major fire. (At the time this fire occurred, code enforcement and the ability to have contractors cease operations were not controlled by the fire department. Today there is joint responsibility for ceasing operations between the fire department and the city's code enforcement bureau.)

Needed Fire Flow

A problem facing the initial Incident Commander at a fire scene is how much water will be needed for effective fire control. This information will impact the incident in terms of determining needed resources and the implementation of tactical operations.

Determining the amount of water needed to extinguish a fire in a specific building is best accomplished during the preplanning stage. This can be attained through a deliberate calculation of the occupancy, considering conditions when establishing the needed fire flow. When preplanned information is available to the initial Incident Commander upon arrival at an incident, strategic and tactical decisions can be made more readily and accurately.

To determine **needed fire flow (NFF)** during preplanning requires the application of a fire flow formula to conditions observed during an inspection of the premises. (See Figure 1-13.)

On many occasions, fire incidents are encountered where fire flow information is not available. Under these circumstances, experienced fire officers are able to determine the NFF based on their experience and knowledge of similar situations they have encountered in the past.

There are occasions when a newly appointed or relatively inexperienced officer lacking the expertise of a seasoned officer must quickly judge the amount of water needed to effectively control a fire. The NFA in Emmitsburg, Maryland, has developed a formula that allows for quick calculations. The formula was derived through a study of fire flows that were successful in controlling a large number of working fires, along with interviews of numerous experienced fire officers throughout the country regarding the fire flows they had found to be effective in various fire situations. The NFA quick-calculation formula can be used as a tactical tool to provide a starting point for deciding the amount of water required at an incident scene. This will permit decisions to be made on the amount and type of apparatus needed to deliver the water and the number of firefighters needed to apply it.

The information developed by the NFA indicated that the relationship between the area involved in the fire and the approximate amount of water required to effectively extinguish the fire can be established by dividing the square footage of the area of fire involvement by a factor of three. This formula is expressed as:

$$\text{Fire flow} = \text{length} \times \text{width} \div 3$$

This formula is most easily applied if the estimated square footage of the entire structure is used to determine an approximate fire flow for the total structure, and the fire flow estimate is then reduced accordingly for various percentages of fire involvement.

The example shown below illustrates how the formula can be applied to a single-family dwelling 60 feet long by 20 feet wide and one story high:

$$(60 \times 20) \div 3 = 400 \text{ gallons per minute (gpm)}$$
$$100\% \text{ involvement} = 400 \text{ gpm}$$
$$50\% \text{ involvement} = 200 \text{ gpm}$$
$$25\% \text{ involvement} = 100 \text{ gpm}$$

The quick-calculation formula indicates that if the dwelling were fully involved, it would require 400 gpm to effectively control the fire. If only half of the building were burning, 200 gpm would suffice, and 100 gpm should be sufficient if one-fourth of the building were involved.

MULTISTORY STRUCTURES

In multistory buildings, if more than one floor in the building is involved in fire, the fire flow could be based on the area represented by the number of floors actually burning. For example, the fire flow for a two-story building of similar dimensions as the previous example would be:

$$(60 \times 20) \div 3 \times 2 \text{ (floors)} = 800 \text{ gpm if fully involved}$$

If other floors in a building are not yet involved, but are threatened by possible extension of fire, they should be considered as interior exposures. Twenty-five (25) percent of the required fire flow for the fire floor should be added as exposure protection for each exposed floor above the fire floor to a maximum of five interior exposures. (In the

60x20 / 3x1 = 400 GPM for 100% Involvement

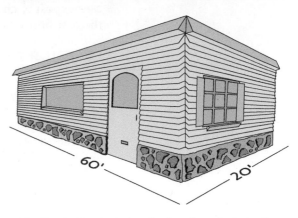

FIGURE 1-13 To calculate fire flow for a single-family dwelling 60 feet long by 20 feet wide. *Used with permission of Michael DeLuca.*

needed fire flow ■ The theoretical amount of water needed to control and extinguish a fire.

previous example, a fire on the first floor would threaten the second floor, and a 25 percent exposure charge should be added. A second-floor fire would probably not threaten the first floor, so no interior exposure would need to be calculated.)

EXTERIOR EXPOSURES

Likewise, if exterior structures are being exposed to fire from the original fire building, 25 percent of the actual required fire flow for the building on fire should be added to provide protection for each side of a building that has exterior exposures.

The following example shows how to apply calculations for exposures to our previous one-story dwelling with exposed exterior structures on two sides of the fire building:

$$(60 \times 20) \div 3 \times 1 = 400 \text{ gpm}$$
$$\text{2 exposures: } 400 \text{ gpm} \times (25\% \times 2) = 200 \text{ gpm}$$
$$\text{Total fire flow required} = 600 \text{ gpm (for 100\%}$$
$$\text{involvement of the original fire area)}$$

If the exposure becomes involved in fire (either additional floors of a multistory building or adjacent structures), the exposure(s) should then be treated as a separate fire area and calculated separately, and then added to the required fire flow for the original fire area.

In using the quick-calculation method to determine required fire flows, it is important to remember that the answers provided by this formula are approximations of the water needed to control the fire. The formula is geared to an offensive attack, and its accuracy diminishes with fire involving more than 50 percent of a structure and with defensive operations. Don't forget that you are estimating both the area of the building and the amount of fire involvement within the building.

Since firefighting is not an exact science to begin with, the use of the quick-calculation formula cannot be expected to determine the exact gpm that will be specifically required for full fire control. It has been found that as the amount of involvement reaches a stage where a defensive attack is necessary, the NFF will be found to be slightly greater than predicted. (See Figures 1-14 and 1-15.)

AVAILABLE WATER SUPPLY

Available water flow must be known. The type and location of the water supplies should be specified. Provide exact locations if using hydrants or drafting sites. If a tanker/tender operation is to be used, determine how many will be needed to ensure a constant water supply.

DETERMINING TYPE OF ATTACK

Once the required fire flow has been determined, the capability of available resources will determine the strategy and tactics needed to control the incident. If the fire flow capability of available resources exceeds the required fire flow, an interior attack on the fire can usually be made. However, before this decision is implemented, the Incident Commander should consider:

■ Do existing conditions allow sufficient safety for the firefighters on an interior attack?
■ Are there sufficient firefighters on scene?

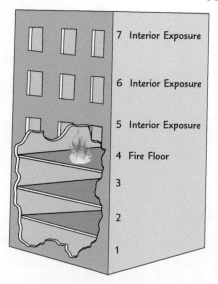

FIGURE 1-14 A charge of 25 percent will be calculated for each floor above the fire floor up to a maximum of five floors. *Used with permission of Michael DeLuca.*

FIGURE 1-15 If exterior exposures are being threatened by fire from the original fire building, 25 percent of the actual required flow for the building on fire should be added to provide protection for each side of a building that has exposures. *Used with permission of Michael DeLuca.*

- Is the fire area accessible?
- How many hose-lines and firefighters are needed?
- Where is the best location from which to attack the fire?
- What support activities are needed (ventilation, forcible entry, search and rescue, accountability, rapid intervention crew)? (See Figure 1-16.)

If the fire flow requirements exceed the fire flow capability of available resources, a defensive mode of operation is usually required. In these situations, larger hose streams, more apparatus, more equipment, and more personnel may have to be requested. Situations will occur where fire is attacking lightweight structural components, and though there is a sufficient water supply and resources, the conditions will be too dangerous for an offensive attack. The Incident Commander must also recognize that there will be situations where nothing can be done with the available resources to save the involved building. In these circumstances, exposure protection becomes the primary objective.

FIGURE 1-16 A flammable liquid fire in an auto garage heavily involves the building and vehicles. A quick search will be required to ensure that all occupants have evacuated the building. *Used with permission of Brian Feeney.*

SELECTION OF HOSE-LINE

Recognizing that a 1½-inch or 1¾-inch hose-line flows 125–175 gpm and a 2½-inch hose-line flows approximately 250 gpm, we can estimate the number of hose-lines and resources needed to control the fire.

While the NFF formula (fire flow = length × width ÷ 3) will provide the Incident Commander with a starting point to determine how much water may be needed for an effective fire attack in normal situations, common sense and good judgment are required to evaluate the effect of the water on the fire as it is being applied. There may be unforeseen factors impacting the situation, such as barriers that prevent the water from reaching the seat of the fire or building contents that cause unexpected fire behavior.

If control is not achieved within a reasonable period of time, the amount of water may have to be increased or a defensive attack may need to be implemented. If immediate knockdown of the fire takes place, the amount of water being applied can be reduced to minimize water damage to the structure and contents.

The Fire Officer

COMPANY OFFICER

The company officers are among the most critical members of a fire department. They may be selected by a civil service test, appointed by the fire chief, or elected by the members. The responsibility is the same regardless of the method of promotion or whether they serve a career or volunteer department. The company officer is the direct link for the firefighters between middle and executive management, and he or she must maintain a critical balance between them. Officers must accomplish the goals of the department while looking out for the well-being of their firefighters. This can often be a demanding challenge.

A company officer has to get work accomplished through others. This may be achieved in different ways. Whether an officer is an authoritative or congenial type is usually determined by his or her personality. The style is not as important as the fact that there must be consistency. Inconsistency occurs when attempting to be a hard-line type of officer one day and overly friendly another day. This change in style will lead to frustration on the part of the firefighters. It is better to find a style that suits you and be consistent.

Company Leader Firefighters look to the company officer as their leader. Good leaders lead by example and gain the respect of subordinates and superiors alike. Those who need to rely on the adage "do as I say, not as I do" will have a more difficult time getting their orders carried out than the officer who leads by example. By sharing personal fire and emergency response experience with the members, the officer allows them to grow within their department and prepares them for promotion.

Respect and admiration is gained through many individual qualities. Company officers must have the ability to adapt to changing situations, be quick thinking, and have good communication skills. They must remain calm in critical situations, frame orders clearly and concisely, and give, not yell, the orders. This will foster teamwork and cooperation.

They should seek a variety of ways to achieve personal development. This can include seminars, training, and especially pursuit of educational degrees in fire science, management, and public safety fields.

Common Sense A trait that some people seem to be born with is common sense. It can also be developed if one works at it. We must think before we act. If we allow ourselves to think of how to handle a situation rather than rush right in, we can decide on the best possible approach. At an incident scene, time is at a premium and decisions must be forthcoming. This time may consist of only seconds, but some thought is often the difference between a good and a bad decision. Quick decisions that are not thought out can become faulty decisions. Take the time to "Stop, Step Back, and Think" before acting. Good decisions are often associated with common sense.

People Skills The successful company officer is one with people skills. By studying each subordinate, he or she can determine how to bring out the best in each of them. What are their strong points? What areas need to be strengthened? The key ingredient is motivation. They must be personally motivated. A key to motivation is finding out what drives each individual. For some, it will be how they benefit personally. For others, it is praise for a job well done or a pay incentive. The company officer must find these methods of motivation to assist in molding the firefighter as an individual and as a team member.

Company officers can be successful if they praise their firefighters' good behavior publicly and criticize their mistakes privately. When reviewing a misdeed with a firefighter, the discussion must focus on the mistake that was made and not become a discussion of personalities. This type of strong supervision creates a positive work environment.

Knowledge of District The officer should do a risk analysis of the unit's response district/community. This will assist in gaining personal knowledge on specific problems. It might include locations where a long response is necessitated due to the presence of limited access highways, railroad crossings that may slow response time, and/or occupancies (hospitals, schools, nursing homes, etc.) that will demand a maximum effort by the responding firefighters. Special attention should be given to all target hazards, and their preplans should be reviewed at the scene when performing building inspections. Officers should draw upon the experience of their firefighters. Previous responses to these locations can provide invaluable information. This sharing of information can be initiated at the scene. A good officer will take notes and continue the discussion after returning to the station. Information should be gathered on what support may be expected from plant personnel and the types of fires these installations have had in the past. It should include the positive things that occurred, as well as the problems that arose and how they were overcome. This review not only allows the officer to build a base of knowledge but enables the other firefighters to learn as well. (See Figure 1-17.)

Teamwork Chief officers don't extinguish fires. Chiefs develop overall incident objectives and the basic strategy. The fulfillment of orders to achieve the chief's strategy is the responsibility of the company officers and firefighters. These members must have the courage and fortitude to accomplish the needed objectives. Success occurs from a total team effort. The firefighter operating the nozzle may actually extinguish the fire, but he or she is only one key player in a company or combined unit operation. The company officer must direct the unit and give assignments to ensure success. From the size and type of hose-line, the source of water supply, and how and from where to attack the fire, he or she must be sure to define roles and responsibility. Coordination with other units ensures a balanced attack.

The company officers' duties on an emergency can cover a wide spectrum. Initially, they may be called on to perform the role of Incident Commander. Upon being relieved, they may be assigned to supervise a division or group or return to their company. In the incident's final stages, they may revert to Incident Commander to complete the overhauling stage of the incident.

FIGURE 1-17 A high-pressure natural gas main was broken by a worker who was operating a backhoe. It threatened numerous exposed buildings and destroyed nearby vehicles.

The Incident Commander or Division or Group Supervisor assigns tactics that need to be accomplished. The company officer assigns tasks to his or her members to achieve those goals while keeping the Incident Commander apprised of the progress being made.

Training The competence of officers is most often judged by their ability to command a company. Under their guidance, firefighters train and practice to be ready when called to a fire or other emergency scene. Training is the backbone of every good organization. How well we practice dictates how well we will perform at an emergency. The more we train, the better we become.

Through the use of standard operating procedures or standard operating guidelines, we have the ability to practice the performance of routine tasks that can then be applied at the incident scene.

Riding Assignments Another tool many departments utilize is riding assignments. This means assigning tools and basic tasks to firefighters. Career departments assign these duties at the start of each shift. Some volunteer departments have assignments established by the apparatus seat position. The seats are numbered and specific duties are assigned to the firefighter riding in each seat. This allows the officer to give a general order and reserve time for the vital decisions that are sure to crop up at an incident.

Responsibility The company officer is responsible for many areas. These include training, physical fitness, and mental readiness of the entire crew. There should be an esprit de corps among the members that will drive the firefighters to want to do the best that they can. This is an attitude that must be exhibited by the officer, and it will become contagious for all members.

The company officer determines the route and regulates the speed of the apparatus on an emergency response. All crew members should have their personal protective equipment on and be seated with seat belt secured before the apparatus leaves the station. The officer should be cognizant of intersections where they may meet other responding units. On arrival at the scene, the company officer will direct the positioning of the apparatus, considering the functions that will need to be accomplished. (See Figure 1-18.)

FIGURE 1-18 The company officer will direct the positioning of the apparatus, considering the functions that will need to be accomplished. *Used with permission of Joseph Hoffman.*

The company officer, on arriving at the incident scene, must size up the structure. In addition to life-safety factors, this observation must include anything out of the ordinary. Are doors or windows open? Where is the fire on arrival? Is there more than one fire? Because officers are responsible for initial fire investigations as to the cause of fire, this point will assist them and the fire marshal, fire investigator, or police agency in completing their investigations. (Chapter 3 covers size-up.)

Realizing that size-up is the duty of each member operating at the scene, the officer's thought process should be geared to predicting areas where pitfalls could endanger company members. Observations must include reading a building prior to entering, to consider secondary means of egress. By sharing this information quickly with the firefighters, they can react promptly if they need to use a secondary exit.

Teamwork is enhanced by communication between the company officer and the firefighters and with other company officers. The sharing of information facilitates the accomplishment of assignments and prevents duplication of effort. It may entail the size of ladders needed in the rear of a structure, where ventilation is required, or in which areas a primary search has already been completed.

> Size-up and safety are everyone's responsibility.

Safety Safety is one of the most important responsibilities of a company officer. Incident scene safety doesn't just happen; it must be anticipated and addressed in training sessions. The implementation of safety in practice evolutions will carry over to the emergency scene. Cultivation of good work habits in routine situations develops into good incident scene practices. The company officer must ensure that everyone is wearing full protective clothing when operating at an incident. Inspecting the firefighters' personal protective equipment at regular intervals will uncover any deficiencies. Follow-up is needed to ensure that defective gear is repaired or replaced.

THE CHIEF OFFICER

The chief officer is the leader of a fire department. The position of chief can encompass a variety of ranks: battalion, district, division, deputy, assistant, and chief of department. The titles and duties will vary in day-to-day operations, but there is little difference when operating at an incident scene. At an incident, the duties of an Incident Commander must be all-encompassing, regardless of titles.

There is a difference between a chief and a company officer. A company officer is part of an engine or truck company. Firefighters will confide in company officers to ask questions or seek advice. The rank of chief is the next step on the promotional ladder after company officer. The experience learned as a company officer will be the foundation that chief officers can build upon. A chief must assume responsibility for management and leadership in the department. He or she should lead by example, by making the difficult decisions, not necessarily the most popular ones. The orders given must be based upon what is best for the community and the department as a whole. There should be no inclination toward favoritism to any one person or group. This is a balance that must be maintained to be an effective leader.

Personal Development The chief officer must be personally motivated and build on the skills learned as a firefighter and company officer. Too many fire departments do little training for members who reach the chief officer ranks. Chief officers must continue to expand their knowledge. They must obtain information through academia, training, and at incident scenes. Their knowledge base must include an understanding of fire science, incident scene management, safety, building construction, hazardous materials, supervisory skills, time management, delegation of tasks, and many other areas too numerous to mention.

Knowledge Knowledge of fire science and building construction is fundamental for firefighters. The firefighter needs to recognize how building construction features can affect fire travel through a structure, the proper handling of hose-lines, and how to effect ventilation to achieve control and extinguishment of a fire. These areas will need to be expanded as one is promoted to company officer and again to chief officer.

A company officer must build upon that knowledge to order and place the proper size and number of hose-lines and to understand the duties required of a truck or heavy rescue company in placing ladders or assigning search and rescue crews. Company officers must recognize what additional functions will be required. They will direct operations while coordinating and supervising a division or a group.

The chief officer will assess the scene and assign the necessary units to accomplish the strategies developed. They must continuously size up the incident and ensure that the orders given will accomplish the needed tasks. The chief's knowledge of fire

FIGURE 1-19 A chief officer surveys the rear of a fire building before assuming command. *Used with permission of Joseph Hoffman.*

science, strategy, tactics, and construction is important to predict fire spread. The chief's assessment must determine whether the size of the fire and the type of construction will permit the fire to be controlled, while predicting what additional problems could occur. (See Figure 1-19.)

Management Before being able to manage others, chiefs must control and manage themselves. What do we need to accomplish? A positive way of stating our mission is "to get the job done." But it must be done the right way for both the fire department and for the firefighters who will be doing the job. Since many individual personalities come into play, it is impossible to specify how to accomplish this task. Many chief officers just give the orders, and they are carried out. However, if the tasks are not getting done or not getting done properly, the chief must find out where the breakdown is. It could be in the way the orders are given. It could be the inexperience of the company officer who was given the assignment. A critical factor causing breakdowns is poor communication. The importance of framing orders properly may not be serious in nonemergency situations, but it is critical at an incident scene. There are many methods of ensuring that orders are understood. One way is to have the person receiving the order repeat it to ensure it was properly received.

Delegation Chief officers who fail to manage their own time will be ineffective. Chief officers who refuse to delegate assignments to subordinates will find themselves overwhelmed with minor details and totally ineffective in emergency and nonemergency situations. Delegation is a major part of leadership. Trying to do everything alone is a recipe for failure. However, delegation is a tool that should be used for development, not as a method of avoiding work. Delegation permits subordinates to assume responsibility and to make decisions. It permits a supervisor to assess the skills of subordinates and can lead to suggestions on how they can improve. It is a necessary training process whereby company officers can learn the duties and responsibilities of the chief officer. Naturally, every decision made during this learning process will not be perfect, but by learning from their mistakes, subordinates will gain valuable experience.

An excellent method for a company officer to learn decision making at emergency incidents is with oversight from a chief officer. Should an unsound decision be considered, the chief can interject and ensure that it is corrected. This monitoring/mentoring permits the sharing of knowledge between the chief and company officer. The experience of the chief can be invaluable and can be utilized by the company officer at future incident scenes.

Incident Scene What is expected of a chief officer at an incident scene? This question can have many answers.

- Leadership
- Direction
- Safe operating procedures
- Problem solving
- Common sense
- Unity of command
- Teamwork
- Dependability
- Initiative
- A positive attitude
- Enthusiasm
- Ability to adapt to changing situations
- Professional conduct

At an incident, the chief, after surveying the scene by doing a 360-degree walk-around, should establish a position in front of the fire building from which to command the incident. One of the more difficult aspects of promotion to the chief officer's rank is that the chief must allow the firefighters and fire officers to perform their jobs without interfering. There is a tendency by some newly promoted chiefs after assuming command to continue operating at an incident in ways that they are familiar with, that is, being inside and assisting with the actual fire extinguishment. They must resist the urge to operate within the fire building to get a hands-on feel for the incident. This is a natural desire since firefighters learned the profession by doing, not observing. Initially this will feel awkward, but through experience, a comfort level will be reached. The chief officer must rely on the company officer to be his or her eyes and ears. The progression to chief means depending upon others not only to accomplish the fire suppression but to report the actions being taken and the progress being made. This interaction between the chief and the company officers may take some time to develop. Mistakes will occur in underestimating or overestimating the size of problems by the company officers operating on the interior. This is a learning process. The chief must interpret the verbal reports received and compare them to what is observed. This comparison will allow the chief to decide if satisfactory progress is being made, and it will become the basis for future orders. The chief must let it be known what is expected of the company officer. The areas that should be discussed include basic operations for engine or truck companies and the information the chief requires on arrival from the initial Incident Commander. Post-incident analyses and informal discussions can help to rectify any problems that occurred. These discussions will improve future operations. Remember, it is the responsibility of the chief officer to train fire officers and to assume the blame for their mistakes.

Sectoring As a fire increases in size or complexity, higher-ranking chief officers will respond and often assume command of the fire. This will permit other chief officers to be assigned to supervise various divisions or groups. (See Figure 1-20.) (My personal objective at an incident is to place chief officers in the most critical areas. I rely heavily upon their experience and strongly consider their recommendations.)

Safety Considerations Command decisions must be based upon what is encountered at an incident. The safety of everyone operating at an emergency is an awesome responsibility. The chief must ensure that no one takes unnecessary chances that would endanger themselves or others while attempting to control or extinguish a fire. There can be a fine line

FIGURE 1-20 The Incident Commander can assign chief officers to supervise various divisions or groups. *Used with permission of Marty Griffin.*

between what is acceptable and what presents too much of a risk to the firefighters. If in doubt, we must err on the side of safety.

Specialization   Chiefs need an intricate knowledge of the implementation of a command system to handle the large variety of problems that could occur. They need to understand what can be accomplished with the various apparatus, tools, and equipment at their disposal. Though the chief does not need to know all the idiosyncrasies involved, it is helpful to know the limitations of apparatus and equipment.

With the proliferation of hazardous materials, the threat of terrorism, and other special operations, there will be many demands on the chief at these incidents. The chief officer assuming the role of Incident Commander must delegate the actual handling of specialized areas of an incident to others. This could include a hazardous materials group. Realize that the Incident Commander still retains control and overall responsibility for the incident scene.

Command Presence

The ability to command an incident scene takes preparation and development on the part of the Incident Commander. It is a demanding, autocratic position. The critical nature of the emergency scene does not allow decisions to be made by a committee. There can be only one person in command.

How commanders conduct themselves will influence the conduct of all operating at the incident scene. Leaders who exhibit confidence in themselves will gain the confidence of their subordinates. This characteristic can be referred to as **command presence** or command leadership. The need for command presence is magnified at emergency scenes. High-stress situations demand it. Time constraints placed on the Incident Commander in life-or-death situations require orders to be specific and forthright. When arriving at the incident scene, time cannot be wasted if people are in need of rescue. Immediate action is required.

command presence ■ The traits an Incident Commander exhibits that will influence the conduct of everyone operating at an incident scene.

SELF-DISCIPLINE

Similar to firefighting skills, command presence must be developed. Though easily recognized, it can be difficult to attain. Leadership starts with the ability to possess self-discipline. Before leaders can attempt to control others, they must have control over themselves. They should be take-charge types who can recognize potential incident problems and concentrate on the task of controlling the emergency. A good leader knows what needs to be accomplished and gives deliberate orders that are easily understood. When indecisive orders are issued, they leave doubt in the minds of those on the receiving end and can lead subordinates to question their validity. For this reason, it is important to be precise, deliberate, and decisive.

Leaders will gain the respect of their peers more easily if they have the ability to remain calm. Composure is contagious and will result in a professional operation. Prior to the use of radios, command officers often had to shout to communicate orders. Handheld radios now allow a much better exchange of information. The officer who consistently shouts at an incident scene is frowned upon. Shouting itself denotes a sudden emergency that demands immediate action or it denotes an unsolvable problem. Shouting creates unnecessary excitement that can be contagious and lead to poor decision making and a chaotic incident scene.

VISUALIZE INCIDENT SCENES

Command leadership ability can be tested quite easily. Mentally place yourself in various locations: a single-family dwelling, a manufacturing plant, or a 10-story apartment building. Envision a fire in any one of these structures at 2 P.M. or 2 A.M.

- What strategy and tactics would be necessary in each situation?
- What resources would be needed?
- Where would the resources come from?
- What are the safety considerations for our personnel?
- How would personnel be assigned?

FIGURE 1-21 Though a portion of this dwelling was heavily involved in fire, proper deployment of personnel protected the larger section of the building.

- What would need to be accomplished first?
- How would rescues be made?
- How should the incident management system be implemented?

If we visualize various scenarios and then go over how we would handle the problems that might arise, it enables us to preplan our command moves. As with a building preplan, the more planning we undertake, the fewer critical decisions will have to be made at the emergency scene. Visualizing scenarios allows us to prepare for the eventuality of certain occurrences. It lessens the surprises and allows the Incident Commander more time for other decisions. (See Figure 1-21.)

UTILIZING EXPERIENCE

We can gain experience from previous responses. We must draw on what has worked well in similar situations in the past and project what resources and time frames will be needed to implement the strategy. Experience allows insight as to whether the tactics employed will achieve the strategic goals. Is the attack on the fire accomplishing what was intended? If not, then how must we adjust the strategy? (See Figure 1-22.)

Time factors must be considered when dealing with specific kinds of incidents. A hazardous materials incident may involve a relatively short exposure for firefighters but can result in long-term disabling complications. Situations may also arise when insufficient personnel are available for incident control. Physical abilities and stamina have their limits. Attempting to accomplish too many tasks with insufficient personnel will often fail to achieve the desired goals. A fire officer must consider the safety of the members when making decisions. Firefighters who realize they are a prime consideration of the Incident Commander in his or her decision-making process will often give that extra effort to ensure success of their assignments.

THE LEARNING PROCESS

The saying "We don't live long enough to make all the mistakes ourselves, so learn from the mistakes of others" is most appropriate for the fire service. In addition to knowledge gained through experience, the studying of texts, fire journals, and case studies are methods that can significantly broaden our knowledge. (See Figure 1-23.) The lessons learned from an incident in another jurisdiction are also helpful in enabling us to prepare for a similar occurrence in our area.

FIGURE 1-22 Experience allows insight into whether the tactics employed will achieve the strategic goals.
Used with permission of Joseph Hoffman.

FIGURE 1-23 Realistic exercises enhance the training process.

The career development of a fire officer must encompass the basic aspects of everyday organizational rules. This includes an appropriate span of control to keep from being overwhelmed and to allow proper control of the incident. The delegation of authority will ensure that areas are not overlooked and reserve sufficient time for important decisions that must be made by the Incident Commander.

Firefighter Life Safety Initiatives

Firefighter safety must be an attitude in every fire department. It needs to be stressed in every aspect of fire department operations: training, station activities, and especially at the incident scene. The attitude of safety must be initiated at the fire chief's level and be fully understood and enforced at all ranks. It must be realized that unsafe acts are not acceptable and will not be tolerated. From everyday basic acts of ensuring that all members are seated and belted in before an apparatus can move, to performing air monitoring prior to removing SCBAs at an incident scene, these actions must become routines that are never violated. Firefighter life safety also means keeping oneself in good physical condition so that every firefighter will be able to perform when called upon. In an attempt to reduce firefighter deaths and bring about change in the fire service, the National Fallen Firefighters Foundation (NFFF) decided to take drastic steps and find ways to accomplish that goal.

On March 10–11, 2004, the first National Firefighter Life Safety Summit was held in Tampa, Florida. It was attended by more than 200 individuals representing every identifiable segment of the American fire service who participated in the process. They focused on how to prevent line-of-duty firefighter deaths. With approximately 100 firefighter line-of-duty deaths a year, they realized that changes were needed.

The NFFF hosted the summit as the first step in a major campaign. In cooperation with the USFA, the foundation has established the objectives of reducing the fatality rate by 25 percent within five years and by 50 percent within ten years. The purpose of the summit was to produce an agenda of initiatives that must be taken to reach those milestones and to gain the commitment of the fire service leadership to support and work toward accomplishing this goal.

The summit marks a significant milestone: it is the first time that a major gathering has been organized to unite all segments of the fire service behind this common goal. It provided an opportunity for all of the participants to focus on the problems, jointly identify the most important issues, agree upon a set of key initiatives, and develop the commitments and coalitions that are essential to move forward with their implementation.

The first National Firefighter Life Safety Summit produced 16 major initiatives that will give the fire service a blueprint for making changes.

FIREFIGHTER LIFE SAFETY INITIATIVES

1. Define and advocate the need for a cultural change within the fire service relating to safety, incorporating leadership, management, supervision, accountability, and personal responsibility.
2. Enhance the personal and organizational accountability for health and safety throughout the fire service.
3. Focus greater attention on the integration of risk management with incident management at all levels, including strategic, tactical, and planning responsibilities.
4. Empower all firefighters to stop unsafe practices.
5. Develop and implement national standards for training, qualifications, and certification (including regular recertification) that are equally applicable to all firefighters, based on the duties they are expected to perform.
6. Develop and implement national medical and physical fitness standards that are equally applicable to all firefighters, based on the duties they are expected to perform.
7. Create a national research agenda and data collection system that relate to the initiatives.
8. Utilize available technology wherever it can produce higher levels of health and safety.
9. Thoroughly investigate all firefighter fatalities, injuries, and near misses.
10. Ensure that grant programs support the implementation of safe practices and/or mandate safe practices as an eligibility requirement.
11. Develop and champion national standards for emergency response policies and procedures.
12. Develop and champion national protocols for response to violent incidents.
13. Provide firefighters and their family access to counseling and psychological support.
14. Provide public education more resources and champion it as a critical fire and life safety program.
15. Strengthen advocacy for the enforcement of codes and the installation of home fire sprinklers.
16. Make safety a primary consideration in the design of apparatus and equipment.

Firefighter Life Safety Initiatives ■ The first National Firefighter Life Safety Summit produced 16 major initiatives that will give the fire service a blueprint for making positive changes.

On March 3–4, 2007, fire service leadership gathered for the 2007 National Firefighter Life Safety Summit in Novato, California. The 2007 summit was developed and delivered under the aegis of the NFFF and the **Firefighter Life Safety Initiatives** (FLSI) program. Their aim was to persist in developing solutions to the continuing problem of firefighter line-of-duty deaths, and by extension, firefighter line-of-duty injuries. At the summit's conclusion, the participants proffered more than 100 recommendations. These recommendations will become the material from which action plans for preventing line-of-duty deaths will be constructed and made available to all fire departments.

Following the goals of the USFA to reduce line-of-duty firefighter fatalities by 50 percent by the year 2014, the NFFF, partnering with fire organizations and fire service leaders from around the United States, has created pathways and programs to prevent line-of duty firefighter deaths and, by extension, serious injuries. These are the "Everyone Goes Home" program and the "16 Firefighter Life Safety Initiatives," created from the first National Firefighter Life Safety Summit in 2004, and six subsequent mini-summits held between 2004 and 2007. (Reports from the first national summit and the mini-summits are available at the Everyone Goes Home website).

The chart below shows the progress that has been made since 2007 in reducing firefighter deaths. Though the intended goal of a 50 percent reduction has not yet been achieved, some progress has been made.

Firefighter Fatality Statistics from the US Fire Administration

Year	Fatalities
2014	92
2013	106
2012	81
2011	83
2010	87
2009	91
2008	120
2007	119

Source: FEMA

There have been inroads made in reducing line-of-duty firefighter deaths. The trend has been downward, but some tragic incidents have taken numerous firefighters. In 2013, an explosion in West, Texas, took the lives of nine firefighters and one EMS responder, and a wildfire near Yarnell, Arizona, took the lives of 19 wildland firefighters.

For the 16 Firefighter Life Safety Initiatives to work, they must be adopted and implemented by fire departments.

Safe Operation of Fire Department Apparatus

Though time is critical in firefighting, so is safety. It is imperative that firefighters safely arrive if they hope to have a positive impact on the outcome of a fire event. The driver of a fire apparatus is responsible for the life safety of the public and those riding on the apparatus. The other firefighters trust the driver with **safe operation of fire department apparatus**. As a team member, the driver must protect the other firefighters just as they all do when they enter a burning structure and when they rely on each other to ensure safety and completion of assignments.

safe operation of fire department apparatus ▪ A shared responsibility by the driver and the officer to ensure safe driving procedures whenever a fire department apparatus is in motion.

The legal responsibility placed on a fire department and especially the firefighter driving the fire apparatus is closely scrutinized. This can be seen in charges and lawsuits against many apparatus drivers who have been involved in accidents. Examples of firefighters and police involved in serious accidents while responding to and returning from incidents include the following:

- A firefighter driving a fire apparatus was arrested and negligent homicide charges were lodged against him for allegedly speeding through a stoplight and striking another vehicle, killing its two passengers.
- A fire department apparatus was involved in an accident at a traffic light–controlled intersection, and three members in a passenger car were killed. Initial reports stated that the fire apparatus failed to stop for a red light.
- A state trooper struck the rear of a car at a high rate of speed, killing the driver and injuring her child. The instrumentation on his vehicle showed that just prior to the collision he had reached 127 mph on a two-lane roadway. It was later learned that he was not needed at the incident to which he was responding. Naturally we don't know that when we receive a call, but he never arrived. He was subsequently fired from his job.
- A firefighter died as a result of an apparatus rollover. The engine he was driving was responding to a car fire in a residential garage.

Excessive speed does not guarantee arriving in less time, but it can lessen the possibility of arriving at all. The only way to gain time once a response is dispatched is in the station, by quickly donning your personal protective equipment and getting seated and belted in.

The use of sirens and warning lights does not automatically give anyone the right of way. State laws typically require that emergency vehicles be operated in a safe manner. This means stopping for red lights to ascertain that all traffic has seen and/or heard you and has pulled over or stopped to allow you passage. An accident while responding places at risk:

- Those on the apparatus
- Pedestrians
- The drivers and passengers of other vehicles
- Those who placed the initial call for help

 ON SCENE

As a new firefighter, my first captain's initial discussion with me emphasized that I could be the best firefighter in the world, but in order to accomplish anything, I needed to be there. He reinforced this in two ways: punctuality when reporting for duty, and ensuring that when driving, I would respond in a safe manner. The captain's message stuck with me throughout my career.

DRIVER TRAINING IN AMERICA

In 1917, United Parcel Service (UPS) gave its bicycle messengers, and those who drove its Model T delivery vehicles, their first defensive driving handbook. Private industries, and UPS in particular, value safe driving. That first driving handbook expounded on five "Seeing Habits."

1. *Aim High in Steering.* Look as far down the road as possible to uncover important traffic information and make appropriate decisions.
2. *Get the Big Picture.* Maintain the proper following distance so you can comfortably determine the true hazards around your vehicle. Don't tailgate others.
3. *Keep Your Eyes Moving.* Scan—don't stare. Constantly shift your eyes while driving. Active eyes keep up with changing traffic conditions.
4. *Leave Yourself an Out.* Be prepared. Surround your vehicle with space in front and at least on one side to escape conflict.
5. *Make Sure They See You.* Communicate in traffic with your horn, lights, and signals to establish eye contact with motorists and pedestrians. Be reasonably sure of people's intentions.

With more than 200 million licensed drivers in the United States and 250 million vehicles, this information is as applicable today as it was in 1917. It is especially useful to drivers of fire department apparatus or any emergency vehicles. We can take this information a bit further and add other concerns as firefighters.

Aim High in Steering

When we aim high in steering, we should be looking not only at where we are but where we will be, in one-block or two-block segments. On a highway, our view should encompass as far ahead as we can see down the highway. How is the traffic stacked up ahead? Could all lanes be shut down due to an accident? Which lane would be the easiest to safely navigate the roadway?

Since you are already looking ahead, you will see potential problems and need to find ways to avoid them. One method to avoid trouble may be changing lanes to find an open roadway and continue to respond or, if not on a response, to be able to exit the roadway to find a more suitable route. When responding, an emergency vehicle changing lanes too often can confuse the drivers trying to get out of your way. If possible, select a lane and try to fully utilize it. This allows the other drivers time to formulate how they can get out of your way. When given a choice, it is best to be in the passing lane, since more options are available.

No matter how loud the sirens or air horns, or how visible the flashing lights, traffic may be gridlocked and have nowhere to go to allow you to pass. You may need to consider a parallel roadway or alternate route, or have another company dispatched from the other direction.

A consideration of all responding units is the intersection where you could expect to meet adjacent responding companies. Should a unit's members not be responding on the normal route from their station, they should notify the other responding units of this change of response route.

Get the Big Picture

To get the big picture is to know at all times where other vehicles are around you. Read the anticipated actions of the other drivers. In normal driving, we must anticipate the actions of aggressive drivers who weave in and out of traffic. They often cut in front of other cars or trucks. This aggressiveness can cause accidents between the aggressive driver and an innocent driver who reacts or overreacts to that driver's actions. In some instances, the aggressive driver infuriates other drivers, and it can become a deadly game of "Chicken" between the aggressive driver and other drivers.

Your life is more precious than the value aggressive drivers place on their lives with the dangerous maneuvers they engage in. In these cases, ease off the gas, give yourself as much room as needed should an accident occur in front of you, and be prepared to stop. Knowing where all of the vehicles are behind and alongside of you allows for more options should you need to act defensively.

ON SCENE

An example of misapplied reasoning involved Jon Corzine, then governor of New Jersey. In a highly publicized accident, a New Jersey state trooper was driving the governor's state vehicle in a caravan of two vehicles. They had their emergency lights flashing and were driving on New Jersey's Garden State Parkway at speeds up to 91 mph in a 65-mph zone. The lead vehicle containing the governor was involved in an accident, and he was not wearing a seat belt. The governor was thrown from the front to the rear seat and sustained a broken leg, collar bone, a dozen ribs, and other injuries. At the time, he was heading toward an inconsequential meeting, which he never attended. A state police spokesman noted that "speed was not a factor in the accident," a statement that is difficult for a rational person to comprehend. As the governor of New Jersey, Corzine fully understood the speed of his vehicle and had the authority to instruct his driver to slow down. In any case, he was as responsible for the accident as the driver.

Keep Your Eyes Moving

Move your eyes up and down and from side to side, using all of your mirrors. Be aware of those on the sidewalk or near crosswalks, and try to anticipate their movements. Elderly pedestrians may have impaired hearing and/or vision and might attempt to cross the street, observing that the traffic has stopped for a traffic signal but not hearing or seeing the responding apparatus. Attempt to make eye contact with pedestrians and other drivers at intersections. This helps to ensure that they see you and react in a reasonable manner.

Children may run toward the street, attracted by the sight of the responding fire engines. Whether on an emergency response or during routine driving, you may encounter children chasing a ball or a family pet into the street from between parked cars. Situations like this require that you have control of your apparatus at all times and be prepared to stop. There is no difference between dead wrong and dead right—in both cases someone is dead.

Leave Yourself an Out

Ensure that there is enough space around your vehicle to avoid an accident. By being fully aware of your apparatus' capabilities and braking distances, you should know the safe distances to come to a complete stop at any speed. The faster a vehicle's speed, the greater the distance that must be covered to stop, due to the driver's reaction time and the stopping distances needed once the brakes are applied.

When closely following other responding apparatus, a sudden stop by the lead apparatus to avoid an accident will force a similar action by the following vehicles. This can create a situation in which following too closely can cause the responding vehicles to collide. Conversely, if the second or third responding apparatus are lagging too far behind when the first apparatus has cleared an intersection, then cars, trucks, and pedestrians may resume their normal activities, which will place you and them in danger. When traffic is entering a highway into your lane, know what vehicles are around you. This allows you to change lanes, permitting the traffic to safely enter the roadway.

Professional drivers drive a minimum of 35 hours a week. A firefighter's actual time driving a fire vehicle can be 10 hours a week, but typically is much less. When faced with inclement weather, firefighters need to read the road conditions and adjust their driving accordingly. Rain, snow, fog, or other inclement weather needs to be considered and adjustments made. This will usually mean reduced speeds, since stopping distances will be greatly increased on wet or icy roadways.

Make Sure They See You

It is hard to believe, given the size of fire apparatus, that you need to make sure they see you, but it is true. People are often caught up in their daily lives, and distraction carries over to their driving.

Warning devices are a great tool for alerting drivers and pedestrians of an approaching emergency vehicle. The problem is that sirens are directional, meaning that the greatest sound intensity is straight ahead and not to the side or rear. The siren's range will be reduced by tall buildings, intersections, and curving roadways. Some fire departments rely heavily on air horns. Though they have great value at times, their effectiveness can diminish if sounded continuously.

The cars on highways today are engineered to keep offensive sounds from entering the vehicle. Sound systems make it highly unlikely that drivers will hear your warning devices from a distance. Some drivers employ headphones for music or cell phone use, minimizing their ability to hear anything except through their headphones. Other concerns are drivers who talk on cell phones or text message while driving. Drivers may even be driving with one hand as they hold a phone or text a message with their other hand. When they realize that you are right behind them with sirens blaring, they may panic, stop dead, or make a radical move, endangering themselves and others.

Intersection Accidents

Most serious accidents occur at intersections. Fire department standard operational guidelines (SOGs) for responding to emergencies should specify that all apparatus come to a complete stop at a red light and only proceed when safe to do so. This means when there is no traffic, or traffic has stopped to allow you safe passage.

When entering blind intersections, where the intersecting traffic is not visible to the apparatus driver, even with a green light drivers should remove their foot from the accelerator, be prepared to stop if necessary, and proceed only when it is safe to do so.

 **ON SCENE**

Safe driving is a serious concern for the fire service. In one eight-year period, 175 firefighters died while responding to or returning from alarms. We need to eliminate this type of firefighter death. The only way to achieve that goal is to make safe driving a priority, for every fire company and for every member of those companies. Every fire officer and every firefighter who drives an apparatus needs to recognize the enormous responsibility that is placed on their shoulders.

Fixed Objects

Drivers frequently strike fixed objects. This can occur when making turns in tight areas or backing up. Whenever backing up an apparatus, the driver must be guided. All other members should dismount the apparatus and ensure that the backing procedure is performed safely to clear any impediments (street poles, hydrants, cars, or pedestrians) and then remount the apparatus when it is safe to proceed. Whenever an apparatus is placed in reverse to back up, use guide persons. This policy still holds true of apparatus equipped with onboard cameras.

Fixed objects include mirror to mirror contact. This kind of accident is common on narrow streets. Some mirrors are adjustable and can be collapsed against the apparatus. Drivers should also be cognizant of the entire width and length of their vehicle, to ensure safe clearance distances when passing similar vehicles. If in doubt, stop the apparatus and utilize guide persons to safely navigate past the difficult areas.

Height must be considered when encountering low bridges and tree limbs that overhang streets. Drivers should know locations with inherent driving hazards in their response district where apparatus may need to use extreme caution.

Arriving at the Incident Scene

As the apparatus nears the dispatched location, slow down. It is necessary for the engine driver to check for hydrants as they near the fire building. They can take a wrap with a hose-line and stretch it with the apparatus, or the driver will need to stretch back to the hydrant once they are positioned near the front of the fire building. When stretching hose-line, always attempt to keep it to the side of the roadway to permit access for additional vehicles.

The address of the incident must be found quickly, and slowing down allows the firefighters to locate it. On a fire response, placement of the initial units must be considered.

The front of the fire building should be reserved for the first-due ladder truck company. If the engine and ladder truck company are responding one behind the other, the engine will need to pull past the front of the fire building to allow access and placement of the ladder truck company. Should they enter the street from opposing directions, the engine company will need to stop short of the fire building.

The ladder truck should be spotted directly in front or on the corner of the fire building, depending on the intended use for placing the main ladder or for using a platform.

Seat Belt Usage

There is always the possibility of firefighters being killed or seriously injured in accidents. One area that UPS did not address in 1917 (but certainly has in the ensuing years) was the need for seat belts. In the past, firefighters have been seriously injured and killed due to not wearing seat belts. No fire apparatus should move until the driver and all passengers are properly seated with their seat belts fastened. Ensuring this fact is a dual responsibility of both the company officer and the apparatus driver. No excuses are acceptable: buckle up every time, arrive safely, and be prepared to address the nature of the emergency call.

Responsibility

Our goal as firefighters should be to become courteous, defensive drivers. We should never assume or demand the right of way. It must be realized that the company officer is responsible for the route to be taken and for regulating the speed of the apparatus. Proper supervision and taking command of the route and speed sets the tone for a safe response and overall driving safety. Remember, the emergency is happening to someone else, and we are coming to solve their problems. Don't compound the situation by having an accident and not arriving.

Dealing with drivers who ignore us or cannot hear us can be frustrating, but as professionals we have to exhibit control at all times. The repercussions of not arriving in a timely manner can impact the lives of those who are counting on us for help. It could be a heart attack victim who, deprived of a first responder company or a paramedic unit, could suffer irreparable damage or die. A person trapped in a burning building may succumb if having to await the arrival of the second-due company.

Summary

Understanding the behavior of fire and the different classes of fires will assist firefighters in their attempt to handle an incident scene. Further knowledge of how a fire spreads and through what mediums can help in forecasting strategies, tactics, and tasks. Knowing the stages of fire growth and the potential conditions of rollover, flashover, and backdraft will aid in incident scene safety.

Remember the six Ps: "Proper Prior Planning Prevents Poor Performance."

Prior knowledge is the foundation for information that a well-informed Incident Commander can utilize for handling specific problems. The knowledge of the amount of fire flow needed is a tool. As more data becomes available, it can be added to the preplanned information. The better the planning, the more professionally the scene will be managed. The more decisions made in the preplanning stages, the fewer that have to be made at the emergency.

If we don't plan for emergencies, we cannot deal with them effectively. Preplanning allows a fire department to be forewarned; to be forewarned is to be prepared. Once problems are identified, the way has been paved for successful emergency operations.

Aspiring firefighters looking to future advancement have a lot of work ahead of them. A firefighter desiring a position of command will need to do a great deal of studying and personal development. Remember, a promotion alone does not ensure leadership and command presence traits. They are acquired through hard work, persistence, and perseverance.

Though we do not usually associate the position of leader with guts, mettle is certainly needed to properly handle some of the tasks that come our way. There are times when we have to make unpopular decisions. They may come in the form of countermanding a previous order given by a subcommand or changing a fireground strategy to bring a situation under control.

We can learn from our mistakes as well as build on our successes. We must realize that what has worked in the past will probably continue to work in the future. Since the methods used to accomplish assignments vary, the company officer is responsible for the readiness of his or her company as a unit and for the members individually. Successful fire departments are those with a strong core of company officers. Likewise, chief officers are the leaders. Their actions will reflect strongly on their fire departments. A successful fire department needs strong leaders at all levels.

The 16 Firefighter Life Safety Initiatives must be adopted and enforced by all fire departments. Only through a combined effort can firefighter injuries and deaths be reduced. Ignoring these safety areas will directly impact brother and sister firefighters and their families.

Review Questions

1. What is the fire tetrahedron?
2. What are the five classes of fires? Define each class and describe the symbol, its coloring, and alphabetical letter that depicts each class.
3. What are the methods of heat transfer?
4. List the five stages of fire.
5. Define rollover, flashover, and backdraft.
6. What various types of training are used in your fire department?
7. Who benefits the most from training—the firefighter, the officer, or the fire department? Explain.
8. Discuss the different types of preplans and decide which type would be most effective in your department.
9. In your department, what buildings' preplans should include needed fire flow, or should all buildings that are preplanned have this information? Explain.
10. In your department, what are the requirements for becoming a company officer? Chief officer?
11. Do the requirements for becoming both company and chief officer accomplish the needed goals in your department? Why? Why not?
12. What traits do you recognize in effective fire officers? List them.
13. Who is the best fire officer you have served under? Why?
14. How does your department assign tools and fireground responsibilities prior to arrival on the incident scene? Is this method effective or could another method be more effective? Explain.
15. What is the goal of the 16 Firefighter Life Safety Initiatives?
16. Select two of the life safety initiatives and explain how your fire department has implemented them. If you are not a fire department member, summarize how in your estimation two of these initiatives could be implemented.

Suggested Readings, References, or Standards for Additional Information

US Fire Administration (USFA)

Routley, J. G. "The Value of Preincident Planning for Effective Emergency Management." Technical Report 051.

Routley, J. G. "Three Firefighter Fatalities in Training Exercise." Technical Report 015.

For statistics on firefighter fatalities in the United States, see the USFA website: www.usfa.fema.gov. Suggested search titles: "firefighter fatalities in the United States" or "annual report on firefighter fatalities in the United States."

For reports on the summits and mini-summits, see www.everyonegoeshome.com.

National Fire Protection Agency (NFPA)

Standards:

1041: Standard for Fire Service Instructor Professional Qualifications

1035: Standard on Fire and Life Safety Educator, Public Information Officer, Youth Firesetter Intervention Specialist, and Youth Firesetter Program Manager Professional Qualifications

1021: Standard for Fire Officer Professional Qualifications

1002: Standard on Fire Apparatus Driver/Operator Professional Qualifications

1001: Standard for Fire Fighter Professional Qualifications

1000: Standard for Fire Service Professional Qualifications Accreditation and Certification Systems

Related Courses Presented by the National Fire Academy, Emmitsburg, Maryland

Practical Applications of Fire Dynamics and Modeling

Training Operations in Small Fire Departments

Leadership I: For Fire and EMS: Strategies for Company Success

Leadership II: For Fire and EMS: Strategies for Personal Success

Leadership III: For Fire and EMS: Strategies for Supervisory Success

Executive Development

Executive Leadership

Leadership and Administration

Executive Planning

Strategic Analysis of Community Risk Reduction

Training Program Management

Challenges for Local Training Officers

21st Century Training for Fire and EMS Training Officers

New Fire Chief

New Fire Chief I: Challenging Issues

New Fire Chief II: Administrative Issues

New Fire Chief III: Contemporary Issues

Related Courses Presented by the Emergency Management Institute, Emmitsburg, Maryland

Managing Classes and Classrooms

Leadership and Influence

Development and Evaluation of Training

2
Management Tools

> " It is not the critic who counts: not the man who points out how the strong man stumbled or where the doers of deeds could have done them better. The credit belongs to the man who is actually in the arena, whose face is marred by dust and sweat and blood, who strives valiantly . . . who, at the worst, if he fails, at least fails while daring greatly, so that his place shall never be with those cold and timid souls who knew neither victory nor defeat. "

—President Theodore Roosevelt

A command post will benefit the Incident Commander and outside agencies at the incident. *Used with permission of Joseph Hoffman.*

KEY TERMS

incident scene management, *p. 36*

US Department of Homeland Security (DHS), *p. 41*

emergency operations center (EOC), *p. 45*

Incident Commander (IC), *p. 53*

incident scene control, *p. 72*

status report, *p. 75*

OBJECTIVES

Upon completion of this chapter, the reader should be able to:

- Understand the origin of the incident command system.
- Understand the implementation of the National Incident Management System.
- Understand the positions of an incident management system.
- Discuss the position of Command.
- Know the different types of status reports and the information required in each type of report.

Command and control is necessary at an emergency scene if the problems found are to be mitigated. Organization requires the implementation of an incident management system. Chapter 2 discusses the positions/roles in an incident management system. It also discusses implementation of the command position at an incident.

This section looks at the need for and types of status reports that are necessary at an incident scene and the information needed in the various types of status reports.

Incident Scene Management

The variety of emergencies that confront firefighters daily demands that fire departments have and use an incident command system for **incident scene management**. In 1970, multijurisdictional fires consumed large portions of wildland areas and structures in southern California and were the impetus for the development of a new incident management

incident scene management ■ The utilization of an incident command system to achieve command, control, and coordination at an incident.

FIGURE 2-1 In addition to routine fires or emergency medical incidents, the incident command system can handle unusual incidents, such as a plane crash. *Used with permission of William J. Shouldis.*

system (IMS) known as the incident command system (ICS). ICS is the result of a federally funded project called FIRESCOPE. This group consisted of federal and state forestry service personnel along with city and county fire departments. FIRESCOPE's charter was to examine various aspects of interagency response to incidents. These agencies saw the need to document a system that allowed them to work together toward a common goal in an effective and efficient manner.

FIRESCOPE derives its name from **FI**re **RES**ources of California Organized for Potential Emergencies. The FIRESCOPE ICS is primarily a command and control system delineating job responsibilities and organizational structure for the purpose of managing day-to-day operations for all types of emergency incidents. While originally developed for wildland incidents, it was found that the system could be easily applied to day-to-day fire and rescue operations. It is also flexible enough to manage catastrophic incidents involving thousands of emergency response and management personnel. This system is often referred to as the National Fire Academy Model Incident Command System. (See Figure 2-1.)

NATIONAL INTERAGENCY INCIDENT MANAGEMENT SYSTEM (NIIMS)

A group of federal agencies, which included the Bureau of Land Management (BLM), the Bureau of Indian Affairs (BIA), the US Fish and Wildlife Service (USFWS), the US Forest Service (USFS), representatives of the National Association of State Foresters (NASF), and the National Park Service (NPS), found ICS to be adaptable to a wide range of situations, including floods, hazardous materials incidents, earthquakes, and aircraft crashes. This was accomplished by introducing minor terminology, organizational, and procedural modifications to FIRESCOPE ICS. These modifications created the National Interagency Incident Management System Incident Command System (NIIMS ICS), which was adaptable to an all-hazards environment. While tactically each type of incident may be handled somewhat differently, the overall incident management approach still utilizes the major functions of the ICS.

NIIMS consists of five major subsystems that collectively provide a total systems approach to risk management. The subsystems are:

- ICS, which includes operating requirements, eight interactive components, and procedures for organizing and operating an on-scene management structure
- Training that is standardized and supports the effective operations of NIIMS
- A qualification and certification system that provides personnel across the nation with standard training, experience, and physical requirements to fill specific positions in the ICS

- Publications management that includes development, publication, and distribution of NIIMS materials
- Supporting technologies such as orthophoto mapping, infrared photography, and Multi-agency Coordination Systems (MACS) that support NIIMS operations

In the ensuing years FIRESCOPE and NIIMS were blended. The National Wildfire Coordinating Group (NWCG) was chartered to coordinate fire management programs of the various participating federal and state agencies. The FIRESCOPE agencies and the NWCG have worked together to update and maintain the Incident Command System Operational System Description. This document would later serve as the basis for the National Incident Management System (NIMS) ICS.

Incident Command System

ICS is a management system designed to control personnel, facilities, equipment, and communications throughout an emergency operation. It is intended to begin developing from the time an incident occurs until the requirement for management and operations no longer exists. The structure of the ICS can be established and expanded depending upon the changing conditions of the incident.

The system can be used for any type or size of emergency, ranging from a minor incident involving a single unit, to a major emergency involving several agencies. ICS allows agencies to communicate using common terminology and operating procedures. It also allows for the timely combining of resources during an emergency.

ICS is designed to be used in response to all hazards: fires, floods, earthquakes, hurricanes, tornadoes, tidal waves, civil disturbances, hazardous materials, or other natural or human-caused incidents.

ICS has management capabilities for:

1. Single jurisdiction/single agency involvement (Single Incident Commander)
2. Single jurisdiction/multi-agency involvement (Unified Command)
3. Multijurisdiction/multi-agency involvement (Unified Command)

ICS allows emergency responders:

- An organizational structure adaptable to any emergency or incident to which response agencies would be expected to respond
- A system applicable and acceptable to users throughout the country
- A system readily adaptable to new technology
- The ability to expand in a logical manner from an initial attack situation into a major incident
- Basic common elements in organization, terminology, and procedures
- Implementation with the least possible disruption to existing systems
- Effectiveness in fulfilling all management requirements costs

The system is intended to be staffed and operated by qualified personnel from any emergency services agency and may involve personnel from a variety of agencies.

Components of the Incident Command System   ICS has eight interactive components. These components provide the basis for an effective ICS concept of operation:

- Common terminology
- Modular organization
- Integrated communications
- Unified Command structure
- Consolidated action plans
- Manageable span of control
- Designated incident facilities
- Comprehensive resource management

Common Terminology   The need for common terminology in any ICS is essential. Major organizational functions and units are pre-designated and titled in the ICS, and the system's terminology is standard and consistent. To prevent confusion when multiple incidents occur within the same jurisdiction or on the same radio frequency, each one should be named. (A fire located on Broad Street would be named Broad Street

Command. A fire occurring at the same time on Market Street would aptly be named Market Street Command, alleviating any possible confusion between the two incidents.)

Common names are established and used for all personnel, equipment, and resources conducting tactical operations within the ICS, as well as for all facilities in and around the incident area. When units are given a designated function, they will no longer use their standard call letters. They will adopt their new designation for all communications, i.e., Engine 1's officer, assigned as Division 1 Supervisor, will then use Division 1 for all radio communications and not Engine 1. If Battalion 1 assumes command, it will use the title Command or Broad Street Command when communicating on the incident scene. Likewise, facilities will be designated as Base, Staging, Command Post (CP), etc.

Modular Organization   The ICS organizational structure develops in a modular fashion from the top down at any incident. The functional areas, which are implemented as the need develops, are Command, Operations, Logistics, Planning, and Finance/Administration. The command function is always established. The specific ICS organizational structure for any incident is based on the incident's management needs. For example, a simple incident does not require staffing sections to manage each major functional area. The operational demands and the small number of resources do not require the delegation of management functions. However, a complex incident may require staffing sections to manage each major functional area, and the number of resources committed may require delegating management functions.

Modular expansion can be implemented at an incident such as a mass casualty or hazardous materials incident. These particular incidents create specific types of problems that need to be addressed. A mass casualty incident may require triage, treatment, transportation, and other functions to be initiated. Similarly, a hazardous materials incident will require trained technicians to mitigate the problems through product identification, intervention, and mitigation. By assigning a competent individual as a Group Supervisor or Branch Director, these critical types of incidents can be well managed. The use of a branch instead of a group to handle mass casualty or hazardous materials incidents readily allows the Branch Director to expand the branch without span of control concerns while responding to the myriad of problems that he or she may confront.

Integrated Communications   Integrated communications involves managing communications at an incident through the use of a common communications plan. Standard Operating Guidelines (SOGs) should be established using common terminology and clear text. Effective two-way communication is essential to effective incident management. Not only is it important that messages are received, but it is also important that they are properly acknowledged. Interoperability of communications is needed for diverse agencies to be able to effectively interact at an incident scene. This includes the ability of fire, police, and other agencies to have access to common communication networks.

Unified Command   The command function within ICS may be conducted in two general ways. Single command may be applied when there is no overlap of jurisdictional boundaries or when a single Incident Commander (IC) is designated by the agency with overall management responsibility for the incident. Unified Command may be applied when the incident is within one jurisdictional boundary, but more than one agency shares management responsibility (i.e., a civil disturbance requiring a large commitment from fire and police departments where civil unrest and multiple fires are involved). Unified Command also is used when an incident is multijurisdictional in nature and more than one agency needs to share overall management responsibility (i.e., a wildland fire involving federal, state, and local fire departments, or a natural or manmade disaster involving multiple agencies).

Incident Action Plan   Every incident needs an Incident Action Plan (IAP). On most incidents the IAP will not be written; rather, it usually is a thought process of the IC. Written IAPs should be considered whenever:

- Multiple jurisdictions are involved in the response
- The incident is of a long duration or will involve multiple operational periods

- A number of ICS organizational elements are activated (typically when all General Staff Sections are staffed)
- It is required by agency policy
- Hazardous materials are involved in the incident
- Special operations would involve confined space rescue, trench rescue, etc.

The plan should cover incident objectives, strategies, tactics, and support activities needed during the operational period. This can be accomplished on long-duration incidents where the incident management team (IMT) will develop the upcoming operational period IAP. The IMT normally has a formal planning meeting to establish an IAP for the next operational period. This plan will be its blueprint to handle the current and foreseeable problems. (IAPs are discussed in further detail in Chapter 3.)

An operational period is usually a set period of time that an IMT will be supervising an incident; it can be a 6-, 8-, 12-, or 24-hour time period during the response phase, and weeks or months during a recovery phase.

Manageable Span of Control Span of control refers to the number of personnel reporting to any given individual. Optimal span of control in the ICS is five, with an acceptable spread of two to seven. With a situation that is not yet under control, typically no one operating under the ICS should have more than five personnel reporting to him or her (the rule of five).

Span-of-control ratios can be driven by a number of factors. They are:

- *Training/experience level of subordinates.*   Poorly trained or less experienced personnel require more direct supervision, thereby lessening the number of subordinates one can manage effectively. Likewise, highly trained and experienced personnel require less direct supervision.
- *Complexity of the incident.*   The more complex the incident, the less the number of personnel that can be effectively managed. A hazardous materials incident may require more mental concentration, thereby leaving less time available to supervise personnel.
- *Type or time frame of the incident.*   The speed in which an operation is progressing may influence the span of control. A fast-moving incident may require a tighter span of control with fewer divisions/groups in place; in a slower-moving operation such as overhaul, the supervisor is less pressed for time and decision making and therefore can manage more personnel.

For span-of-control purposes, the following functions are not counted against the rule of five as reporting to a supervisor: Safety Officer, Liaison Officer, PIO, Rapid Intervention Crew, and Staging Area Manager.

By dividing the incident into manageable segments, the IC is able to reduce the number of individuals directly reporting to him or her and is able to properly manage the incident.

Command officers must anticipate span-of-control problems and prepare for them, especially during rapid buildup of incident organization. Effective management is difficult if too many people are reporting to one supervisor.

Designated Incident Facilities Designated incident facilities—such as a Command Post (CP), an Incident Base, or a Staging Area—can be established based on the requirements of the incident. The IC or Logistics determines when these facilities are established and where they are located.

The CP is the location from which all incident operations are directed. Only one CP should be established per incident. (The establishment of only one CP is critical when operating with multiple agencies. In the past, each agency wanted to establish its own CP, which led to confusion and a weak Command at these incidents.) Cooperating and assisting agency representatives are assigned to the Liaison Officer. Those agencies with a statutory responsibility for incident outcome will be part of a Unified Command. An incident base can be implemented for high-rise fires, hazardous materials incidents, and wildland fires. It is an area where primary logistics functions are coordinated and administered. (Whereas staging is a location from which units can be rapidly deployed within three minutes, the location of base will be a safe distance from the incident scene and units will not be readily deployable.)

Comprehensive Resource Management Comprehensive resource management may be accomplished using three different methods, depending on the needs of the incident.

1. *Single resources* include individual engines, squads, ladder trucks, rescues, crews, etc.
2. *A task force* is a group of resources, having common communications and a leader, that may be pre-established and sent to an incident, or formed at an incident (e.g., two engines, a truck, and a chief officer formed as a task force for a specific assignment).
3. *A strike team* is specified combinations of the same kind and type of resources, with common communications and a leader (e.g., in many areas, five engines and a chief officer would form a strike team of engines).

Comprehensive resource management, when performed effectively, should maximize resource use, consolidate control of large numbers of single resources, and reduce the communications load.

Major Functional Areas The ICS has five major functional areas:

1. Command
2. Operations
3. Planning
4. Logistics
5. Finance/Administration

Use of the ICS improves safety by providing proper supervision, accountability, coordinated efforts, and improved communications. Effective incident management also minimizes freelancing and can reduce the department's or jurisdiction's liability, as well as the financial impact of emergencies on the community.

US Department of Homeland Security

In the aftermath of the attacks on the United States of America at the World Trade Center and the Pentagon on September 11, 2001, the president of the United States and Congress created the **US Department of Homeland Security (DHS).**

DHS's responsibility of protecting the citizens of the United States encompasses many areas. One major function is coordination of all emergency responses. On March 1, 2003, the Federal Emergency Management Agency (FEMA) became part of DHS. FEMA's continuing mission within the new department is to lead the effort to prepare the nation for all hazards and effectively manage federal response and recovery efforts following any national incident.

HOMELAND SECURITY PRESIDENTIAL DIRECTIVES

On February 28, 2003, the president issued Homeland Security Presidential Directive-5 (HSPD-5), which directs the secretary of Homeland Security to develop and administer a NIMS. It states in part:

> This system will provide a consistent nationwide approach for Federal, State, and Local governments to work effectively and efficiently together to prepare for, respond to, and recover from domestic incidents, regardless of cause, size, or complexity. To provide for interoperability and compatibility among Federal, State, and Local capabilities, the NIMS will include a core set of concepts, principles, terminology, and technologies covering the incident command system; multiagency coordination systems; unified command; training; identification and management of resources; qualifications and certification; and the collection, tracking, and reporting of incident information and incident resources.

US Department of Homeland Security (DHS) ■ Its mission is to lead the unified national effort to secure the country and preserve our freedoms. While created to secure our country against those who seek to disrupt the American way of life, its charter also includes preparation for and response to all hazards and disasters.

The reason for and mandates of HSPD-5 are:

[in order] to prevent, prepare for, respond to, and recover from terrorist attacks, major disasters, and other emergencies, the United States Government shall establish a single, comprehensive approach to domestic incident management. The objective of the United States Government is to ensure that all levels of government across the Nation have the capability to work efficiently and effectively together, using a national approach to domestic incident management.

All federal departments and agencies are required by Homeland Security to adopt the NIMS and to make NIMS adoption by state and local organizations a condition for federal preparedness assistance.

NIMS has six components:

1. Command and management
2. Preparedness
3. Resource management
4. Communications and information management
5. Supporting technologies
6. Ongoing management and maintenance

Command and Management The NIMS standardizes incident management for all hazards and across all levels of government. The NIMS–standard incident command structures are based on three key components:

- *Incident Command System* ICS is a standardized on-scene emergency management designed to provide for the adoption of an integrated organizational structure that reflects the complexity and demands of single or multiple incidents, without being hindered by jurisdictional boundaries. ICS is the combination of facilities, equipment, personnel, procedures, and communications operating with a common organizational structure, designed to aid in the management of resources during incidents. ICS is used for all kinds of emergencies and is applicable to small as well as large and complex incidents. ICS is used by various jurisdictions and functional agencies, both public and private.
- *Multi-agency Coordination Systems* On large- or wide-scale emergencies that require higher-level resource management or information management, MACS may be required. MACS is a combination of resources that are integrated into a common framework for coordinating and supporting domestic incident management activities. These resources may include facilities, equipment, personnel, procedures, and communications. The primary functions of MACS are to:
  - Support incident management policies and priorities
  - Facilitate logistics support and resource tracking
  - Make resource allocation decisions based on incident management priorities
  - Coordinate incident-related information
  - Coordinate interagency and intergovernmental issues regarding incident management policies, priorities, and strategies

  MACS includes emergency operations centers (EOCs), locations from which the coordination of information and resources to support incident activities takes place, and in certain multijurisdictional or complex incidents, Multi-agency Coordination Entities, which typically consist of principals from organizations with direct incident management responsibilities or significant incident management support or resource responsibilities. Direct tactical and operational responsibility for the conduct of incident management activities rests with the Incident Commander. (EOCs are discussed later in this chapter.)
- *Public Information Systems* The Public Information Officer (PIO) is a member of the command staff and operates within the parameters established for the Joint Information System (JIS). The JIS offers an organized, integrated, and coordinated mechanism for providing information to the public during an emergency. Key elements of a JIS include interagency coordination and integration, developing and delivering coordinated messages, and support for decision makers. To ensure coordination of public information, a Joint Information Center (JIC) can be established.

Using the JIC as a central location, information can be coordinated and integrated across jurisdictions and agencies, and among all governmental partners, the private sector, and nongovernmental agencies.

Preparedness   The NIMS establishes specific measures and capabilities that jurisdictions and agencies should develop and incorporate into an overall system to enhance operational preparedness for incident management on a steady-state basis in an all-hazards context.

The operational preparedness of our nation's incident management capabilities is distinct from the preparedness of the individual citizens and private industry.

Resource Management   The NIMS defines standardized mechanisms to describe, inventory, track, and dispatch resources before, during, and after an incident; it also defines standard procedures to recover equipment once it is no longer needed for an incident.

Communications and Information Management   Effective communications, information management, and information and intelligence sharing are critical aspects of domestic incident management. The NIMS communications and information systems enable the essential functions needed to provide a common operating picture and interoperability for incident management at all levels.

Supporting Technologies   The NIMS promotes national standards and interoperability for supporting technologies to successfully implement the NIMS, as well as standard technologies for specific professional disciplines or incident types. It provides an architecture for science and technology support to incident management.

Ongoing Management and Maintenance   The NIMS Integration Center was established by the secretary of Homeland Security to provide "strategic direction for and oversight of the NIMS, supporting both routine maintenance and the continuous refinement of the system and its components over the long term." The NIMS Integration Center is a multijurisdictional, multidisciplinary entity made up of federal stakeholders and state, local, and tribal incident management and first responder organizations. It is situated in the DHS's FEMA.

See Appendix F in www.pearsonhighered.com/bradyresources for the ICS Organization Chart.

DIFFERENCES BETWEEN FIRESCOPE AND NIMS

NIMS adapted the FIRESCOPE Model ICS for its own use and made two minor changes.

1. Under NIMS, the position of Intelligence/Investigations is created. This position is used to assist the IC(s) in developing sufficient data in the event of a terrorist event or other activity involving the need for police intelligence. It is a position that will normally be staffed by local, state, or federal police officials. The Intelligence/Investigations function may be organized in one of the following ways:
   - Officer within the Command Staff
   - Unit within the Planning Section
   - Branch within the Operations Section
   - Separate General Staff Section
2. Under NIMS, the Command Staff position previously known as "Information Officer" will be known as "Public Information Officer" (New Hampshire Department of Safety).

INCIDENT MANAGEMENT TEAMS

Throughout the United States, IMTs have been created at the national, state, and local levels. These IMTs consist of individuals who are trained in the all-hazard approach of managing various levels of incidents based upon their size and complexity. The IMT members are trained in the various sections and positions of the ICS. When a team is requested by federal, state, or local organizations, it can utilize its leadership and communication skills at complex incidents. Its area of expertise is the implementation of incident objectives, strategies, and tactics, and the initiation and management of IMTs. These IMTs are trained for all-hazard incidents to assume any role, position, or specific responsibilities required by the situation.

A concept that has been adopted in some areas is the creation of incident support teams (IST). It was recognized that many fire departments and other agencies do not have the staffing to handle the many roles needed at a major incident. Often, the calling of additional resources will bring additional companies, and the officers responding will be needed to manage those resources. The premise of the IST is to support and mentor the local IMT and not to take over an incident. These IST can support a local team and offer their expertise to provide various levels of documentation. A common arrangement is that the members of the IST report to the IC, and work in a support role for Planning, Logistics, Staging, Public Information Officer, Safety Officer, and other areas serving at the IC's pleasure. These IST members generally receive training to the same level as a Type 3 or 4 IMT.

INCIDENT TYPES

The US Fire Administration has created a typing of incidents to denote the level of expertise that is needed to manage the incident, based upon incident size. The types are listed from Type 5, which is the smallest or routine incident size, to Type 1, which is the most complex. By identifying the type of incident, an appropriate response of an IMT can be dispatched, and decisions can be made on resource requirements. Incident types are based on the following five levels of complexity.

Type 5
- The incident can be handled with one or two single resources with up to six personnel.
- Command and General Staff positions (other than the IC) are not activated.
- No written IAP is required.
- The incident is typically contained within an hour or two after resources arrive on scene.
- Examples include a vehicle fire, an injured person, or a police traffic stop.

Type 4
- Command Staff and General Staff functions are activated only if needed.
- Several resources are required to mitigate the incident, possibly including task forces or strike teams.
- The incident is typically contained within one operational period in the control phase, usually within a few hours after resources arrive on scene.
- The Agency Administrator may have briefings to ensure the complexity analysis and that the delegation of authority is updated.
- No written IAP is required but a documented operational briefing will be completed for all incoming resources.
- Examples may include a major structure fire, a multiple vehicle crash with multiple patients, an armed robbery, or a small hazmat spill.

Type 3
- When capabilities exceed initial attack, the appropriate ICS positions should be added to match the complexity of the incident.
- Some or all of the Command and General Staff positions may be activated, as well as Division/Group Supervisor and/or Unit Leader level positions.
- A Type 3 IMT or incident command organization manages initial action incidents with a significant number of resources, an extended attack incident until containment/control is achieved, or an expanding incident until transition to a Type 1 or Type 2 team.
- The incident typically extends into multiple operational periods.
- A written IAP is typically required for each operational period.
- Examples include a tornado touchdown, earthquake, flood, or multiday hostage stand-off situation.

Type 2

- This type of incident extends beyond the capabilities for local control and is expected to go into multiple operational periods. A Type 2 incident may require the response of resources from out of the area, including regional and/or national resources, to effectively manage the operations.
- Most or all of the Command and General Staff positions are filled.
- A written IAP is required for each operational period.
- Many of the functional units are needed and staffed.
- Operations personnel normally do not exceed 200 per operational period and total incident personnel do not exceed 500 (guidelines only).
- The Agency Administrator is responsible for the incident complexity analysis, Agency Administrator briefings, and the written delegation of authority.
- Typically involve incidents of regional significance.

Type 1

- This type of incident is the most complex, requiring national resources to safely and effectively manage and operate.
- All Command and General Staff positions are activated.
- Operations personnel often exceed 500 per operational period and total personnel will usually exceed 1,000.
- Branches need to be established.
- The Agency Administrator will have briefings and ensure that the complexity analysis and delegation of authority are updated.
- Use of resource advisors at the incident base is recommended.
- There is a high impact on the local jurisdiction, requiring additional staff for office administrative and support functions.
- Typically involve incidents of national significance.

Emergency Operations Center

A large-scale incident can occur anywhere and at any time. Every community, large or small, has the likelihood of an overwhelming emergency event and is potentially at risk. Since these incidents can transcend jurisdictional boundaries, emergency planners need to assess the capability of all available resources to ensure community readiness. Responsibility for taking command, management, and preparedness actions is held by local government. The roots of a community's all-hazard preparedness program hinge on the departmental leadership, intergovernmental relations, and a detailed planning process that includes external assistance agreements. Emphasis is on technical information, based on an accurate assessment of risk, vulnerability, and capabilities.

The **emergency operations center (EOC)** is a tool to enhance emergency management functions and to identify a meeting place/location. Since resources will be scarce during a large-scale incident, the EOC can identify and obtain additional assets that are not always available to the on-scene IMT. The EOC mobilizes people and equipment to handle low-frequency/high-risk emergencies that are typically outside the ability of any single agency to resolve. The purpose of the EOC is to ensure that departmental response capabilities are maintained and authoritative information is disseminated to the general public. Capturing disaster intelligence at an EOC provides senior officials with real-time data to set strategic directions, establish priorities, allocate resources, and, under extreme circumstances, declare a disaster. These actions by an EOC allow field commanders to focus on the incident scene while the EOC staff takes responsibility for coordination and support endeavors.

The EOC is an integral component of the larger MACS. The entire MACS consists of Initial Dispatch, On-Scene Command, Coordination Resource Centers, Coordination Entities, and the EOC. Often major incidents require the activation of all components of the MACS during the response and recovery phases.

The establishment of an EOC will result in building a stronger public safety system when resource decisions are critical and there is a dire need for decisive actions. The EOC

emergency operations center (EOC) ■ The purpose of the EOC is to ensure that departmental response capabilities are maintained and authoritative information is disseminated to the general public. The EOC allows field commanders to focus on the incident scene while the EOC staff is responsible for handling coordination and support endeavors.

serves as a vital link between the IMT and resource centers. Under ideal conditions, the EOC is centrally located where representatives from various agencies gather during an emergency and perform tasks that require active communication among key officials. The EOC becomes the link for organizations that work behind the scene to coordinate and support the field forces in protecting the population at risk during a disaster. The delivery of information to the general public is a NIMS requirement and involves a JIS based on structure, plans, and procedures.

The integrated effort that it takes to efficiently run an EOC will increase operational readiness. SOGs and emergency operating plans (EOPs) will remove much of the guesswork when confronted with cascading risk and hazards. Proper planning will ensure a smooth transition during a disaster, from response to recovery to rebuilding.

WHAT MAKES THE EOC WORK?

- A good concept of operation
- Good teamwork
- Good space
- Good technology
- Good communication
- Regular practice

TYPES OF EMERGENCY OPERATIONS CENTERS

Fixed Facility

The primary type of EOC is a central location with permanent equipment. It is usually found in an existing multipurpose government building that will reduce delays in the arrival of senior officials and staff personnel. It should be located in a secure area, since chaos would be created if, during an emergency event, the EOC were endangered. It is important for financial and logistical reasons to identify potential alternate sites for reliability and suitability as part of an overall community risk management plan. Local government should follow the federal government guidelines that suggest a need for continuity of government (COG) and continuity of operations (COOP).

The Mobile EOC

When a disaster is widespread and the decision makers have an urgent need to see the situation, a mobile EOC can be utilized to transmit pictures or video to the Policy Group from the vicinity of the incident. The flexibility of a mobile EOC allows for optimal interaction between the IMT and the Policy Group when reviewing critical information. A tractor/trailer style vehicle can offer a degree of comfort, while a full communication center with software can tie together radio frequencies of different agencies. Vehicles should have linkage to the Internet and multiple phone connections. Most have a security camera system that monitors all six sides. The mobile EOC allows for interagency planning from various locations and can provide valuable insight for executive decision makers in a stable situation with short-term recovery challenges.

The Virtual EOC

This innovative concept has combined with new technology to replace the traditional brick-and-mortar EOC facility. Virtual EOCs are a state-of-the-art emergency management solution that exist solely or partially in cyberspace, with private networks and satellite communication. Remote locations will reduce the reflex time and limit the vulnerability of having all senior officials assembled at one central location. With a virtual EOC, authorized individuals have anytime, anywhere access, via a user name and password, to multiple secure servers with the required software applications at high-tech terminals.

Coordination and Support

The EOC is responsible for all organizations to work together and be aware of each other's missions. In any type of EOC, whenever a complex incident or widespread disaster strikes a community, the efforts of traditional and non-traditional first responders must be closely coordinated and fully supported. To manage the numerous resources from different disciplines, EOC and IMT must build a solid reporting relationship that avoids conflicts, confusion, and duplication of efforts. EOC staff will need different competencies, based on the kind of damage or destruction caused by a manmade or natural disaster.

Without adequate and accurate information, the EOC will fail in its unique mission. The EOC staff and IMT need sufficient time to create a meaningful plan of action based on factual data. History has shown there can be a serious disconnect between the IMT and EOC when a joint planning process is not practiced. After-action reports of major incidents clearly reveal that local resources will be overwhelmed and will require external assistance as well.

For effectiveness at the emergency scene, a single EOC facility that is well-designed and sufficiently staffed will improve coordination and support endeavors. A strong partnership between the EOC and IMT will protect lives and property, as well as minimize the impact of response and recovery problems. (See Figure 2-2.)

ESSENTIAL FUNCTIONS OF AN EOC

The EOC must maintain a comprehensive view for continuity of government. Planning enables agency representatives to focus on vital services to maintain the safety and well-being of the public-at-risk. Factoring in all of the essential functions will serve as a decision making tool to fairly distribute necessary resources during a crisis.

A workable EOP requires a commitment to sufficient staffing levels to accurately chart the critical information. This allows time for the decision makers to gain an understanding of the documented data. Understanding the impact of these essential functions will provide strategic guidance and could resolve any conflicting agency policies and/or procedures.

The proper time to develop and practice an EOP is before a community suffers a low-frequency/high-risk incident. Essential functions that must be documented during EOC activation are:

FIGURE 2-2 Mock drills and tabletop exercises allow the EOC members to reinforce essential functions and duties. *Used with permission of Jim Smith, Jr.*

- Damage assessment reports
- Resources acquired and used
- Media and public's request for assistance
- Accounting of public safety expenses
- Response plan for providing evacuation and sheltering
- Support services for food, water, ice, and utilities
- Incident log of occurrences and actions

 ON SCENE

In 2005, during Hurricane Katrina, much of the infrastructure to support communication endeavors in the Gulf Coast region was heavily damaged. This made the transfer of information difficult or impossible for some time, which caused information delays, raised the risk to responders, and created inferior emergency notifications to the public.

DEVELOPMENT OF AN EOC

The EOC development process begins with a formal assessment and defining the area's requirements for readiness, based on likely threats. The nature of the incident and resource requirements will determine the agencies that will be represented in the EOC.

An EOC facility may range from dual-use conference rooms to a complete stand-alone facility. Either type must meet life safety regulations, have instant access to governmental records, and be designed with enough space. Providing for basic human comforts for long periods of time is critical to success.

ANALYZING ACCESSIBILITY

Research must be done to determine if there has been a population shift or changes in the configuration of the bridges, rails, and roadway systems that could cause congestion and unreasonable delays in the arrival of key officials, suppliers, vendors, or support personnel to the EOC.

A periodic hazards analysis of access to an EOC could reveal modern threats and changing vulnerability to a geographic area. The high vulnerability for extensive damage by an earthquake should be a clear warning that any investment of an EOC in a low-rise or high-rise district may not be prudent. A history of flash floods in low-lying areas of a community could significantly extend travel and increase the potential for cascading events. When constructing an independent facility, the elevation should be set at the 500-year flood mark, while a renovated building should be located above the 100-year flood plain. Feasibility studies should review these access and egress issues.

SITE SELECTION

Construction of a separate and dedicated EOC facility is often beyond the financial reach of most communities. If there are a low number of local-level full activations of the EOC, there are many monetary benefits in collocating it with emergency dispatchers.

Every effort should be made to locate the EOC close to the heart of government offices for ease of travel. Support teams of administrative, maintenance, and security officials may need the convenience of access to vital members of the emergency management operations section, such as law enforcement, fire, emergency medical services, public works, public health, and mass care.

Site selection for an EOC should consider safety, suitability, and security. Whenever possible during activation, personnel should be able to take a comfortable break away from their EOC work space, have an opportunity for physical exercise, and possibly be within walking distance for meals. Locations that can accommodate a helicopter landing can be advantageous when widespread damage needs aerial observation by elected officials or a technical specialist.

 **ON SCENE**

In retrospect, the terrorist bombing in 1993 in New York City at the World Trade Center should have been a strong indicator that it was a targeted area and a poor choice for the Port Authority to locate its main EOC in the World Trade Center's Building #7, which was destroyed in the attacks on September 11, 2001.

STAFFING AN EOC

Because daily use of a full-time staff at the EOC is not cost effective or practical in many communities, emergency duties need to be designated well in advance of an activation. If a plan is to be effective, its content must be known and understood by those who are responsible for implementation. Up-to-date rosters that include current levels of training must be maintained as a means for credentialing. Succession planning found in the COOP Guidance document recommends that each EOC assignment should be three-deep in trained personnel. Contact lists with phone numbers and email addresses should be updated every three months. Staff training should be held on a quarterly basis. A full review of the EOP should be scheduled for each year.

Job aids should allow easy conversions of a normal work space into a cohesive EOC. The goal for changeover should be targeted at 30 minutes. This approach allows competent individuals to take necessary actions and begin to collect information until properly relieved. They can refer to a posted checklist containing detailed steps on the transition procedure. Training will increase the odds of a quick conversion and proficiency by assigned personnel.

ORGANIZATIONAL STRUCTURE OF AN EOC

EOCs exist in many forms; no single organizational structure is correct for all jurisdictions. The key is to maximize the efforts of the staff and allow decision making at the lowest practical level. Presidential Directive #5, which outlines command and management procedures, does not mandate that a community adopt any particular model as its organizational structure. However, in the future, federal grant money may be linked to the specific way an EOC is organized. It should be structured with the intent of facilitating the smooth operation of data collection, documentation, and executive decision making, and to maintain public confidence under ever-changing conditions while acting on routine, priority, and classified information. The EOC organization must have the flexibility to shape itself to the emergency incident. That means it should not be so rigid that it cannot be easily modified, expanded, or diminished as required by the situation.

Local government is entirely responsible for the management of emergency activities within its jurisdictional boundaries. Response priorities focus on getting the right resources to the right place at the right time. This focus promotes total uniqueness in the concept of operations at an EOC. The only accountability requirement is for strict site security. Basic measures are needed to ensure that unauthorized personnel can be excluded, such as a reception area or checkpoints for bag inspection, sign-in sheets, photo identification, and single point of entry.

Typically, there are four ways to organize an EOC facility, each with some recognized coordination and support challenges. Over time, many jurisdictions have used these organizational structures very successfully.

The four organizational structures for an EOC can be based upon:

- Major Management Activities
- Incident Command System
- Emergency Support Function
- Multi-agency Coordinating Group

Major Management Activities Structure

Policy Group Comprised of high-level elected officials and department heads who focus on the overall objectives and priorities for the community. Decisions made by the Policy Group set the direction for implementation by the Coordination, Resource, and Operations Groups.

Coordination Group Personnel responsible for the collection and analyzing of data that an EOC must gather, including advanced predictions based on the essential functions impacted by the damage, resource allocation, public information, and expenditures.

Resource Group Representatives from any agency or organization that provides, or could be asked to supply, resources to the scene. These organizations may include transportation agencies, utility companies, business and industry representatives, and mutual aid partners.

Emergency Management Operations Group Representatives from any agency with responsibility for any portion of the response or recovery. Units within the Operations Group are dictated by the specific incident for a tactical assignment. This can include law enforcement, fire, public works, emergency medical services, and numerous other agencies.

Advantage and Disadvantage of Major Management Activities Structure The advantage of this model is that the organizational structure is relatively simple, with straight lines of communication and chain of command. With this model, all key problem-solvers and representatives from participating agencies can contribute to decision making and resource allocation.

The disadvantage of a Major Management Activities structure is the weak link with on-scene commanders. There is not a one-to-one match between the organization of the on-scene IMT and the EOC organization.

Incident Command System (ICS) Structure

Emergency Program Manager This designated person, who fills the top EOC position and serves a similar role to the Policy Group, makes executive decisions that establish the overall objectives and concept of operation for the EOC. Often under this format, the Emergency Program Manager is viewed as the EOC commander. This military style command structure can increase interaction and heighten the level of situational awareness.

Emergency Management Operations It takes responsibility for coordinating with and supporting on-scene responders. Position titles of branches, divisions, and groups are created and organized as necessary to support the incident.

Emergency Management Planning It is responsible to gather and analyze information, informing decision makers of changes in the use of resources. Technical Specialists may be used to provide special insight and expertise.

Emergency Management Logistics It serves as the single ordering point from the IMT. Coordinating the network of primary and backup communications equipment will assist in any large-scale incident but is especially meaningful during an evacuation or reentry with special transportation and housing needs.

Emergency Management Finance/Administration It takes responsibility for designing financial projections of the incident cost.

Advantage and Disadvantage of the Incident Command System Structure The advantage of this model is the clarity of roles and the functional integrity, which leads to a clear contact point between the IMT and the EOC. The coordination and support of logistical and financial duties relieves the workload at the incident scene and dispatch center.

The disadvantage of an ICS structure in the EOC is the potential for confusion about authority at the incident scene versus the EOC facility, due to the similarity of the titles.

Emergency Support Function (ESF) Structure

WHAT ARE EMERGENCY SUPPORT FUNCTIONS (ESFs)?

For incidents requiring local mutual aid resources, the local EOC typically uses local agreements already in place. For incidents requiring intrastate resources, the local EOC usually goes to the state EOC to fill resource requests. For incidents requiring interstate resources, the local EOC normally reaches out to the state EOC to facilitate resource requests. To acquire federal resources, the local EOC also uses the state EOC to facilitate resource requests.

ESFs provide the structure for coordinating federal interagency support for a federal response to an incident. They are a mechanism for grouping the functions most frequently used to provide federal support to states and federal-to-federal support, for both declared disasters and emergencies.

The roles of the ESFs are:

ESF #1 – Transportation
ESF #2 – Communications
ESF #3 – Public Works and Engineering
ESF #4 – Firefighting
ESF #5 – Emergency Management
ESF #6 – Mass Care, Housing, and Human Services
ESF #7 – Resource Support
ESF #8 – Public Health and Medical Services
ESF #9 – Urban Search and Rescue
ESF #10 – Oil and Hazardous Materials Response
ESF #11 – Agriculture and Natural Resources
ESF #12 – Energy
ESF #13 – Public Safety and Security
ESF #14 – Long-Term Community Recovery
ESP #15 – External Affairs

The Operations Manager The Operations Manager is in charge of the EOC. The Operations area can include branches such as Firefighting, Public Works/Emergency Engineering, Public Health and Medical Service, Urban Search and Rescue, Mass Care, and Law Enforcement. Currently there are 15 branches under the National Response Framework (NRF).

Planning The Planning area includes Situation Analysis, Documentation, Advanced Planning, Technical Services, Damage Assessment, Resources Status, and the geographic information system (GIS).

Logistics The Logistics area includes a service and support branch.

Finance/Administration This area includes Compensation Claims, Purchasing/Procurement, Cost Concerns, timesheets for personnel, and Disaster Financial Assistance, based on legal records such as contracts, accounting records, and property management photos.

Advantage and Disadvantage of the Emergency Support Function Structure The advantage of this model is that where these agencies have sufficient expertise and depth of personnel, it appeals to local and state EOCs because there is a clear one-to-one relationship with the NRF as well as with on-scene ICS organizations.

The disadvantage of an ESF structure is that local EOCs may not have the required depth of personnel, training, and expertise, which means that all positions will not be staffed and the local organization may not correspond directly with the state and federal ESFs.

Multi-agency Coordinating (MAC) Group Structure

A MAC Group is made up of organization, agency, or jurisdiction representatives who are authorized to commit resources and funds. The success of the MAC Group depends on its current membership. Sometimes membership is obvious by the organizations that are directly impacted and have a resources commitment to an incident. Often the organizations that should be members of a MAC Group are less obvious. These organizations may include the local Chamber of Commerce, volunteer groups, the Salvation Army, American Red Cross, faith-based charities, or other organizations with special expertise or knowledge. These groups may not have hard resources or funds to contribute in the response and recovery phase, yet their contacts, political influence, or technical expertise could provide the foundation for a collaborative effort.

The MAC Group Coordinator This is an optional position that provides supervision to the various components. Members of the MAC Group directly distribute the results of their deliberation to their own organizations as well as through the chain of command (MAC Entities, dispatch centers).

The MAC Group Situation Assessment Unit It collects and assembles information needed for the MAC Group to fulfill its role. At times, a MAC Group Resource Status Information Unit will gather information on the status of resources.

The Joint Information Center A JIC is a public information unit that has access to local information sources, governmental entities, and is responsible for coordinating a summary report. Public information must be organized around a JIS that is overseen by Public Information Officers. Intelligence information is routed from the EOC to the JIC for collection, validation, and public dissemination.

Advantage and Disadvantage of the Multi-agency Coordinating Group Structure The advantage of this model is that it works well to ensure coordination among other MAC Entities. It is useful where no system exists to provide short-term multi-agency coordination and decision-making. Typically, a MAC Group fits into a policymaking part of an existing EOC.

Some of the more common MAC Group applications include the following:

- A single jurisdiction may establish a MAC Group as part of its EOC function. In this application, it is important that the jurisdiction broadly define its role because of the impact on other agencies and organizations.
- MAC Groups are frequently defined geographically, especially when an emergency crosses jurisdictional boundaries.

- A MAC Group may be organized functionally. For example, law enforcement agencies at local, state, and federal levels may establish a MAC Group to assist in coordinating a response to a major terrorist activity.
- A MAC Group may be organized nationally. For example, during wildfire season, a national MAC Group convenes at the National Interagency Fire Center in Boise, Idaho. This MAC Group includes representatives from the federal wildland fire agencies, the states, FEMA, and the military.

The disadvantage of a MAC Group structure is a lack of clearly defined, standardized relationships to other MAC Entities. There is no associated implementation of staff, and it is rarely used as a stand-alone structure in an EOC. A generic MAC component can be used at any level of government.

CHARACTERISTICS OF AN EFFECTIVE EOC

Once the reference materials are in place and the SOG and EOP are written, training will be the key element for an effective EOC. It has been repeatedly shown that in a disaster scenario, most workers remember what they practiced rather than what they were told or had read. This is especially true when decision making gets very intense. Panic can set in, and even the simplest task can become difficult to remember. For building an effective organization, it is critical to maintain sustainable skills in data collection, documentation, and in setting parameters for decision making during an EOC activation. Personnel changes, reorganizations, and downsizing can have a dramatic effect on performance. Orientation, mock drills, tabletops, functional, and full-scale exercises all have value in reinforcing essential functions and duties. A yearly review of the development and fine-tuning of an EOC team should be conducted.

COMMUNICATIONS

Effective communication is the strength of any organization. Determining the scope of the disaster is only possible when first responders in the field accurately report conditions to the EOC so that positive actions can be undertaken. A communication network with the ability to transfer information between agencies with sufficient backup equipment is the minimum requirement.

Incident information must be shared:

- Within the EOC
- Between the EOC and IMT
- Between the EOC and the public
- Between the community's elected officials and other jurisdictions
- Between the community and state and federal officials

The EOC is the hub for communications with other levels of government, the private sector, and the general public. NIMS has established interoperability and redundancy as two basic requirements for communication at an EOC. Each can improve the chances that clear and concise reports are received from the IC, to allow the EOC to continually prioritize operational directions and resource allocation.

Interoperability

Public safety providers need to communicate on routine and expanding incidents with the staff from other responding agencies and to exchange voice and data communication, on demand and in real time. Interoperability has been a longstanding challenge in most communities for first responders and various partners who assist in tactical activities. The sharing and management of communication equipment, policies, procedures, and systems has proven troublesome during such diverse emergencies as winter storms, hurricanes, floods, earthquakes, terrorist actions, and nuclear power plant incidents.

Redundancy

Emergency providers need a reliable primary and secondary means to communicate. Having a redundant system is a top priority. Interagency planning is the key to improve reliability. The key to effective communication during a disaster is developing various levels of backup data lines and repeaters. No type of communication is 100-percent reliable and all systems have limitations. For first responders, radios are the main means of communicating during a disaster. It is critical to have emergency procedures for notifying field personnel to switch to a backup communication system. For the public, the use of a series of sirens, a paging system, or broadcasted warning signals have proven effective communication devices.

ACTIVATION AND DEACTIVATION OF THE EOC

The extent of EOC activation depends on established policies and the magnitude of the incident. Procedures for activation take effect when normal channels of direction and control are disrupted. Staff is called into the EOC when needed and increased as the situation warrants. The levels of activation are based on notification and time to react to circumstances. A hazard analysis can help to determine when a limited, partial, or full activation is recommended.

Deactivation levels should be based on current incident status and how much time will be required to satisfactorily complete the incident objectives. An obvious sign for demobilization is when the demand for resources begins to slow down.

FUNCTIONING IN THE EOC

Accountability is mandatory. Personnel must always follow the check-in process to ensure a proper briefing and full understanding of their duties. Succession plans must be posted for providing a clear chain of command for each position.

Being prepared means there is no substitute for planning, training, and exercising. Individuals are not very good at tasks they do not do often. Progressive communities have instituted cross-training to ensure that essential functions of an EOC will be documented, even with a small number of staff. All staff knowing their responsibilities and contacts, having a checklist, and understanding their activation procedures will prove very valuable during a low-frequency/high risk disaster.

It must be recognized that the EOC is the key to minimizing any duplication during the response or recovery phases of a large-scale incident. By having high-level decision makers and policymakers located together; there is a greater chance that all resources will be used to a maximum level of efficiency. The coordination and support by the EOC will assist the on-scene commanders in the accomplishment of strategic and tactical activities. The principles of emergency management—risk assessment, hazard analysis, and capability limitations—will ensure that best practices guide a smooth transition from daily duties to disaster management.

The Incident Commander

The position of Command encompasses every phase of an incident and must be implemented on every assignment. The **Incident Commander (IC)** assumes a high level of responsibility due to the need to protect lives and property while ensuring the safety of firefighters as they attempt to accomplish their tasks. (See Figure 2-3.)

The responsibilities of the IC or Command, as it is commonly referred to, include gathering and evaluating information relative to preplanning and size-up, as well as development and communication of plans.

The successful implementation of a command system requires delegation of authority and responsibility. Orders and decision making must be performed at the lowest level in an organization that has the ability to make those decisions.

Although the IC may delegate functional authority, he or she always retains ultimate responsibility for the incident. If the IC chooses not to delegate authority for one or more functions, then the IC must perform those functions.

The IC must determine the incident objectives and then direct available resources to accomplish incident goals. A responsive organization must be developed to ensure proper incident management by coordination of command, tactical operations, and support functions.

Incident Commander ▪ Has the responsibility for the overall management of an incident.

| Incident Commander |

FIGURE 2-3
Incident Commander.

The IC must be able to communicate effectively within the organization and assess feedback. The use of terms that are understood by all resources is critical to the proper managing of an incident. Gathering and assigning resources functionally and geographically are also included in the IC's responsibilities.

Overall effectiveness of the IAP must be continually evaluated to ensure that the operational decisions are correct and being properly applied. By using this feedback, the IC can modify the IAP if necessary.

COMMAND RESPONSIBILITIES

The command responsibilities of the IC are to:

- Assess incident priorities
- Determine incident objectives
- Develop an IAP
- Determine strategies
- Determine tactics when there is no Operations Section Chief
- Develop an appropriate organizational structure
- Manage incident resources
- Coordinate overall incident activities
- Ensure safety of on-scene personnel
- Coordinate activities of outside agencies
- Authorize the release of information to media

Unified Command

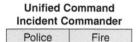

Unified Command Incident Commander	
Police	Fire

FIGURE 2-4 Unified Command.

Unified Command allows agencies with jurisdictional responsibility to participate in the management of an incident. This participation is demonstrated by developing and implementing a common set of incident objectives and strategies that all can subscribe to, without losing or abdicating agency authority, responsibility, or accountability (additional information on the role of Unified Command is contained in Chapter 3 under "Reviewing the Planning 'P'"). (See Figure 2-4.)

Incidents involving hazardous materials, natural disasters, terrorist events, or wildland fires may involve a number of jurisdictions and/or agencies that will need to be involved in the decision making process. This will require the implementation of Unified Command. Instead of one person being in charge of the event, command decisions will be made by a group of individuals appointed by their respective agencies. They will ultimately decide priorities for resource assignment, with overall success for all agencies and jurisdictions being the outcome. (See Figure 2-5.)

Under the Unified Command concept, all involved agencies contribute to the command process. Objectives, strategies, conducting integrated tactical operations, and maximizing the use of all available resources are decided jointly.

Selection of participants to work effectively within a Unified Command structure depends on the location and type of incident. They must support the incident with resources at their disposal. Problems can occur with Unified Command if individuals or groups do not subordinate their personal goals for the overall good of mitigating the incident. Previous training or experience of the individuals as a group will enhance the probability of their success. A Unified Command structure could consist of a key official from each jurisdiction, or representatives of several functional departments within a single political jurisdiction.

FIGURE 2-5 Unified Command may be necessary when multiple agencies are involved. The collapse of this pier used as a nightclub took the lives of three people and required extrications and water rescues of many others. This operation involved the interaction of numerous agencies. *Used with permission of William J. Shouldis.*

Unified Command can be implemented should any of the following occur:

- A fire spreads to include local, county, state, and federal fire agencies, where each fire department will be directly affected (multijurisdiction/multi-agency involvement).
- A hazardous material incident directly impacts numerous agencies within a single jurisdiction. This could involve fire, police, and public health agencies (single jurisdiction/multi-agency involvement).
- A mass casualty incident involves fire, emergency medical services (EMS), police, and other agencies (single jurisdiction/multi-agency involvement or multijurisdiction/multi-agency involvement).
- A terrorist event over a widespread area involves fire, police, and federal and state agencies (multijurisdiction/multi-agency involvement).

To be successful in implementing a Unified Command requires:

1. A written plan that stipulates how the system will be applied
2. Application of the plan in simulated situations
3. Support for Unified Command by all participants at an incident
4. Those serving as part of Unified Command to have the authority to deliver the needed resources from their own organization

(*Note:* Implementing the IAP under Unified Command is the responsibility of the Operations Section Chief.)

Some factors in helping to determine whether an agency should be part of Unified Command are:

1. Does the agency/jurisdiction have a clear legal or jurisdictional responsibility to respond to this incident?
2. Would this agency normally spend money to respond to some aspect of this incident?
3. Does this agency have funds and/or resources to support participation in this response organization?
4. Do the other agencies in Unified Command agree that this agency meets the criteria for Unified Command?

INCIDENT COMPLEX AND AREA COMMAND

ICS is designed to handle various types of emergency situations. To accomplish that task may require implementation of either an incident complex or area command.

Incident Complex

An incident complex is two or more individual incidents located in the same general proximity that are assigned to a single IC or Unified Command to facilitate management. These incidents are typically limited in scope and complexity and can be managed by a single IMT.

An example would be a number of wildland fires, each fire being separate, but in the same general area. Rather than create separate incidents for each fire, one IMT can be employed. An organizational tool to facilitate the handling of these fires would be the assigning of individual branches or divisions for each fire.

Area Command

The ICS is very flexible. To maintain that flexibility, it can expand to handle a very large incident, or area, with multiple IMTs. It accomplishes this management by establishing an Area Command. Area Command can be used any time incidents are close enough that oversight direction is required of multiple IMTs to ensure that conflicts do not arise.

Area Command can be implemented on incidents involving wildland fires, all-hazard incidents, and natural or man-made disasters affecting multijurisdictional areas. The size of these events creates the appropriate situation for the possible use of Area Command.

The criteria for implementing Area Command are:

- Several major or complex incidents of the same kind are in close proximity.
- Critical human or property values are at risk due to the incidents.
- Incidents will continue into the next operational period.
- Incidents are using similar resources and there are limited critical resources.
- Difficulties are encountered with inter-incident resource allocation and coordination.

Area Command is beneficial since:

- Coordination needed between IMTs can be accomplished by the Area Commander.
- Area Command will set priorities between incidents and allocate critical resources according to the priorities established by the Agency Executive (or Agency Administrator).
- Area Command helps the Agency Executive by ensuring that agency policies, priorities, constraints, and guidance are being made known to the respective ICs.

Note that Area Command does not create an IAP, but reviews the IAPs from the IMTs within Area Command. The Area Command organization should be kept as small as possible. Area Command can initiate the following functions as needed:

- Assistant Area Commander Logistics
- Assistant Area Commander Planning
- Area Command Critical Resources Unit Leader
- Area Command Situation Unit Leader
- Area Command Public Information Officer
- Area Command Liaison Officer

Area Command may also add Technical Specialists to provide specific information and expertise depending upon the type of incidents involved. It is important to remember that Area Command does not in any way replace the incident-level ICS organizations or functions. The above positions, if established, are strictly related to Area Command operations.

Defining Command

Leadership goes a long way toward ensuring that actions at an emergency scene will proceed properly. Someone must be in charge. At a fire or other incident, success or failure is entrusted to the IC. The primary duty of Command is taking control of the overall situation. It is the most visible function at the scene.

The IC is responsible for all functions at an incident scene. Every incident must have an IC and that person must follow a sequential process to ensure that nothing has been overlooked.

The National Institute for Occupational Safety and Health (NIOSH) has investigated more than 600 firefighter deaths. Shortcomings in the Command function was a common factor noted in traumatic fireground deaths. At times, transfer of Command was given to one who was not yet on the fireground (passing command), or no actual transfer of command was initiated and Command was accepted by a higher ranking fire officer on his or her arrival. Lack of accountability on the part of Command was another significant factor in firefighters' traumatic fireground deaths.

PROBLEMS FACING COMMAND

Command must be able to predict changes in the incident scene while evaluating the effectiveness of the firefighting efforts. The importance of firefighter safety demands that they be an integral part of a known plan. This will prevent firefighters from inadvertently being placed in dangerous positions.

The problems facing Command are the same regardless of the size of the department. The best solution to handling incident problems is calling for and receiving sufficient resources. When insufficient resources exist, it will impact the plan being formulated. When operating with limited resources, the IC may feel a tendency to become involved physically, which can result in critical areas being overlooked.

Firefighters can be seriously injured or killed if the command function is omitted or not used properly. This factor has been noted in many NIOSH reports on investigations into firefighter fatalities. There are enough dangers to contend with; lack of Command should not be one of them. A department with minimal staffing must look at a system as the best way of maximizing its resources and maintaining safety. Utilization of an IMS may be the single most effective way to increase firefighter safety.

Since an incident must be handled in an effective manner, someone must be in charge. When an emergency occurs, the fire department is called and we respond. When no one knows how to handle a particular emergency, everyone looks to the fire department for direction.

> It is important to realize that we have been called to stabilize the incident—not to be part of the problem, but the solution to the problem.

Constant reevaluation of the incident is necessary to ensure that the incident objectives, strategies, tactics, and tasks are accomplishing the goals of the IC. If not, adjustments will need to be implemented to reach those goals. (See Figure 2-6.)

UNITY OF COMMAND

Before many fire departments implemented a command system, fireground command was similar to an old western movie where Indians would attack by circling a wagon train. Chief officers would continuously circle the fireground, each giving orders. This often created conflicting orders, leading to confusion and frustration on the part of the firefighters.

Unity of command dictates that *no person reports to more than one person and everyone has someone to answer to.* Departments that adhere to the principle of unity of command avoid contradictory orders and their emergency scenes are better organized. Should a conflict occur, the person receiving the latest order should bring it to the attention of the officer giving the conflicting order and be guided by his or her reply.

The IC at most incidents will oversee the tactical operations and sector the scene, ensuring that his or her span of control remains manageable. The basic assignments for a building fire will usually include one company on fire attack, another assigned to search and rescue, another performing ventilation, and a rapid intervention crew. The position of Command will be assumed by the initial company officer or a chief officer.

Large or complex incidents make many demands on the IC. To guard against becoming overwhelmed and to reserve sufficient time to make decisions on problems that the incident presents, the IC must be prepared to expand the IMS based on incident needs and demands. The aim is to control the incident, not to develop a fancy system. As general staff positions are created, the section chiefs delegated these assignments can implement any additional components of the system that are needed to accomplish their goals. This is an important point. Additional positions should be implemented to address the complexity of problems that arise or are anticipated.

INCIDENT ACTION PLAN

An IAP must be developed. This IAP will require that information be gathered pertaining to the incident. On small incidents, the IAP will usually only consist of a thought process by the IC, while on larger incidents it should be written. (IAPs are discussed in detail in Chapter 3.)

FIGURE 2-6 Command must be able to predict changes in the incident scene while evaluating the effectiveness of the firefighting effort. *Used with permission of James H. Bampfield, Jr.*

It is the responsibility of the IC to determine the incident objectives. On a basic structure fire, or a fire that is routinely handled by the responding fire department, the incident objectives typically follow the incident priorities of life safety, incident stabilization, and property conservation. Once the incident objectives that must be accomplished are known, sufficient resources to achieve those goals can be requested. The plan will consider resources at the incident scene and those available upon request. It also must consider the response time and capabilities of the units.

NIOSH FIREFIGHTER FATALITY REPORT F2007–18

On June 18, 2007, nine career firefighters died when they became disoriented and ran out of air in rapidly deteriorating conditions inside a burning commercial furniture showroom and warehouse facility. The first arriving engine company found a rapidly growing fire at the enclosed loading dock connecting the showroom to the warehouse. The Assistant Chief entered the main showroom entrance at the front of the structure but did not find any signs of fire or smoke in the main showroom.

He observed fire inside the structure when a door connecting the rear of the right showroom addition to the loading dock was opened. Within minutes, the fire rapidly spread into and above the main showroom, the right showroom addition, and the warehouse. The burning furniture quickly generated a huge amount of toxic and highly flammable gases along with soot and products of incomplete combustion that added to the fuel load. The fire overwhelmed the interior attack crews and they became disoriented when thick black smoke filled the showrooms from ceiling to floor. The interior firefighters realized they were in trouble and began to radio for assistance as the heat intensified. One firefighter activated the emergency button on his radio. The front showroom windows were knocked out and firefighters, including a crew from a mutual-aid department, were sent inside to search for the missing firefighters. Soon after, the flammable mixture of combustion byproducts ignited, and fire raced through the main showroom. Interior firefighters were caught in the rapid fire progression and nine firefighters from the first-responding fire department died. At least nine other firefighters, including two mutual-aid firefighters, barely escaped serious injury.

Key contributing factors identified in this investigation include: not having a written standard operating procedure (SOP) for a safety and health program, operating without a written IMS, the lack of SOPs identifying requirements for members expected to serve in command roles, the need to ensure that the IC is the only individual with overall authority and responsibility for management of all activities at an incident, the need to ensure that the IC conducts an initial size-up and risk assessment of the incident scene before beginning interior firefighting operations, the need to communicate interior conditions to the IC as soon as possible and provide regular updates, the need to ensure that a stationary CP is established, the need to ensure that the IC maintains the role of director of fireground operations and does not become involved in firefighting efforts, the need to ensure the early implementation of divisions/groups, the need to ensure that the IC continuously evaluates the risk versus gain when determining whether the fire suppression operation will be offensive or defensive, the need to ensure that the IC maintains close accountability for all personnel operating on the fireground, the need to ensure that a separate Incident Safety Officer who is independent from the IC is appointed at each structure fire, the need to ensure that crew integrity is maintained during fire suppression operations, the need to ensure that a rapid intervention crew (RIC) is established and available to respond immediately to emergency rescue incidents, the need to ensure that adequate numbers of staff are available to respond immediately to emergency incidents, and the need to ensure that ventilation to release heat and smoke is closely coordinated with interior fire suppression operations.

There are numerous additional key contributing factors associated with the outcome of this fire that can be accessed at the NIOSH website. **See Appendix B: NIOSH Reports in www.pearsonhighered.com/ bradyresources to read the complete report and recommendations.**

DIRECTING RESOURCES

When implementing the IAP, consider the order of accomplishing tasks: the effectiveness and time needed and doing the right task at the right time. Directing resources or activities is a major function of Command.

Coordinating operations is similar to fitting the pieces of a puzzle together. The proper sequence of events must be followed to complete the right task at the right time. Line placement, forcible entry, search and rescue, and ventilation must be coordinated. The hose-line crew may not be able to gain entry into the fire area until ventilation has been effected. Opening a roof in the wrong location or before attack lines are ready can cause a fire to spread to other areas.

Command must evaluate decisions made prior to his or her arrival. A thorough analysis should ascertain if the plan in place will effectively achieve the goals required. If the proper goals will be reached, though by different means, the new IC must allow the plan to continue. If minor adjustments are needed, they should be done.

Because decisions will be based upon available information, time constraints are a realistic consideration. Use of reports received by subordinate officers will help. Though each situation is different, decisions are based on readily available information. Securing all information before making a decision, even if possible, consumes too much time and normally is not a practical approach at most incident scenes.

A difficult decision must be made when the plan in place will not accomplish what must be done. Whether modification, expansion, or even starting over again is required, the necessary changes must be made. The IC must not live with an intolerable situation. Conditions can change in minutes from a situation that originally required an interior attack to one that now requires an exterior attack. With rapidly changing conditions, orders must be given immediately, since a delay can jeopardize firefighters.

CONTROLLING THE INCIDENT

Effective control of an incident dictates that certain functions be performed. The IC at a major incident becomes a manager. He or she must make use of staff by delegating tactical decisions to subcommands, giving an overall objective, and allowing those who will achieve the objective the latitude to accomplish the task. Specific point-by-point orders become counterproductive. They allow no discretion by the individual officer, and if the task can't be performed exactly as ordered, it frequently necessitates additional questions.

An important point is that ample training is required of all members prior to the implementation of any command system. This includes not only simulations and exercises to gain familiarization but specific training in the command and general staff positions. This training must cross agency and jurisdictional boundaries to prove successful at future incidents. This requirement can be quite time-consuming when training for the positions of Safety, Logistics, Planning, Finance, and Liaison. Numerous situations must be practiced. Each participant should be allowed the opportunity to role-play in each incident command position that he or she will be required to assume at an incident. In fact, multiple opportunities at each position work the best. This training should be done under the direction of a trainer who can point out the correct procedures and the pitfalls that could occur. Through the constant use of the IMS, coupled with training and critiquing of each incident, members will achieve proficiency.

COMMAND STAFF

A command staff can be utilized to assist Command in performing some basic functions. This staff consists of Safety, Public Information, and Liaison Officer positions and can include an Intelligence/Investigations Officer. Each position can be critical, depending upon the type of incident and demands on the IC's time. Realize that Command is responsible for all functions at an incident. On a minor incident, Command can personally address each area of concern. On major incidents, delegation is necessary.

COMMAND STAFF

- Safety Officer
- Public Information Officer
- Liaison Officer
- Intelligence/Investigations Officer

Safety Officer

Because firefighter safety is a prime concern, the appointment of a Safety Officer helps to make it a priority. The Safety Officer reports to the IC to see what specific concerns the

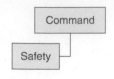

FIGURE 2-7
Command staff—Safety.

IC may have. (See Figure 2-7.) The next step is a survey of the incident scene to look for unsafe conditions. The Safety Officer will discuss safety issues with Division and Group Supervisors and report back to Command with an overview of the scene. Safety concerns will then be presented to Command for action. Should the Safety Officer discover an operation that involves an unsafe act and needs immediate correction, he or she has the responsibility and authority to stop that act. This usually is accomplished through a suggestion to the units involved, though the suggestion can become an order if it is not immediately heeded. (The Safety Officer's duties are covered in Chapter 6.)

Public Information Officer

There is a need and a demand to interface with the media. The Public Information Officer (PIO) is responsible for this interaction with the media and other appropriate agencies. This function is implemented to relieve the IC of working with media, taking him or her away from command responsibilities. Media needs are real and must be met. Reporters need accurate and consistent information. When the IC is not able to handle both the incident and the media, the PIO position should be implemented. (See Figure 2-8.)

A press (public information) area may need to be established. It should be away from the CP and all incident activities. Media representatives need to be made aware of its location and the importance for them to report there.

The media will want tours of the incident and photo opportunities. They need to understand which areas are safe and which areas are off-limits. It is essential that the press not go into unsafe locations, and it is equally important that members of the media not interfere with the incident activities. (Realize that some state laws allow wide latitude to the press, and in many instances they cannot be restricted. It is important to be familiar with local requirements.)

The PIO acts as a central clearing point for the dissemination of information, reducing the risk of providing conflicting information from multiple sources. Large incidents may require the PIO to handle town meetings to keep the public informed of the incident.

A JIC may be created at large or complex incidents, and the PIO may become part of the JIC. The JIC is used to gather incident data, analyze public perception, and keep the public informed. Using a JIC allows multiple organizations or agencies to work together and accomplish the goals of public information.

PIOs must coordinate all releases of significant information with the IC. They will decide on sensitive topics, such as the cause of the incident, victims' names, and any other information that should not (and does not have to) be released immediately to the press.

At some time during the incident, arrangements should be made for the press to talk to the IC. The IC may have to transfer command of the incident for a few minutes to meet with the press.

Realize that if the fire department does not interact and release timely reports, the media will find sources that often are less reliable and can reflect poorly on the fire department.

Liaison Officer

A Liaison Officer (LOFR) is the point of contact for assisting or cooperating agencies. This function is assigned to prevent the IC from becoming overloaded by questions, needs, and demands from the number of cooperating and assisting agencies that some incidents attract. The LOFR position usually is implemented at large or complex incidents. (See Figure 2-9.) It can also be utilized on smaller incidents when a large number of outside agencies respond, such as the police, electric company, gas company, or water department.

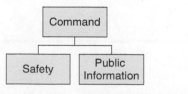

FIGURE 2-8 Command staff—Public Information.

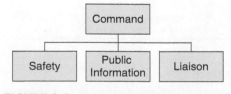

FIGURE 2-9 Command staff—Liaison.

One of the most important responsibilities of the LOFR is to coordinate the management of assisting or cooperating agencies. This is essential to avoid duplication of efforts. It allows each agency to perform what it does best. Liaison management provides lines of authority, responsibility, and communication, and increases the control necessary to provide for the safety of personnel from all involved agencies.

The LOFR is a conduit to the IC for individual stakeholders affected by the incident and for potential problems that will need to be managed. The LOFR is responsible for getting the information required to make proper decisions to the IC. This information includes any political circumstances the IC needs to be aware of. Politics can create explosive situations that need to be handled correctly. The LOFR can keep stakeholders apprised of the situation by holding meetings or forums.

When an agency lacks familiarity with ICS, the LOFR acts as a diplomat. He or she can be particularly useful when agencies lack the joint training necessary to understand their involvement in the incident. Occasionally, it becomes necessary to give strong direction to help agencies understand where and how they fit into the system. This may mean telling instead of requesting.

The agency representatives with whom the LOFR interacts need to have decision-making authority, since the time delay of going through channels to get answers may have a negative effect on the needed coordination.

LOFRs need to identify a location for agencies to report in, work, and communicate with each other.

The appointment of a LOFR eliminates the desire of other agencies to set up separate CPs.

Through the utilization of a LOFR, issues of firefighter safety, authority, responsibility, and communications can be properly managed.

Intelligence/Investigations Officer

The position of Intelligence/Investigations Officer can be initiated as a member of the command staff. (As mentioned above, the NIMS ICS may also use the Intelligence/Investigations function within other areas.) This position is staffed by law enforcement when there may be criminal activity at an incident. It will be a critical position at any incident when acts of terrorism may be involved. (See Figure 2-10.) In addition to assessing potential criminal activity, the position of Intelligence/Investigations can be used to gather information on other all-hazard types of incidents. At a pandemic (widespread outbreak of disease), finding basic information about those involved can assist medical personnel in finding others who are infected and potential areas where spread of the disease may occur, as well as other basic knowledge or intelligence.

STAGING

Staging is an area from which personnel and equipment can be deployed within a three-minute time frame.

It also provides a standard system for resource placement prior to tactical assignments. Staging assists in the control of units, to implement the strategy and tactics of the IC. (See Figure 2-11.) With the exception of the first-due company, units not given

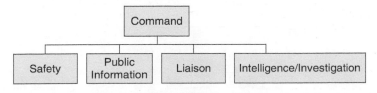

FIGURE 2-10 Command staff—Intelligence/Investigations.

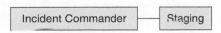

FIGURE 2-11 Staging is a location from which units can be quickly deployed at an incident.

an assignment must report to Staging and await orders. Units that are not needed at an incident can be quickly returned.

Failure to use such a system will result in added confusion on the incident scene as units determine their own tactical assignments. When units bypass Staging and commit themselves prior to being given an assignment, they determine their own objectives and often fail to achieve the goals of the IC. Command can lose track of these resources, resulting in poorly applied resources, priorities being overlooked, the inability to oversee personnel safety, and a general lack of accountability. This form of freelancing can result in the IC calling for more resources to accomplish the necessary tasks and bring the incident under control.

Implementation of the staging function will enhance the overall operation. It has been documented that a tired firefighter is more prone to injury. The rotation of companies from Staging to relieve units who have been operating at the scene provides for well-rested firefighters and enables more firefighters to gain experience by performing various fireground operations.

Someone should be appointed as Staging Area Manager. Staging reports directly to the IC unless an Operations Section Chief has been implemented. In that case, Staging will report to Operations.

Staging can be used in a variety of ways and at various levels. In cities and towns where streets are narrow and mobility is severely restricted, the use of Level 1 Staging has proven quite helpful. This is accomplished by having the responding units go to predetermined sides of the fire building, allowing coverage on all sides of the incident. With the exception of the first-due companies, all other responding companies that have not received an assignment should automatically proceed to Level 1 Staging. Engine companies in Level 1 Staging will then secure a water supply and notify the IC of their arrival and location. They should be prepared to go into operation when and where ordered by the IC. Level 1 Staging allows units to be situated early in the incident. This early deployment has units in place, should they be needed, without having to shut down large-diameter hose-lines and move other apparatus at a later time to position them. (See Figure 2-12.)

Level 2 Staging utilizes a specific location or marshaling area for units to proceed to and await further orders. It should be far enough from the incident scene that it won't interfere with the ongoing operation, yet close enough for immediate deployment of the requested units.

The initial IC determines the level of staging required. If he or she orders Level 2 Staging, a location for the staging area should then be designated. When Level 2 Staging is instituted, a Staging Area Manager should be assigned and will be responsible for managing all activities that occur in Staging. (See Figure 2-13.) The Staging Area Manager does not have to be a command officer. The use of a driver/engineer will suffice on smaller incidents. Larger incidents may require a company officer, with assistance to track the larger number of units.

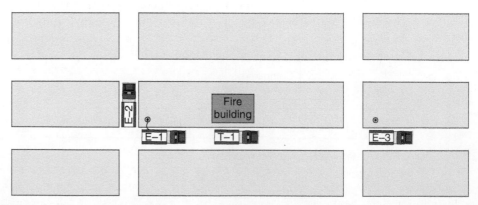

FIGURE 2-12 Level 1 Staging allows units to respond to a prearranged side of the fire building and to be prepared to go into operation if ordered by the Incident Commander. *Used with permission of Michael DeLuca.*

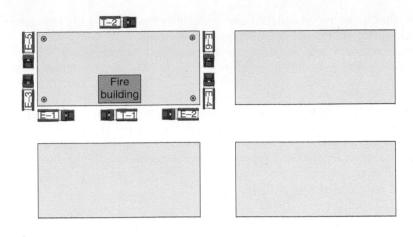

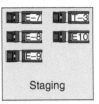

Staging

FIGURE 2-13 Level 2 Staging utilizes a specific location or marshaling area for units to proceed to and await further orders. *Used with permission of Michael DeLuca.*

TITLES FOR ICS POSITIONS

POSITION	TITLE
Incident Commander	Command
Deputy Incident Commander	Deputy
Command Staff: Safety, PIO, Liaison	Officer
Assistant Safety, Assistant PIO, Assistant Liaison	Assistant
General Staff: Operations, Logistics, Planning, Finance/Administration	Section Chief
Deputy Operations, Deputy Logistics	Deputy
Deputy Planning, Deputy Finance/Administration	Deputy
Branch	Director
Division or Group	Supervisor
Strike Team or Task Force	Leader
Resource Unit or Medical Unit	Leader
Staging Area	Manager

GENERAL STAFF FUNCTIONS

In addition to Command, the general staff, sections, or major incident management positions include Operations, Planning, Logistics, and Finance/Administration. (See Figure 2-14.) As discussed earlier, Intelligence/Investigations, at the discretion of the IC, could be assigned as a general staff position. Let's look at the general staff positions.

OPERATIONS

An Operations Section Chief should be designated when there is a great demand on the IC's time. (See Figure 2-15.) Operations should not be appointed when doing so would mean that the only person reporting to Command would be that one individual, unless it is in preparation for an expanding incident. After Operations is implemented, the duties of the IC are modified. The IC will be responsible for the development of the incident objectives and strategies and the communication of that information to the Operations Section Chief.

The Operations Section Chief will run the operations that are implemented to mitigate the on-scene emergency. When ICs initiate this position, they will review with Operations

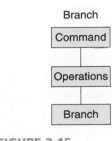

FIGURE 2-14 Operations.

FIGURE 2-15 Operations—Branch.

the incident objectives that they have established, the strategy and tactics already in place, and their expectations, anticipated problems, and specific concerns.

The Operations Section Chief is responsible for the direction and coordination of all tactical operations. As a part of this overall responsibility, the Operations Section Chief also:

- Assists the IC in the development of strategies and tactics for the incident
- Directs and coordinates the overall tactical operations
- Develops operational plans
- Develops tasks and assigns units
- Develops a command structure that will report to Operations to handle the current and anticipated problems
- Requests or releases resources through the IC
- Consults with the IC about the overall IAP
- Ensures that the incident objectives are achieved
- Keeps the IC informed of situation and resource status within Operations
- Supervises the Staging Area Manager
- Supervises the Rapid Intervention Crew

The most common reason for staffing Operations is to relieve span-of-control problems for the IC. Such problems occur when the number of branches, divisions, or groups, coupled with Planning and Logistic elements, exceeds the IC's ability to manage effectively. The IC may implement the Operations Section to reduce the span of control by transferring the direct management of all tactical activities to the Operations Section. The IC then is able to focus attention on the overall management of the entire incident as well as interact with the command and general staffs.

The Operations position is usually delegated to someone already operating at the scene. A natural progression to this position can occur when a higher-ranking officer assumes Command. The officer relieved of Command can be assigned Operations.

For many departments, the position of Operations is a radical change. Before adopting an IMS, the chief was always the IC and made all the decisions. With an IMS, once the position of Operations has been created, the IC must give that person the latitude to handle those duties.

BRANCH

A branch is an organizational level between divisions/groups and the IC or Operations. A branch can also be established under Logistics. Branches are managed by a Branch Director. A branch can be utilized as a functional, a geographic, or an organizational level for major parts of the incident operations. (See Figure 2-15.)

At large-scale or complex incidents, the number of divisions and/or groups may create a significant span-of-control problem. When this occurs, the implementation of branches should be considered. Branches in wildland fires are usually identified by Roman numerals, such as Branch I, Branch II, and so on. In the all-hazard world, branches can be used when a large number of resources are committed to a specific functional activity and designated to carry out that function. A good example is a fire incident that also involves mass casualties. A Medical Branch can be implemented to manage this problem. Other examples include a Hazardous Materials Branch, or a branch for a geographic area (Branch Alpha, Branch Bravo, etc.) of the incident scene. Another method is to specify the various disciplines as branches, such as the Fire Branch, Law Enforcement Branch, Public Works Branch, and so on. Once a branch is implemented, the Branch Director will report to the IC, unless Operations or Logistics have been implemented.

An incident involving various agencies can be handled with functional branches. An example is a natural disaster involving fire, police, and public works. In this scenario, a Fire Branch, Police Branch, and Public Works Branch could be established. Good working relationships and training between the various agencies can allow integration of the various disciplines into geographical branches. This possibility is often dependent on the demands of the incident as well as other considerations.

A fully expanded Logistics Section could exceed the rule of five, and the Logistics Section can implement a Service Branch and a Support Branch to prevent exceeding their manageable span of control.

The specific responsibilities of a Branch Director include:

- Implementing the portion of the IAP appropriate to the branch function
- Coordinating the activities of the units within the branch
- Evaluating goals and objectives and requesting additional resources, if needed
- Keeping his or her supervisor (either the IC, Operations Section Chief, or Logistics Section Chief) informed of the status in the branch's area of responsibility
- Assigning specific tasks to divisions, groups, or units within the branch
- Resolving logistical problems associated with the units deployed in the branch

RESOURCES

Though national standards call for a minimum of one officer and three firefighters for a fire apparatus, resources will depend on many factors. In most areas of the country, resources will consist of engine and possibly truck (ladder) companies, heavy rescue companies, or other specialized units. These individual companies are commonly referred to as a single resource. The staffing of these units can vary from one firefighter to an officer and five firefighters. Naturally, the amount of work that can be accomplished will be determined by a combination of the personnel and their capabilities. The response of a limited number of firefighters will make it difficult to safely accomplish assignments at an incident. Some volunteer and combination departments permit their firefighters to respond directly to the incident scene. They assemble from two to six firefighters and classify them as a crew. The number of personnel should not exceed the recommended span-of-control guidelines. A common method of accountability is to list either a member's name or the function that will be performed for the crew (e.g., Crew O'Malley or Vent Crew).

The National Institute of Standards and Technology (NIST) conducted a study to investigate the effects of varying crew size, first apparatus arrival time, and response time on firefighter safety, overall task completion, and interior residential tenability, using realistic residential fires. It was based on "NFPA 1710: Standard for the Organization and Deployment of Fire Suppression Operations, Emergency Medical Operations, and Special Operations to the Public by Career Fire Departments." The results showed that four-person crews operating on a low-hazard structure fire completed all the tasks on the fireground (on average) seven minutes faster—nearly 30 percent—than two-person crews. The four-person crews completed the same number of fireground tasks (on average) 5.1 minutes faster—nearly 25 percent—than three-person crews.

STRIKE TEAM

A strike team is specified combinations of the same kind and type of resources (apparatus and staffing) and has common communications with a Strike Team Leader. (See Figure 2-16.) These can be any type of unit, though fire service apparatus typically consist of engines. At wildland fires, strike teams of dump trucks or bulldozers are common. A typical arrangement is to assemble five fully equipped and staffed engines with a chief officer as the Strike Team Leader. The Strike Team Leader will report to the IC or Operations

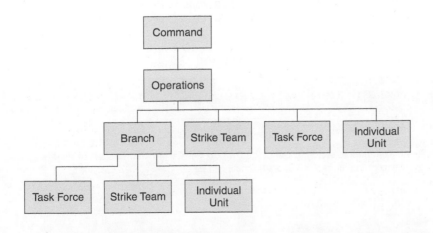

FIGURE 2-16 Operations—Strike Team/Task Force/individual unit.

for an assignment. This enables the IC/Operations to utilize five companies while needing to speak to only one person. At a large fire, three strike teams of engines could be operating in a division under Operations with only the three strike team leaders conferring with Operations. This would allow 15 engine companies to work in one area without exceeding the span of control of any supervisor and permitting common communications between the operating resources.

Realize that in locales with limited resources, a strike team could consist of three engine companies with a Strike Team Leader.

TASK FORCE

A task force is a group of resources having common communications and a leader that may be pre-established and sent to an incident, or formed at an incident. The task force should not exceed five units. A common arrangement for structural firefighting is two engines and one or two truck companies with a chief officer as the Task Force Leader. Similar to the strike team, task forces allow many companies to operate within a division or group and still maintain an appropriate span of control while enhancing communications.

DIVIDING THE FIRE: ASSIGNING BRANCHES, DIVISIONS, AND GROUPS

By dividing an incident scene into sectors or areas, you are taking a problem and breaking it down into manageable segments. There is both geographic and functional sectoring. The terms can include branches, divisions, or groups.

Geographic sectoring allows the IC to take an incident scene, divide it, and place someone in charge of each area. Those individuals will then be assigned units to accomplish the tasks. They will be responsible for requesting additional resources from the IC or Operations and for giving timely progress reports. The model ICS refers to geographic sectors as divisions. (See Figure 2-17.) The NIMS mandates the use of alphabetical sectoring. The alphabetical system starts with the front of the building being Division A.

Going clockwise around the building: Division B would be the left side as you face the building, Division C would be the rear, and Division D the right side. Some departments, to avoid confusion of letters sounding alike during radio transmissions, have adopted the use of Alpha, Bravo, Charlie, and Delta. Exposures can be referred to as exposure Alpha or exposure Bravo.

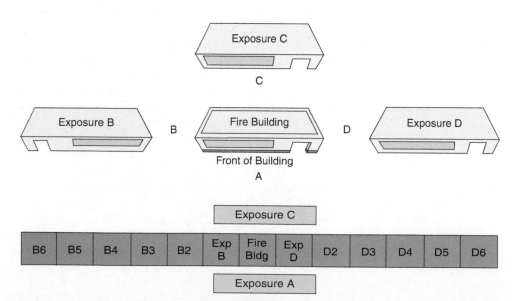

FIGURE 2-17 Geographic sectoring can utilize Division A-B-C-D to identify sides of the fire building or exposures. Adjacent buildings, such as a strip mall, can be denoted as D-2, D-3, or B-2, B-3, etc.
Used with permission of Michael DeLuca.

When dealing with multistory structures, each floor can be designated a division. (See Figure 2-18.) This is especially effective when dealing with high-rise structures. The 15th floor would be Division 15, 16th floor Division 16, and so on. The units assigned to these divisions would accomplish all the needed tasks within their geographic area, including fire attack, ventilation, search and rescue, and overhauling.

At high-rise incidents, the number of floors that could be affected by a fire in a building can easily cause span-of-control problems, should each floor be designated as a division. A 30-story building with problems on many floors would quickly exceed the span of control of an Operations Section Chief. One method that can be utilized is the inclusion of branches. Chapter 9 has further discussion on creating branches at a high-rise incident.

Functional sectoring is the establishment of a group to accomplish a specific function. Functions could include, for example, ventilation or search and rescue. These teams would use the radio-call letters Ventilation Group or Search and Rescue Group.

As noted earlier, a branch can be utilized as a functional, a geographic, or an organizational level for major parts of the incident operations.

A Branch Director, Division Supervisor, or Group Supervisor would be assigned the responsibility of accomplishing the assigned tasks.

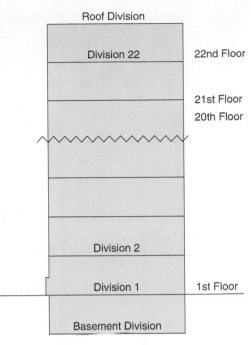

FIGURE 2-18 When dealing with multistoried structures, each floor can be designated a Division. *Used with permission of Michael DeLuca.*

PLANNING

The Planning Section Chief is a member of the general staff and is responsible for the collection, evaluation, dissemination, and use of information about the development of the incident and the status of resources. The tracking of all personnel is critical for ensuring firefighter safety.

At a complex or rapidly escalating incident, the IC may need assistance in planning how to deal with the many and varied problems that may develop. An important function of the Planning Section Chief is the development of the IAP or modification of the present plan. (See Figure 2-19.)

The Planning Section Units are Resource Unit, Situation Unit, Documentation Unit, Demobilization Unit, and Technical Specialists.

Resource Unit

The Resource Unit is responsible for the tracking of units committed to the incident and anticipating resource needs.

On most day-to-day assignments, Planning typically assists by maintaining the situation and resource sheets. These tactical worksheets, or boards, track units as operating at an incident or in Staging, and prevent critical details from being overlooked. The sheets can be separate or combined to show the relevant data. They should contain a rough drawing or plot plan of the fire area and any threatened exposures. This creates a picture for the IC that reflects the distribution of resources and shows where additional resources may be needed. A list of each unit dispatched and its operating location should be maintained. This is helpful when units are rehabbing or in staging. The list states who is the supervisor of each group or division, and the units assigned to each area. This ensures accountability of every unit. While reviewing situation and resource sheets, the Planning Section Chief may recognize that a need for rotation of personnel will be required or that sufficient resources are not available at the incident. Providing this information to the IC will assist in strategic development.

The charts should contain command assignments that allow you to see what functions have been delegated and those still retained by the IC. This can be reflected in the form of an organizational chart starting with the IC and continuing through each area of the IMS that has been instituted, showing functional responsibilities. This chart gives the IC the ability to make assignments or adjustments of needed tasks. If a Planning Section Chief is assigned, that person will be responsible for seeing that these charts are maintained. (See Figure 2-20.)

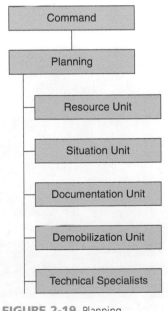

FIGURE 2-19 Planning.

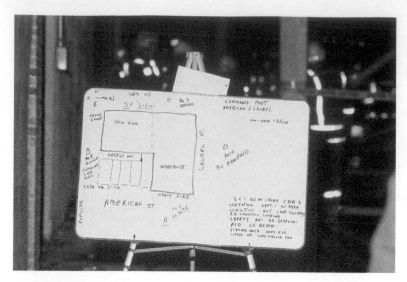

FIGURE 2-20 Fireground charts are a tool. They require functionality of a discernible sketch, not the perfection of an artist's drawing. *Used with permission of Joseph Hoffman.*

Charts assist in the transfer of command to a higher-ranking officer during the escalation of a fire, in descending order when an incident has stabilized and is reverting to lower-ranking officers, or at the change of shift in career departments. Because the information has been committed to paper, there is less likelihood of important information being overlooked.

On complex incidents, the role of the Planning Section can become quite involved, including the preparation of an IAP. (See Chapter 3's coverage of the IAP for additional information.)

Situation Unit

Major incidents of a technical nature (protracted hazardous material incidents, prolonged high-rise fires, wildland fires, etc.) may necessitate the need for predictions of likely outcomes and anticipated problems by the Planning Section. Though the number of times this is required is minimal, when needed it is a vital position. This will require the assigning of a senior officer.

Concerns of the Situation Unit are:

- What has happened?
- What is currently happening?
- What may happen?

Documentation Unit

The Documentation Unit records and protects all documents relevant to the incident. This includes incident reports, communication logs, injury claims, and situation reports. These reports are often utilized to present data at critiques.

Demobilization Unit

The Demobilization Unit is responsible for developing a plan for the demobilization of the resources committed to an incident and assisting in the implementation of that plan. This plan allows an orderly and logical method of releasing units from the scene. It should consider the length of time units have been at the scene, the physical condition of the personnel, and any mutual-aid agreements that need to be considered. Proper demobilization is dependent on adequate planning.

Technical Specialists

The Planning Section Chief may interact with technical specialists who can provide insight on an operation. (See Figure 2-21.) Technical specialists are individuals with skills or knowledge that may be applied to support incident operations. Examples include building maintenance engineers, meteorologists, industry representatives, or private-sector chemists. Information obtained by the technical specialists should be shared with the IC.

LOGISTICS

The Logistics Section Chief is a general staff position. (See Figure 2-22.) The responsibility of the Logistics Section Chief is crucial in keeping units supplied during a major incident. Complex assignments can present a variety of demands, often within restricted time frames. The number of units needed for logistical assignments can match or

FIGURE 2-21 The Planning Section Chief may interact with technical specialists who can provide insight on specific operations.

exceed those assigned to suppression or mitigation efforts. The Logistics Section Chief must work closely with the IC and the Operations Section Chief.

Routine incident scene needs can include:

- Water supply
- Foam at a flammable liquid fire
- Resupplying of air for self-contained breathing apparatus (SCBA)
- Refueling or repairs to apparatus
- Caches of emergency medical supplies at a mass casualty incident
- Specialized equipment

Wildland fires and all-hazard incidents can place a high demand on logistics to provide facilities, services, and materials for the incident, including setting up large camps with sleeping, eating, and bathing facilities. The wildland fire may require the acquisition of vehicles to provide water for both ground and air operations.

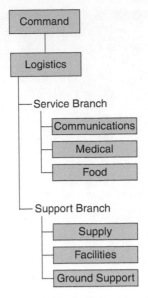

FIGURE 2-22 Logistics.

> A ratio of two logistic companies for each unit involved in suppression at a working high-rise fire is a reasonable consideration.

Since a fully expanded Logistics Section can include Communications, Medical, Food, Supply, Facilities, and Ground Support, this expansion exceeds the accepted rule of five persons under supervision, as dictated by ICS. For this reason, the Logistics Section can be divided into two branches, the Service Branch and the Support Branch. Another approach that Logistics Section Chiefs (LSCs) often use is to create a Deputy Logistics Section Chief (DLSC), in lieu of creating branches. The DLSC can then be utilized by the LSC to handle any area(s) of concern with span-of-control issues, potential problems, or close supervision required by the LSC.

Service Branch
The Service Branch is responsible for service activities at incidents. These activities include communications, medical treatment for incident personnel, and provisions for feeding operating forces. The Service Branch is managed by a Branch Director and contains three functional units: Communications, Medical, and Food.

Communications Unit Large incidents involving numerous mutual-aid companies may require a communications plan. The Communications Unit develops this plan and can utilize ICS Form 205 (Incident Radio Communications Plan) to document and distribute the plan.

If necessary, they can distribute communications equipment, supervise the communications network, and maintain/repair communications equipment.

Medical Unit The Medical Unit is responsible for providing emergency medical treatment for emergency personnel. This unit does not normally provide treatment for civilians.

There is often confusion that the Medical Unit comes under Logistics and not under Operations. ICS was designed originally for wildland fires. There was normally not a civilian problem, and medical needs were for the emergency responders. Realize that on structure fires or other types of incidents, an EMS group may be assigned to treat civilians and will report to Command or Operations if established.

Responder rehabilitation (rehab) will be provided by the Medical Unit. It includes medical evaluation and treatment, food and fluid replenishment, and relief from extreme climatic conditions for emergency responders, depending on the circumstances of the incident.

Large incidents may require a Medical Plan. The Medical Unit develops this plan and can use ICS Form 206 (Medical Plan) to document and distribute the plan.

Food Unit Providing meals and drink for personnel involved in an incident is the responsibility of the Food Unit. This may be a significant logistical task at major incidents, and could be necessary at relatively minor incidents during extended operations.

FIGURE 2-23 Many demands can be placed on Logistics. Large-scale incidents will require resupplying of air cylinders for self-contained breathing apparatus. *Used with permission of Joseph Hoffman.*

Support Branch

The Support Branch is responsible for providing the personnel, equipment, and supplies to support incident operations. These activities include supply, provision of fixed incident facilities, and ground support (such as fueling and maintenance of equipment). The Support Branch is managed by a Branch Director and can contain three functional units: Supply, Facilities, and Ground Support.

Supply Unit The Supply Unit orders the equipment and supplies required for incident operations and maintains ongoing inventory and control of these resources. Equipment and supplies may include additional SCBA cylinders, specialized equipment required for a hazardous materials spill, or expendable supplies, such as breathing air or foam concentrate. (See Figure 2-23.)

Facilities Unit The Facilities Unit provides fixed facilities for an incident. Most often, fixed facilities are required for incidents of long duration, and may include the Incident Base. The Base serves several functions: It is the location where primary support activities are performed, and it serves as a reporting and marshaling area for resources. Base is not commonly used at structure fires. However, it may be used during wildland fires, hazardous materials incidents, or high-rise incidents. (See Figure 2-24.)

Other fixed facilities include feeding and sleeping areas, sanitary facilities, and a formal CP.

Ground Support Unit The Ground Support Unit is responsible for fueling and maintenance or repair of vehicles, transportation of personnel and supplies, and preparation of an Incident Traffic Plan. The Traffic Plan, if necessary, is used to facilitate the flow of apparatus and equipment within the incident area and can be included in a written IAP.

FINANCE/ADMINISTRATION

The Finance/Administration Section is a part of the general staff. (See Figure 2-25.) The Finance/Administration Section Chief's position can be implemented at a major incident or one where a major cost recovery is evident. This position ensures the tracking of expenditures and the proper documentation of anticipated costs. A major function of this section is to document the financial costs of the incident to keep the IC apprised. The person most qualified to handle this assignment may be the city or county finance officer. In addition to wildland fires and hazardous materials incidents, there have been attempts to recover the costs of firefighting when a fire occurs due to violation of a code or ordinance.

Finance/Administration can include the following units: Time Unit, Procurement Unit, Compensation and Claims Unit, and Cost Unit.

Time Unit The Time Unit is responsible for equipment and personnel time recording. This will allow for the proper payment for these services including overtime accrued.

Procurement Unit The Procurement Unit is responsible for administering all financial matters pertaining to vendor contracts, leases, and fiscal agreements.

FIGURE 2-24 Base should be a safe distance from the incident scene, and units will not be readily deployable. *Used with permission of Joseph Hoffman.*

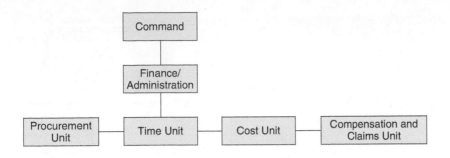

FIGURE 2-25
Finance/Administration.

Compensation and Claims Unit The Compensation and Claims Unit is responsible for the overall management and direction of all administrative matters pertaining to compensation for injury and claims-related activities (other than injury) for an incident.

Cost Unit The Cost Unit is responsible for collecting all cost data, performing cost effectiveness analyses, and providing cost estimates and cost-saving recommendations for the incident.

IMPLEMENTATION OF THE SYSTEM

An IMS can be correctly implemented in numerous ways. It is a system that should and must be used by firefighters on routine or everyday types of incidents. This will ensure that implementation becomes second nature when confronted with a major incident.

A vehicle fire response might only consist of a single engine company. The command structure for this incident would require the engine company officer to assume command and notify Dispatch that Engine 1 will be 7th Street Command. In this instance, the only people being supervised by Command would be the firefighters in Engine 1. Should there be an injury and people trapped, then a rescue company can be dispatched along with a medic unit and police. Engine 1's officer would retain Command and have the Rescue, Medic Unit, and Police Officer reporting to him or her. (See Figure 2-26.)

Realize that there are many correct methods of implementing an IMS and being fully compliant with the system. But being fully compliant and designing a system that is workable on an incident can be two different things. That is especially true when fire departments implement only groups and not divisions for structural fires. A common arrangement is a fire-attack group, a search-and-rescue group, and a ventilation group. On a one-story structure, this system will work fine. If you never have any structures taller than one-story, it will probably not need to be changed in any way. The problem shows up in areas where not all structures are one-story, and if the next fire response is to a three-story dwelling or a three-story motel, it will not be as successful. Fire attack groups are a function and are not defined geographically, so in reality, sending a fire attack group into the burning three-story motel with fire burning to some degree on all three floors would make them responsible for all floors. Supervision of firefighting on three floors is unworkable for one supervisor.

Fire departments could initiate different systems depending upon the type of response (groups for one-story structures and divisions for multistoried structures). The problem is that we are taught to be consistent, and the system that we use daily is the system that we will try to employ at most structure fires. We may recognize on a high-rise response that we need to adjust our system, but most fire departments will try to implement the system that they use on a daily basis.

A much better way to implement the system is to create a division for each floor and a ventilation group. The typical structure fire in a one-story dwelling will utilize a very basic system. (See Figure 2-27.) There will be an IC, a Division 1 that will handle everything on the interior, and a Ventilation Group to handle ventilation. The Division 1 Supervisor would be assigned sufficient resources for fire control and search and rescue. He or she would need to coordinate both operations.

A fire in a three-story nursing home that requires rescues, fire attack, and ventilation demands an expanded system to handle the problems. This incident may necessitate a Division 1, Division 2, Division 3, and Ventilation Group. The fire attack and

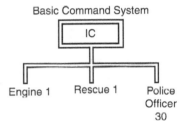

Basic Command System

FIGURE 2-26 A command structure for a vehicle accident with an injury could include the engine company officer as the IC and the rescue and police officer.

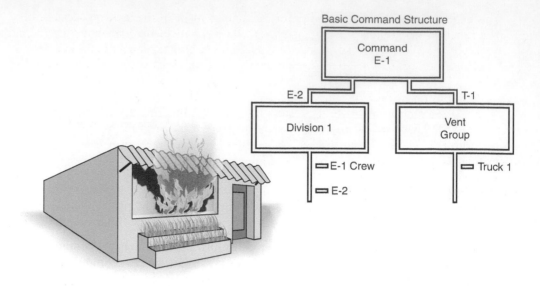

FIGURE 2-27 A typical structure fire will utilize a very basic command structure. *Used with permission of Michael DeLuca.*

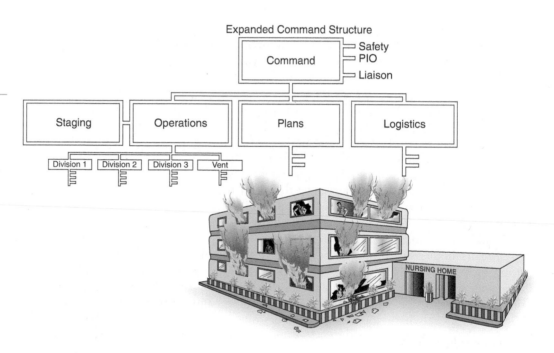

FIGURE 2-28 A fire involving a complex situation will demand an expanded command structure, as shown in this nursing home fire. *Used with permission of Michael DeLuca.*

search-and-rescue are handled by the individual divisions. Command needs to provide sufficient resources to each division or group to enable the completion of their assignments. A fire of this magnitude may include the implementation of Operations, Planning, Logistics, Safety, and other positions. (See Figure 2-28.)

Incident Scene Control

For the IMS to function properly, there are certain steps that are required to ensure a successful outcome. The first arriving officer at an incident scene must assume Command. This is called **incident scene control**.

ESTABLISHING COMMAND

When Command is first established, Dispatch should be notified. An example is "Dispatch from Engine 1. Engine 1 will be Broad Street Command." This accomplishes a number of tasks. It notifies Dispatch and other units that Command has been established, it tells which unit is in Command (Engine 1), and where Command is positioned (Broad Street).

incident scene control ■ A method used to assure that all units are operating within a plan that ensures a safe and successful conclusion to the emergency.

This action places the responsibility for the management of the incident scene on the IC. The function of Command must be present at every incident. Someone must be in charge or chaos will result. The absence of Command allows for indecision and duplication of effort. Certain fireground tasks can be overlooked. Units will establish their own priorities.

Command must be initiated at all incident scenes. The everyday practice on minor fires allows for a smooth transition to a major incident. Once Command is established, its continuity must be maintained. A system must be in place to allow for Command to be transferred to a higher-ranking officer. Some systems require that the senior officer must assume Command upon arrival at an incident. A better method is allowing the higher-ranking officer the latitude to take Command or let the current officer maintain it.

ASSUMING COMMAND

As higher-ranking officers arrive on the scene, they must decide whether to assume Command. Upon arrival and prior to transferring Command, that officer should visually size up the incident scene. This size-up should be conducted as if that officer were the first officer to arrive.

- What do you see?
- How would you attack the fire if you had been the initial IC?
- What fire department actions are taking place?
- Are these actions consistent with the problems observed?

The officer, after performing the size-up and discussing the scene with the current IC, will decide whether to assume Command. An expanding incident may require a higher level of experience, necessitating a transfer of Command.

Whether the higher-ranking officer assumes Command or allows it to remain with the present officer, he or she still assumes the responsibility of the scene. Some have the misconception that if Command is not formally transferred, then neither is responsibility. This is erroneous; you cannot disavow responsibility.

TRANSFER OF COMMAND

If a decision to assume Command is made, then there should be a formal transfer of Command. (See Figure 2-29.) This change should not occur prior to a transfer of critical information and will require notifying Dispatch and units via radio. The officer assuming Command must have a handle on:

- What has happened
- What is occurring now
- The anticipated problems

Depending on resource availability, the IC being relieved may stay at the CP for a time after the transfer of Command. This assures the new IC that if any information has been overlooked during the transfer of Command, it is readily available.

The reasons for not assuming Command can vary. There would be no need to transfer Command if all major decisions have been made and the current IC has the situation well under control. A complex incident often necessitates much time in the transfer of information. In this instance, the higher-ranking officer can assume Command and keep the officer who has been relieved to assist until he or she gains familiarity with the incident.

This knowledge or transfer of information must occur before a good transfer of Command takes place. It should be noted that it is unfair to allow a low-ranking officer or an officer with little experience to remain in Command of a very complex incident.

Permitting a junior officer to remain in Command with supervision from a higher-ranking and more experienced officer is an excellent way of

FIGURE 2-29 If a decision to assume Command is made by a higher-ranking officer, then there should be a formal transfer of Command. *Used with permission of Joseph Hoffman.*

developing fireground command skills and future chief officers. If a problem occurs, the junior officer can make the decision or "call the shots." If he or she is about to make a critical mistake or seems befuddled, the senior officer can give assistance.

When transfer of Command does occur, there should be a standard procedure to be followed. The IC should use situation and resource forms to allow a much easier transfer, with a review of the command structure and the units assigned to each function, the strategies that are in place, and the tactics and tasks that have been implemented.

For an IMS to function at peak efficiency, each firefighter involved must contribute to the overall success of an operation. The first-arriving officer at an incident, after sizing up the situation, has many decisions to make. He or she must decide whether to set up a CP and assume Command or if active participation in the incident can have a greater impact.

If the first-arriving officer decides that getting physically involved in the initial stages can save a life or prevent a minor fire from becoming a major incident, he or she can notify Dispatch of this decision and pass Command to the next-arriving officer.

PASSING COMMAND

Passing Command is an option of the first-arriving officer. Should the first officer take this option, Dispatch should be notified and Command must be assumed by the next officer to arrive at the scene. This must occur to ensure that this vital position is staffed.

Passing Command allows some latitude to the first-arriving officer. When units are responding from a great distance, it will take considerable time for the next unit to arrive on scene. The first-arriving officer may need to become physically involved. Under fire or emergency conditions, it will be impractical if the first engine to arrive is staffed with only two or three firefighters, and one of them sets up a CP and assumes Command while a person in need of immediate rescue is not tended to.

MOBILE COMMAND

Some fire departments, due to closely aligned stations and rapid response of a chief officer, permit the first-arriving company officer to assume Command, yet remain mobile.

To accomplish this, the first-arriving officer:

- Gives an initial status report
- Gives orders for the incoming units (either specific orders or those units will go to Level 1 Staging)
- Identifies and assumes Command

Mobile Command works well for fire departments where the chief officer arrives practically at the same time, or a minute or two after the arrival of the initial unit. Realize that until the officer who is a Mobile Command transfers Command to another fire officer, or chief officer, he or she still has the responsibilities of any IC.

In fire departments that utilize Mobile Command and the chief officer will be delayed, the chief can notify Dispatch of his or her delayed status, and the next-arriving company officer will assume Command as if Command was passed.

COMMAND POST

A CP provides the IC a stationary position from which to command the incident. (See Figure 2-30.) It is essential when multiple companies are operating and its importance increases as the incident grows in complexity. Its location should allow the IC to be easily found. It gives the IC a location to assemble staff and other resources. It is a place where management functions occur to bring an incident under control.

FIGURE 2-30 A Command Post provides the Incident Commander with a stationary position from which to command the incident. *Used with permission of Joseph Hoffman.*

The CP collects and disseminates information. It is a place where decisions are made, strategies developed, and orders given. It offers shelter from the elements, light for night-time operations, and a place where reference books can be utilized.

The actual size of the CP will be dictated by the needs of each incident. As functions and incident demands increase, the CP will increase accordingly. Caution must be taken to prevent overloading. This can occur if unneeded personnel are allowed to gather there.

The selection of where to locate the CP should consider a vantage point from which to view the incident. If possible, a good location is in front of the fire building or incident scene with a view of the front and the more critical of the two sides of the structure or the direction toward which a fire may spread. This will be predicated on a number of factors, such as the accessibility of the command vehicle as well as fire and smoke conditions encountered. If apparatus are operating at this location, a high noise level may prohibit use of this site for the CP.

A vantage point allows the IC to compare reports received from the divisions or groups with his or her personal experience and knowledge. This is especially helpful when receiving conflicting reports, such as an interior division reporting that a fire is knocked down and a roof division reporting continual or increasingly heavy smoke and heat conditions. Conflicting reports need to be assessed by the IC, and a view of the building can assist in this evaluation.

A CP can utilize a chief's vehicle, a special communications vehicle, or a nearby building. The location of the CP should be communicated to Dispatch, to the units operating at the scene, and to incoming units. The CP location should be indicated by a special recognizable designation. This could be a flag or special-colored light.

Status Reports

The successful organization is one that can quickly assimilate into the various types of emergency situations that fire departments are called to daily.

> We have found that what starts out well usually ends well. Status reports enhance our fireground response.

INITIAL STATUS REPORT

When responding to an emergency, a duty of the first-arriving company officer is to quickly assess conditions and give a report of the fire conditions to Dispatch. Some departments keep a plastic laminated card with the required information for the initial and subsequent reports in the cab of all apparatus. This card is always in clear view of the officer.

This initial **status report** should start off with a repeat of the address where the fire is located. This can differ from the dispatched location. Accurate addresses need to be given. Most cities and towns have a standard practice in which addresses are assigned to specific directions. For instance, all odd-numbered addresses will be on the north and east sides of the streets; all even-numbered addresses will be on the south and west sides. This allows responding units to determine which side of the street the fire building is located on, thus assisting proper response and apparatus placement.

status report ■ Report prepared by the IC to assess fire conditions for Dispatch; helps fulfill initial and ongoing demands of constantly sizing up and reporting incident scene conditions.

Initial status reports differ from department to department. A comprehensive report accomplishes many goals. It forces the first-arriving officer to review the basic points of size-up, such as life safety, number of stories, and size and construction of the building, as well as fire conditions on arrival. It may be a requirement of the initial status report to contain orders for all responding units. This could be detailed orders of specific duties or general orders, such as for units to proceed to Staging.

An example of an initial status report is, "Dispatch from Engine 4. We are on location at 123 Maple Street. We have a two-story frame dwelling approximately 75 feet by 25 feet. There is heavy fire involving the first floor on Charlie side. There is a report of people trapped. Have the first-due truck prepare for rescue. Have the second-due engine stretch a supply line and a 1¾-inch hose-line to back up Engine 4. Engine 4 will be Maple Command. Strike out a second alarm and dispatch a medic unit. Staging will be located one block east on Maple."

When giving the dimensions of a structure, always give the width and then the depth of a building. This standardized procedure allows responding units to better visualize the property.

A comprehensive report benefits everyone. It allows Dispatch to get a full picture of the incident scene. The responding chief officer can start initiating strategy en route, since a thorough report allows the chief to envision the magnitude of the problem. If the situation *may* require additional resources, they should be requested. When it's not clear whether more resources will be needed, the axiom "It's better to have them and not need them, than to need them and not have them" dictates. Firefighters realize the importance of being prepared for the nuances of a changing fireground situation. They know the need for staging and the necessity of having rested personnel readily available to relieve other crew members or in the event that other problems should occur.

Dispatch can utilize the status report to prepare for move-ups or cover-ups of companies, to notify mutual-aid companies of the situation, or to pass along information to senior officers who may have response duties. It also establishes command at the scene and informs everyone who is in command.

Responding units will be given specific assignments en route or report to Staging. This allows certain assignments to be known prior to arrival with no loss of time. It prevents freelancing and assists in firefighter accountability and overall incident scene safety. Firefighters can adjust their thought process to the particulars of the incident. Knowing what orders are being given to responding units, they can anticipate areas still needing to be addressed. When the units in Staging are needed, they can then move to the correct location to perform their assignments.

> Move-ups or cover-ups occur during a working fire, when Dispatch relocates other units to those stations that have been vacated by the fire response in preparation for additional dispatches in those now vacant areas, or to prepare for additional alarms should they be requested by the IC at the working fire.

Typical Initial Status Report Data
- Address
- Type of building
- Size and stories
- Fire conditions
- Disposition of the occupants
- Exposures
- Orders for the responding units
- Identify Command
- Assume Command
- Request additional units if needed
- Announce a location for Staging

ONGOING STATUS REPORT

Just as size-up is an ongoing process that demands constant evaluation of changing conditions and adjustment of strategies and tactics accordingly, the ongoing status report should be a part of the SOGs for incident response. Requiring a status report at timed intervals is a method of alerting the IC that a specific amount of time has elapsed since the previous report and that if the size-up hasn't been updated, he or she should do it now. A good time frame for the required status reports is at 10-minute intervals until the fire is placed under control. Dispatch can notify the IC every 10 minutes with a message: "Broad Street Command, be advised that you are 10 minutes into the incident, we need a progress report." (A similar message should be given to Command every 10 minutes, updating the message as: "Command, you are now 20 minutes into the incident," etc.). The ongoing status report can follow the basic format of the initial report, without the preliminary information about the site and construction of the fire building. An ongoing status report may sound like this:

"Dispatch from Maple Command. Battalion 2 is now Maple Command. We have heavy fire conditions on the Charlie side of the fire building. It appears at this time that we will be able to confine the fire to that area. We are in the process of opening walls and ceilings to expose hidden fire. All occupants have been accounted for. One has been transported via Medic Unit 2 to the hospital. We are now operating with two engines and two trucks and have one engine in Staging. Dispatch another medic unit to the scene."

Ongoing status reports allow ICs to review their size-up and determine if the present strategy and tactics are accomplishing what was hoped. If not, they can make modifications. By addressing what companies are committed, they know how long these units have been operating and whether they need relief.

ICs must remain proactive in their approach to handling the incident. Resource needs may change due to the extension of fire or prolonged operations requiring rotation of personnel.

Typical Ongoing Status Report Data
- Any information that was not available for the initial report
- Current conditions of the fire and the disposition of the occupants
- Specific units operating at the scene
- Units in Staging

FINAL STATUS REPORT

A final status report should be given at the time a fire is placed under control. The best description of "fire under control" refers to a situation in which the life and safety of civilians has been assured, no additional extension of fire will take place, and normally, no additional firefighting forces will be required to respond at emergency speed. This could be a scene that still requires a lengthy time commitment from the fire department, including the relief of the personnel at the scene, but "fire under control" means that the scene is stabilized.

The final status report will give an overall report of current fire conditions and how long the remaining units will be on the scene, if that information is known. A typical report is: "Dispatch from Maple Command. Place the fire under control. All visible fire has been knocked down. Engine 17 is being relieved by Engine 4. Engine 4, Truck 2, and Medic 5 will be remaining at the scene for overhauling and salvage. They will be here approximately one hour. Engine 4 will be Maple Command. Battalion 2 is available for service."

Status reports help us fulfill the initial and ongoing demands of constantly sizing up and reporting conditions at the incident scene. When done in a professional manner, they help improve our overall operation.

> The term *fire under control* refers to a situation where the life safety of civilians has been determined, and that no additional extension of fire will take place, and normally no additional firefighting forces will be required to respond at emergency speed. This could be a scene still requiring a lengthy time commitment for the fire department, including relief of the personnel at the scene, but it means that the scene is stabilized.

Typical Final Status Report Data
- Current conditions
- Final disposition of occupants
- Units remaining at the scene
- Units preparing to leave the scene
- Approximate length of time units will remain on the scene
- Transfer of Command to a lower-ranking officer

COMMUNICATIONS

Communications is a critical component of every operation. Communications is the giving and receiving of information, the backbone of any emergency response organization.

Because our decisions are based on the information that we receive, we must attempt to obtain accurate data as quickly as possible. This includes feedback on the progress being made by Division and Group Supervisors. Are they being effective in what they are attempting to accomplish? Do they need more resources? Can they complete their assignments? Has the situation changed, whereby their present course of action is no longer feasible? When communications are planned for and carried out properly, the overall incident will benefit. The IC must control communications and request regular updates from each division and group.

Reporting Problems and Solutions

A breakdown in communications seems to be consistent with seemingly unsolvable operational problems. An invaluable part of communication occurs when problems are discovered. The person reporting those situations must not only describe the problem but also offer potential solutions. An example is: "Command from Charlie Division. We have fire extending to the exposure on the Charlie side. I have Engine 30 stretching a line into the exposed building. I will need another engine and truck company to assist them." This message tells Command what the problem is and what resources are needed to solve it. If only the first part of the message is given, Command will then need to ask questions to establish the extent of the problem and the needed resources.

An excellent method for the IC to determine conditions is to receive CAN reports from units operating on the incident scene which contain:

- **C**onditions – what conditions they have found in their respective areas
- **A**ctions – what actions are currently being performed and what is required to mitigate the conditions found
- **N**eeds – what resources are needed

Call Letters

Communication is enhanced through the utilization of appropriate call names. Many departments still use their assigned unit number after arriving at the incident and being given an assignment. Once designated as a Division or a Group Supervisor, the radio call sign must be the area or function assigned. Using the area or function (and not the unit number) allows everyone operating on the appropriate radio frequency to know which functions have been implemented and the conditions being reported from that assignment. An example is: "Command from Vent Group. We have completed vertical ventilation," or "Command from Division 2. The primary search is completed, and we are making good progress in controlling the fire." Both of these messages describe either a location (the roof or the second floor) or activity (ventilation or search/rescue and confinement). It does not matter who the units are, but the important fact is that everyone monitoring the radio knows what is happening and where it is occurring.

Face-to-Face Communications

A good communicator can read the body language of a person giving or receiving a message if face-to-face with that individual. These nonverbal messages can be expressions or gestures and must be interpreted properly by the receiver. An example of body language or gestures that a firefighter can make when receiving an order is a frown or quizzical look, indicating that he or she does not understand the order or has reservations about the order being given. The repeating back of the order by the firefighter receiving the order ensures that it has been received. Further questions can clarify if it has also been fully understood. If it has been understood, yet body language is expressing a hesitation on the part of the firefighter, it may be wise to ask if he or she foresees a problem carrying out the order or has a better method to accomplish the task. Naturally, this is dependent on time factors and in no way implies that giving orders at an incident scene is better performed by participative management practices. This whole process should take only a matter of seconds.

Departmental Radio Procedure

A good radio procedure should encourage the use of face-to-face dialogue when possible and should restrict unnecessary radio communications. This allows sufficient time for broadcasting important messages.

Radio usage policy can include:

- The receiver will acknowledge the receipt of an order.
- When an assignment has been completed, this information will be given to the person originally issuing the order.
- If unable to complete an assignment, the reason will immediately be given to the person issuing the original order (insufficient personnel, safety condition preventing it, etc.).
- Be prepared to give a progress report, if requested.
- Immediately report to the IC if a safety factor is discovered that can impact on the overall operation or firefighter safety (serious structural crack in a wall, a buildup of water on a floor in a building, etc.).

Emergency Transmissions

Large incident scenes will still demand a high degree of radio traffic. A method must be in place to clear a radio band if an important message has to be given. This may be necessary if a company or firefighters find themselves in jeopardy or if a serious occurrence happens that the IC must be made aware of immediately. One method that can be used is by requesting "emergency traffic" or stating "Mayday." An example is: "Emergency traffic! Command from Engine 1!" Command can then recognize the message and have Engine 1 proceed with its message. The term *Mayday* is used to indicate that a firefighter is in trouble. These messages alert everyone to clear the frequency to allow transmission of the important message. Another way to get an important message recognized can be through the use of the word *priority*. This allows a unit using this signal to receive prompt recognition of its call letters; the message is of importance but not an emergency situation. An example is: "Priority! Command from Engine 1." When "Mayday," "emergency traffic," or "priority" is used, all other units must keep off of the radio to allow the important message to be given.

The use of numbered codes has drawbacks, especially on interagency responses, and should not be used. Messages should be concise. Effective communications allow other agencies to understand our specific needs.

It must be assumed that an order given via radio that has not been acknowledged has not been received. The receiver must confirm messages and seek clarification if an order is unclear.

Radio Protocol

Many fire departments have changed how they send radio messages. In the past, most departments stated their own call letters and then the call letters of the person they were calling ("Engine 1 to Command," or "Command to Dispatch"). They found that the person receiving the call heard his or her own call letters but many times did not hear who was calling them, resulting in additional radio traffic. A more effective method in use today places the unit being called first and then the unit sending the message ("Dispatch from Engine 1," or "Command from Division 1"). The unit being called is alerted and listens for who is trying to make contact. This method has been found to reduce unnecessary radio communications.

Cellular Phones

Communications can be enhanced by the use of cellular phones. They allow ready access to Dispatch on a relatively unmonitored line. Cellular phones can be used to contact outside agencies, such as the Chemical Transportation Emergency Center (CHEMTREC), to provide technical information on transportation accidents. Some departments are now using them as a command channel on major incidents. They allow the IC an avenue to talk to Division and Group Supervisors without interfering with radio transmissions.

Summary

Utilization of an incident management system is an endeavor that must be planned and practiced. The Incident Commander must develop a structured organization to address the many and varied problems that occur.

For a system to be effective, everyone must be familiar with it. Familiarity can be accomplished only through adequate training sessions for all members. Training should include using scenarios that allow expansion of the system from a minor incident to a major disaster. Ensuring that proper procedures become instinctive under the stress of emergencies can occur only through training and constant application of the system. Operations will be better organized. Assignments will be well defined. Communications will flow up and down the organizational structure. All members will know what is expected of them. Problems will be recognized and handled, and a strong emphasis will be placed on firefighter safety.

When a command system is expanded and assignments are given, those individuals do not operate independently. Ensuring firefighter safety demands coordination and communications. There must be discussion on the progress being made and the problems that must be overcome. Through these discussions, solutions can be found and required adjustments to strategies made.

A department that communicates well eliminates the problems of units duplicating assignments. Good communications allow for the assigning and tracking of personnel and assist the IC in recognizing when units will need to be relieved.

Build on your strengths, and work on weak areas to develop them into strengths. Recognize what is possible. Consider the resources available and their capabilities.

Real organization will be found in the acts of those in Command and the coordinated efforts of those being commanded.

Review Questions

1. How well does your fire department interact with your mutual-aid companies when implementing an incident management system? Are you in agreement in all areas? Or are there some areas in which your departments operate differently?
2. Discuss a past major incident in your department. What areas of your incident management worked well? Which areas have room for improvement? What specific steps would you take to improve these areas?
3. What factors should the Incident Commander consider when expanding the positions in an incident management system?
4. What are the eight basic components of the incident command system that provide the basis for an effective ICS concept of operation?
5. Describe how Unified Command should be handled at a major incident that involves your fire department and a number of outside agencies at a passenger plane crash in your response district.
6. List the general staff and command staff positions in the incident command system.
7. What are the six components of the National Incident Management System?
8. What are the responsibilities of the Incident Commander?
9. What are the responsibilities of the Operations Section Chief?
10. What are the responsibilities of a fire officer who is a Mobile Command?
11. How does unity of command assist in ensuring firefighter safety?
12. List the positive aspects of utilizing Staging at an incident scene.
13. In your department, who could assume the position of Finance/Administration if needed at an incident?
14. What information is required for the transfer of Command to be effective?
15. List the considerations for establishing a command post. Who should be located at the command post? How should the command post be marked in order to be readily identifiable?
16. What alternative methods of communications could be utilized by your department at an incident scene?
17. Who benefits from a comprehensive initial status report?
18. Name the three types of status reports.
19. At an incident, how well does your department communicate with mutual-aid departments? Do you operate on the same frequencies? Are there methods of improving your current communications at an incident? How?

Suggested Readings, References, or Standards for Additional Information

New Hampshire Department of Safety. "Emergency Planning Municipalities: National Incident Management Incident System (NIMS): Frequently Asked Questions." Homeland Security and Emergency Management. Available at: https://www.nh.gov/safety/divisions/hsem/Planning/planning_muni_nims_faq.html#A9.

US Fire Administration (USFA)

"Incident Command System 100" (Self-study course).
NIST Technical Note 1661. April 2010. Report on Residential Fireground Field Experiments. This report on crew size available at: http://www.nist.gov/fire/ssr1.cfm.

National Fire Protection Agency (NFPA)

Standards:
1600: Standard for Disaster/Emergency Management and Business Continuity Programs
1561: Standard on Emergency Services Incident Management System
1221: Standard for the Installation, Maintenance, and Use of Emergency Services Communications Systems

Related Courses Presented by the National Fire Academy, Emmitsburg, Maryland

Emergency Medical Services: Incident Operations
Introduction to Unified Command for All-Hazard Incidents
Communications for Emergency Services Success
ICS-100—An Introduction to Incident Command System
ICS-200—Incident Command System
ICS-300—Intermediate Incident Command System for Expanding Incidents
ICS-400—Advanced ICS Command and General Staff, Complex Incidents, and MACS for Operational First Responders

Related Courses Presented by the Emergency Management Institute

EOC/IMT Interface
NIMS ICS All-Hazards Public Information Officer
Introduction to the Incident Command System
NIMS ICS All-Hazard Finance and Administration Section Chief
NIMS ICS All-Hazard Safety Officer Course

3
Decision Making

❝Plans are nothing; planning is everything.❞

—Dwight D. Eisenhower

Fast-moving fires in commercial buildings can threaten numerous exposures. *Used with permission of Marty Griffin.*

KEY TERMS

Incident Action Plan, *p. 83*
Planning "P", *p. 86*
incident scene decision making, *p. 90*
size-up, *p. 93*

strategy, *p. 107*
tactics, *p. 108*
tasks, *p. 108*
crew resource management (CRM), *p. 112*

firefighting ventilation practices and tactics, *p. 113*
modes of fire attack, *p. 119*

OBJECTIVES

Upon completion of this chapter, the reader should be able to:

■ Describe an Incident Action Plan.
■ Understand the difference between cue-based and classical decision making.
■ Identify and discuss the 13 points of size-up.
■ Discuss strategy, tactics, and tasks found in the classical decision making process.
■ Discuss the modes of fire attack.

Incident scenes are dynamic events. Because there is the ever-present possibility of change, as Incident Commanders we must attempt to predict these changes and be prepared to be one step ahead of these problems.

Overall success involves preparation, as discussed in Chapters 1 and 2. Once we arrive at the emergency scene, our training and experience must take over. Chapter 3 looks at the importance of developing an Incident Action Plan (IAP). Having a practical IAP will ultimately lead to better incident scene decision making. A critical part of the IAP is identifying problems, which we accomplish through our size-up and determining incident objectives. We can then achieve our incident objectives by implementing strategies, tactics, and tasks. This chapter also reviews the offensive, defensive, and transitional modes of fire attack.

In addition, this chapter reinforces the need to use our training and experience to assist in our constant goal of improving firefighter safety. Having a workable IAP at every incident will enhance safe operations and our decision making process.

Incident Action Plan and the Planning "P"

The key to successful incident scene operations is proper planning. A major component of the planning process is the use of an **Incident Action Plan** (IAP). The IAP can be just a thought process of the Incident Commander (IC) on a minor incident, or it can evolve into a comprehensive written document on larger incidents.

Incident Action Plan ■ Contains measurable incident objectives to be achieved in a specific time frame.

IAPs ensure that everyone is working in concert toward the same goals. They include measurable incident objectives to be achieved in a specific time frame called an Operational Period. IAPs provide a coherent means of communicating the overall incident objectives for both operational and support activities; they provide all supervisory personnel with direction for actions to be taken. They may be verbal or written, except for hazardous materials incidents and other special operations, where IAPs must be written. When written, they are prepared by the Planning Section.

For simple incidents of short duration, the IAP will be developed by the IC and verbally communicated to subordinates. The planning associated with this level of complexity does not demand any formal Planning Meeting or written instructions.

Under certain conditions, such as additional lead-time, increased staff, and cascading consequences at the scene, Command may need to engage in a more formal process. On larger incidents, the Incident Commander/Unified Commanders can receive assistance in developing the incident objectives and the IAP.

A written IAP should be considered whenever:

- Multiple jurisdictions are involved in the response.
- The incident is of long duration or will involve multiple Operational Periods.
- A number of ICS organizational elements are activated (typically when all General Staff Sections are staffed).
- It is required by agency policy.
- Hazardous materials are involved in the incident.
- Incidents involve special operations, such as confined space or trench rescues.

An IAP can be prepared for planned events, such as major training exercises, conventions, parades, or concerts, as well as emergency incidents.

Emergency incidents will demand immediate planning and organizational actions to ensure safe operating activities. The big difference between a planned event and an emergency incident is that the planned event can be prepared for in advance without time pressures in which decisions have to be made, while the emergency incident that occurs without the benefit of preplanning can demand that life safety measures be immediately applied.

Incident scene planning involves:

- Evaluating the situation
- Developing incident objectives
- Selecting a strategy(ies)
- Deciding which resources should be used to achieve the objectives in the safest, most efficient, and cost-effective manner

WHAT A WRITTEN INCIDENT ACTION PLAN DOES

An IAP formally documents incident goals, Operational Period objectives, and the response strategy, as defined by the IC during the planning process. It contains work assignments (tactics) to achieve goals and objectives within the overall strategy, while providing important situational information on event and response parameters. Equally important, the IAP facilitates dissemination of critical information about the status of response assets (apparatus, personnel, supplies, and equipment). Because incident scenes are constantly evolving, action plans must be revised on a regular basis (at least once per Operational Period) to maintain consistent, up-to-date guidance across the system.

The following should be considered for inclusion in an IAP:

- Incident objectives (where the response system wants to be at the end of a response)
- Operational Period objectives (major areas that must be addressed in the specified Operational Period to achieve the goals or control objectives)

- Response strategies (priorities and the general approach to accomplish the objectives)
- Response tactics (methods developed by Operations to achieve the objectives)
- Organization list (with an ICS chart showing primary roles and relationships)
- Assignment list (specific duties with work assignments and special instructions)
- Critical assessments (situational updates)
- Composite resource status updates (accountability of all personnel)
- Health and safety plan (to prevent responder injury or illness)
- Communications plan (how functional areas can exchange information)
- Logistics plan (procedures to support Operations with equipment, supplies, etc.)
- Responder's medical plan (providing direction to care for responders)
- Incident map (diagrams/drawings of incident scene)
- Traffic plan (how responders will move around the scene)
- Additional components as indicated by the nature, complexity, and scalability of the incident

OPERATIONAL PERIOD

An IAP is designed for a specific period of time. This time frame is determined by Command along with input from the Command and General Staffs. The Operational Period can be for 6, 8, 12, or 24 hours during a response phase, and weeks or months during a recovery phase, as would occur from a natural disaster. After Hurricane Katrina, weekly IAPs were created to address demolition and debris management issues.

The considerations for an Operational Period would be:

- Safety of responders and victims
- Work assignments of resources on scene
- Availability of additional personnel
- Future involvement of additional agencies and jurisdictions
- How environmental conditions will impact the incident scene

Documentation for the planning process is achieved through the use of standard NIMS-approved, incident command system forms. These forms are used to document basic areas of the incident scene. The incident scene documentation starts with Form 201 Incident Briefing (The 201 is a formalized tactical worksheet or command chart). It shows a drawing of the incident scene, a summary of the current actions being performed, the current incident objectives, an organizational chart that has been created to address the problems found, and a list of all resources dispatched to the scene. This structured form contains the same basic information that an IC would quickly document, in a less formal manner, on tactical worksheets. Though Form 201 Incident Briefing is not part of the IAP, the information that it contains allows for a smooth transfer of command at an incident scene and has the basic information needed to initiate the IAP.

The other forms that make up the foundation of the IAP are:

- ICS Form 215 Operational Planning Worksheet
- ICS Form 215A Safety Officer's Report

ICS Form 215 is developed by the Operations Section Chief as a result of the Tactics Meeting to provide insight into resource requirements, work boundaries, and tactical assignments.

ICS Form 215A is completed by the Incident Safety Officer (ISO). It addresses each operational aspect of the incident and specifies safety mitigation measures for identified hazards in the divisions/groups.

ICS Forms 201, 215, and 215A are not part of the formal IAP, but are the foundation pieces. (See Figure 3-1.) A strong foundation will support the structure of a building. Likewise, these preparatory forms, when properly utilized, will give the IAP a solid platform. Because they address the incident objectives and are used for briefing, resource requirements, and personnel safety, there is a greater chance of a positive outcome. (See the ICS Forms Booklet in Appendix J in www.pearsonhighered.com/bradyresources.)

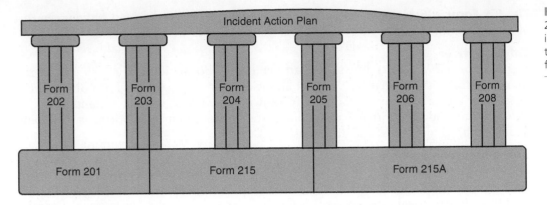

FIGURE 3-1 Though Forms 201, 215, and 215A are not included in a written Incident Action Plan, the information they contain is the foundation of the plan.

COMPONENTS OF AN INCIDENT ACTION PLAN

ICS Forms 202, 203, 204, 205, 206, and 208 are the basis of the IAP and the preparation of these forms is delegated to the different ICS sections or officers. The main components of any IAP include:

- What do we want to do? (ICS Form 202)
- Who will be responsible for doing it? (ICS Form 203)
- How will it be done? (ICS Form 204)
- How will we talk to each other? (ICS Form 205)
- What happens if someone gets hurt? (ICS Form 206)
- How can we protect responders? (ICS Form 208)

Planning Section
The Planning Section is responsible for preparing ICS Forms 202, 203, and 204.

ICS Form 202 Incident Objectives provides a general situational awareness that may include:

- The incident objectives that are received from the IC(s)
- The projected weather forecast for the upcoming Operational Period
- A safety message from the Safety Officer (see ICS Form 208 Safety Message/Plan)

ICS Form 203 Organizational Assignment List contains the names of everyone in the incident management system who has received an assignment, such as the IC, section chiefs, branch directors, supervisors, and unit leaders.

The ICS Form 204 Assignment List denotes the divisions, groups, strike teams, task forces, and crews with their individual work assignments or tactics. The work assignment (tactics) is taken from Form 215 Operational Planning Worksheet. For strict accountability, a separate Form 204 is created for each division and group. This form is important because it spells out the duties and special instructions that are required from division and group supervisors and that must be performed to achieve the incident objectives for the Operational Period. It includes safety considerations and the radio frequencies to be used by each division or group.

Logistics Section
The Logistics Section is responsible for preparing Forms 205 and 206.

ICS Form 205 Radio Communications Plan The use of a common communications plan is essential for ensuring that responders share information with one another during an incident. Communication equipment, procedures, and systems must operate across jurisdictions and have interoperability between agencies. Developing an integrated voice and data communications system must occur prior to an incident. The ICS Form 205 contains designated radio frequencies for Dispatch, tactical, and support functions. A Form 205A may be prepared listing all methods of contact: radio frequency, phone, cell phone, or pager numbers of the essential individuals in the incident management organization. Since the IAP is a public document, many incident management teams will create the 205A but not include it in the IAP. They will distribute it separately to keep that contact information from possible mass distribution.

The ICS Form 206 Medical Plan contains information on medical aid stations, means of transportation, and hospital information for responders who are injured.

Safety Officer

The ICS Form 208 Safety Message/Plan contains information on how to protect responders from hazards that may exist at an incident. It may be in the form of a general message or a detailed plan. It may specify known hazards and specific precautions to be observed. It should be written in clear concise statements and include priorities, key command emphasis, decisions, and directions.

Other Forms in an IAP

Additional information can be included in the IAP. Common reports that are created by the Logistics Section include a facility map and a traffic plan. The traffic plan can be used to designate certain streets or roads for the delivery of supplies, or to establish drop-points for equipment. It can indicate if certain roadways are being utilized as one-way streets to facilitate movement of apparatus/vehicles.

The Planning Section/Situation Unit Leader can create maps that delineate division boundaries and the location of helispots, which are used as landing sites for helicopters to load and unload supplies or personnel.

The Safety Officer can include a detailed safety plan for clarification of specific hazards and conditions.

Additional information and components of the IAP are at the direction of the IC. The intent of the IAP is to provide flexibility and scalability as a tool for decision making based upon the on-scene conditions.

> *Helispot* is a term that originated in wildland firefighting and is a temporary location at the incident where helicopters can safely land and take off to deliver supplies and personnel. Multiple helispots may be used.

PLANNING "P"

Planning "P" ■ Is a practical tool that is used in the planning process and for the development of the Incident Action Plan.

The **Planning "P"** is a practical tool that is used in the planning process and for the development of the IAP. (See Figure 3-2.) It is a blueprint of actions and meetings that are used to facilitate the handling of an incident scene and the planning process. The leg of the "P" describes the initial response period and the needed actions once the incident/event begins. The specific steps are:

- Notifications
- Initial Response and Assessment
- Incident Briefing (using ICS Form 201)
- Initial Incident Command (IC)/Unified Command (UC) Meeting (if Unified Command)

At the top of the leg of the "P" is the beginning of the first operational planning period cycle. In this circular sequence there are nine steps:

1. IC/UC Develop/Update Objectives Meeting
2. Command and General Staff Meeting
3. Preparing for the Tactics Meeting
4. Tactics Meeting
5. Preparing for the Planning Meeting
6. Planning Meeting
7. IAP Prep and Approval
8. Operations Period Briefing
9. Execute Plan and Assess Progress

The final step, execute the plan and assess progress, completes the subsequent cycle and begins a new Operational Period; decision makers then begin another review of the Planning "P" for the next Operational Period.

Reviewing the Planning "P"

Initial Response and Assessment Planning begins with problem identification (on-scene size-up). This is the observation of conditions and circumstances that provide the information, cues, prompts, or indicators necessary to make initial management decisions.

Agency Administrator Briefing (If Appropriate) This briefing allows interaction between the incident management team and the agency administrator. It allows the IMT to learn information about the incident that they have been brought in to manage. It can be used to give a delegation of authority to the IMT and typically sets parameters of authority, release of information to the media, and allowable spending for the IMT in handling the incident. This meeting would not be held for incidents within an IMT's response district, but for teams that are brought in from outside the area to manage the incident.

Incident Briefing Using ICS Form 201 The ICS Form 201 contains critical information about the incident situation and the resources allocated to the scene. This form serves as a permanent record of the initial response to the incident and is used by Command in the transfer of Command so the incoming IC can rapidly gain situational awareness. Preparation of the Incident Briefing Form may be done by the initial IC if he or she has sufficient time, or it may be prepared by the oncoming IC as part of conferring on the transfer of information of the incident. The form's preparation could also be delegated to the oncoming Planning Section Chief, or the oncoming Operations Section Chief. Regardless of who actually prepares the form, it will be shared with all of these individuals.

Initial Incident Commander/Unified Command Meeting (If Unified Command)

Unified Command is a collaborative team-effort approach. When Unified Command is established, the Unified Incident Commanders meet and decide upon a designated spokesperson, location of a single incident command post (ICP), and timeframes for the creation of a single IAP. This meeting allows the Unified Incident Commanders to discuss and concur on important issues and allows each agency stakeholder to present jurisdictional limitations, concerns, and restrictions. During this meeting, participants need to agree on the following points:

■ The organizational structure
■ Selection of an Operations Section Chief
■ The General Staff positions that will be filled
■ A procedure for resource ordering
■ A policy for sharing the cost of the incident
■ Directions for the release of incident-related information
■ Other pertinent matters

The Start of Each Planning Cycle

The Incident Commander/Unified Commanders establish the incident objectives. Incident objectives are broad statements of guidance and direction that will be used to develop appropriate strategies. The incident objectives state what is to be accomplished in the Operational Period. It should be noted that not all incidents require detailed written plans. When a detailed written plan is needed, a Planning Meeting is one method that can facilitate the process.

The cyclical planning process is designed to use the overall incident objectives to create tactical assignments and achieve them during the specific Operational Period. It

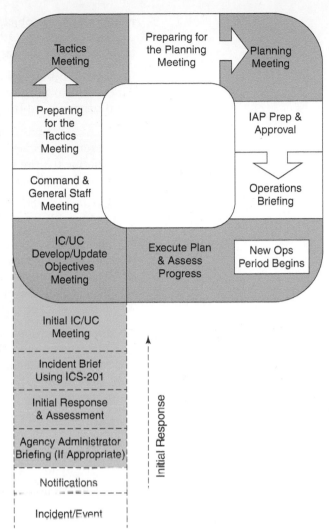

FIGURE 3-2 The Planning "P" is a guide to the process and steps involved in creating a written Incident Action Plan for an incident.

is important that objectives on health and safety of personnel are maintained during the course of the event. Other incident objectives can focus on specific steps to reduce operational problems during a single Operational Period.

Incident Objectives

An example of incident objectives, at a fire in a large two-story motel where many lives are threatened and fire has control of the first floor, could be:

- Provide for the safety of all responders and civilians throughout the incident
- Ensure safe removal of all occupants from the fire building within five minutes
- Confine the fire to the first floor within 15 minutes
- Provide feedback to the media within one hour

Incident objectives should be written in a *S.M.A.R.T.* format. This implies that they should be:

- **S**pecific—Is the wording precise and unambiguous?
- **M**easurable—How will achievements be measured?
- **A**ction-oriented—Is an action verb used to describe expected accomplishments?
- **R**ealistic—Is the outcome achievable with given available resources?
- **T**imely—What is the timeframe? (if applicable)

Command and General Staff Meeting

The Incident Commander/Unified Commanders may meet with the Command and General Staff to gather input and ideas or to provide immediate direction that cannot wait until the planning process is completed. This meeting occurs as needed and should be short and concise.

Preparing for and Conducting the Tactics Meeting

The Operations Section Chief, Safety Officer, Logistics Section Chief, and Resources Unit Leader attend the Tactics Meeting. The Planning Section Chief may also attend, but all others are by invitation only. The Operations Section Chief and the Safety Officer have major roles. Documentation is made on ICS Form 215 Operational Planning Worksheet and ICS Form 215A Incident Safety Analysis.

The creation of ICS Form 215 can build upon the current organizational structure that is already in place on the incident scene, or it can establish branches, divisions, and groups to achieve the incident objectives for the next Operational Period. The basis for a written IAP is to list the objectives, determine the strategies needed to achieve the objectives, and develop the necessary tactics to achieve life safety, incident stabilization, environmental protection, and property conservation endeavors. (Tactics can be referred to as "work assignments" or "control operations" in some NIMS documents/forms.)

The purpose of the Tactics Meeting is to review the work assignments, resource requirements, and reporting locations for the next Operational Period's resources. This information is then placed onto Form 215. The Operations Section Chief develops the tactical measures that will support the proposed strategies and objectives provided by the IC/UC.

The Tactics Meeting will provide the following information:

- Assignments of resources to implement the tactics
- Methods for monitoring tactics and resources to determine if adjustments are required (e.g., different tactics, different resources, or a new strategy)
- Reporting location and time for oncoming resources

The Tactics Meeting will ensure adherence to the unity of command principle, span of control guidelines, and accountability. It will also ensure that the resource assignments consist of the needed resources by kind, type, and numbers to achieve the objectives mandated for the next Operational Period. If the required tactical resources are not available, then an adjustment must be made to the work assignments being considered. It is very important that the availability of tactical resources be determined prior to spending a great deal of time selecting strategies and developing tactical activities that realistically cannot be achieved.

Preparing for the Planning Meeting

Following the Tactics Meeting, preparations are made for the Planning Meeting, to include the following actions coordinated by the Planning Section:

- Review the ICS Form 215 developed in the Tactics Meeting
- Review the ICS Form 215A Incident Safety Analysis (prepared by the Safety Officer), based on the information contained on ICS Form 215
- Assess the current operational and logistical effectiveness
- Gather information to assist in Command decision making; the plan will already have been agreed upon

Planning Meeting

The Planning Meeting provides the opportunity for the Command and General Staff to review and validate the operational decisions as proposed by the Operations Section Chief. Attendance is required for all Command and General Staff. Additional incident personnel, public officials, and private sector representatives may attend at the request of the Planning Section Chief or the IC. The Planning Section Chief conducts the Planning Meeting, following a fixed agenda, and it should be completed in 30 minutes or less.

The Operations Section Chief delineates the amount and type of resources needed to accomplish the objectives of the Operational Period. The Planning Section's Resources Unit works with the Logistics Section to accommodate the request for resources.

At the conclusion of this meeting, the Planning Section Staff will indicate when all elements of the plan and supporting documents are to be submitted for duplication, collation, and distribution for the Operational Period Briefing.

IAP Preparation and Approval

The next step in the Incident Action Planning process is preparation and approval of the plan. The written plan is comprised of a series of standard forms (202, 203, 204, 205, 206, and 208) and supporting documents that convey the IC's intent and the Operations Section's direction for the accomplishment of incident objectives during the next Operational Period.

Once the IAP is completed, it must be approved and signed by Command.

Operations Period Briefing

The Operations Period Briefing may also be referred to as the Operational Briefing or the Shift Briefing. This briefing is conducted at the beginning of each Operational Period and presents the IAP to branch directors; division and group supervisors; and strike team, task force, and crew leaders. The Operations Section Chief will discuss the strategies and tactics needed to accomplish the incident objectives, and each division and group supervisor will review their individual ICS Form 204. This form, which is contained in the IAP, addresses their work assignments, accountability, communications, and safety information.

Following the Operations Period Briefing, supervisors will meet with their assigned resources for a detailed briefing on their respective assignments. This will include any information that has been received from off-going supervisory personnel from the previous Operational Period or any new developments that have occurred since the IAP was finalized.

Execute Plan and Assess Progress

The Operations Section directs the implementation of the plan. Supervisory personnel in all sections are responsible for accomplishing various duties during the Operational Period.

The plan is evaluated at various stages in its development and implementation. The Operations Section Chief may make adjustments during the Operational Period to ensure that the objectives are being met and effectiveness is assured.

The cyclical process of the Planning "P" enables it to be used to continually evaluate the situation and adjust for changes that may occur at the incident scene.

A planning cycle should be initiated for each Operational Period. An assessment must be made of the current situation and whether it is stable or dynamic. If changing:

- Is it increasing in size or complexity?
- Are the objectives effective, or is a change needed?
- How long until the objectives are completed?
- What is the current status of resources?

The IAP and the Planning "P" facilitate effective management at both planned events and unplanned emergencies. Proficiency in the planning process requires training and exercises. Regular practice will ensure that the initial responders are properly prepared to handle any size or type of incident in their community.

ON SCENE

The story of a deadly refinery fire that took the lives of eight Philadelphia firefighters and critically injured two others shows how one comment can save a life. Crews were operating in an area of poor drainage, and water had built up to a height of over two feet. Unbeknownst to the firefighters, leaking crude oil was floating on the surface of the water beneath a layer of foam. As firefighters were refilling a foam unit, they broke the foam layer, and a foam unit's muffler ignited the crude oil.

At the time of the fire, I was the captain of Engine 49 on another platoon. One of the injured firefighters was assigned to my company, and while visiting him later at the burn center, our discussions led to how he was able to survive this disaster. He stated that at the time he was engulfed in flames, two thoughts entered his mind: "Get the hell out of there" and "Don't breathe." The thought about not breathing was something that had been told to him as a new recruit at a fire academy 15 years earlier. During an informal discussion, a lieutenant said that if caught in flames without a mask, breathing in the superheated vapors would cause internal damage, probably resulting in death. The injured firefighter could not recall ever thinking of this discussion in the previous 15 years until the moment that he needed it most. This cue saved his life.

(See Appendix H: Case Studies online for a full description of the Gulf Oil Refinery Fire.)

Incident Scene Decision Making

incident scene decision making ■ The two basic methods are the classical method and the *naturalistic,* or *recognition prime,* or *cue-based* method.

There are two basic methods of **incident scene decision making.** The system most often discussed is the *classical method.* In this system, an individual proceeds step-by-step in a long and precise process that in and of itself is time-consuming. This method has some application at an incident, but usually in a minor role.

The system used by most fire officers is referred to by a variety of names: *naturalistic, recognition prime,* or *cue-based* decision making. This is a process in which extremely fast decisions are made. The process is dependent on the experience and training level of the decision maker.

CUE-BASED DECISION MAKING

The fire officer, through personal experience of training exercises, incident scene responses, and study, has built a base of knowledge. This knowledge is embedded in the brain and is readily recallable. At an incident, the brain will automatically associate past experiences with current developments. This tying together of past and present events lets the IC recognize signs or cues that are present and determine strategies, tactics, and tasks to mitigate the emergency. (See Figure 3-3.) A recognition factor can occur through sight, hearing, or smell. The brain automatically reacts to remember or visualize past associations. It can be thought of as reliving a dream or happening. (It is similar to looking at a photo album of snapshots of a long-past family outing and recalling the events that took place the day the photos were taken.) These snapshots of past events may or may not have been occurrences that fully impressed the fire officer at the time they originally occurred. In many post-incident debriefings and critiques that I have participated in, many firefighters did not seem to remember these happenings as having been significant occurrences at the time.

FIGURE 3-3 Fires in three-deckers are routine operations in some areas. Previous experience of the Incident Commander can make cue-based decision making extremely rapid and can be easily applied to this type of fire situation. *Used with permission of Joseph Hoffman.*

Critical Cues

The decision maker relies on reading critical cues from the incident, analyzing them, and reacting to those critical cues in a manner previously witnessed or learned. For the highest efficiency, the decision maker must know which critical cues are important for the specific situation and the most effective responses to each critical cue. (For example, if we learn the correct critical cues but associate an incorrect response to those critical cues, the decisions made will be flawed and the correct and most efficient solutions will not be applied.)

Even if the firefighter has not been exposed to a particular type of fire situation, he or she can often see occurrences that will automatically be associated with similar events. Though there may not be enough information to fully solve the problem, it usually allows a starting point. The fire officer can then monitor the situation and look for additional signs to continue to mitigate the problems as they develop.

Cue-based decision making is extremely rapid and is the desired method for emergency operations when the experience level allows the IC to implement it. Without a sufficient knowledge base, decisions will need to be analyzed through the classical method.

CLASSICAL METHOD OF DECISION MAKING

The classical method of decision making is used for training exercises, development of preplans, or incidents with cues that require a set of reactions that the decision maker has not experienced or learned before. An example would be an overturned gasoline tank truck involved in fire on an interstate highway. If the decision maker has never encountered or trained for a similar event, he or she may need to refer to the command sequence for assistance. In this instance, the decision maker must process information through the following steps:

- Read the cues.
- Compare those cues to what has been learned from similar situations.
- Review the command sequence.
- Determine the actual problems, the incident objectives, and the strategies to solve those problems.
- Evaluate and select the most effective tactics.
- Implement the tasks.
- Arrive at a conclusion or result by hypothesizing.

While going through the above steps and possibly receiving additional information at the scene, the decision maker may at any time feel that he or she has a handle on the problems found and decide on a plan of action. Having completed this process, and if the resulting actions are successful, the decision maker will simply use the cue-based knowledge gained from this experience to reach the desired conclusions and results in future similar situations.

THE COMMAND SEQUENCE

Fireground operations must be handled with a systematic approach. To respond to a scene with no prior thought and expect the overall operation to run smoothly is being overly optimistic. The implementation of an IAP by the first-arriving fire officer is critical. That officer has to determine the incident objectives. With structural firefighting, the incident priorities are typically the basis for the incident objectives. The officer assuming Command must use a system to perform a proper size-up of the situation. The command sequence can be utilized to assist the IC in decision making. It gives the IC a standardized and sequential thought process to ensure that nothing has been overlooked. The command sequence consists of five levels:

- Level 1: Incident Priorities
- Level 2: Size-Up
- Level 3: Strategy
- Level 4: Tactics
- Level 5: Tasks

Let's look at each level of the command sequence:

Level 1: Incident Priorities
The best method of remaining proactive is through the use of a logical thought process. The determination of incident priorities, in conjunction with size-up, assists in the

FIGURE 3-4 Property conservation includes fire damage and any resultant damage associated with the firefighting efforts. Good overhauling procedures can minimize damage. *Used with permission of Greg Masi.*

development of strategy and tactics. Incident priorities are the foundation of the command sequence. All of our actions are based upon three priorities: life safety, incident stabilization, and property conservation. This list follows the order of importance placed on each priority.

Priority 1—Life Safety Life safety is always our Number One consideration. It includes endangered civilians, responding firefighters, and other public safety personnel operating at the incident scene.

Priority 2—Incident Stabilization We hope to mitigate the problems encountered. The actions taken will differ depending on the type of incident.

■ At a structure fire, our aim is to confine the fire to as small an area as possible. Realistically, this could be one room, one floor, an entire building, or several buildings. The factors that must be considered are numerous: construction, resources at the scene, resources responding, and the many other points of size-up performed at every incident.

■ On a medical response, it is important to stabilize the patient.

■ A hazardous materials incident could include stopping a leak and containing the spilled material.

The accomplishment of Priority 2 cannot expose firefighters to undue risk and thereby violate our first priority of life safety.

Priority 3—Property Conservation A fire department's efforts should be directed toward minimizing property damage. (See Figure 3-4.) This can be achieved through quick extinguishment of fire and employing good salvage methods. Property loss reduction benefits the building owner, occupants, the community, and the fire department through good public relations. Property conservation includes fire damage and any resultant damage associated with the firefighting effort. Proper ventilation and the minimizing of water damage can reduce the total loss.

Priorities never change, though implementation may. Quick extinguishment of a fire on a lower floor of a multistoried building will protect the lives of occupants above. By implementing Priority 2 (incident stabilization), we accomplish Priority 1 (life safety).

Size-Up

NIOSH FIREFIGHTER FATALITY REPORT F2013-17

On May 20, 2013 a 51-year-old male career firefighter (the victim) was conducting a primary search for occupants after the fourth alarm at a fire in an apartment complex. He was killed inside the building when it collapsed. The victim and his partner were in the first floor hallway knocking on doors to the apartments, which were inset from the hallway by small vestibules. The victim's partner was in the vestibule knocking on the third door to the left, and the victim was in the hallway going to knock on the third door on the right. In an instant, the second floor walkway and possibly the third floor walkway collapsed into the first floor hallway, killing the victim. The victim's partner was trapped in the inset of the doorway.

Key recommendations based on contributing factors included adequate size-up and risk assessment before beginning interior firefighting operations; communicating critical benchmarks to IC; clear procedures for operational modes and coordinate changing modes; clear predesignated positions (IC at command post, Safety Officer, assignments on arrival); inspections to ensure working sprinkler system; firefighter training in situational awareness, personal safety, and accountability; training to understand operation elevated master streams and effects on structural degradation; and training all personnel in risks and hazards related to structure collapse. **See Appendix B: NIOSH Reports in www.pearsonhighered.com/bradyresources to read the complete report and recommendations.**

Level 2: Size-Up

Size-up lets the IC gather information for the development of strategic goals. It is a mental process, weighing all factors of the incident against the available resources. Size-up can be looked at as solving a problem or as a puzzle that requires putting the pieces in their correct place by gathering and interpreting the available information. It is an evaluation process that reviews all the critical factors that could have a positive or negative impact on an incident.

Size-up starts in the preplanning stages. Preparation allows for a better overall operation and makes information readily available to the IC. At the time of the alarm, the previously gathered information will be invaluable as our size-up is initiated. It has been said that size-up is anticipation en route and habitation on arrival. While responding, we are wondering what we will find. What type of building? What is the fire load? What is the fire location, and are there any exposed buildings? On arrival,

FIGURE 3-5 In the initial size-up, a chief must do a 360-degree walk-around to ensure a full survey of the fire area. Otherwise, the chief's view limits his or her knowledge as to the size of the structure and the amount of fire involvement. *Used with permission of Joseph Hoffman.*

we shift into a higher gear and must immediately address the life safety of civilians. The size-up en route will be based on information given to Dispatch and relayed to the IC as he or she responds. It should include personal knowledge of the structure, information the driver or engineer may possess, and any preplans that may be in place. Size-up information can come from occupants; bystanders; or placards on buildings, tanks, or vehicles containing hazardous materials.

size-up ■ Identifies problems at an incident scene.

On arrival at the fire scene, it is best to do a 360-degree walk-around of the fire building or incident. (Some large-area fires or hazardous-materials incidents make this impractical, in which case reports from individual divisions and groups will have to suffice.) (See Figure 3-5.) In some cases, the rear or sides of a building can be viewed from the apparatus or chief's vehicle as it arrives on scene. This 360-degree walk-around permits a view of the entire structure. It allows us to develop a personal size-up assessment and later, when receiving status reports from division and group supervisors, to establish whether the situation has improved or deteriorated by comparing it to the conditions observed on arrival. (See Figure 3-6.)

Company Officer versus Chief Officer Size-Up Size-up in preparation for assuming Command falls into two categories: the company officer's and the chief officer's. There is a distinct difference between these two size-ups (assuming that the company officer is first to arrive). The company officer has a need for action. He or she has to make an

FIGURE 3-6 The 360-degree walk-around enables the officer doing the size-up to see the initial conditions. As the incident progresses and division and group supervisors are assigned, Command will know from their progress reports if conditions are improving or worsening. *Used with permission of Michael DeLuca.*

immediate decision about the volume and intensity of the fire and determine the initial strategies and tactics to be deployed. In this growing emergency, when the life hazard will be most severe, he or she will have no prior assessment of interior conditions to assist in making decisions.

The chief officer has certain advantages while performing size-up. There is slightly more time to make decisions. The life hazard will, most likely, already have been assessed, so the chief will have the benefit of an interior assessment on which to base decisions.

OBSERVATION THROUGH THE USE OF OUR SENSES

Those operating on the incident scene need to know many size-up factors. Though size-up is a primary responsibility of the IC, it must be performed by all firefighters at an incident. A supervisor who has received a report or performed a 360-degree walk-around should share any necessary facts with the company members. A wall crack or an unsafe floor is a safety factor that everyone operating in those areas needs to know about. Size-up allows us to handle an emergency incident in a professional manner, and it is facilitated through our observations. Size-up requires the use of the senses of touch, smell, hearing, and sight. What do you feel? What do you smell? What do you hear? What do you see?

Sense of Touch

The sense of touch enables us to determine the weather conditions or temperature differences when confronted with the heat of a fire:

- Icy conditions underfoot
- The temperature of runoff water
- The condition of a roof or a floor—is it spongy or firm?
- The wind blowing in our face allows us to determine how a fire might be affected by the weather conditions.

Sense of Smell

Certain odors are quite common and identifiable. These include:

- The smell of natural or liquefied petroleum gas
- The obvious odor of a misfiring oil burner
- The strong odor of gasoline fumes or paint thinners, indicating the possibility of explosive mixtures
- The distinctive burnt smell of unattended food left on a stove

Yet some serious problems are odorless, as in the case of carbon monoxide. A detector or monitor is needed to warn us of the presence of high concentrations of this deadly gas.

Sense of Hearing

Communication is one of the most critical parts of any successful operation. Our sense of hearing and understanding of the spoken word is indispensable. Yet the noise associated with an emergency scene can create a problem for firefighters. The blare of radio messages, fire officers giving orders, apparatus diesel engines being accelerated to increase pump pressures or to supply power to an aerial-equipped apparatus, are all part of the sounds found at an incident scene. Added to that noise might be the excited occupant or neighbor who attempts to give information on the location of those trapped within a fire building.

Firefighters can be the recipients of misguided information. Being told that everyone is out of the fire building may only refer to that individual's family or specific business occupancy and not other occupants within a building containing multiple apartments or businesses. Even at a private residence, the report that everyone is out of the building can be faulty. A visiting relative or friend of a child who is at the house may be forgotten.

There are those who will meet firefighters at an incident scene and feel compelled to say something. An example is the person begging you to rescue the children. This beseeching of our help would lead one to believe that children must be trapped and in need of immediate rescue. In many cases, what the person is actually saying is that the family living in the fire building includes children. They may have no idea if anyone is home, but they tell you of the need for rescue just in case they are.

Verbal reports from civilians at the scene will often dictate firefighting tactics. The fact that someone is reported trapped within a burning building will lead firefighters to concentrate on protecting the area where they are reported to be, for their safety and that of the firefighters attempting to rescue them.

Firefighters need to question anyone giving them information about occupants to clarify with certainty that someone is within the fire building. Receiving good information will assist in determining the risk/benefit factor: the amount of risk that firefighters take in attempting to save a life, or in fighting a fire. Reports of people trapped within a fire building can lead to fighting a fire differently than if everyone is safely out of a building. Misinformation can endanger firefighters and delay searching other areas where occupants could be trapped. It can place firefighters in dangerous areas where they may not have otherwise gone.

Language Barriers

Another problem arises when language barriers interfere with our communications. Asking occupants who have escaped questions on whether anyone is still within the fire building might be answered with an affirmative nod of the head. This seeming "yes" answer may be to a question that was not understood by the occupant due to a lack of understanding, if English is not his or her first language. It can lead to the firefighter receiving inaccurate data. This situation is especially true when rapid-fire questions are being asked by firefighters who know that seconds count and want an immediate response. Language barriers may be overcome by consulting with bilingual firefighters, police officers, or neighbors who can assist in interpreting for the fire officer.

Other Sources of Information

Listening starts at the time the alarm is received and continues throughout an alarm. Words and other sounds that are indicators or cues include:

- Initial dispatch data and the reports received en route
- Messages received in face-to-face communications
- Reports that are given either directly or indirectly during radio transmissions
- Progress reports of the headway being made or the problems encountered that are hindering progress
- The striking of additional alarms by the IC, or the request for more units to respond to the scene, which indicates a growing emergency
- Calls of warnings in the form of Maydays or urgent messages
- Progress reports given by the IC
- Cries for help from a trapped person, or pleas for the rescue of a family member who is still in a burning building
- Power saws operating on a roof, signaling vertical ventilation operations being performed
- Generators used to operate extrication or forcible entry tools

All of these auditory cues that firefighters may hear at an incident assist them in their individual size-up process.

Sense of Sight

Sight permits us to gather many of the size-up factors needed to effectively analyze the incident scene. What is seen on arrival? Can we be easily distracted? Are we fixated on fire pushing or blowing from a window? Has a person calling for help from a rooftop taken all of our attention? Can a complete visual observation of the structure and the surrounding buildings be performed?

Sight allows us to identify factors such as the location of smoke and fire, its volume, and energy. The flame color could indicate a higher temperature. The color and action of smoke can indicate a growing fire situation or a fire being brought under control through the fire department's efforts.

The initial view from the exterior rarely gives a complete picture of what must be accomplished. The exception is when the building is fully involved and leaves little doubt that the life safety of the occupants has already been compromised and that a defensive attack and protection of exposures is the only strategy that can be employed.

FIREFIGHTER SIZE-UP

In addition to the size-up data required by the fire officer who will assume Command, size-up information should be gathered by each firefighter. Size-up includes general areas for everyone at an incident scene as well as specific areas that are dependent on their individual assignments.

The size-up information needed by a firefighter assigned as a driver/engineer on an engine or pumper will be different for a driver/engineer of a truck company. Likewise, a firefighter assigned to an engine and responsible for control and extinguishment will have a different size-up than a firefighter assigned to a truck company and concerned with rescue, ventilation, and forcible entry.

As various structures are sized up, specific concerns of those occupancies will need to be considered. The size-up factors could continue to grow depending upon the slightest variations of circumstances, such as life safety, occupancy, contents, time, area, and height.

An excellent preplanning exercise is to select a building and list the size-up considerations for the IC and the factors that should be identified for each firefighting assignment.

There are general and specific size-up factors for various assignments. Let us look at some size-up areas for consideration at a structure fire.

General Size-up Considerations for All Firefighters

There are some general size-up factors that are important for everyone to know:

- Information given by Dispatch on the alarm (type of structure, report of people trapped, hazardous materials incident, etc.)
- Supplemental information given by Dispatch
- The size of the building: How many stories? Type of structure: private dwelling, apartment building, commercial structure?
- If an apartment building, how many units? Count the mailboxes, doorbells, electric or gas meters.
- The specific type of building construction and how fire could affect it
- The number of stories of the fire building and the exposed buildings
- The location of the fire within the building. The location of the fire will impact life safety and fire attack, such as a top floor fire versus a fire on a lower floor. The top floor fire can take longer to reach and will have longer hose-line stretches, whereas the fire on the lower floor will threaten the occupants on the floors above the fire floor.
- Accessibility into the fire building, to the upper floors (fire stairs, elevators), to the rear of the structure, to the roof, etc.
- The smoke conditions
- Offensive or defensive attack? Hand lines or master streams?
- Electric hazards: Are there any downed electric wires or arcing wires in danger of falling?
- Time factors: What is the lag time or reflex time? (the time it takes from receiving an order to accomplishing it)
- What life safety cues are present? For example, a car parked in the driveway of a dwelling, or curtains on the windows of the floors above a store.
- Changing conditions that could affect any of the firefighters operating at an incident scene. This could include collapse indicators, rapid fire spread, or dangerous conditions (i.e., holes in floors or the presence of hazardous materials). Should any of these conditions be found, the Division/Group Supervisor and/or the IC should be immediately notified.
- The locations for ground ladders or main ladders that are raised to windows or roofs, should they be needed as a secondary means of egress. After you enter the fire building, check to ensure that those ladders are still at the same location.
- Any lower roof locations that can be used as a secondary means of egress, from windows or from adjacent roofs.

General Size-Up Considerations for Drivers of Fire Department Apparatus Responding to a Fire Assignment These could include:

■ The exact address of the incident.
■ The route to take to the incident scene. Are there any street detours that could affect the response?
■ Whether other fire department units on the dispatch might use the same intersections while responding. Who has the right-of-way?
■ Are there any parked vehicles or other obstructions (such as trash dumpsters) that will affect apparatus placement?
■ Placing the apparatus in a way that allows for easily stretching hose-lines while keeping streets open for other apparatus.
■ Location of overhead electric wiring that, if attacked by fire, could fall onto the apparatus and electrify it.
■ Safe placement of the apparatus, considering immediate and anticipated collapse or safety zones. In placing the apparatus, the operator must consider the potential for an expanding fire that could increase the size of the collapse zones.

Specific Size-Up Considerations by the Firefighter or Engineer Driving an Engine or Pumper These could include:

■ Is the address located in a hydrant area? If so, where is the closest hydrant located?
■ Are the water mains of sufficient size to provide an adequate water supply?
■ At a major fire, is there a point where the water mains would be overtaxed?
■ Is the closest hydrant accessible or obstructed by illegally parked vehicles?
■ In nonhydrant areas, will a tender be assigned? Are there other sources of water available to use as a water supply, such as a lake, stream, or swimming pools?
■ If using an alternate water source, is there suitable apparatus access to set up a drafting operation?
■ Where is a good location for dump tanks?
■ What is the type of structure? Will it necessitate the use of 2½-inch or larger hose-lines? Can a large diameter hose line (LDH) be utilized for either supply lines or firefighting?
■ If not the first-due engine, what assignment has your company been given? Will this impact your apparatus placement?
■ If a truck company is responding, is it coming from the same direction as the engine company or a different direction? This must be known so the engineer can reserve the front of the fire building for the placement of the truck company.
■ Will ground ladders be needed from the first-due engine company? If yes, will they be needed to perform rescue operations before hose-lines are stretched?
■ Will the engineer initially need to supply hose-lines or assist in raising ladders for the rescue of occupants?
■ If the second-due engine company's responsibility is to provide a continuous water supply for the first-due engine, what is the best way to accomplish it?
■ What will the needs of the company be? Predict and be prepared to meet those needs. Should a fire not be darkened down, will another attack hose-line be required?
■ Are there standpipes or sprinklers that should be pressurized to ensure an adequate flow of water?

Specific Size-Up Considerations Performed by Firefighters Responsible for Handling Hose-Lines These could include:

■ In what location(s) is anyone at a window in need of rescue?
■ Are the occupants physically able to self-evacuate? Are they mentally or physically challenged? Can they understand instructions? Will they require fire department assistance?
■ Are there smoke detectors or carbon monoxide (CO) detectors sounding?
■ Are coded alarms sounding to indicate a specific fire location in a large building?

- Are there indicator panels giving the location of the alarms received?
- Has someone from the building met the fire department and informed them of the reported location of the fire or other pertinent data?
- What is the best way to attack the fire while protecting the civilians and firefighters and minimizing the spread of the fire?
- What size hose-lines will be needed? How many hose-lines? Placement of hose-lines?
- In what location are portable ladders already placed? Can they be used as a secondary means of egress from the fire building?
- In what location(s) are other units and hose-lines operating in a fire building? Ensure safety and teamwork in attacking the fire.
- Where are the fire escapes located? Are occupants using them to escape the building? Can they be used to advance hose-lines to the upper floors?
- Are buildings set back a distance from the street, or are buildings deeper than normal? This could indicate that pre-connected hose-lines might not be of sufficient length to reach the fire.
- What is the best method of getting a hose-line to an upper floor in a building with no standpipe?
- Are there standpipes in a building that can be utilized? Is the system looped? Are there individual risers that need to be specifically pressurized? How are the individual risers marked or identified?
- Are there any impediments that would prevent egress, such as window air conditioners or steel bars on windows?

Remember:

- The job of hose-line crews is to protect search and rescue crews operating around and above the fire area.
- When required, back up a hose-line with a hose-line of the same size or larger.
- Ensure that before proceeding to an upper floor to fight a fire a hose-line(s) is in place on the floors below, attacking fire at those locations.

Specific Size-Up Considerations by the Driver/Engineer of the Truck or Platform
These could include:

- Have you sized up all sides of the fire building to help identify locations for the placement of apparatus, aerial devices, and portable ladders?
- What assignment has your company been given? Will this impact on your apparatus placement? Apparatus placement must consider the use of the main ladder or platform, spotting the turntable for rescue at an offensive attack, or placing the apparatus at the corner of the building for a defensive attack.
- What is the best way to rescue those in need? Main ladder, platform, portable ladders, interior rescue?
- Are there any obstructions (trees, electric wires, etc.) that will impact on raising portable ladders, main ladders, or platforms?

Specific Size-Up Considerations by the Firefighters on the Truck Who Will Perform the Rescue, Laddering, Forcible Entry, and Ventilation   These could include:

- In what location is the fire? Is it safe to operate on the roof or the floors above the fire?
- Has the fire extended to any exposed buildings that may need to be evacuated?
- What is the location of smoke and visible fire within the fire building? What information can be gleaned from the radio reports? Where are the hose-lines being stretched? What reports are being given to the other units that are on the scene?
- Is there any information on people in need of rescue? From where? Is anyone visible at windows or on rooftops? If people are at windows, who is in the greatest danger?
- Can a search be performed under the protection of a hose-line?
- What is the location of the fire within the building and the best way to achieve ventilation: horizontal, vertical, positive pressure, hydraulic, or utilization of the building's heating, ventilation, and air-conditioning (HVAC) system?
- Will the wind affect the spread of the fire? In which direction should the ventilation be performed?

- Are there impediments to ventilation? Bars on windows, window air-conditioners, glass block windows, stationary awnings over windows, etc.?
- What exterior ventilation openings do the truck members recommend making to the interior crews? Once those recommendations are made, the interior crews will coordinate with the exterior crews on where the ventilation openings should be made and when.
- Is forcible entry needed? If so, where and how? What tools are needed in addition to the normally assigned tools? Are there any special concerns, such as special occupancies with highly advanced security systems, guard dogs, etc.?
- Have ladders already been placed by engine companies or other truck companies?
- Where should portable ladders be placed? For what use: rescue, ventilation, access, or as a secondary means of egress?
- Are there fire escapes that can be used for access, egress, ventilation, rescue, etc.?
- How should the primary search be conducted? What is the most critical area? Second most critical area?
- What exposures are there and what duties must the truck members perform, either in or on them?
- Is there a possibility of fire spread? Can the truck members perform duties to minimize fire spread?
- Can the truck members discover hidden fire? In buildings that have been converted from large single-family dwellings to apartments, expect numerous interconnecting void spaces where a fire can burn undetected, making it difficult to locate.
- Are there interconnected attics or cocklofts?
- Can thermal imaging devices be used to determine the location of people trapped and the possible spread of fire?
- What cues are there on the roof that the IC needs to know about? Dead loads in the form of air conditioner units? Heavy smoke or fire conditions?
- What overhauling will be needed after the fire has been knocked down?
- Will salvage be needed?
- What are the contents of the building, such as flammable contents? What are the life safety considerations at a school, an orphanage, nursing home, or hospital? How can they impact on the overall operation?
- What is the status of the building's HVAC system?
- What is the status of the elevators?

The following is an overview of the basic areas every good size-up should consider.

Wallace Was Hot   There are an endless number of mnemonic devices that are used for remembering the different size-up factors. One that covers the 13 points of size-up is Wallace Was Hot. Each of the letters represents a different size-up factor. (See Box 3-1.)

BOX 3-1: WALLACE WAS HOT

Water	**W**eather	**H**eight
Area	**A**uxiliary appliances	**O**ccupancy
Life hazard	**S**pecial matters	**T**ime
Location, extent		
Apparatus/personnel		
Construction/collapse		
Exposures		

Water   Assuming that sufficient resources are available, water supply will dictate whether a fire can be controlled. Units should have a system to quickly determine hydrant location, size of water mains, drafting sites, or other sources of available water. If a hydrant system is in place, is there a sufficient number of hydrants supplied by adequate-sized water mains? Operations can be hindered when dealing with dead-end or undersized water mains.

Water supply and available pressure should be monitored. It may be necessary to have the water company increase the pressure on hydrant systems. There may be a private hydrant system available. In addition to their domestic hydrant systems, some cities have secondary hydrant systems that use ocean water or untreated river water to protect high-value districts and industrial areas. The salt in ocean water may cause damage to apparatus and, if used, flushing of the apparatus must be performed as soon as possible after use.

In non-hydrant areas, water needs to be brought to the scene. If drafting sites are available, the use of large-diameter hose-lines, in conjunction with relay pumping, can ensure a continuous supply of water. (See Figure 3-7.) Relay pumping allows the movement of water by using multiple apparatus to pump water from a source to the fire scene. The use of large-diameter hose-line can facilitate this operation. Relay pumping is usually performed when the available water supply at the incident scene is insufficient or nonexistent. The water source can be a lake, pond, or municipal fire hydrants. This operation demands organization and management to ensure the proper size hose-lines, sufficient apparatus, and their correct placement. The IC must realize the amount of time required to set up a lengthy relay and what effect this will have on the fire.

A water tender operation can be utilized when a remote water supply is used. A dump tank(s) is set up at the fire scene. The water tenders can quickly dump their water supply and return to the source of the water supply to refill. This quick release of thousands of gallons of water by multiple units can ensure a continuous supply of water to be drafted by the pumpers at the fire scene. When implementing a water tender operation, the time involved and water available must be considered. Terrain, travel time (the effect of weather on road conditions), and access both to and from the fire area will determine the fire department's ability to ensure a continuous water supply for the fire scene.

FIGURE 3-7 Large-diameter hose-line can be utilized as an aboveground water main to relay water to the fire scene. *Used with permission of Joseph Hoffman.*

Area Knowing the size of the building and the exposed area will assist the IC in determining the amount of water needed for the potential fire flow. Area must consider the building's layout. It may be difficult to ascertain the size of irregularly shaped structures when located in areas of high building density. The fire building may be completely surrounded by other structures, severely impacting the IC's ability to do a 360-degree walk-around. In these situations, the preplan is invaluable. By referencing the preplan, the area of the building can easily be determined. Realize that the larger the structure, the greater the potential magnitude of the problem. The area involved in fire (as well as what is actually burning), where the fire is going, and what is in its path must be determined. By identifying these factors, the IC can determine the level of resources that will be needed.

FIGURE 3-8 Life safety of occupants and firefighters and the fire's location, intensity, and extent will dictate how the fire will be fought. *Used with permission of Joseph Hoffman.*

Life The most important factor of size-up is always life safety. Everything else is secondary. On arrival at an incident, the life factor must be determined. Who is in the building? What is their location(s)? What is the most expedient way to protect or rescue them from danger? Can they evacuate themselves or should firefighters escort them out of the building? Do ladders need to be placed and rescues made? Immediately exposed occupied structures, spectators, and responding firefighters must all be considered. (See Figure 3-8.) When operating at the incident scene, firefighters must wear full protective gear. This ensures a reasonable degree of personal safety during firefighting operations.

Trapped citizens have no protective gear, they are untrained to handle the stress associated with a fire, and they will most likely try to leave the fire building the same way they entered, regardless of the danger. The most endangered will be those in the immediate vicinity of the fire area and those directly above the fire. (Consideration must be given to everyone above the fire floor.) Ventilation and the placement of hose-lines must take into account fire confinement and protection of trapped occupants.

Location The fire's location, intensity, and extent will dictate how the fire will be fought. Location will determine the possible travel of a fire. For example, in a cellar fire, all floors above are potential interior exposures. On the other hand, a fire occurring on the top floor of a multistory building will not normally cause a severe interior exposure problem or threat to life on the floors below the fire floor.

Smoke can hide the location of the fire by banking down and permeating a building. Ventilation will remove pent-up combustion products and allow the incoming air to fuel the fire, assisting firefighters in locating it.

When dealing with a large structure or a complex of interconnected buildings, it is helpful to glean information from the preplan or have an engineer from the complex at the command post to answer questions. Knowledge of the location of fire walls and fire doors and whether they are kept closed will be useful in predicting fire spread and placement of units to contain the fire.

Narrow aisles, high-piled stock, and heavy smoke conditions can delay finding the seat of the fire, allowing it to expand. High ceilings can cause fires to go undetected for a longer period of time and allow fire extension. After the fire's location is identified, the best way to prevent it from spreading horizontally and vertically must be determined.

Apparatus/Personnel Incident control can be assisted through the use of specific apparatus and personnel. It is important to consider which aerial equipment or special apparatus can be used for rescue, fire attack, or protection of exposures. This includes tower ladders, elevated platforms, ladder pipes, communications vehicles, fireboats, and lighting vehicles. When sizing up the area, the best use of a certain piece of apparatus should be determined so that when called, the unit can be given specific orders on which location to respond to and the duties it is to perform. The quantity of water needed to control the fire and the ability of the apparatus pumps to deliver that amount must be considered. This information can be incorporated into the preplan.

What resources will be needed to control the incident? Departments with a large number of resources or those with strong mutual-aid agreements will be better able to handle large or unusual fires. When resources are limited, ICs must adjust their priorities to match what can actually be accomplished. When assessing the resources required, it is necessary to know the capabilities of personnel and equipment. Personnel ability is directly related to the level and amount of training received.

A request for additional resources to an incident, in the form of extra alarms, strike teams, task forces, or specific units, should be based on the current and anticipated conditions. The requested units responding can be utilized to:

- Accomplish a specific assignment
- Relieve units already operating at the scene
- Remain in Staging for anticipated problems

Some fire departments use different call signs for special apparatus. Philadelphia has multiple call signs for engine companies, depending on the type of apparatus. Foam pumpers are assigned as front-line engine companies, but contain a foam tank and additional cans of foam. Their call sign changes from Engine 18 to Foam 18. Squrts contain an elevating boom and their call signs are changed from Engine 38 to Squrt 38. Certain pumpers carry a greater amount of large-diameter hose-line, and their call letters change from Engine 52 to Pipe Line 52. (This is a reference to pipe line companies in the middle of the last century; they were special apparatus that carried larger-diameter hose-line for use at multi-alarm fires.) Engine companies with special monitors, capable of delivering master streams exceeding 2,000 gallons per minute with reaches that far exceed standard master stream devices, are identified as Deluge 49. Similarly, ladder companies also have special designations. The call signs for Tower Ladders or Snorkels are changed to signify that they are special apparatus. This change in designation allows the IC to know what special apparatus is responding. Should a company's special apparatus be out of service and it is responding with a reserve apparatus, then that unit's call sign would revert to Engine 38 or Ladder 28.

Career departments should have a method in place for recalling off-duty personnel. These members can provide additional staffing at the scene or staff reserve apparatus to provide coverage for those units operating at the fire.

Volunteer and combination departments should have a method of informing the IC of the number of firefighters responding with an apparatus. Some departments do not allow an apparatus to respond unless a minimum number of personnel are aboard. Other departments may respond with a driver only. The IC bases his or her strategies on the ability to perform specific duties and place a certain number of hose-lines into operation. These assignments require a minimum number of personnel. If the number of units called for does not supply sufficient personnel, the IC will need to summon additional units. One method of notifying the IC of responding staffing is for the company officer or engineer to announce to Dispatch the number of firefighters aboard each unit as they acknowledge their response.

Personnel can also include resources needed from outside agencies. The astute IC makes good use of outside agencies to assist in commanding and controlling the incident. At a mass casualty incident, the Red Cross can notify the next of kin, assist them in securing transportation, and keep them abreast of the condition of their loved ones. The Red Cross can also set up temporary shelters and make housing arrangements for those displaced by fire or evacuation. The Environmental Protection Agency (EPA) will respond to incidents involving hazardous materials. It has the authority to initiate cleanup procedures and can engage private contractors to do so. There may be a need to call the Coast Guard, utility companies, police, public works, or other agencies that can assist in alleviating incident problems. Each agency can help in its specific area—for example, by shutting down endangered overhead power lines or closing down shipping lanes on exposed waterways.

When outside agencies respond to an incident and the IC is not familiar with all the potential resources of that agency, a good procedure is to have the Liaison Officer discuss with the representative what actions or resources that agency can provide to mitigate the current problems. Though ideally this information should be known in the preplan stage, in too many instances outside agencies have stood by and the potential of their agencies was not utilized due to the IC's lack of knowledge of their capabilities.

FIGURE 3-9 This photo shows an opening that existed in the wall to the previously attached structure. The breakdown of a wall that is utilized as a fire-stop can allow fire to extend to an adjoining property.

Construction/Collapse   Good construction practices assist in controlling a fire, while poor methods allow a fire to expand and spread from structure to structure. (See Figure 3-9.) A prime example of poor construction practices is the continued use of wood-shingled roofs in many areas of the country. They have proven to be one of the major reasons for conflagrations.

Plot planning that prescribes proper spacing of building lots and clear spacing between buildings can prevent fires from extending from one building to the next. Likewise, town houses erected with proper fire-stops offer firefighters a better chance to save a life, confine a fire, and preserve property. Each area of the country has specific types of construction germane to its communities. Efforts must be made to be aware of these methods and how they can affect operations, should a fire occur.

As fire attacks structural components, the potential for collapse must be a strong consideration for the IC. The type of building and the building components will vary, as will the collapse potential. The importance of knowing the strengths and weaknesses of the various building components is critical to firefighters. (See Chapter 5 on construction and Chapter 6 on collapse.)

Exposure   To predict the probable spread of the fire and thus develop a strategy, it is necessary to know what is in the fire's path that can burn. The exposure problem concerns two basic areas: internal and external. With internal exposures, contents in the immediate and adjoining areas must be considered. When dealing with exterior exposures, the direction in which a fire may spread and to which other structures must be considered. The extension of fire to another building will impact operations and cause additional problems.

When deciding upon the most critical exposures, knowledge of construction, proximity, contents, threat, and life safety factors is required. Consideration of all factors is assisted by utilizing the preplan and available on-scene information.

What is the life safety factor in the exposed buildings? Schools, nursing homes, hospitals and other high-profile target hazards involve at-risk populations. The exposed buildings may contain hazardous materials. An explosion or a release of poisonous materials could impact upon a large segment of the surrounding community. The construction type and the proximity of the exposed buildings can help to determine the severity of the problems that the IC will be confronted with. Attached structures, or structures built with little spacing between buildings, can be severely threatened, because fire can easily extend to them from the original fire building. The spread of fire can occur due to:

FIGURE 3-10 Exposed structures require the protection of cooling water streams. *Used with permission of Deputy Chief Thomas Lyons.*

- Interconnected or common areas, such as common attics, cocklofts, or cornices
- Floor and roof joists abutting in common wall sockets, which could allow fire to extend to the exposed buildings by conduction through the joist
- A breakdown of fire-stops in concealed spaces

A fire occurring in closely constructed buildings may severely endanger nearby exposures. Wood frame dwellings with combustible siding will easily ignite under intense heat from the original fire building and possibly cause the fire to spread to other exposed structures. (See Figure 3-10.)

Commercial buildings may be of ordinary or heavy timber construction with masonry exterior walls and wired glass windows. The wired glass windows will crack from high heat, but the glass will remain in place due to the wire imbedded in the glass. If the glass can be kept cooled with protective water lines, it will stay in place. At 1600 degrees Fahrenheit, the glass will drop out and leave the window opening unprotected.

Some structures may contain exterior sprinklers that provide protection above the exposed windows. These sprinklers may function automatically or they may need to be manually activated. When activated, they are highly effective and will provide a downward inverted V spray, protecting the window opening.

 ON SCENE

In addition to individual sprinklers over exposed windows, larger areas of exposed buildings may receive exterior exposure protection from deluge sprinkler systems. In a warehouse fire that I commanded, fire had taken control of all floors in a large eight-story vacant building. On the C side was a 13-story fire-resistive building that was constructed in the 1950s for Bell Telephone, located approximately 20 feet away. The architect designed an exterior deluge sprinkler system for the exposed wall. Fire impingement on the Bell Building was an early concern, and I assigned a second alarm company to pressurize the system. As fire lapped up the face of the building, the sprinkler system activated and knocked down the flames, fully protecting the building. This is an excellent example of how properly designed and functioning fire protective systems can be highly effective, even many years after their installation.

Weather Temperature, wind, precipitation, and humidity can all play a part in an incident. Freezing weather can increase response times because firefighters will need to don extra clothing, and apparatus will be slowed if ice or snow is on the ground,

FIGURE 3-11 Icy conditions can impact fireground operations. *Used with permission of James H. Bampfield, Jr.*

increasing the chance of an apparatus accident. (See Figure 3-11.) The impact of winter weather will hinder operations. Falls on slippery surfaces will slow and can injure firefighters. It becomes necessary to ensure that portable ladders are secure and not sitting on ice. Fire escapes and exterior stairs may be slippery and dangerous, necessitating more firefighters to affect a rescue. Freezing weather can cause hydrants to freeze, or they may be hidden by snow.

During hot weather, some cities experience reduced hydrant pressure and water availability because of children illegally opening hydrants. Hot weather and high humidity will quickly drain the strength of firefighters. Extreme heat means more frequent relief will be needed for the operating forces to maintain fresh personnel. If firefighters work too long under these conditions without adequate rest, they will reach a point of exhaustion and be of little use at the incident scene.

The speed and direction of wind often determines how a fire will travel and at what speed. High winds make for a fast-moving fire that is not only difficult to control but breaks down fire streams, decreasing their reach and limiting their effectiveness. Firefighters operating on the leeward side of a fire must be constantly aware of their dangerous position and be prepared to move to a safer location at a moment's notice.

Auxiliary Appliances Auxiliary appliances are any built-in fire protection appliances present at the scene. There are a variety of suppression systems that can be encountered: sprinkler, standpipe, CO_2, or even foam flooding systems at chemical or bulk storage plants. Know the different types of systems in place and determine how they can best be used. (Chapter 4 contains additional information on sprinkler and standpipe systems.)

Special Matters Special matters cover everything from topography or urban interface when dealing with a wildland fire, to elevated highways, railways, or bridges. (See Figures 3-12 and 3-13.) Remember that special matters can involve a multitude of areas. It is important to consider impediments that may interfere with a normal operation, such as:

- Excavations adjacent to a fire building that restrict access
- Narrow streets or alleyways
- The potential of a backdraft or flashover
- The need for medical aid and requests for the appropriate level of assistance, be it basic life support, advanced life support, or medical team response

Height In any structure more than one story, height will be a consideration. The floors above the fire will mean possible vertical spread of fire and pose a threat to life. Attached or adjacent structures of equal or greater height must be considered as immediate exposures. A building of greater height can be exposed on an upper floor from radiant heat. Adjoining structures of greater height may have windows above the roofline that could be exposed should a fire break through the roof.

Moving firefighters and equipment to upper floors requires time, sufficient resources, and proper management. Some areas restrict the height of structures to be built in the community. This is based on the fire department's ability to safely protect the people living or working

FIGURE 3-12 Narrow alleyways, fences, and overhead electrical wires can hinder operations. *Used with permission of Joseph Hoffman.*

in a building. A common maximum height in many areas without aerial ladder trucks is 35 feet, which often coincides with the longest ladder carried by the fire department.

FIGURE 3-13 Fences and other obstructions can limit access to the rear of fire buildings.

Occupancy To know how to fight a fire, we must determine the contents of the fire building. Signs on the exterior of the building can be helpful. Highly combustible stock produces a high rate of fire spread, and certain manufacturing processes create situations allowing flash fires or rapid spread.

Contents can be of high value, such as items in a jewelry or computer store. Museums or libraries present special challenges to responding firefighters. Historic properties need to be identified and a value placed on them. Some structures have been fully restored and the building itself is the valuable asset that needs to be protected. In other cases, the building has been rebuilt and the contents are the things of value.

There are a number of historic buildings in the Germantown section of Philadelphia that were involved in the Revolutionary War. In one particular property, British soldiers carved their initials in the window glass. Should a fire occur in this structure, the firefighters would need to find other means of ventilation in order to protect those windows. Conversely, the Betsy Ross House where the first American flag was created is a structure that has been rebuilt and contains no historical significance, yet the furnishings are all authentic and need to be protected.

Time A daytime fire in an office or commercial building presents a threat to the lives of employees working in the building, but it allows early discovery and (we hope) early notification of the fire department. The exception to this rule is when employees try to fight the fire themselves prior to notifying the fire department. Many fires that are costly in terms of loss of life and financial loss seem to be related to either delayed discovery or delayed notification of the fire department. A nighttime fire in the same building would present little threat to life with the exception of a skeleton crew and the responding firefighters.

Nighttime residential fires tax firefighters the most. Responding firefighters, realizing the life factor involved, give that extra effort. They go a little deeper into a structure and stay a little longer, listening for the sound of a child in distress. Firefighters operating at night encounter more locked doors, necessitating forcible entry. Darkness restricts vision, requiring the assignment of resources to illuminate the area.

Time of day can affect our response time. Delays must be anticipated during rush-hour traffic in and around our cities. Seasonal shopping can also delay response in and around shopping districts and malls. Seasons of the year can mean increased fire loading, due to additional holiday stock or the threat from specific dangers such as wildland fires.

ON SCENE

The fire department may not be aware of situations that can arise at a facility. One evening, a battalion chief noticed approximately 1,000 children accompanied by adult guardians entering a planetarium with sleeping bags under their arms. A sleepover had been planned in the building as part of a science program. The panic of 1,000 people confronted with something as minor as a burning sleeping bag as they groped for an exit in the dark would have caused major problems for any responding department. The building was not meant to accommodate the campers, and the fire department had not been notified of the situation. An immediate inspection by the fire marshal set up ground rules to be followed for that night. The next day brought lengthy meetings to plan for future sleepovers.

Size-Up Concerns Summarized Through size-up, we must attempt to learn:

1. Where is the fire located?
2. Is it confined?
3. Where is it going?
4. What is the life hazard to civilians and firefighters?
5. What is the type of construction?
6. What are the inherent dangers with this type of construction?
7. What are the fire conditions?
8. Is there a potential for a backdraft or flashover?
9. What are the immediate and long-term problems?

The answers to these questions assist the IC in determining the incident objectives that will be needed for a successful operation. Incident objectives are broad goals that should include the safety of responders and civilians. Once the objectives are established, strategies and tactics can be developed.

Size-up is an ongoing process. When used properly, it assists in the handling of an incident scene. It is the basis for establishing the incident objectives and developing strategy and tactics. Size-up is not reserved for the IC only. Each individual operating at an emergency scene must size up the incident. As firefighters perform their various operations, they constantly have to be on the alert for situations that can arise or conditions that have changed. Changing conditions can mean danger for firefighters. Weakened floors, a sagging roof, or a raging fire in a void space containing lightweight structural components can be a threat to firefighter safety. When a proper size-up is performed, important points will not be overlooked and the incident scene will be run in a professional manner.

Objectives: Strategy, Tactics, and Tasks

Size-up identifies problems that we must solve by implementing the necessary strategies to achieve the incident objectives. The systematic deployment of strategy should be considered a tool and utilized along with knowledge, experience, and training. Knowledge is needed to evaluate the information gathered. Experience lets us draw upon actions that have been successful in the past. Training allows us to be proficient in the performance of our duties, eliminating unnecessary actions and quickly accomplishing tasks.

Unsatisfactory operations in one area can lead to an entire plan going awry.

Improper ventilation or no ventilation can stall or endanger a hose-line crew. Loss of water on a lower floor can allow a fire to accelerate and may necessitate the withdrawal of firefighters operating on the floors above.

Many firefighters group strategy, tactics, and tasks together. Strategy should be viewed as overall goals or *what* you want to accomplish. Good examples of strategies would be "rescue of trapped occupants" or "fire confinement."

Tactics can be viewed as *how* you are going to achieve your strategies or goals. If rescue were the strategy, then "performing a search and rescue of a fire area" would be a tactic to help achieve that strategy. If fire confinement were the strategy, "stretching a hose-line to confine the fire to a specific area" is an example of a tactic to achieve that strategy.

Tasks or actions stipulate *who* will do which step and *when*. To accomplish the aforementioned strategies and tactics, an example of a task is "Truck 1, on arrival, conduct a primary search of the second floor." This addresses the strategy of rescue and the tactic of performing a search and rescue of a fire area. It also states who will do the task (Truck 1) and when it will be done (on arrival).

Another example of a task is "Engine 1, on arrival, stretch a 1¾-inch hose-line to the first floor via the front door and confine the fire to the rear rooms." This addresses the strategy of fire confinement and the tactic of stretching a hose-line to confine the fire, as well as who will do the task (Engine 1) and when they will do it (on arrival).

As you can see, we use our incident priorities as a basis for decision making. We then review our size-up factors to determine the problems. We select the strategies to address the problems found. Then we determine the proper tactics and assign units to accomplish the tasks to resolve the situation.

Level 3: Strategy

RECEO-VS The seven basic strategies are referred to as RECEO-VS. (See Box 3-2.) These letters stand for rescue, exposures, confinement, extinguishment, overhaul, ventilation, and salvage.

BOX 3-2: RECEO-VS

Rescue   **V**entilation
Exposures   **S**alvage
Confinement
Extinguishment
Overhaul

ICs must prioritize their strategies based on the incident priorities. The strategies of rescue, exposures, confinement, extinguishment, and overhaul are in priority order of consideration at an emergency. Ventilation and salvage strategies can be implemented at various times. The initial strategy implemented may be ventilation, but a lack of resources may relegate salvage to a lesser role until the arrival of sufficient personnel.

A critical point to remember is that an initial responder will basically be confronted with only four of the seven strategic considerations. *Rescue, exposures, confinement*, and *ventilation* will be the initial concerns. This is not to minimize the importance of extinguishment, overhaul, or salvage, but they will be considered either later in the incident when extinguishment and overhaul are accomplished, or when sufficient resources arrive to accomplish salvage. Limiting the initial considerations to these four areas assists the IC in decision making. Let's look at each area.

Rescue The considerations for rescue are the location of the occupants and the best way to protect or rescue them. Can a quick extinguishment of the fire protect those who are still in the building? This is one of the most difficult things to determine. There can be a wealth of concurring or conflicting information. When a discrepancy occurs, the IC must quickly decide the course of action to take. If there is no on-scene information available, actions will be based on the probability that there are occupants, and search and rescue will be a priority. (Chapter 4: Company Operations has additional information on rescue.)

Exposure Exposure protection considers the potential for extension of a fire to involve internal or external exposures. (See Figure 3-14.) The protection of these areas is vital in containment and control efforts. What is exposed in adjacent areas? What is on the floors above? Can fire extend to these areas through open doorways or open stairways? Through air or light shafts? To adjacent buildings?

The importance of protecting exposures cannot be emphasized enough. Companies must enter and inspect all possible areas of fire entry and, when necessary, utilize a thermal imaging camera or open those areas suspected of containing hidden fire. With properties that are attached, common attics or cocklofts should be checked for fire spread. Some new fire officers mistakenly fear damaging the exposed property. In the absence of a thermal imaging camera, they may be tentative about opening ceilings and walls to check for hidden fire. Experience teaches us when it is necessary to open an area: look for discolored or blistered paint, hot surfaces, or smoke pushing from moldings.

Once an attic or cockloft is opened, a thorough check should be made. Indications of fire spread should be sought. The presence of cobwebs is one indicator that there is no fire extension to that area. Explaining to the occupant the reason for the damage caused by opening areas where suspected fire may be hidden is an important duty after the fire.

FIGURE 3-14 A fire on the second floor is extending to the exposure on the Delta side of the fire building. *Used with permission of Joseph Hoffman.*

Confinement The goal is to confine the fire to as small an area as possible. The exact location and extent of the fire may be difficult to determine. The path it will most likely travel must be considered. This allows the IC to predict what problems must be overcome to achieve confinement. Determining these factors will dictate the number of hose-lines and their placement. Orders can be given to shut fire doors, supply sprinkler systems to augment them, and check for extension of fire. Attacking a fire in a single room or small area often accomplishes both confinement and extinguishment at the same time. (Chapter 4: Company Operations has additional information on confinement.)

Extinguishment Extinguishment involves knocking down all visible fire and hidden fire exposed during the overhauling stage. Extinguishment calls for the judicious use of hose-lines. Indiscriminate use of water causes needless damage. When a fire has been controlled, sprinkler systems should be shut down and minor spot fires handled with handheld hose-lines. (Chapter 4, Company Operations, has additional information on extinguishment.)

Overhaul Overhauling ensures that all fire has been extinguished. Areas where fire could extend must be checked with a thermal imaging camera or opened and checked. Smoldering contents must be removed to the exterior. The overhaul stage is a critical step in complete extinguishment. When poor overhauling techniques are used, rekindling can occur.

Overstuffed furniture that has been involved in fire should be removed to the exterior and thoroughly wet down. Deep seated fire burning in a mattress, overstuffed chair, or sofa is difficult to fully detect. Leaving these furnishings within the fire building can often lead to rekindles. Realize that these items are damaged and of no use to the homeowner and will need to be replaced, so the fire department is doing the occupant a service by removing them to the exterior.

Electricity should be shut down to any damaged circuits. Other utilities should be shut down as necessary. Fire protection systems should be restored. Overhauling operations also involve determining the cause of the fire. Telltale signs of origin and cause can be uncovered. If arson is suspected, fire investigators may request special assistance during their investigation. (Chapter 4, Company Operations, has additional information on overhauling.)

Ventilation Ventilation and salvage are utilized whenever they are needed during the course of firefighting and not at one set juncture. Ventilation allows intervention of hose-line crews to effect extinguishment and reduce damage. Ventilation may be performed during the rescue stage to draw fire and smoke away from a trapped occupant. (Chapter 4, Company Operations, has additional information on ventilation.)

Salvage Salvage lessens the amount of the loss by protecting items from damage caused by smoke and water. The structure should be left in as good a condition as possible, considering the seriousness of the fire itself. Salvage can start as soon as enough personnel are available and the area is safe to operate in. (Chapter 4, Company Operations, has additional information on salvage.)

Level 4: Tactics

tactics ■ The way strategies or goals will be achieved.

Tactics achieve the selected strategies. Tactics are *how* you are going to achieve the strategies. If ventilation is the intended strategy, then the tactic could be "horizontal and vertical ventilation over the fire area." Multiple tactics may be needed to achieve a strategy. For example, to enter the building for confinement or search and rescue, it may be necessary to use forcible entry, place ladders to upper floors, or stretch hose-lines.

To determine when tactics have been accomplished, it is necessary for them to be measurable and specific. To give the order "place water on the fire" is vague and could be improperly accomplished by placing a stream through a window. It is better to give the order "place the hose-line in the first floor and confine the fire to the area involved." The person giving the order will know when the hose-line is in place and the fire is confined. The achievement of tactical objectives will depend upon on-scene conditions and resources available.

Level 5: Tasks

tasks ■ Stipulate who will do which step and when.

Tasks or actions are implemented by giving orders to the units that will carry out the tactical operations. Tasks describe *who* is going to do a task and *when* they will do it.

The IC (either a company officer or chief officer) must decide whether there are sufficient resources to handle the current and anticipated problems. This assessment must include the resources needed at the present time and those that may be necessary for future incident control. (See Figure 3-15.) How many companies will be required for relief? One word of caution: Many times at an incident scene, it is recognized that the fire will require numerous additional resources (a minimum of a third or fourth alarm or, by comparison, eight to ten more companies). If all the resources are called for immediately, standard operating guidelines must dictate that the units proceed to Staging unless given specific orders en route. This prevents companies from determining their own course of action, alleviates congestion on the incident scene, and ensures that units address the strategy developed by the IC.

FIGURE 3-15 Heavy fire conditions on arrival demand the immediate request for additional resources. *Used with permission of Greg Masi.*

Constant reevaluation of the incident is necessary to ensure that the strategy, tactics, and tasks are accomplishing the goals of the IC. If not, adjustments will need to be made to reach those goals.

AN EXAMPLE OF IMPLEMENTING THE STRATEGIES, TACTICS, AND TASKS

There should be a flow to an incident scene. If we arrive on the scene of a two-story frame dwelling at 0200 hours with a car parked in the driveway, fire visible through the first-floor windows, and an occupant at a second-floor window, with moderate smoke conditions throughout the building, how would we handle this situation?

First, we establish the strategies. If people are trapped, then an initial strategy (*what* we want to do) will be rescue. The next step is to see which tactics (*how* we are going achieve the rescues) will be needed to accomplish the rescues. More than one tactic may be needed to achieve our strategy. In this instance, when a person is reported trapped on the second floor, it may be necessary to implement search and rescue procedures to locate the victim or other victims. Forcible entry or laddering the building may be required. Another strategy would be confinement. One possible way to accomplish confinement is by preventing the upward spread of fire via the interior stairs.

If ventilation is identified as a strategy, then a decision on the best method of ventilation should be made. The tactic could be to employ horizontal, vertical, or positive pressure ventilation. (Though ventilation is a strategy, it can also be implemented as a tactic.)

Now that we have established *what* we want to do, or our strategies, and *how* we want to do it, our tactics, we then implement our tasks. This will involve *who* will do the tactic and *when* it will be accomplished. Examples of typical orders for different strategies and tactics could be as follows.

To achieve rescue: "Truck 1, on arrival, provide forcible entry of the front door and perform a right-hand search of the second floor. Truck 2, on arrival, place portable ladders to the second floor windows for possible rescue."

To achieve confinement: "Engine 1, on arrival, stretch a 1¾-inch hose-line via the front door and contain the fire to the first floor."

To achieve ventilation: "Truck 2, on arrival, perform horizontal ventilation of the first and second floors to support fire attack and search and rescue operations."

The sequence in Box 3-3 graphically outlines the flow of utilizing the command sequence. It is an excellent training tool. The actual decisions needed for the tactics and tasks are not all made by the IC. In actual use, the decisions on the type of search or location of ground ladders, for example, will be made by the ones accomplishing those tasks.

BOX 3-3: STRATEGY, TACTICS, AND TASKS

STRATEGY (WHAT)	TACTICS (HOW)	TASKS OR ACTIONS (WHO & WHEN)
Rescue	Implement search and rescue procedure	Truck 1, on arrival, perform a right-hand search of the second floor.
Confinement	Forcible entry	Truck 1, on arrival, force the front door.
	Contain fire to first floor	Engine 1, on arrival, stretch a 1¾-inch line via the front door and contain the fire to the first floor.
Ventilation	Horizontal ventilation of the first and second floors	Truck 2, on arrival, ventilate via the windows on both floors.
		Truck 2, on arrival, place portable ladders to the second-floor windows to support the search and rescue crews.

SUCCESSFUL OPERATIONS

Through strategy, tactics, and tasks, we achieve the goals determined by the incident priorities. Attaining these actions can be demanding. The initial IC may be faced with limited resources. Minimal information may be available on the status of the occupants or the location of the fire, and time constraints may prohibit the confirmation of this data. The demands of the scene must be prioritized to ensure the maximum utilization of available personnel.

The development of strategy will depend on the available resources. Strategy can be accomplished in different phases. Certain actions can be achieved with the limited resources initially available. When adequate personnel arrive, implementation can be expanded to cover all aspects of the IAP.

Strategy and tactics are not an exact science. We cannot say that if a certain problem arises, the solution will be handled by Plan A, or if another situation develops, then Plan B will be the answer. We must constantly reevaluate the situation. We must learn from our mistakes and try never to make the same mistake twice.

APPLYING INCIDENT MANAGEMENT TO SOLVE PROBLEMS

After we have identified problems through size-up, we can solve those problems with our strategies, tactics, and assignments of tasks. The next step that should be a natural progression at an incident is the implementation of an incident management system. The Strategy Prompter is a tool that can be useful to achieve that goal. By utilizing the strategies necessary to bring an effective solution to the problems found, we can also determine how our incident management system should be structured. (See Figure 3-16.)

Let's continue with the same scenario of the two-story frame dwelling at 0200 hours with a car parked in the driveway, fire visible through the first-floor windows, and an occupant at a second-floor window, with moderate smoke conditions throughout the building. Let's assume that three engines, one truck, one squad, and one chief are responding. Each engine and truck company has four personnel, and the squad company has two firefighter/EMTs. With these facts, we can go to the next step, which is developing our incident management system. Earlier we identified the strategies that needed to be implemented: rescue, confinement, and ventilation. Our incident management system might look like this:

- The officer of Engine 1 on arrival will assume the role of IC.
- Engine 1's crew will operate in Division 1, indicating that they will operate on the first floor and accomplish all tasks needed there, including fire confinement and search and rescue.
- Engine 2's officer will become Division 2's Supervisor with Engine 2's crew. Division 2 will perform search and rescue, check for extension of fire to the second floor, and conduct any other duties needed on the second floor.

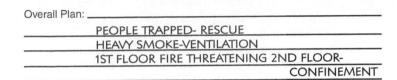

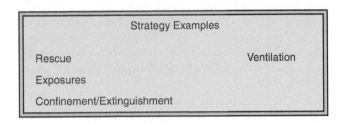

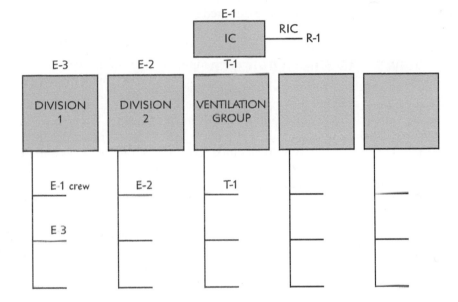

- Engine 3's officer will become Division 1's Supervisor with the crews of Engine 1 and Engine 3.
- Truck 1's officer will become Ventilation Group Supervisor with Truck 1's crew. The Ventilation Group will effect the ventilation necessary at the incident.
- Squad 1 will become the Rapid Intervention Crew.

When the chief officer arrives, the chief can assume Command or permit the officer of Engine 1 to retain Command.

As you can see, by identifying the needed strategies, it becomes a natural progression to initiate our management system to handle the problems found. The actual unit assignments may vary. Naturally, there are other ways that the management system could be structured to handle this assignment. Some departments may form a Suppression Group, a Search and Rescue Group, and a Ventilation Group. Because of the small size of the structure, another method could be to assign only an Interior Division and a Ventilation Group, or to assign a Division 1, a Search and Rescue Group, and a Ventilation Group. As you can see, there are many right ways of implementing the management system. The important points to note are:

- The Strategy Prompter can assist in the quick development of a management system, and by using divisions and groups from the beginning, additional units can be assigned quickly to those divisions or groups as needed.
- The IC, by speaking to a maximum of three individuals (the supervisors of Division 1, Division 2, and Ventilation Group), can get immediate progress reports and maintain a manageable span of control.

Crew Resource Management

Crew resource management (CRM) was implemented by the airline industry to optimize a crew's interactions at times of high stress. It was found that human error was the cause of most airline accidents. The airline industry recognized that in a number of airline crashes, members of the flight crew knew that problems had developed and, in fact, had brought the problem to the captain's attention, only for him not to act on it. CRM was originally called cockpit resource management, but was changed to crew resource management to incorporate all members of the flight team. CRM uses training to ensure that the members of the flight crew are bold and assertive so that the captain, who has the final say, gets the critical information and crew input for safe flight operation. This input can come from the first officer, flight attendants, maintenance personnel, or air traffic controllers. The implementation of CRM has drastically reduced airline disasters from twenty a year to around one to two a year.

CRM recognizes that humans behave predictably, and if certain behaviors are learned and practiced, operations can be made significantly safer. It can be seen as knowing and understanding what is going on, and knowing how to react to certain situations. The program met with such success that it was adopted by the US military, and they too found it quite beneficial.

AIRLINE CREWS AND FIRE SERVICE CREWS

Many NIOSH reports that investigated firefighter line-of-duty deaths have listed communication failures, poor decision making, lack of situational awareness, poor task allocation, and leadership failures as contributing factors in those deaths. These were the same factors that were listed in numerous air crashes and led to the adoption of CRM by the airline industry. There are numerous other similarities between flight crews and fire company crews. Crews are structured with a leader or company officer. The crew or company works best as a team. They can spend many hours on nonemergency tasks, such as maintenance, prevention, and training, and then be dispatched to an incident and act quickly to address a life-or-death situation. Another similarity is that some crews/companies may work together quite often, while others may be assembled on short notice.

CRM IN THE FIRE SERVICE

With the number of deaths and injuries that occur yearly, the fire service can benefit from adopting CRM. As with the airline industry, the chief or company officer has the final say, but input from firefighters to their company officer or a company officer to a chief officer can be beneficial for everyone.

CRM is not management by committee. It is a tool created to optimize human performance by reducing the effect of human error through the use of all resources that are available. It is similar to the way many progressive fire officers have operated during their careers; they have welcomed input from firefighters and fire officers alike. They recognized that multiple sets of eyes and ears, as well as the experience of their personnel, can only be beneficial.

Examples of how CRM could be beneficial in the fire service include:

■ Input while on a response from the apparatus driver, telling the captain that he is familiar with the building in which the fire is reported. It may be a daycare center that his child attends. His knowledge may include the fact that the building is not sprinklered and is constructed of lightweight material.

■ On arrival at the scene, another firefighter notices smoke coming from the building's attic vents under pressure and sees flames at that location. He may recommend to the captain checking the building with a thermal imaging camera.

These minor statements may seem needless, yet they make up a body of information that the captain can use in his or her size-up. Viewing the flames at the vent could have gone unseen by the captain while giving an initial report to Dispatch. Verbalizing these factors ensures that what was seen by one is shared with the company officer. The captain can process the data that he or she has seen and heard from others. In the first example, the captain can then deduce that children could still be inside the building, and that fire may be in the attic area of a lightweight-constructed building, which will drive his or her decision making.

CRM is based on five factors:

- *Communications*—focuses on speaking directly and respectfully
- Situational awareness—stresses attentiveness and the effects of perception, observation, and stress on personnel
- *Decision making*—stresses the amount of information needed to be able to evaluate the risk/benefit analysis at an incident scene
- *Teamwork*—stresses the need and value of teamwork for successful, safe operations
- *Barriers*—recognizes the effect of barriers on the first four factors and how to neutralize them

The fire service can only benefit by learning more about CRM and implementing training for their personnel to improve incident operations while reducing firefighter deaths and injuries. (See References at the end of this chapter to access IAFC's comprehensive manual on CRM for the fire service.)

 ON SCENE

Miracle on the Hudson. On January 15, 2009, a commercial passenger flight from New York City to Charlotte, NC, was disabled three minutes after its initial takeoff from LaGuardia Airport by a flock of geese. The airplane lost almost all thrust to both engines, which resulted in the pilot being forced to make an emergency landing. Realizing that he would be unable to land the plane at an airport or in the heart of Manhattan, the pilot brought the plane down into the Hudson River just three minutes after losing thrust. All 155 occupants safely exited the plane and were quickly rescued by nearby watercraft. The NTSB released a statement that credited the accident outcome to the fact that the aircraft was carrying safety equipment in excess of that mandated for the flight, and excellent CRM among the flight crew.

Firefighting Ventilation Practices and Tactics

Underwriters Laboratories (UL) and the National Institute of Standards and Technology (NIST), in collaboration with various fire departments, including the Fire Department of New York, Chicago Fire Department, and Spartanburg SC Fire Department, conducted tests to scientifically determine how ventilation affects a fire burning in one- and two-story dwellings and the tactics needed to fight those fires, in coordination with ventilation.

The following information has been gleaned from UL and NIST reports that were created after testing horizontal and vertical ventilation during live burns and fires in dwellings constructed for these projects, and data from Governors Island testing in acquired dwellings slated for demolition. Realize that I have condensed hundreds of pages of important information into a few pages. This brief overview will not supply the reader with all of the information needed to implement changes in suppression techniques. At the end of this chapter, I have included seven website locations where these reports can be accessed. I fully recommend that the reader use these websites to read the entire reports in their exact words, in addition to my abbreviated report and personal comments, and through training, that we implement the changes needed to enhance firefighter safety.

Firefighting ventilation practices and tactics ■ Studied by UL and NIST to scientifically determine how ventilation affects a fire burning in one- and two- story dwellings.

WHY WAS THE TESTING DONE?

Ventilation is frequently used as a firefighting tactic to control and fight fires. In firefighting, ventilation refers to the process of creating openings to remove smoke, heat, and toxic gases from a burning structure and replacing them with fresh air. If used properly, ventilation improves visibility and reduces the chance of flashover or backdraft. If a large fire is not properly ventilated, not only will it be much harder to fight, but it can build up enough poorly burned products of combustion to create a backdraft or smoke explosion, or enough heat to create flashover. Poorly placed or timed ventilation may increase the fire's air supply, causing it to grow and spread rapidly. Realize that any opening in a structure including opening the front door creates a ventilation opening. Used improperly, ventilation can cause a fire to grow in intensity and potentially endanger the lives of firefighters who are between the fire and the ventilation openings. For many years, we have been aware that the fires fought today are different than fire fought before the prolific use of plastics in the home, workplace, and practically everywhere else. The burning of these plastics creates a much hotter fire with deadly smoke concentrations, and the fire is more likely to become ventilation-limited than previously experienced with fires involving Class A materials. The desired result of the testing was to find better firefighting tactics, which would result in fewer firefighter injuries and deaths.

VENTILATION IS A TOOL

Ventilation is like any tool; there are right ways and wrong ways it can be used. Effective use of ventilation can create a safe working environment for firefighters. When used improperly or not coordinated with fire suppression activities, it can imperil anyone in a fire building. Though fire departments have used ventilation for years, many aspects of ventilation were not fully understood until recently. It was always thought that a hoseline directed into a window opening would push fire and heat into a building, making conditions unbearable. Repetitive tests with repeatable results have now conclusively shown that is not the case.

The introduction of water into burning dwellings from the exterior was conducted via windows and doors. The results of each test were measured and documented using sophisticated scientific equipment. In one example, a live burn in a two-story dwelling had fire showing from the second-floor front bedroom window. A hose-line was directed from the outside through a window and onto the ceiling of the fire room and, delivering a relatively small amount of water (25 gallons), it was able to knock down the fire. The temperature in the fire room decreased from 1792 degrees F to 632 degrees F in 10 seconds. The bedroom door was open and the temperature in the hallway decreased from 273 degrees F to 104 degrees F in 10 seconds. This tactic would not only slow the fire's progression, but make it a more tenable operation for the interior hose-line crew to accomplish final extinguishment.

HORIZONTAL AND VERTICAL VENTILATION TESTING

Horizontal and vertical ventilation tests were performed on one- and two-story dwellings that were built of legacy construction (solid beam) in laboratories. The intent of the tests, which were conducted multiple times, was to investigate the effects of both horizontal and vertical ventilation on a fire when the contents were involved (not to test the construction of the building). These were independent tests performed at different times, starting with horizontal ventilation and followed later by vertical ventilation. The testing was conducted scientifically, to develop empirical data on horizontal and vertical ventilation, how ventilation impacts the growth or extinguishment of a fire, suppression practices and the resulting fire behavior, and the impact of different suppression techniques.

All of the experiments began with the exterior doors and windows closed and all of the interior doors in the same location either open or closed. One bedroom door was kept closed alongside a bedroom door that was left open. This was done to see the effects of the fire should an occupant stay in the bedroom with the door closed. Instrumentation and video were set up to show heat flow through the structures; air speed and direction in and out of the fire compartment; air movement; concentrations of oxygen, carbon monoxide, and carbon dioxide; and other factors.

The fires were ignited either in the living room, family room, bedroom, or kitchen, and allowed to grow until ventilation operations were simulated. The one-story home was ventilated at eight minutes after ignition. The two-story house was ventilated 10 minutes after ignition. This timing was determined based on three factors: time to achieve ventilation-limited conditions in the house; potential response and intervention times of the fire service; and window failure times from previous window failure experiments. The horizontal ventilation was performed by single openings of either the front door or window in the immediate fire area, or a window remote from the fire area, or a combination of openings. The vertical ventilation was accomplished by various combinations of roof opening, front door opening, and window opening(s).

After ventilation, the fire was allowed to grow until flashover or perceived maximum burning rate, based on the temperatures, observation of exterior conditions, and monitoring of the internal video. (The time frames that the fire was allowed to burn after the ventilation differed for each scenario, since the experiment looked to understand the impact of ventilation on the fire). Once the fire had maintained a peak for a period of time, with respect given to wall lining integrity, a firefighting hose stream was flowed in through an external opening. The experiment was terminated approximately one minute after the hose stream was applied, and suppression was completed by a deluge sprinkler system and/or the firefighting crew.

VENT, ENTER, ISOLATE, SEARCH

It was found that the tactic of Vent, Enter, Search (VES) should be changed to Vent, Enter, *Isolate*, Search (VEIS). VEIS is the rapid search of a targeted area where there is a high probability of locating a victim. Entry into this targeted area can be accomplished by directly entering the space from the exterior via a window or door. It is important to remember that this access point is a ventilation opening and can draw a flow path through the area to be searched.

Isolate means that a firefighter entering a bedroom window from the exterior to check that room for occupants should immediately shut the bedroom door to isolate that room from the rest of the building, to protect the firefighter and any occupants of the room. Tests showed that by shutting the door, air from the window did not reach and increase the fire. Isolation created a safer environment to conduct the search, made the searched space more life supporting, and at times increased visibility. (There is additional information on VEIS in Chapter 4.)

FLOW PATH, FLOW PATH CONTROL, AND DOOR CONTROL

The testing documented how air movement occurs within these structures under fire conditions, and the need to control air flow to prevent a rapidly expanding fire and possible flashover. To control the air movement, one has to understand flow path, controlling the flow path, and how door control can assist in that venture.

Flow path is the movement of heat and smoke from the higher pressure area of the fire towards the lower pressure areas of open doors, windows, and roof openings. As the heated fire gases move away from the fire area towards the low pressure areas, the energy of the fire pulls in fresh air from these low pressure areas. Based on varying building design and the available ventilation openings, there may be several flow paths within a structure. Operations conducted in the flow path (such as firefighters entering the front door to attack the fire or perform search and rescue) will place members at significant risk due to the increased flow of fire, heat, and smoke toward their position. The opening used to gain access to conduct fire attack or search and rescue is a ventilation opening and could potentially pull the flow path in your direction.

Flow path control is the tactic of controlling or closing ventilation points, which will:

- Limit additional oxygen into the space, thereby limiting fire development, heat release rate, and smoke production.
- Control the movement of the heat and smoke from the fire area to the exterior and to other areas in the building not yet exposed to heat, fire, and smoke damage.
- Improve victim survivability by controlling the flow path. Door control and less oxygen will equal lower temperatures.

Findings show that not controlling the amount of ventilation can increase air inlet paths to the fire and/or exhaust paths from the seat of the fire; these flow paths will result in fire growth and spread.

Door control refers to ensuring that the door providing access to the fire area is controlled and closed as much as possible. If necessary, a tool can be used to prevent the door from locking behind the entering members. By controlling the door, we control the flow path of fire conditions from the high pressure fire area towards the low pressure area on the other side of the door. Door control also limits fire development by controlling the flow path of fresh air from the lower level of the open door towards the seat of the fire. Door control is a means to control ventilation to a fire by interrupting the flow path.

GOVERNORS ISLAND TESTING

This fire research project developed empirical data from full-scale house fire experiments to examine ventilation, suppression techniques, and the resulting fire behavior. Tests on Governors Island were conducted to determine the effects of flow paths and suppression tactics on conditions within the acquired townhouses/row houses. Fires were set in the basement, living room, kitchen, and bedrooms. Their findings include the following:

1. Water applied via exterior attack is effective and does not push the fire.

A key finding of the experiments showed that the common belief about exterior fire attack (offensive-exterior attack) pushing the fire is unfounded and that innovative fire attack tactics can improve the safety and effectiveness of firefighting efforts.

2. Rather than making conditions more hazardous, applying water directly into the fire compartment as soon as possible results in the most effective means of suppressing the fire.

The tests conclusively documented an immediate and drastic reduction in the temperature and conditions of a compartment involved in fire when you apply water from the exterior via a window or door into the fire room. This reduction occurs when using a solid or straight stream, in a direct or indirect attack, through a window or door, into the fire room. Delivering a relatively small amount of water knocked down the fire and reduced temperatures throughout the structure.

Even in cases where front and rear doors were open and windows had been vented, the application of water through one of the vent openings improved conditions throughout the structure.

Experiments showed that exterior fire attack increases the potential survival time for building occupants and provides safer conditions for firefighters performing search and rescue. In fact, the experiments demonstrated that the traditional practice of increasing ventilation openings to a ventilation-limited structure fire, by opening additional doors, clearing windows, or cutting the roof without interior coordination and prior to mounting an attack on the fire, increased the fire hazards and the potential for a rapid transition to flashover.

3. While the attack should be commenced from the exterior to improve conditions for firefighters and building occupants, it must be finished inside.

The offensive-exterior attack can improve conditions by slowing the fire's progression and cooling the compartment, but it will not completely extinguish the fire. Applying water as soon as possible from the outside to a fire that is venting from one or more openings softens the target and helps firefighters to gain the upper hand. After the initial knockdown from the exterior has made safer conditions inside the structure, continuing the attack from the inside increases the speed and effectiveness of fully extinguishing the fire.

Coordinated Fire Attack

Every effort should be made to coordinate and communicate efforts between fire suppression and ventilation. Ventilation tactics that are properly coordinated with fire suppression can:

- Limit fire spread
- Control fire development and improve conditions inside the building
- Reduce potential injuries of members on the interior conducting search and rescue or fire suppression activities

Uncoordinated ventilation can have a negative impact on the fire. Feeding oxygen to a ventilation-limited fire before water is ready to be applied can intensify the fire and create a ventilation-induced flashover. Consideration should be given to controlling the horizontal and vertical ventilation openings until a charged hose-line is advancing to the seat of the fire. Communication with interior crews prior to initiating vertical or horizontal ventilation allows the interior crews to assess the timing and placement of ventilation from the inside.

4. Water applied via exterior attack does not push the fire.

The anecdotal experience of firefighters "that water introduced into a structure via a window will push a fire" can be explained by one of the following scenarios: a) A flow path is changed by ventilation and not by water application; b) A flow path is changed by water when the thermal layer is disrupted and steam moves ahead of the hose-line, elevating the level of heat and creating the impression to those downstream that the fire is being pushed; c) Turnout gear becomes saturated with heat, which begins to pass through to the firefighter; if this occurs about the time a hose-line is opened, it might appear that the hose-line caused the rapid buildup of heat; d) One room is extinguished, which allows air to flow into another room and causes that room to ignite, burn more intensely, or reach flashover.

BASEMENT FIRES

The Governors Island tests showed the benefit of softening the attack from the outside. It worked better to begin with a transitional offensive-exterior attack through a basement window or rear basement door, followed by an interior attack. These basement tests reinforced that fire attack through a window or exterior door will not push fire. They also showed that fire and heat travelled upwards through pipe chases in the kitchen, collapsing the kitchen floor, rather than through the open stairwell door to the first floor. Flowing water from a hose-line at the top of the basement stairs had limited impact on a fire. Tests showed that the best method of entry for a basement fire was via an exterior ground level entrance into the basement.

Findings of Interest

The purpose of this study was to improve firefighter knowledge of the effects of ventilation and the impact of different suppression techniques. The researchers recognize that all fireground circumstances cannot be analyzed. They anticipate that the data from this testing can be used to complement firefighters' previous observations and experiences.

- Significantly, the tests show that a relatively small amount of water (25 gallons from a straight stream) directed into the fire room from the exterior drastically reduced the interior temperatures as it knocked down the fire. This approach allowed a safer entry for firefighters to access the fire area and complete extinguishment, while simultaneously performing search and rescue.
- A coordinated fire attack is needed to synchronize size-up, entry, ventilation, search and rescue, and the application of water.
- When using ventilation openings in small rooms containing 8-foot ceilings, the water applied was directed at the ceiling. Visible flames were extinguished in less than 5 seconds, steam did not reduce smoke layer height, and 15 seconds after water application, the smoke was beginning to lift and conditions were improved.
- In larger volume spaces, such as a family room or great rooms with 17-foot ceilings, testing showed it is important to put the water on what is burning, rather than direct the hose-line to the ceiling.
- The experiments did not simulate water being applied from inside the structure by an advancing hose-line, which happens on most fires.
- Measurements taken of air moving from the ventilation openings through the structure showed that both window and door openings allowed hot air to escape through the top of each opening, while cool air flowed into the dwelling through the bottom of the openings.
- Testing reinforced that it is critical to use thermal imaging cameras (TICs) at modern fires, due to the heavy smoke conditions now produced. Units equipped with TICs must carry and use them at all structural fires.
- Test results recommend that the first-arriving fire officer use a TIC to do a 360-degree walk-around of the fire building. A TIC can help to locate the fire if no visible flames show on the exterior.

- If an attack starts on the outside of a structure, what matters is the direction you're moving in. Whether you call it a transitional attack, offensive-exterior attack, softening the target, blitz attack, quick water, hitting it hard from the yard, or something else, the direction you are moving in is the deciding factor. If you are moving forward, you are deploying an offensive fire attack.
- Fire spread is largely a pressure-driven phenomenon. Fires spread along a flow path, the path of least resistance in the lower pressure space between an inlet for fresh air (a ventilation opening) and the fire, and the path of least resistance in the higher pressure space between the fire and the outlet for hot gases and smoke (a window or roof opening).
- In basic ventilation training, firefighters are taught to ventilate as close to the seat of the fire as possible. This approach seeks to release the heat and smoke from the fire and to localize fire growth to the area of origin. Ventilating farther away from the seat of the fire creates the potential to spread the fire to uninvolved parts of the house by creating a flow path and source of oxygen from that uninvolved area.
- Ventilation of a fire building must be a controlled operation. It should not be an attempt to open all of the windows and doors throughout a structure. All interior and exterior ventilation tactics must be controlled, communicated, coordinated, and approved by the crews operating inside the fire area to be vented. Before ordering any horizontal or vertical ventilation, the interior crews must evaluate the impact this tactic will have on interior conditions.
- If too long a time frame exists from the time of ventilation until the final extinguishment on the interior, the ventilation can introduce enough oxygen to cause the fire to flashover and rapidly expand and spread throughout the structure.
- Wind is a powerful force in influencing the direction of a flow path. A window opening, either caused by the fire department or by a window failing due to attack by fire on the windward side of a building, will create a rapid influx of air and can have a tremendous impact on a fire.
- Fog streams should not be used on an offensive-exterior attack. Straight streams or solid streams should be used to minimize air entrapment. Keep nozzle movement to a minimum and aim the streams at the ceiling of the fire room. The nozzle should be moved deliberately and slowly, from side to side. This should allow smoke and gases to continue exiting through the opening.

Summary

Some of this scientific data seems contrary to my personal experience, particularly to direct hose-lines into windows at incidents. In my experience of using that tactic, the interior conditions deteriorated rapidly and it seemed that the hose-lines pushed the heat toward the firefighters operating on the interior. Now, with this scientific data and the accompanying explanations, we know why we made these conclusions, and must realize that those conclusions were wrong.

The presence of flow paths and how air movement occurs within these structures has been documented scientifically. But the information provided by these studies still requires firefighters to utilize our own experience and expertise in determining the problems that exist at an incident scene and finding methods to solve those problems. We can utilize scientific data along with our knowledge of tactical operations, and after thorough training, we can better apply these principles at incident scenes and experience safer fireground operations.

The common issue in fighting any fire is the availability of resources and their capabilities. Fire departments staffed with two or three personnel and having to wait 10 to 20 minutes for additional resources to arrive will be challenged to fight any substantial fire, compared to a large fire department responding with four engines and two truck companies staffed with a minimum of four personnel per apparatus.

One factor does not change: Every fire is different. The differences range from controlled live burns with everything set up to all the unique situations found on a fireground. The one constant thread is to control the ventilation openings as part of a coordinated fire attack, and that final suppression efforts still need to be accomplished from the interior.

Modes of Fire Attack

There are four basic **modes of fire attack** for structural firefighting:

modes of fire attack ■ The types of attack for structural firefighting are offensive, defensive, and transitional, which can be offensive/defensive or defensive/offensive. The use of nonintervention mode will also be a consideration when encountering hazardous materials incidents.

1. Offensive
2. Defensive
3. Offensive/defensive
4. Defensive/offensive

Fire has destroyed many structures because a fire department hesitated to initiate a prompt attack on arrival. Implementing an offensive attack maximizes control efforts by allowing a quick attack on the fire. With an offensive attack, water is delivered to the seat of the fire and rescues can be made under the protection of interior hose-line crews. Realize that a major part of the decision making process is weighing the risk being taken by the firefighters versus the gain to be achieved. High-risk situations with little or no gain must be avoided. (Risk versus gain is discussed under Incident Scene Safety, Chapter 6.)

Decisions on the mode of attack must be made upon arrival of the first-due company officer. Other questions to be decided include the following: What size hose-line will be needed? Where should the first hose-line be placed? What about the second and third hose-lines? These initial orders can mean the difference between success and failure.

OFFENSIVE MODE

Offensive attack means that hose-lines are moving forward in an offensive manner, whether deployed through a window or door opening, or from the interior of the structure. The initial IC at a dwelling fire will need to decide whether to make an offensive-exterior attack or offensive-interior attack.

The first hose-line has to protect civilians who are still exiting the building or in need of rescue. This hose-line should be placed between them and the fire. If the first hose-line is able to extinguish the fire, the rescue will be a simple task. If the amount of fire is too great for one hose-line, the first priority is to protect the trapped occupants and the firefighters attempting to rescue them. If fire involves or threatens an open stairway, the same hose-line should be positioned to also protect the stairs, or a second hose-line can be given this priority to allow for egress of the occupants.

Gaining control of the stairs is the first step in gaining control of the building. (See Figure 3-17.) Protected stairs can be used by firefighters to enter the structure and by

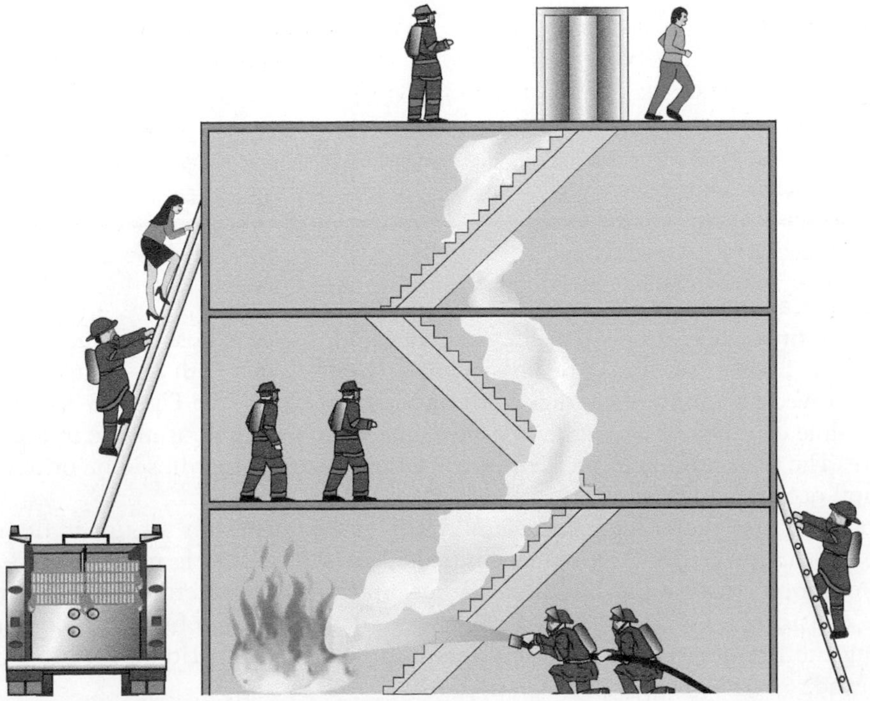

FIGURE 3-17 Gaining control of the stairs is the first step in gaining control of the building. Proper hose-line placement can reduce the heat of the fire extending to the upper floors via the stairs. *Used with permission of Michael DeLuca.*

FIGURE 3-18 The rear of a fire building must be checked. My experience has shown that a major loss of life in single-family and multi-occupancy dwelling units occurs in the rear of these properties. *Used with permission of Michael DeLuca.*

civilians to leave it. Proper hose-line placement can reduce the heat of the fire extending via the stairs to the upper floors. While occupants are exiting a building, any ventilation needs to be limited to ensure that oxygen supplied to the fire will not accelerate it and draw the fire towards the fleeing occupants. Firefighters operating on the interior must advise outside ventilation crews on what ventilation openings are needed and when to initiate those openings.

If an offensive-exterior attack is initiated, the second hose-line should be prepared to enter and attack the fire on an offensive-interior attack once the exterior attack is completed. If the first hose-line has initiated an interior attack, the second hose-line should back up the first hose-line. If fire involves more than one floor, the second hose-line can be placed on the floor above to control the fire there.

The third hose-line is taken to protect any secondary exits. In my experience, a major loss of life in single-family and multiple-occupancy dwelling units occurs in the rear of these properties. (See Figure 3-18.) This hose-line will ensure that anyone who might be trapped initially and seeks escape via another exit will receive the protection of a hose-line. The placement of the hose-line at this location will have to be monitored carefully. Coordination is required to assure that when hose-lines are operating from different directions, they do not oppose one another. The rear of the fire building has the potential for rapid undetected fire spread and life safety considerations. A response to the rear must be ensured by at least one engine and one ladder or truck company to assist in rescues, laddering, ventilation, and forcible entry. If a ladder company is not available, a second engine company or a rescue or squad company can assist in accomplishing these duties.

 **ON SCENE**

A fire that I worked in an old mill building that had been converted into condominiums forced the companies to take a 50-foot ladder over a 35-foot wall to make rescues in a courtyard. The normal entrance was rendered useless because heavy fire cut off the entry. Sixteen people were rescued from upper floors via the courtyard.

Hose-Line Placement in Unoccupied Buildings

In structure fires where no civilian life is endangered, the first hose-line should attempt to cut off the fire spread. It may be possible to accomplish this with an offensive-exterior attack followed by an interior attack with the first hose-line. (See Figure 3-19.) The second hose-line can then back up the first hose-line or go to the floor above to attack any fire there. The third hose-line can then be placed to back up either hose-line or to contain horizontal or vertical fire spread.

Safety dictates that a hose-line should not be operating on an upper floor until the fire on the lower floors is either controlled or a hose-line is placed in operation on those floors. Because the building is unoccupied, the life factor of the firefighters must be considered. Operating on a floor above an uncontrolled fire, with no hose-line attacking the fire below, places firefighters in a high-risk situation with little gain to be achieved.

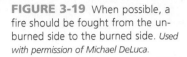

FIGURE 3-19 When possible, a fire should be fought from the unburned side to the burned side. *Used with permission of Michael DeLuca.*

Hose-Line Usage

When stretching hose-line for interior operations, company officers must keep in mind that when more than two hose-lines are pulled through the same doorways or up the same stairs, it becomes very difficult for any of the hose-lines to advance. If a third hose-line is needed, the third hose-line should be pulled through a window or other doorways. This will assist the units in overall incident scene efficiency.

If preconnected hose-lines are utilized, you must consider the length of the hose-line that will be needed and whether additional lengths of hose-line will be required.

Probably no other component is as important in achieving success at an incident scene as the proper placement of the hose-lines. Our ability to control and extinguish a fire is often solely dependent on this factor. The size of the fire will dictate the number and size of hose-lines required. When feasible, the preferred mode is an offensive attack on the fire. The company officer, when assigned to an offensive attack, must realize the importance of this assignment. For an interior attack to be successful, units must be able to advance the hose-line to the seat of the fire. Stopping the fire should be the aim, rather than pushing or following the fire. If an offensive-exterior attack is initiated and the exterior fire is knocked down, that hose-line or another hose-line must be prepared to advance through the interior to the fire area as quickly as possible to complete extinguishment.

Doorways allow quicker entry into a building than windows and a faster exit, if needed. Therefore, hose-lines should be run first through doorways, then through windows.

Advancing a deflated hose-line is faster and easier than advancing a charged hose-line. When the deflated hose-line reaches the fire area, the water should be ordered. Hose-line pulled up the exterior face of a building or up the stairwell opening uses less hose-line and allows companies to go into operation much quicker than if laid on the stairs. (See Figure 3-20.)

A rule of thumb for hose-line pulled up the exterior of a building or up a stairwell is one length for every three floors. (See Figure 3-21.) Hose-line laid on the stairs will require one length per floor and additional firefighters for advancement. We should allow a minimum of two lengths for the fire floor.

To keep below the heat, a firefighter advancing a hose-line should keep as low as possible, even resorting to a low crawl on knees and elbows until the fire area is reached. After the fire has been knocked down and the hose-line is being advanced into the fire area, the floor should be swept with the hose-line to clear away burning debris. The nozzle-person should operate from a crouched position, allowing him or her to keep below the heat and take advantage of what little visibility exists. It also prevents contact with hot surfaces or embers, which can cause burns to the leg area through bunker pants or boots.

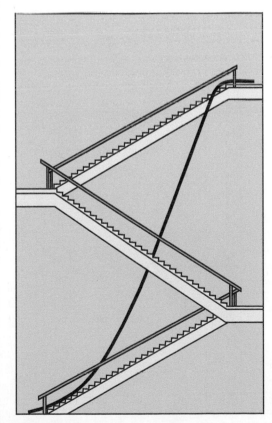

FIGURE 3-20 Hose-line stretched up a stairwell opening uses less hose-line and allows companies to get into operation faster than hose-line laid on the stairs. *Used with permission of Michael DeLuca.*

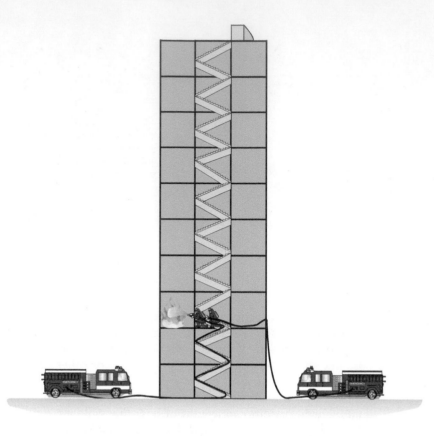

FIGURE 3-21 A rule of thumb for hose-line stretched up the exterior of a building is one length for every three floors. Hose-line laid on the stairs should allow one length per floor. *Used with permission of Michael DeLuca.*

Ventilation allows the hose-line crew to advance into the fire area. If ventilation has not occurred, the opening of the hose-line into a closed room or area on fire will create steam, making it untenable to fight the fire and driving back those operating the nozzle. When fire is encountered, a solid stream should be played on the ceiling and quickly rotated around the room. (See Figure 3-22.) The tendency to strike the fire directly must be resisted. The heat accumulates in the ceiling area. Remember that what goes up will come down. The water played onto the ceiling will come down and knock down the fire in the room quickly and efficiently. The exception is large volume rooms, where the water should be directed onto the burning material.

When the fire has been darkened down, the shutoff should be closed, and either that stream or another hose-line backing it up should be placed on a fog pattern to clear the room of smoke. A single stream with an adjustable nozzle can first knock down the fire and then be switched to fog for hydraulic ventilation. When using a fog stream to ventilate a room, it works best if the entire pattern is discharged through the window or door opening to the exterior of the building, with allowance for some opening around the stream. This enables the stream to pull the smoke from the area to be ventilated. This ventilation will reduce the temperature of the room, allowing firefighters to enter and complete the extinguishment and overhaul of the area.

The loss of water to hose-lines on a lower floor may seriously endanger the crews operating on the floors above. An interrupted water supply can occur due to pump failure or a burst length of hose-line. Safe operating procedures mandate that backup hose-lines be placed to guard against this kind of failure. Many departments have rules that if more than two hose-lines are operating and an adequate hydrant system is available, hose-lines should be supplied by more than one pumper.

FIGURE 3-22 When fire is encountered, a solid stream should be played on the ceiling and quickly rotated around the room. *Used with permission of Michael DeLuca.*

DEFENSIVE MODE

Exterior or defensive operations should be initiated when:

- A fire is beyond the control of hand-lines.
- There are heavy fire conditions and no civilians in the fire building.
- There are insufficient resources on scene.
- There is an insufficient water supply.
- A fire involves lightweight structural components.
- The risk to the firefighters outweighs the gain.

The first-arriving officer's observations and ensuing orders are critical. This person must evaluate conditions in relation to resources at the scene. Can the personnel present adequately protect exposures while mounting an attack on the initial fire building? Should the initial forces be deployed to concentrate on protecting exposures and later-arriving units be utilized to attack the main body of fire?

FIGURE 3-23 A second alarm has already been struck. If you were this chief officer approaching the fire building, what would be your immediate problems and considerations? What would be your long-term concerns and considerations?

Exposures

Master streams must be placed where they are most needed. This can result in placing many of the initial streams between the fire and a seriously endangered exposure. Water can be played onto the exposed building to keep it cool.

The ability to protect exposures will be in direct proportion to the distance between the exposure and the fire buildings. Frame structures create the greatest challenge due to the large amount of combustible material. Where adjoining buildings are encountered, the possibility of extension is greatly increased.

The first hose-line for a defensive mode of attack should be placed to protect the most endangered property. This helps to achieve incident stabilization and property conservation. If life is threatened within the exposures, the one most critically endangered will be protected first. The second hose-line will be placed using the same criteria. If the first hose-line cannot adequately protect the primary exposure, the second hose line should back it up. The placement of additional hose-lines will follow.

If no life is endangered, the priority of placement will be determined by the value of the property. A business employing many people would need to be protected. A business utilizing a hazardous process that could endanger a large segment of the community if involved in fire requires immediate attention. Exposures in the rear areas of a property must not be overlooked in hose-line placement.

Hose-lines placed to protect exposures can accomplish this goal by directly extinguishing the original fire. (See Figure 3-24.) Sometimes the first hose-lines can be placed to protect the exposed structure from extension of the fire while fixed appliances or aerial devices are being positioned to allow direct extinguishment of the original fire building.

Fire Travel

When a strictly defensive operation is apparent, the first-responding officer must make a reasonable assumption of what the fire will consume and where it can be stopped. The IC must be prepared to write off parts of buildings or even entire buildings if they are past the ability of a fire department to save. Failure to give up a building that is already lost and concentrating resources on it will result in chasing a fire. This objective usually sets up a losing situation. On the other hand, making realistic goals on what can be saved and where the fire can be stopped will minimize losses.

FIGURE 3-24 The ability to protect exposures will be in direct proportion to the distance between the exposure and the fire buildings. *Used with permission of Michael DeLuca.*

FIGURE 3-25 Continued heavy smoke conditions can indicate fire burning within void spaces in a building.

The first-arriving officer must predict the path and speed that a fire is traveling. Then he or she must estimate the amount of time required to place a sufficient number of master stream devices into operation to control the fire. It must be determined where the fire may spread to during the time it takes to set up master stream devices. If the master streams are set up at the wrong location, fire may have already burned past the streams' effective reach. Factors that will directly impact these decisions include:

a. The extent and type(s) of fuel that is burning
b. The proximity of the immediate exposures
c. The amount and capabilities of the resources
d. The available water supply
e. Other factors (See Figure 3-25.)

Exterior operations must consider the correct appliance to deliver the stream: tower ladders, ladder pipes, elevating platforms or booms, deluge guns, or mounted monitors. Consideration must be given to stream placement, the maximum effectiveness anticipated, and the provision of sufficient pressure and volume of water to the appliance to place the water onto the fire.

A stream that is not hitting the fire is ineffective. It steals water from other appliances, causes unnecessary damage, and can further weaken the structure.

After the fire appears to be knocked down, exterior hose-lines should be shut down and the area examined to determine whether it is safe for units to operate on the interior to overhaul hot spots. If this is not feasible due to water accumulation, structural damage, or other unsafe conditions, the areas still needing attention should be noted and exterior hose-lines can be repositioned to effect complete extinguishment.

Apparatus placement for a defensive attack must consider current and potential collapse zones. Company officers must predict the current and anticipated incident needs in positioning engine and truck companies. The relocation of apparatus after

FIGURE 3-26 Once life safety is addressed, the next priority is exposures. High heat has damaged the exterior siding of this exposed structure, but cooling water kept the fire from the building's interior. *Used with permission of Joseph Hoffman.*

FIGURE 3-27 Heavy fire conditions on the fifth floor initially trapped occupants. After the use of mobile hose-lines assisted their escape, a defensive attack was initiated. *Used with permission of Jim Smith, Jr.*

master streams are operating becomes quite involved. Apparatus may be boxed in by other apparatus and charged hose-lines, and to move this apparatus will necessitate the shutting down of streams and the movement of multiple apparatus.

TRANSITIONAL MODES

Transitional modes of fire attack consist of a combination of offensive and defensive attacks.

Offensive/Defensive Mode

Situations can develop where certain positions or locations must be defended, knowing that relocation of the attack will be necessary after rescues have been made, or if the fire is found to be beyond the control of an offensive attack.

A structure fire may have control of the better part of a building, yet rescues can still be accomplished under the protection of a hose-line. In this particular instance, the initial IC will have the first-arriving units go into an offensive-interior attack mode utilizing hand-held hose-lines to protect the search and rescue crews. On completion of the rescues, all units will be removed from the structure and a defensive attack will be implemented.

Changing fireground situations can dictate a need to change from an offensive to a defensive mode. (See Figure 3-27.) Our goal is to predict and be prepared for changes prior to their occurrence, but this is not always possible. As changes occur at the scene, we must be prepared to adapt to the situation.

If withdrawal from a building is necessary during an offensive attack, the officer giving the order should stipulate whether it is an immediate and emergency withdrawal necessitating the abandonment of the hose-line, or if the hose-line will be backed out of the building.

Defensive/Offensive Mode

A defensive/offensive attack can be used when the amount of fire is beyond the control of hand-lines or when there are not sufficient resources at the scene to mount an offensive attack. If the fire is beyond the ability of hand-lines to control, master streams can be utilized to knock down the fire in a blitz attack. (See Figure 3-28.) After knockdown, the building can be checked for structural stability, and if sound, an offensive-interior attack can then commence.

FIGURE 3-28 In an unoccupied building, a blitz attack can knock down a large body of fire prior to an offensive attack. *Used with permission of Joseph Hoffman.*

If the defensive attack was initiated due to insufficient staffing at the scene, the IC can start knocking down the fire from the exterior until the arrival of sufficient resources. Upon their arrival, the IC can switch to an offensive attack.

ON SCENE

The coordination of all activities is essential. One particularly difficult hotel fire prevented hose-lines from advancing to a fully involved large front room on the second floor. The floor areas in the hallway leading to the room were weakened from fire. Bringing a portable ladder to the upper floor to bridge the area was time-consuming. An engine company operating in the Delta exposure heard the problems being encountered and notified the Division 2 Supervisor of their ability to play a stream on the fire. Units in the fire building were pulled back and the stream was played into the side window and proved highly effective in knocking down the fire. When the ladder arrived, it was put into place and companies overhauled the area.

The location of the exposure line made it possible to operate into the fire building without forcing the heat onto the other units because the fire was able to vent through the front windows. This is the exception to normal fireground operations, in which outside and interior streams do not mix well.

EVALUATING THE INCIDENT SCENE

Determining the correct mode of attack to be deployed at an incident will result from the IC's size-up. As higher-ranking officers arrive on the scene, their assessment of the incident must consider whether to:

- Continue the strategies and tactics that are already in place
- Modify these strategies and tactics as required to reach the desired goals
- Change the mode of operation to reflect the conditions on their arrival

Realize that ever-changing fireground conditions must be considered and the proper decisions made accordingly. If the strategies and tactics employed upon the arrival of the chief officer are correct, then his or her job is a little easier. However, the chief still has to go through a complete size-up procedure to ensure that all areas are being considered.

The difficult decision for the senior officer is whether to continue or change the strategies and tactics that are already in place. This decision will be based on his or her size-up, knowledge, and experience. At times, modifications have to be made to accomplish goals. After considering all the options and weighing them, if you, as the IC, decide to withdraw everyone from the structure, then do it. If the collapse you anticipated does not occur, it does not mean you were wrong. Decisions have to be made on what might happen and the disastrous results that could occur if the firefighters remained in the structure. The safety of personnel is a prime consideration.

Withdrawing interior streams and setting up a defensive attack is one of the most difficult jobs of Command. This is especially difficult when the initial IC's actions were correct at the time of his or her arrival but because of changing conditions are no longer feasible.

This decision should in no way be seen as a reflection on the initial strategy employed. Changing fireground conditions dictate alterations in strategy and tactics. The change to a defensive attack may now be warranted. Conditions can change rapidly. Heavy fire development or the notifications of collapse indicators (see collapse indicators in Chapter 6 Building Collapse) are but a few signs. The companies operating on the interior may not always have the big picture or all the information available to the IC. Most aggressive firefighters feel that if left alone, they can put out any fire. The problem is they may not be weighing all the factors involved.

ICs must avail themselves of whatever information is available to make proper fireground decisions. The responsibility of changing strategies when necessary will impact directly on firefighter safety and incident stabilization. (See Figure 3-29.)

FIGURE 3-29 Decisions on where a fire can be stopped must be made with realistic goals in mind, considering available resources, water supply, and the construction of the fire building and surrounding structures. *Used with permission of Marty Griffin.*

 ON SCENE

At one particularly difficult fire, I found the top floor of a three-story building well involved on my arrival. The first-in company officer had ordered all companies to operate with 1¾-inch lines. I felt the call was correct and checked with the captain of the first-due truck company to brief me on interior conditions. He reported that the engine company had just arrived at their location with charged hose-lines, and they were starting to advance to the third floor.

A few minutes passed, and seeing no appreciable difference from the exterior, I again contacted the truck captain. He reported that they were making no progress and conditions had worsened; the fire was now starting to drop down onto them. At that point, I ordered the companies from the building and directed all incoming companies to operate with large-diameter hose-line and utilize master streams. The strategy paid off. Within a short time of the evacuation, the roof and the sidewall of the structure collapsed.

Though the initial orders for an offensive attack were correct in my estimation, continuing size-up, along with additional information (interior assessment) and changing fireground conditions, necessitated the change to a defensive mode.

CHANGING ATTACK MODES

There must be clear direction from the IC when an attack mode is changed. All units must acknowledge the order to change the attack and immediately start initiating the change. Interior and exterior operations are rarely compatible. Exterior streams must not be directed into windows or doors while firefighters are operating in the building unless fully coordinated by either the IC or crews operating on the interior. With few exceptions, these operations should not be mixed. When going from an offensive to a defensive mode, there is a tendency to continue operating handheld hose-lines that have been withdrawn to the exterior, along with master streams. This should be allowed only to protect exposed areas until master stream devices can be placed. Handheld hose-lines place firefighters in danger zones. A defensive attack should be performed with master stream devices.

Summary

Organization and decision making do not just happen at an incident scene; planning and training are required. The implementation of an Incident Action Plan will assist the Incident Commander in accomplishing the needed actions for handling an incident scene. At a major incident it will be a written plan; at a minor incident it will usually occur in the mind of the IC. The use of the Planning "P" and the Planning Meeting will facilitate incident scene safety by developing a written IAP, determining incident objectives, and ensuring that they are safely accomplished.

A major portion of the IAP is the decision making process that is required for handling the problems found. The experienced IC can achieve control through the implementation of cue-based decision making. The inexperienced IC will need to rely on reviewing the command sequence, in which case strategy, tactics, tasks, and implementation of the appropriate ICS will need to be decided.

Crew resource management is a tool that progressive fire departments can utilize to enhance operations and safety.

Underwriters Laboratories and the National Institute of Standards and Technology have contributed to the fire service a good amount of scientific data. Their test results and recommendations, particularly in the area of using ventilation, need to be reviewed by all firefighters, and training needs to be initiated to improve firefighting operations.

The implementation of the correct mode of attack at a structure fire will contribute to firefighter safety. Whether implementing an offensive, defensive, or transitional attack, it must be accomplished in a safe operation.

Review Questions

1. List the considerations for a written Incident Action Plan.
2. Though not a part of the Incident Action Plan, what are the three ICS forms that make up the foundation of the IAP?
3. Discuss the difference between classical decision making and cue-based decision making.
4. What are the five levels of the command sequence?
5. List the incident priorities and discuss how they are utilized in the command sequence.
6. List the points of size-up discussed in this chapter.
7. What is the benefit of the 360-degree walk-around of the incident scene?
8. What do we hope to learn by sizing up an incident scene?
9. List the seven basic strategies.
10. List the modes of fire attack.
11. What is the benefit of assigning divisions and groups to the initial arriving companies?
12. What are the five factors that crew resource management is based on?
13. In what other situations than the ones listed in the text would you consider initiating a defensive mode?

Suggested Reading, References, or Standards for Additional Information

Klein, G., et al. 1998. *Rapid Decision Making on the Fireground*. Technical Report 796. Alexandria, VA: US Army Research Institute for the Behavioral and Social Sciences.

Underwriters Laboratories (UL) Reports on Ventilation Testing

"New Science Fire Safety Article: Innovating Fire Attack Tactics." 2013. Available at: http://newscience.ul.com/wp-content/themes/newscience/library/documents/fire-safety/NS_FS_Article_Fire_Attack_Tactics.pdf.

"Scientific Research for the Development of More Effective Tactics: Governors Island Experiments." 2012. Self-guided course format, available at: http://www.firecompanies.com/modernfirebehavior/governors%20island%20online%20course/story.html.

Kerber, S. 2012. "Analysis of One- and Two-Story Single Family Home Fire Dynamics and the Impact of Firefighter Horizontal Ventilation." Available at: http://newscience.ul.com/wp-content/uploads/2014/04/Analysis_of_One_and-_Two_Story_Single_Family_Home_Fire_Dynamics_and_the_Imapct_of_Firefighter_Horizontal_Ventilation.pdf

Kerber, S. 2013. "Study of the Effectiveness of Fire Service Vertical Ventilation and Suppression Tactics in Single Family Homes." Available at: http://ulfirefightersafety.com/wp-content/uploads/2013/07/UL-FSRI-2010-DHS-Report_Comp.pdf.

"Fire Service Summary Report: Study of the Effectiveness of Fire Service Vertical Ventilation and Suppression Tactics in Single Family Homes." UL Firefighter Safety Research Institute. Available at: http://ulfirefightersafety.com/wp-content/uploads/2013/07/2010-DHS-FD-Summary.pdf.

National Institute of Standards and Technology (NIST)

For fire and government research on ventilation testing, search for the following titles on the NIST website: www.nist.gov.
"Firefighting Tactics." 2013. Summary of hose stream characteristics.
"Spartanburg, SC, Fire Experiments Will Test New Firefighting Tactics." 2014.

International Association of Fire Chiefs (IAFC)

Crew Resource Management. 3rd ed. Fairfax, VA: IAFC, 2003. Available at: http://www.iafc.org/files/1SAFEhealthSHS/pubs_CRMmanual.pdf.

US Fire Administration (USFA)

Routley, J. G. "Detroit Warehouse Fire Claims Three Firefighters." Technical Report 003.

National Fire Protection Agency (NFPA)

Standard:
291: Recommended Practice for Fire Flow Testing and Marking of Hydrants

Related Courses Presented by the National Fire Academy, Emmitsburg, Maryland

Command and Control of Fire Department Operations at Target Hazards
Decision Making for Initial Company Operations
Preparation for Initial Company Operations

4

Company Operations

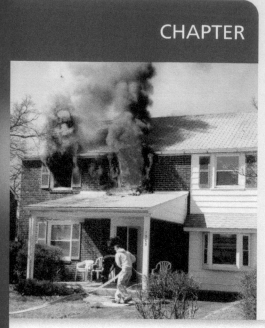

> ❝ Those who cannot remember the past are doomed to repeat it! ❞
>
> —George Santayana

Minimum staffing will challenge firefighters to perform a primary search and control a growing fire. *Used with permission of Brian Feeney.*

KEY TERMS

engine company operations, *p. 132*   **truck company operations,** *p. 141*

OBJECTIVES

Upon completion of this chapter, the reader should be able to:

- Discuss engine company duties.
- Discuss hose-line placement considerations.
- Discuss sprinkler and standpipe operations.
- Discuss truck company duties.

We have discussed the decision making process of the Incident Commander (IC). To solve the problems that are identified and to achieve the goals that are established by the IC requires actions.

Chapter 4 discusses the implementation of our strategies by engine and truck companies. To accomplish their goals, firefighters must mitigate the problems that they are confronted with. Success can occur only by predictability. This requires ample training and set procedures. Knowing each individual's duties allows every member to integrate his or her actions into a team approach. The training required is best accomplished through the engine and truck company concept. Though each company is a separate unit, it must be part of a larger organization that is established at each emergency scene.

The proper implementation of engine and truck company assignments is essential for successful operations. (Some departments refer to truck companies as ladder companies. The duties are the same.) The teamwork needed at an emergency demands close coordination of efforts. To be effective, firefighters must interact. Through training, they learn to get the right thing done at the right time. In fire departments that staff truck companies, there is a distinct difference between the duties of those assigned to an engine or a truck. In areas where no truck is assigned, another unit must assume the duties normally performed by the truck company. These assignments are usually given to a rescue company or another engine company.

The use of specialized apparatus—for example, Quints, tower ladders, elevating platforms, and water towers—demands that specific duties be assigned. Personnel operating this equipment may be required to perform either engine or truck duties or a combination of those duties.

Recruit training and ongoing training should have as a goal that firefighters have the basic skills to perform all fireground functions, regardless of the unit to which they are assigned.

A typical operational guideline that many fire departments use would have the first-due company, Engine 1, make forcible entry and advance a hose-line into the burning building. Engine 2 would secure a water supply for Engine 1 and perform ventilation. Engine 3 would place a backup hose-line to support Engine 1 and perform search and rescue. Rescue or Squad 1 would assume the duties of the rapid intervention crew.

Should a truck company be assigned on the response instead of a third engine, the changes would be minimal. Engine 1's duties would remain the same. Engine 2 would secure the water supply and place a hose-line to back up Engine 1. Truck 1 would perform ventilation and search and rescue. Rescue or Squad 1 would be the rapid intervention crew.

An operational guideline is based on a number of factors. A prime consideration is the amount of staffing on each apparatus. Career departments with minimum staffing requirements can implement guidelines that are specifically written with their staffing as a consideration. If staffing fluctuates, as it does in many volunteer and combination departments, the IC will need to prioritize the assignments.

Some areas of the country have heavy rescue companies, rescue companies, squads, and medic units to share the incident scene duties. The types of apparatus, tools, equipment, and staffing vary greatly, and for that reason, I have limited the specific duties discussed here to engine and truck companies, though the terms *rescue* or *medic unit* are sometimes used. The specialized types of apparatus can include:

- A *heavy rescue unit* can cover a wide category of apparatus. It can be equipped to perform the duties of an Urban Search and Rescue (USAR) team and be staffed with an officer and five firefighters. It can be utilized as a staffed unit that responds for specialized rescue operations, or it can be utilized as additional staffing to supplement units at a working fire. In some fire departments, the rescue unit consists of two firefighters who may be assigned full-time to the rescue unit or may have other duties, such as assignment to a truck company. (If the rescue unit is dispatched, they then respond with that apparatus.)

- The *medic unit* typically responds to medical calls. It may transport patients or give initial medical treatment until the arrival of an ambulance to transport the patient. Some areas refer to these units as rescue squads. Staffing may consist of paramedics, emergency medical technicians, or firefighters. Medic units may provide advanced life support or basic life support, depending upon the equipment carried and the qualifications of the assigned individuals. In many departments, these personnel are cross-trained as firefighters who assist in firefighting. Their duties will be dictated by the standard operating guidelines (SOGs) of the individual fire department. Today, many departments that staff these units with firefighters utilize them as rapid intervention crews.

- *Quints* are apparatus that are a combination of both engine and truck companies. The apparatus contains some form of main ladder or elevating platform and portable ladders, along with a pump and a sufficient supply of hose-line to suffice as an engine company. The Quint concept can be successful in fire departments that have no main ladder or platforms available and purchase the Quints to gain increased flexibility. The Quint concept has been instituted by some fire departments to address staffing problems. The idea is that all members receive cross-training in both engine and truck company operations, and on a response some units perform engine company duties and other units perform truck company duties. Prior to adopting the Quint concept, a fire department should review the context in which it will be utilized and decide the benefits and drawbacks of the concept to ensure that firefighter safety is not compromised.

- *Squad companies* or Special Operations Companies (SOC) have been instituted in some fire departments to provide a host of duties. They can provide specialized rescue in areas involving high angle, trench rescue, vehicle rescue, and others. They may be cross-trained as hazardous-materials technicians to supplement the hazardous-materials unit in stabilization, mitigation, and decontamination. Many fire departments fully utilize these units at working fires to complement their staffing with specialized personnel. The staffing can vary from five firefighters in large metro fire departments to two firefighters in other areas. These units can also be utilized as the rapid intervention crew.

Engine Company Operations

engine company operations ■ Provide the water to control and extinguish fires.

Engine company operations provide firefighters with the water to control and extinguish fires. The engine company typically consists of an apparatus with a pump, a hose-bed for storage of hose-line, and compartments to carry a variety of tools and equipment. There are also specialized apparatus that can be utilized as an engine company. These can consist of a water tower or articulating boom that is attached to the apparatus and stowed in the hose-bed area and can be elevated for a master stream operation during a defensive fire attack. This water tower apparatus is sometimes referred to as a *Squrt*.

The staffing on an engine company can vary greatly. National standards recommend a minimum staffing of four firefighters for career fire departments. The actual staffing is strictly dependent on local factors.

Let's examine the duties of an engine company. The engine company is responsible for:

- Locating the fire
- Confinement
- Extinguishment
- Securing and delivering a water supply to the incident scene
- Utilizing the water supply for control and extinguishment

FIGURE 4-1 Locating this fire was easy. Determining how large an area was involved was another problem. *Used with permission of Joseph Hoffman.*

LOCATING THE FIRE

Locating a fire can be a simple task. It may be unattended food on a stove, an oil burner that has misfired, a smoldering mattress removed to the exterior by the occupants, or rubbish burning on the highway. (See Figure 4-1.)

Location can become complex when confronted with a multistoried building heavily charged with smoke and numerous smoke detectors activated on several floors.

The first part of locating a fire is the initial information given to Dispatch. This can be in the form of an address and additional data. The dispatch center will then transmit this information when the companies are dispatched. When arriving on scene, the address will be confirmed, or changed if it is not the correct location of the fire. Units will gain additional information at the scene through size-up. This information can come from visual indicators, such as fire visible in a window, door, or roof area; from occupants or bystanders who could have critical data; from alarm systems that can pinpoint the fire areas; or from sprinkler systems that have been activated or have water flowing.

It is recommended that the first-arriving company officer at an incident scene do a 360-degree walk-around of the fire building, utilizing a thermal imaging camera (TIC) if available. This action will assist in overall size-up and help in locating a fire if no flames are visible.

CONFINEMENT

The aim is to confine a fire to as small an area as possible. Confinement must consider the six sides of the fire. This includes the four walls and above and below the fire.

We have to consider what we are attempting to do against what can be accomplished. What staffing is available? What is the risk to firefighters? In determining our strategy, a reasonable assumption must be made on the size of the area to which the fire can be confined. (See Figure 4-2.) Adequate incident analysis through a comprehensive size-up allows the IC to assess the current situation. Then he or she must

FIGURE 4-2 In determining strategy, a reasonable assumption must be made on the size of the area to which the fire can be confined. *Used with permission of Joseph Hoffman.*

utilize past experience to forecast anticipated problems and occurrences. It must be a proactive approach, considering the potential for immediate fire spread. What can we reasonably accomplish? What resources are needed? After we determine our current situation, we must make some predictions. When confronted with a minor fire, we must predict what the conditions will be in 5, 10, or 15 minutes. As the complexity of an incident increases, our time frame considerations can increase to 10-, 20-, or 30-minute time frames. On a major incident, our predictions will involve time ranges of 30 or 60 minutes or longer.

The contents of the fire building, the type of construction, the presence of fire protection features within a building, and the resources available all impact confinement. A properly installed and maintained sprinkler system is beyond a doubt the best method of fire protection available.

Construction features, such as fire walls and fire doors, assist in containment. Other building characteristics, such as balloon-frame construction, air shafts, and unprotected horizontal and vertical openings, permit extension of fire to other areas of the building.

A fire occurring in an area of a building that is compartmentalized allows containment by utilizing the construction features in conjunction with good extinguishment methods.

The Underwriters Laboratories (UL)/National Institute of Standards and Technology (NIST) studies on firefighting ventilation practices and tactics (discussed in Chapter 3) found that an effective tactic to confine fires in one-and-two-story dwellings is a coordinated offensive-exterior attack, immediately followed by an offensive-interior attack to achieve full extinguishment.

EXTINGUISHMENT

Containment is followed by extinguishment. The time required to accomplish this function is directly related to the size of the fire area and the amount of overhauling required. A room-and-contents fire could simultaneously be confined and extinguished. Complex incidents could require a fireground detail that takes weeks before complete extinguishment is achieved.

SECURING AND DELIVERING A WATER SUPPLY TO THE INCIDENT SCENE

Securing a water supply is an initial concern of the engine company. If operating in an area that is supplied by fire hydrants, locate the nearest hydrant that will provide an adequate supply to accomplish this. Though many SOGs are written so that the second-due engine company will provide the first-due engine with a water supply, the first-due engine, when possible and with enough time, can secure a hydrant to obtain their own water supply. This initiative will ensure a continuous water supply and, with the second engine company securing their own water supply, will provide multiple water sources at the incident scene. In nonhydrant areas, securing a drafting site or utilizing a water tender operation will be required. It may be possible to set up a relay operation from an adjoining hydrant area or a drafting site.

Large-diameter hose-line allows large volumes of water to be moved from the source to the fire. (See Figure 4-3.) This can be considered an aboveground water main. If long distances are encountered, a relay pumping operation involving multiple engines in the relay can be used.

An evaluation of the available water supply must be made. If accomplished before the assignment through a preplan, the plan can be implemented. If there is no preplan or prior assessment

FIGURE 4-3 Large-diameter hose-line allows large volumes of water to be moved from the water source to the fire. *Used with permission of Brian Feeney.*

of the water supply, it must be an immediate concern. Assignment of a Logistics Section Chief will assure that this important facet is given adequate attention. That individual will determine the amount of water needed by conferring with the IC. The fireground should be surveyed for available water supply. In areas not serviced by hydrants, it must be determined if a relay operation from a nearby lake or river can be set up. If strictly a water tender operation, dump tanks should be set up in locations allowing easy access and egress for the water tenders. The number of water tenders available must be considered, along with the needed fire flow.

The available flow can be calculated by determining water main size, available hydrants, and the proximity of rivers, lakes, and other large supplies of water. Adding the volume of the water tenders and multiplying the time needed for these units to load and unload their water will determine the amount of water available. (Five 3,000-gallon water tenders requiring a round trip of 10 minutes to load and unload could theoretically supply 15,000 gallons of water in 10 minutes, or 1,500 gallons per minute. This leaves no margin for error, and attempting to fully utilize 1,500 gpm would be a short-sighted approach.)

UTILIZING THE WATER SUPPLY

The amount and type of fire will dictate the size of the hose-line used to attack the fire. The 1½-inch or 1¾-inch hose-line is the workhorse of the fire service. Its weight and maneuverability in confined spaces and its ease of handling have proven overwhelmingly successful on fires occurring in residential properties. It is seldom necessary to stretch a larger hose-line into dwelling units.

These hose-lines can, however, prove to be ineffective or less effective when fighting fires in commercial properties containing a heavy fire load or large noncompartmentalized areas. There is a tendency to use 1½-inch or 1¾-inch hose-lines on all interior operations, regardless of the size of the structure or its contents. But fully relying on 1½-inch or 1¾-inch hose-line may mean the difference between the success or failure of an offensive attack. The reluctance to utilize 2½-inch hose-line seems to stem from the fact that staffing levels for engine companies in most departments necessitate the combining of two companies to place a 2½-inch hose-line into operation, and this takes more time. The benefit of using the 2½-inch hose-line is that the larger volume of water provides a greater cooling effect and quicker knockdown.

ICs should consider their options when dealing with a fire in a large-area structure, or one containing a heavy fire load, when the fire is still controllable by an interior attack. Size-up must be used to evaluate the incident and determine if 1½-inch or 1¾-inch hose-line can make a difference, and if there is enough time to utilize 2½-inch hose-line as the initial attack line. The IC may order 1½-inch or 1¾-inch hose-line for the first-alarm companies, to allow them to quickly apply water and attempt to control the fire. The second and greater alarm companies can then utilize 2½-inch hose-line to back up or replace the smaller hose-line. This tactic gets the smaller and easily handled hose-lines into operation quickly enough for containment. The speed of a quick attack may make the difference between gaining initial control of the fire or having to use an immediate defensive attack due to the longer time that is needed to deploy the 2½-inch hose-line. Because of the 2 ½-inch hose-line's infrequent use as an attack line in many departments, regular training sessions to maintain skill levels can reduce the time required for its deployment.

Hose-lines can be stretched from standpipes in high-rise or commercial structures that are so equipped. This will facilitate units getting into operation. Standpipes should be considered a hose-line already laid.

Nozzles

A variety of nozzles can be utilized. Much discussion has occurred on straight-tip or smooth-bore nozzles versus automatic nozzles. I personally feel there is a need for both types. As with any other tools, there is a right one for the job. Using the wrong tool may accomplish the task but not as efficiently. The straight-tip or smooth-bore nozzle delivers a greater flow and has better knockdown than an automatic nozzle. The automatic nozzle offers the versatility to quickly deliver a straight stream or a fog stream,

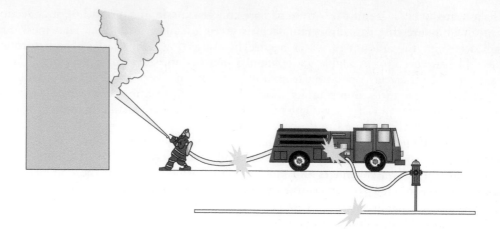

FIGURE 4-4 Closing a nozzle too quickly can produce water hammer, which can damage the hose-line, apparatus pump, and water mains. *Used with permission of Michael DeLuca.*

usually with a twist of the nozzle. The automatic nozzle does require a higher nozzle pressure to be effective. These higher pressures may not be practical in some standpipe systems, though current standards require a minimum of 100 psi from the highest outlet in a standpipe.

Nozzles should be opened and closed slowly. Opening a nozzle too quickly can cause the firefighter to lose control of the line. Closing a nozzle too quickly produces water hammer. (See Figure 4-4.) This can:

1. Cause the hose-line to burst
2. Damage the apparatus
3. Damage water mains

Hose-line advancement requires sufficient ventilation in front of the nozzle, and the hose-line must be of sufficient size to control the fire. A hose-line that does not advance is ineffective. Firefighters operating a hose-line from a fixed location will take more punishment and cause more water damage than if they push it forward and extinguish the fire. Size-up and experience will determine if a hose-line is of sufficient size to control a fire or whether a second hose-line is needed before advancement can occur. Generally, the hose-line must be able to knock down the fire as the firefighter operating the nozzle moves through the fire area. If this does not occur, it will impede advancement. Firefighters must know the reach of their hose-line. There is no need to enter an area that is involved in fire when the hose-line stream can reach all areas of the room/s from the doorway. Safety dictates operating from safe areas whenever possible.

The best position for the company officer is behind the nozzle-person. Fire conditions can be observed, as well as the effect of the stream on the fire. The officer can supervise an inexperienced firefighter and order the line to be advanced as knockdown is achieved. An experienced firefighter on the nozzle will know when to advance the hose-line. If a hose-line cannot advance, the company officer must determine the reason and rectify it. If there is inadequate ventilation, the officer should order it corrected. If the hose-line is too small, the officer should have a larger one brought to the fire.

Keeping Stairs Open Firefighters should not be allowed to gather at the nozzle. They should be spaced along the hose-line and be prepared to feed or push the line forward when ordered. Likewise, stairways must be kept open. There is a tendency to crowd or gather on stairways when fires are above the first floor. The stairs must be kept open to allow firefighters to remove victims, to descend when relieved, to advance additional hose-lines, to allow search and rescue teams access, and to bring equipment or tools to the fire floor.

The debris of the firefighting effort—for example, plaster and drywall—can accumulate on the stairs. Firefighters being relieved have a limited view from their masks as they start to climb down the stairs. The smoke and steam created by the fire and the firefighting efforts, combined with hose-lines stretched over the stairs, produce a dangerous condition, making ascent or descent difficult, and can contribute to firefighter injuries.

A shovel or the blade of an axe can quickly clear the stairway of debris, alleviating a potentially dangerous situation.

Line Placement Considerations The selection of the number and size of hose-lines must consider the type and size of the fire, the life that is threatened, the mode of attack, available resources, and water supply. (See Figure 4-5.) The number of lines and their exact location will be based upon the following considerations:

- Protection of life
- Performing an offensive-exterior attack
- Reaching the seat of the fire
- Protection of the stairs and preventing upward extension of fire
- Protection of search and rescue crews
- Prevention of horizontal spread of fire
- Protection in the rear of the structure
- Protection of roof ventilation crews, and controlling roof fires and fire extension to other buildings due to roof ventilation
- Protection of exterior and interior exposures
- Backup lines stretched for any area above
- Master streams set up as a precaution should the interior lines need to be abandoned

FIGURE 4-5 Do a size-up on this fire considering the initial response your department would send on this assignment. Would the resources be sufficient to handle the problems you have identified? How would you address life safety of the occupants? The firefighters? What orders would be given in regard to ventilation? What water supply would be needed? *Used with permission of Joseph Hoffman.*

Master Streams A master stream is capable of water flows in excess of 300 gallons per minute, too powerful for handheld hose-lines. An exterior operation involving the use of multiple master streams requires a large volume of water and maintenance of adequate nozzle pressure. If sufficient water supply is available, utilization of large-sized tips will deliver greater quantities of water to control and extinguish the fire. When operating in an area where there are large-sized water mains and a sufficient number of hydrants, the initial pumpers may have available outlets that can be utilized by the later-arriving units. This allows shorter hose-line leads and less friction loss, and units can move into operation more quickly.

When confronted with a limited water supply, control of a large fire will be difficult. The placement of streams must be adapted to meet the prevailing conditions. One mistake that must be avoided is attempting to set up more master streams than can be supplied. This can lead to a number of ineffective streams. If this occurs, the IC must see that streams are combined to obtain effective ones.

Master streams can be delivered from a number of appliances: (See Figure 4-6.)

- Deluge or deck guns
- Mounted monitors
- Ladder pipes
- Tower ladders
- Elevating platforms
- Articulating booms or water towers

Each has benefits and limitations. The proper device to use will depend on availability and specific incident needs. Where there is limited access to the fire building, placement of special apparatus that do not respond on initial alarms will be difficult. Large-diameter hose-lines restrict mobility of apparatus. Placement of specialized apparatus will necessitate the use of hose bridges or temporarily shutting down the pumpers supplying the large-diameter hose-lines. Some fire departments handle this situation by dispatching special units on the first or second alarm, and the IC will typically give priority placement to those units.

FIGURE 4-6 Master streams can be delivered from a number of appliances. *Used with permission of Joseph Hoffman.*

FIGURE 4-7 Heavy fire conditions and a lack of resources or water supply can mandate a defensive attack.

Sprinkler and Standpipe Systems

The first arriving engine company should have knowledge of the protective systems in a building. Effective firefighting is enhanced through the use of a building's systems. A properly designed and installed sprinkler system is the most effective method of fire control and life safety in any structure. Additionally, the presence of a building standpipe system will allow units to stretch hose-lines shorter distances and to place water from handheld hose-lines onto a fire much faster.

Automatic Sprinkler Systems

Automatic sprinkler systems are an excellent means of protecting life and property. The inherent value of a sprinkler system is that it immediately delivers water directly over the fire in specific patterns and volume to extinguish or hold a fire in check until the arrival of the fire department. Sprinkler systems consist of a network of properly sized piping with sprinkler heads attached at specific locations.

There is a large variety of types and designs of sprinkler heads. The sprinkler head contains a cap or valve that is held in place by a fusible link or frangible bulb. The fusible link is secured by solder containing various metal alloys, of which the exact melting point is known. When the temperature from the fire reaches its melting point, the cap or valve is opened, the sprinkler system is activated, and water is released onto the fire below. The frangible bulb contains a liquid that when heated expands, and at a set temperature breaks the glass, releasing the valve cap and allowing the water to flow through the opened sprinkler head.

Water supply for sprinkler systems can be provided either by:

- Connection to a public water system that can provide adequate volume, pressure, and reliability
- Properly constructed and placed gravity tanks that can be relied upon for an adequate water supply
- A pressure tank that is situated on the top floor or roof of the building. It is usually filled two-thirds with water and pressurized with air.

Types of Sprinkler Systems

The National Fire Protection Association (NFPA) Standard 13 provides rules for the design, installation, and acceptance testing of sprinkler systems. It identifies four types of sprinkler systems: wet pipe, dry pipe, preaction, and deluge.

Wet Pipe Sprinkler Systems These systems contain water at all times. Activation delivers immediate water spray onto the area below the activated sprinkler head(s). The flowing of water through a wet pipe sprinkler system will activate an alarm that is connected to the main riser. This type of system is preferred, yet it should not be installed in buildings in which the temperature could fall below 40 degrees F.

Dry Pipe Sprinkler Systems These systems are filled with air or nitrogen under pressure in lieu of water. It contains a dry pipe valve that prevents water from entering the system until a sprinkler head is activated. These systems are designed for a small amount of air to hold back a much greater water pressure. This air pressure is usually between 15 and 50 psi. The air pressure can be supplied from a building's system or from a compressor and tank made specifically for the sprinkler system. When a fire activates a sprinkler head(s), the air is expelled through the open head(s) and water is released into the system to the open heads.

These systems are needed in buildings where temperatures could fall below 40 degrees F and there is a possibility of the water freezing. A dry pipe sprinkler system may be a subsystem of a wet pipe system for an area exposed to freezing temperatures, such as an unheated loading dock.

Preaction Sprinkler Systems These systems contain a deluge type valve, fire detection devices, and closed sprinkler heads. They are similar to the dry pipe sprinkler system in that they are charged with air or nitrogen, not water. The air pressure is generally lower than a dry pipe system. In this system, the water is held back by a preaction valve and has a supplemental detection system that, once activated, releases water into the sprinkler system. Once a sprinkler head(s) is activated, it will deliver water onto the fire. This system is similar to a wet system once the preaction valve operates.

Preaction sprinkler systems are monitored for any drop in air pressure that could be caused by a leak. But the building is also protected from accidental water damage, since water is only released into the system by an activated fire detection system.

Deluge Sprinkler Systems These systems utilize open sprinkler heads and a deluge valve. Fire detection devices are arranged in the same area as the sprinkler heads. When fire is detected, the water is introduced into the system and through all of the open sprinkler heads. These systems are typically used in highly hazardous occupancies or where a fast spreading fire may occur; large open areas such as airplane hangars; or dangerous occupancies such as a plant producing explosives.

Sprinkler System Ratings
NFPA Standard 13 defines fundamental hazard levels for sprinkler systems that are based upon the building's contents, not the construction materials. The three hazard levels are listed as light, ordinary, and extra hazard. The ordinary hazards and extra hazards levels are further broken down into Group 1 and Group 2 categories. These classifications of hazard levels require the authority having jurisdiction (AHJ) to ensure that buildings that are sold or rented comply with the contents requirements for which the sprinkler system was intended. If the property is to be used for an enterprise that has a greater hazard than the sprinkler system is intended to protect, the system will need to be upgraded so that proper coverage is in place.

Firefighters' Actions at Sprinklered Buildings
When operating at a fire in a sprinklered building, an initial consideration should be how to provide an ample water supply via supplemental hose-lines into the fire department connection(s) (FDC). This is especially important if a large number of sprinkler heads have been activated.

Preplans should indicate the location of sectional stop valves. These properly spaced shutoff valves can be used to control water flow to all or part of the sprinkler system. Once a fire is controlled, the nearest sectional stop(s) in the sprinkler system should be shut down. This action minimizes water damage and allows for a closer inspection of the fire area to ensure complete extinguishment. Shutting down an entire system should only be done if no sectional stops exist. (Additional information on sprinkler operations can be found in Chapter 9, Commercial and Industrial/Commercial Buildings and Warehouses.)

Extra sprinkler heads should be kept on the premises to replace any heads that are activated or damaged in order to immediately restore the system. Sprinkler heads have temperature ratings that are color-coded for ease of identification. When replacing a sprinkler head, it must be with one that is identical to ensure continued protection. Some fire departments routinely replace sprinkler heads that have been fused as part of their customer service to their community.

Standpipe Systems
The purpose of a standpipe system is to deliver water for firefighting purposes. The standpipe can be designed for full-scale firefighting, first-aid firefighting, or both. Standpipe systems are designated Class 1, 2, or 3.

There are combined piping systems that supply water to both the standpipe and the sprinkler systems. These are either a Class 1 or 3 standpipe system that also supplies water to the sprinkler system, and the only difference may be that larger piping is provided to supply both systems.

Class 1 standpipe systems provide 2½-inch hose-line connections at designated locations in a building for use by firefighters for full-scale firefighting. They

FIGURE 4-8 A Class 2 standpipe outlet with hose-line attached. *Used with permission of Jim Smith, Jr.*

are typically required in buildings of more than three stories and in malls, due to the long hose-line stretches that would be needed.

Class 2 standpipe systems provide 1½-inch hose-line connections at designated locations in a building for first-aid firefighting. These systems can contain a hose-line (commonly referred to as a house-line), nozzle, and hose rack. They are not intended for occupant use, but can be used by fire brigades prior to the arrival of the fire department. Many jurisdictions have eliminated the requirements of having hose-lines located with these systems and discourage the use of any installed house-lines by occupants. Firefighters can use these systems, but when they do, it's important that the firefighters remove the house-line and connect their own hose-line to the outlet. The use of the fire department's hose-line will provide an added measure of safety for firefighting crews. In the past, these house-lines were prone to dry rot and often failed once pressurized. Class 2 systems are often required in large unsprinklered buildings. (See Figure 4-8.)

Class 3 standpipe systems combine both the Class 1 and 2 systems: they provide full-scale and first-aid firefighting capabilities and are intended for both firefighter and fire brigade use. These systems are provided with both 2½-inch and 1½-inch hose-line connections. The connections may be separate or can be connected by using a 2½-inch hose valve and an easily removable 1½-inch adapter attached to the valve by a reducer. Class 3 systems are sometimes installed when both Class 1 and Class 2 systems are required and it is not beneficial to install separate systems.

Types of Standpipe Systems

In addition to the classes of standpipe systems, which define their use, they are also identified by type. The types designate whether the system is filled with water (a wet system) or not (a dry system), and they denote whether the water supply for the system is automatic, semiautomatic, or manual.

- *Automatic-wet systems* contain piping that is filled with water at all times and have a water supply that can meet the demand needed for firefighting.
- *Automatic-dry systems* contain piping that is filled with pressurized air and contain supply valves that release the water into the standpipe system when a valve is opened. The available water supply with sufficient pressure is automatically available for firefighting.
- *Semiautomatic dry systems* contain piping that is normally filled with air and may or may not be pressurized. These systems are arranged to allow water into the system when a remote actuation device, such as a pull station, is operated. They require a preconnected water supply that can ably support the firefighting demands.
- *Manual-dry systems* contain piping that is filled with air and do not have a water supply connected to them. A fire department connection is provided so water can be manually supplied to the system for firefighting. Dry systems can be looped into one interconnected system or contain individual risers. Systems with individual risers will necessitate coordination with the apparatus engineer or chauffeur to ensure pressurization of the correct fire department connection. It should be noted on the preplan the type of system and the location of the connection(s) where the fire department can supply or add additional pressure to the standpipe system. Some bridges and highway systems have manual dry standpipes that do not contain air and are used to move water from a lower level to elevated roadways.
- *Manual-wet systems* contain piping that is filled with water to alleviate the time required to fill the system in the event of a fire. The water supply comes from a connection to a domestic water supply, though the supply is not adequate to provide sufficient pressure for firefighting. The fire department will need to pressurize the system through the fire department connection for firefighting purposes.

Truck Company Operations

Though truck company firefighters perform many tasks, the primary duties assigned as part of **truck company operations** are:

1. Rescue
2. Laddering
3. Forcible entry
4. Ventilation
5. Overhaul
6. Salvage

truck company operations ■ Rescue, laddering, forcible entry, ventilation, overhaul, and salvage are the primary duties assigned to a truck company.

If there is sufficient staffing on a truck company, the firefighters can be divided into teams and assigned specific duties. A team or crew consists of two or more members. One method is initially to assign one team from the first-arriving truck to perform outside duties, which includes raising ladders and ventilation. The interior or inside team can assist in forcible entry and perform search and rescue. After completion of these assignments, additional duties can be assigned. If the staffing level is only two or three firefighters, multiple companies will need to be utilized to perform the needed functions.

RESCUE

In relation to life safety, fireground commanders and firefighters alike want to give trapped occupants the benefit of the doubt. If there is any chance they are alive, a concerted effort to rescue them will be made.

We must find out where the people are located, who are the most threatened, and what is the quickest and safest way of reaching them: through the interior stairs or from portable or main ladders? Other questions include: What is the safest way of protecting occupants, and can those trapped be protected in place? The latter is often a better option in a fire-resistive building than in a two-story dwelling, but it may be a consideration.

Search and Rescue Operations

Preplans or quick-action plans can provide data on the normal times that a building is occupied, the number of people usually in the structure, a floor layout, and any building protective systems. Yet, in one- and two-story family residences there are normally no preplans, and occupants should be anticipated at all times, which requires a primary search at the minimum.

Dispatch information can be invaluable to the responding units. Dispatchers routinely ask what is burning and if everyone is out of a building. Some departments routinely update responding units, but other departments transmit this additional data only if requested from the incident scene. A procedure must be established to transmit all pertinent information to the IC.

The occupants can be a good source of information of the probable location of anyone still in a building. The obvious drawback to this approach is that, due to their excitement and confusion, inaccurate information can be given. Any information gathered should be collated to get the big picture. This responsibility should rest with one person: either the IC or a person appointed by him or her.

The fact that many reports of people being trapped in a burning building turn out to be false is due to various factors. The fire prevention message for those practicing home fire drills to have a designated place to assemble on the outside is usually not followed. There is also the tendency of people to call their pets by name. People may refer to their cats and dogs as children. Terms such as "baby" or "honey" seem to indicate that a person rather than an animal is still within a burning property. This is not to imply that a pet is not important. If the rescue of a pet can be accomplished, that's fine. Firefighters take many risks to save human life, but a firefighter should not be at risk attempting to save a pet.

Routine visual size-up can indicate if anyone is hanging out of windows, as well as smoke and fire conditions, which will impact any rescue attempts. Window stickers have been used to indicate a child's bedroom. These stickers can be misleading, however; rooms may be changed during renovations, children can grow up and move out, or houses can be sold and the stickers remain in place. All rooms must be searched thoroughly for victims.

Primary and Secondary Search

The search is broken down into two distinct phases: primary and secondary. The primary search is certainly the most important; it is performed initially and must be as thorough as possible. Any search will be greatly enhanced when performed with a TIC. Every area should be checked.

Primary Search   A primary search is always necessary. The information that no one is at home can prove faulty. The excitement of a fire can cause memory lapses of a child's friend or a relative spending the night. Would-be rescuers—whether neighbors, passersby, or even police officers—could be overcome and may also need to be rescued.

A search of the interior must be performed in a coordinated and systematic way. The area around the fire and directly above the fire should be checked first. This will allow removal of the most seriously endangered occupants. Also, it can give the IC a good interior assessment of the conditions in these areas. If fire has extended to one of these locations, it will allow the IC to adjust his or her tactics to meet the growing problem. The company officer ordinarily will assign firefighters particular areas of the building to be searched. Firefighters performing search and rescue must not enter a building without the knowledge of their company officer. This notification is necessary for accountability of members.

A one-directional search of a room should be made by keeping close to the walls and reaching out as the room is encircled. (See Figure 4-9.) The searcher continues in the same direction until returning to the starting point. The use of a pattern ensures that all areas are checked (referred to as a right-hand or left-hand search).

Listening for sounds can help to locate victims. Children crying, moaning, or even coughing can allow firefighters to pinpoint their location. Thermal imaging allows search and rescue crews to locate victims and to maintain contact between crew members. Like any piece of equipment, the TIC needs to be checked for proper operation. If this check is not performed prior to a response, then it must be done either en route or on arrival and prior to leaving the apparatus, to ensure its functionality. Proficiency of using a TIC comes with practice. There have been instances when a firefighter was either unfamiliar with the operation of the TIC or the batteries were not charged, and the piece of equipment was useless at the incident. That is unacceptable.

Furniture must be checked under and behind. Large bureau drawers can be children's hiding places during playtime, and a child may seek shelter there during a fire. If a bureau drawer is open, it should be checked. After checking along the walls, the interior of the room should be checked. Closets and bathrooms often become hiding places for victims. In closets, they seek shelter under clothing. Water from a shower often is turned on for protection. When an area has been searched, the IC should be notified of the results. The room should be marked to indicate that a primary search has been made.

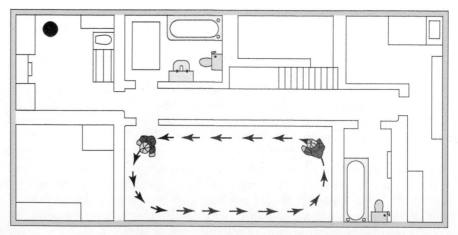

FIGURE 4-9 A one-directional search of a room should be made by keeping close to the walls and reaching out to check the center of the room. The searcher continues in the same direction until returning to the starting point. A thermal imaging camera will enhance the search. *Used with permission of Michael DeLuca.*

At one particular fire, we were searching with no success for two children who slept in a third-floor bedroom. Neither the primary nor secondary searches revealed them. They finally were located in a knee wall area. They had entered this space through a hole in the plaster wall in their closet. It was discovered later that the area often was used as a play area. The children sought refuge from the fire below but to no avail.

Secondary Search The secondary search is done to ensure that no stone is left unturned. It will cover the entire structure, both on the interior and exterior. It also will allow for the examination of all debris for the possibility of a victim. It is a good habit to rotate crews for the primary and secondary searches. This ensures that different teams search all areas.

Vent, Enter, Isolate, Search

A system to search rooms where there are suspected occupants can begin from the outside through a window. The steps in the system are to vent, enter, isolate, and search (VEIS). The window to the room will need to be vented by breaking out the glass and fully clearing the opening, to allow fast egress if necessary. This venting of the window opening will increase the air flow into the building and could provide additional oxygen to the fire. At the same time, the heat, smoke, and fire will be drawn to the window from the fire area, creating a flow path to the opening and to the firefighter. Since the firefighter will be operating at great risk, he or she must decide if an occupant of the room may still be alive, considering the interior conditions. Once the firefighter decides to enter the room, he or she must coordinate with interior crews so they will know his or her location. As soon as the room is entered, the firefighter must shut the door to the room to isolate the room from the rest of the building. This action will protect the firefighter and any occupants of the room by interrupting the flow path and reducing the oxygen flowing to the fire. Now the firefighter can perform a search of the room. If a victim is found, then the firefighter's company officer must be notified to decide on the safest method of removing the victim.

Exterior Search

The operation should not be considered complete until a thorough search is made of the exterior of the structure. (See Figure 4-10.) Incidents have occurred where fire debris thrown from a window covered individuals who had jumped from an upper floor, unbeknownst to firefighters.

Hose-Line Protection

The optimum search is made under the protection of a hose-line. Realistically, this does not always occur, due to the many areas that must be searched in as short a time as possible. And it often means that the search and rescue teams operate under extreme conditions. The importance of full protective equipment with a self-contained breathing apparatus is obvious.

FIGURE 4-10 A thorough search of the exterior of the fire building must be made. Occupants may have jumped prior to the arrival of the fire department and could be lying obscured on the ground behind shrubbery or debris. *Used with permission of Michael DeLuca.*

FIGURE 4-11 Steel bars on windows can trap occupants and firefighters. *Used with permission of James H. Bonner.*

If possible, each member of a search and rescue team should carry a hand tool, light, portable radio, and a TIC. When search and rescue crews are dependent on a hose-line to contain a fire to a specific area, they should make sure that the hose crew is aware of their location and their dependence on the hose-line. Coordination between hose-line crews and rescue teams is imperative. It is critical when a crew goes to an upper floor via an interior stairway before the fire is controlled on a lower floor. If for any reason the hose-line cannot contain the fire and a backup line has not arrived yet, or members are ordered out of the building, the rescue crews operating above must be notified immediately. This withdrawal of the hose-line can occur due to loss of water, building or floor collapse, or any number of reasons. When a search is completed and the search and rescue crew leaves the endangered area, the hose-line crew must be informed. This must be done when they exit where they entered, but especially if they leave by another exit.

Alternate Escape Routes

Constant thought must be given to an alternate escape route, no matter where the search and rescue team is located in the building. This is an individual thought process that must start with arrival on the scene. All outside features of the building should be memorized, including the number of stories and location of fire escapes. Different roof levels that would allow for easy exit from a window should be noted, as well as the presence of window air-conditioners or steel bars on windows that could prevent exit. (See Figure 4-11.) With the information gathered, firefighters should try to envision their approximate location within the building at all times.

 ON SCENE

I investigated a fire in a single family two-story row house that took the lives of two civilians and left a firefighter severely burned. My role in the investigation was to determine the fire department's actions at the incident and whether they conformed to accepted firefighting standards.

The firefighter who was injured had just climbed the interior stairs to the second floor and was in the hallway when heavy fire roared up the stairs. The amount of fire was unbearable and he was engulfed in flames. A Mayday was called and in response to the Mayday, a hose-line fought their way into the dwelling. While one firefighter was able to knock down the flames and protect the injured firefighter's location, another firefighter assisted him from the fire building.

Understand that flames impinging on your body will immediately change any rational thought process to just one thought: to find a way of protecting yourself. It may be by running down the stairs and out the front door through the flames, or jumping from a window, or curling up in a fetal position on the floor trying to keep below the flames and heat of the fire. A situation like this shows the importance of situational awareness and individual size-up, as discussed in Chapter 3. Had the firefighter sized up the situation, starting with his arrival at the incident, and been constantly thinking of a secondary means of egress should he find himself in trouble, he might have realized that the quickest avenue to his safety would be to open the door to the middle bedroom, which he was directly outside of, and to enter that room and close the door behind him.

As a firefighter operating within a burning building, if you have not given prior thought to what actions you will take at all times, it's likely that you will not be able to react in the safest manner possible should a life-threatening situation arise. The injured firefighter sustained severe burns but is still alive, due to the quick-thinking, brave, and aggressive actions of his fellow firefighters. Thankfully, he was able to recover and return to work after a year of treatments. The lesson for all firefighters is to have a mental picture of where you are in a fire building at all times. Know what you would do if you cannot exit the building the same way you entered. Consider your options should you find yourself in a life-threatening situation with only a split-second to react. The simple act of seeking refuge in a room and closing a door behind you may save your life.

Victim Removal

When victims are located, it is important to get them medical treatment as quickly as possible. First consider whether they can be removed through the interior of the building in relative safety. More people can be evacuated quickly and safely this way than by using ladders. Often it is easiest to remove someone who has been overcome and is seriously endangered by the fire by dragging him or her along the floor. This keeps them below the heat and smoke and is more expedient for the rescuer.

The use of ladders in performing rescues can cause problems. People already are in a state of shock. Young children have to be carried. The fear of an unknown person with a self-contained breathing apparatus may cause a struggle. If a victim is unconscious and suddenly regains consciousness while being carried down a ladder, he or she may react by struggling and possibly fall or cause the rescuer to fall. Fear of heights must be anticipated. People may be sick, incapacitated, handicapped, or even in a full body cast. The removal of the aged, or an obese individual who you realistically would not be able to carry down a ladder, can pose problems. In these types of cases, the person trapped would need to be protected in place until it is safe to proceed through the interior, and if required, use an oversized stretcher or backboard to carry them out of the building, or slide the stretcher along the floor to the exterior. Conditions need to be assessed for the safest and best method of protecting the individual.

With all the drawbacks, there are still many instances where rescue by ladder will have to be performed. When many rescues are required, ladder operations tend to be slow because they provide exit from only one window. In this case, the most endangered must be attended to first.

Ladder placement dictates that the head of the ladder be even with, or slightly below, the windowsill. This allows for ease of entry and exit, especially when removing an unconscious person. It also allows a firefighter who gets into serious trouble to make an immediate exit from the room. A firefighter caught in a flashover may attempt to leave the room in a headfirst dive through the window. His or her visibility will be impaired, and he or she will be starting the dive from the floor of the room. If the ladder occupies part of the window area, it will restrict the exit. (Diving out of windows certainly is not a recommended or suggested procedure. The firefighter should recognize the danger in advance of the flashover and leave.)

Latchkey Children

A door that is locked from the hallway side in a private residence must be checked. On more than one occasion, young children have been found locked in a bedroom after their parents left the house. At one incident, a fire started by one of two children in a locked bedroom nearly took their lives. Their parents locked them in the room every day while they both worked at full-time jobs. The neighbors told us that no one was home. A proper search and valiant efforts by paramedics saved the lives of the two youngsters.

Rescue Safety

Search and rescue should be performed in teams. Team members must rely on each other for their safety. Read the fire building. How much time is there to perform rescues, considering the amount of fire and smoke conditions? Is there imminent danger to the search team? Will the engine company be able to quickly control the fire? Does the company officer know where the search team will be operating within the fire building?

Critical Decision Making

The initial IC arriving on a fire scene where many people are endangered will have some critical decisions to make. He or she will have to decide where his or her efforts will do the most good, considering staffing, what additional resources are responding, and what their anticipated arrival time will be. These decisions may entail dedicating all personnel to rescues. Command may decide that a hose-line must be pulled to contain a fire to a specific area as a lifesaving tactic.

FIGURE 4-12 For defensive operations, apparatus should be set up at the corner of a building to keep it out of the collapse zone. This location will offer accessibility to two sides of the fire building and protect those operating at that location. *Used with permission of Joseph Hoffman.*

FIGURE 4-13 Overhead electrical wires can electrocute firefighters raising portable ladders or operating master streams. *Used with permission of Joseph Hoffman.*

There can be more than one right way. The conditions encountered, along with available resources, are part of the consideration. A large metro department that has a response of six or more units with predetermined duties and adequate staffing can more easily predict specific duties and handle multiple-rescue fires. A small department with a limited number of units responding and a minimum number of firefighters on each apparatus may need to use different strategies. The solution to handling large rescue assignments is resources and training in search and rescue procedures.

LADDERING

The front of the fire building should be reserved for the truck company. Depending upon the direction of response of the truck company, the first-due engine should either stop before reaching the front of the fire building or pull past it to permit access for the truck. This positioning will permit ready access of the main ladder or platform if needed. It also places the truck in a position for ready access of the portable ladders or tools that may be needed.

Trees can interfere with raising ladders. A main ladder can sometimes be maneuvered through openings in the tree branches to reach windows or roofs. The operator must check that overhead electrical wires are not running through the tree branches. The presence of wiring unseen by the firefighters operating or climbing the main ladder could energize the ladder and apparatus or shock a firefighter coming into contact with the wire. (See Figure 4-13.) Always operate under the premise that overhead electrical wires are hot at all times. In one incident, an apparent miscommunication between an electric company official and a fire officer on whether the power to the overhead electric lines had been de-energized resulted in a fire officer's electrocution.

 ON SCENE

A main ladder being raised for use as a ladder pipe in the front of a fire building touched an overhead wire and electrified the apparatus. The rapid intervention crew, which was assembled adjacent to the apparatus, received an electric shock, sending five members to the hospital. Fortunately, all recovered from their injuries.

 NIOSH FIREFIGHTER FATALITY REPORT F2005-07

On February 13, 2005, a 36-year-old male career captain (the victim) was electrocuted while working at the scene of a three-alarm residential structure fire. The captain was checking on one of his crew members when he walked under a tree and came in contact with a 12kv power line. The line had burned through early in the fire with one section landing on the ground to the south and the other lodged in a tree near the northwest corner of the fire building. It is believed the victim knew of the downed power line that had fallen to the south. However, it appeared to witnesses that he was unaware of the power line that was hanging in the tree, and possibly did not see the caution tape or hear the warning of a firefighter who was in the vicinity. He walked directly into the power line and collapsed to the ground. A nearby firefighter used an axe handle to secure and hold the power line off of the victim while firefighters pulled him away from the line to a safe area. Advanced life support was administered immediately by emergency medical personnel who were at the scene. The victim was transported to a local hospital where he was pronounced dead.

See Appendix B: NIOSH Reports in www.pearsonhighered.com/bradyresources to read the complete report and recommendations.

Portable Ladders

Truck companies carry a variety of portable ladders. In bygone days, the hook-and-ladder company initially marked the fire building by placing a ladder against it. This established that the company was staking a claim to the fire. There are certainly better uses of the portable ladder.

No one ground/portable ladder can best meet the many and complex needs of the fire service. Consequently, there are different sizes and types of ground ladders in use. They include straight, extension, extension with tormentor poles, pompier, roof, hook, closet, and folding ladders.

In addition to being used for rescuing trapped civilians and entry to and from a building, ladders can be utilized for other functions, including:

- To bridge openings between buildings
- To prop open overhead doors to fire buildings, ensuring a ready exit for firefighters
- As a battering ram to break out plate glass windows. The length of the ladder will keep firefighters a distance from the opening as the falling glass and pent-up heat is released.
- As a stretcher to remove someone injured from upper floors or below-grade locations by securing him or her to the ladder. (This can be especially helpful when dealing with an obese person who will not fit on a stretcher.)
- Placement over floor or roof areas that have been weakened or burnt through to allow firefighters to operate in a safer manner
- Placement over roof openings during overhaul to identify their location

Selecting the correct length and type of ladder comes through training and experience. The company officer should ensure that sufficient personnel are assigned to raise the ladder. Teamwork gained through practice will increase the company's proficiency.

Obstructions on the ground can limit ladder usage. Likewise, the terrain can impact operations if firefighters are confronted with sloping ground or hilly areas. Overhead impediments (e.g., signs, trees, electrical wiring) can also interfere with operations.

Basic Ladder Positions

The assignment to be accomplished determines where the top or head of the ladder should be located. (See Figure 4-14.)

Rescue   When a ladder is placed for rescue from a window, the top of the ladder must be even with the windowsill or slightly below. A ladder protruding into the window opening will impede the removal of occupants.

Entering a Window   A ladder placed for a firefighter to enter a window can be alongside the window and about three rungs above the windowsill, allowing easy access for the firefighter to climb onto the sill and enter the room. If the window is wide enough, the ladder can protrude into the window opening. (In either case, the opening must be cleared of broken glass or window parts with a tool. This makes a doorway out of the window opening and allows safe entry and exit.)

Roof Access   A ladder placed to gain access to a roof should extend at least three rungs above the roof. This permits firefighters to safely mount or dismount the ladder. It also allows those operating on the roof to locate the ladder.

To Play a Stream into a Window   A ladder placed to play a stream of water from a hose-line into a

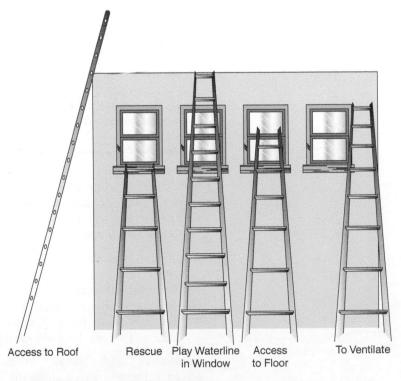

FIGURE 4-14 The basic ladder positions for access to a roof, to rescue, to play a line into a window, to access the building, or to ventilate a window. *Used with permission of Michael DeLuca.*

Access to Roof   Rescue   Play Waterline in Window   Access to Floor   To Ventilate

FIGURE 4-15 The correct positioning of the heel or base of a portable ladder is 75 degrees. This allows a firefighter to climb the ladder erect. *Used with permission of Michael DeLuca.*

window should have the head of the ladder resting on the wall above the opening. The firefighter can then be in a position on the ladder slightly below the center of the window while playing the stream into the building.

For Ventilation A ladder placed to ventilate a window should be placed on the windward side of the opening, slightly above the top of the window.

Placing/Raising

The correct positioning of the heel or base of the ladder will ensure that a proper climbing angle is achieved. The desired angle is 75 degrees. This allows firefighters to climb the ladder erect. (See Figure 4-15.) This angle can be determined by placing the heel or base of the ladder one-fourth of the distance of the height from where the ladder will rest against the building. When the base of the ladder is set too close to the building, the top of the ladder will tend to pull away from the building as the climber nears the head or top of the ladder. If the base is set too far from the building, the base will tend to creep or walk away from the building. Securing the ladder either at the top, bottom, or both ensures that the ladder will not be affected by movement of firefighters on the ladder or by high winds. The base of the ladder should take advantage of cracks in concrete or natural barriers to stabilize it. If a ladder is placed into soft earth, the heel of the ladder can sink, causing it to lean to one side.

Basic Rules When Using Ladders

Climbing a ladder improves with practice. There is an initial tendency to keep your body near the rungs. The proper method is to extend your arms and climb the ladder steadily on the balls of your feet, using your legs to do the climbing. Firefighters should develop a rhythm for smooth ladder climbing. By climbing in the center of the ladder, you will avoid a wobble in the ladder that could cause the heel of the ladder to walk or move away from the building. If a ladder becomes slippery because it is wet or icy, firefighters can improve their footing by placing their feet farther in on the ladder rungs as they climb or descend.

When dismounting a ladder onto a roof while carrying a tool, the tool should be used to locate and feel out the roof before dismounting the ladder. Then place the tool onto the roof to free both hands in order to safely climb onto the roof. Likewise, when mounting a ladder from the roof, place the tool where it can be retrieved after mounting the ladder to descend.

Some additional rules concerning ladders include:

- A firefighter working from a ladder should use a leg lock or a safety belt as a safety measure.
- Care should be taken when ventilating lower floor windows that the escaping gases and fire do not endanger firefighters operating on ladders above.
- Firefighters operating in or on a fire building should look for a secondary means of egress. This can be a fire escape, an adjoining roof, or a main or ground ladder.

Innovative Ladder Usage Fireground situations can arise that require firefighters to be innovative. People trapped above the reach of an aerial ladder will need to be rescued from the interior. If this is not possible due to fire conditions, an option is to lash a ground ladder to the top of the main ladder to extend its reach. (Remember that this is not a routine operation and should be discussed and performed during training in case it is needed at an incident.)

FIGURE 4-16 Trees and overhead electrical wires can interfere with placing and raising ladders.

Ladder Safety The most important factor to remember about ladders is to use them. There never seem to be enough ladders raised at a working fire in a multistory building. One consideration for ladder placement is to provide firefighters operating within or on a structure another means of egress, should conditions deteriorate. Main and portable ladders are an emergency exit for civilians trapped in a building and a secondary means of egress for firefighters.

Firefighters involved in roof operations should ensure that there are at least two independent means of egress from the roof:

1. Two ladders raised by the fire department
2. A combination of a raised ladder and a fire escape or a fire tower
3. A raised ladder and the roof of an adjoining building

Many factors affect safe ladder operations. Safety starts with choosing the right ladder for the assignment and having a sufficient number of firefighters to raise it.

As firefighters enter a building, they should note the location of adjacent roofs and raised ladders. When operating in a structure where a ladder was seen prior to entering the building, the firefighters should do a visual check to ensure that the ladder has not been relocated. Many fire departments paint the tips of their ladders with fluorescent paint to allow firefighters to easily locate the ladders. Firefighters should also make a mental note of the next closest means of egress.

We must ensure that there are no overhead obstructions to snag or interfere with raising the ladder. (See Figure 4-16.) Overhead electrical wires in contact with a metal ladder or a wooden or fiberglass ladder that is wet can electrically charge the ladder. Identifying the location of overhead wires and ensuring that a safe operating distance is kept from them is the responsibility of the firefighters raising the ladder.

 ON SCENE

At one incident, a portable metal ladder was energized on a drizzly night by electricity jumping from overhead wires a span of about four feet. Firefighters were removing the ladder from the building and bringing it to a full upright position in preparation for lowering it. The firefighters lost control of the ladder as the electricity was conducted through it, which threw the firefighters to the ground. Fortunately, as the ladder was electrified, it fell away from the wires, and severe injury was avoided. In other instances, firefighters have not been as fortunate, and electrocution has resulted.

FIGURE 4-17 Cutting roll-down overhead doors will allow access and assist in ventilation. *Used with permission of Joseph Hoffman.*

FORCIBLE ENTRY

With the exception of commercial properties that had sturdy locking systems, forcible entry in the past meant prying a basic door lock to gain entry. Today, forcible entry can be extremely challenging in both residential and commercial buildings. (See Figures 4-17 and 4-18.) Firefighters can be confronted with:

- Multiple locks
- Metal bars on windows
- Barbed wire
- Razor wire
- Guard dogs
- Roll-down doors
- Lexan windows
- Other sophisticated and unsophisticated security systems

In residential properties, forcible entry is usually made through the front door. This often allows access to every part of the building, and it is the way by which most occupants will attempt to escape the fire.

The first rule in forcible entry is to try to open the door. We can all tell horror stories about the door and door frame getting wrecked on a door that had been unlocked. By trying the front door, we can also give a push with our shoulders to see how much resistance or spring is in the door frame.

We should observe which way doors open. In residential buildings, they usually open inward. In public assemblies, exit doors should open outward. The hinges can ascertain the way a door opens. If they are visible, then the door will open toward you; otherwise, the door will open away from you.

Most living areas have doors that open inward, as in bedrooms and hotel rooms. Interior doors that open outward usually contain closets in dwellings and service areas or utility closets in commercial buildings. Unless fire is suspected behind an interior door that opens outward, it should not be forced open in the initial stage of search and rescue. It is a waste of the firefighters' time and effort. When forcing a door that opens inward, it may need to be shut quickly. A small utility rope slipped over the doorknob prior to forcible entry can accomplish this.

FIGURE 4-18 A firefighter cutting the hinges on a steel door to gain access. *Used with permission of Joseph Hoffman.*

At a fire that I encountered in a high burglary area, all visible means to gain entry into a burning basement were sealed off. The outside metal doors placed in the sidewalk had concrete poured into the opening. A search of the smoke-filled first floor revealed no apparent means of entry. As a last resort, I ordered a hole cut in the wood flooring to gain access. The underside of the floor joist had also been filled with concrete, thwarting our efforts. The fire burning unabated in the basement entered the walls and broke through the roof, destroying the building.

Forcible Entry Safety

When using hand tools for forcible entry, a clear operating zone should be established. A firefighter should be assigned to maintain this clear area or safety zone. This is important when operating power saws. As one firefighter operates the saw, another should be behind that firefighter, guiding him or her and keeping a clear safety zone. Safety dictates eye protection and gloves for forcible entry operations.

VENTILATION

To a firefighter, the term *ventilation* means the removal of heat and smoke from a fire building. Horizontal and vertical ventilation prevents or stops the mushroom effect, which occurs when heat and smoke reach the highest level within a fire area or building and start banking downward. (See Figure 4-19.)

Ventilation creates flow paths through which heated gases flow away from the high pressure area of the fire toward the low pressure areas created by door or window openings. As the heated gases move away from the fire to the lower pressure areas, additional oxygen moves in from the door or window openings to the higher pressure fire area, providing oxygen and increasing the burning rate of the fire. A door opening that lets air reach the fire through a flow path will increase the fire and allow the heat and smoke to flow back towards that same open door. As fresh air enters from the bottom half of the door opening, the products of combustion will vent through the top of the door opening. Dependent upon the number of openings in a building, there may be several flow paths occurring within a structure.

The practices of limiting ventilation openings and using door control allow firefighters to control the flow path on the other side of the door. Limited openings limits fire development by controlling the lower flow path of fresh air from moving towards the seat of the fire. The less air, the less fire growth, and the less fire growth, the less water needed to control and extinguish the fire.

FIGURE 4-19 The smoke and heat from a fire will mushroom at the highest level if ventilation is not performed. *Used with permission of Michael DeLuca.*

Ventilation is an action at a working fire that, when performed properly, can have an immediate positive effect. It is initially used to remove the heated smoke and gases, which allows for the rescue of trapped occupants and enables firefighters to quickly locate and approach the fire with hose-lines for confinement and extinguishment. It can also prevent additional damage or further spread of fire. Ventilation improves interior conditions, creates better visibility for search and rescue operations, and can prevent smoke explosion or backdraft situations.

The UL/NIST studies show that an initial ventilation opening in dwellings may be used by a hose-line to hit the fire in an offensive-exterior attack to knock down the fire, allowing entry of the interior hose-line crew to the fire area to complete extinguishment.

Ventilation is a must for structural firefighting. At the incident scene, specific ventilation assignments must be given. Ventilation must be controlled by the crews operating on the interior. Exterior ventilation crews must receive the go-ahead from the interior crews prior to creating any ventilation openings. Addressing ventilation as an afterthought leads to poor fireground operations.

If no civilians are endangered, ventilation should be done after a charged hose-line is laid. This will permit a quick knockdown and minimize damage. Venting too early can cause a fire to spread, quickly involving a much greater area. Once the ventilation opening(s) are created, an immediate attack on the fire should commence.

Ventilation can be as basic as firefighters opening doors and windows to allow free circulation of outside air. This is the preferred method when smoke conditions are light and no life is threatened. Damage is minimized, and the fire is quickly extinguished.

Firefighters must consider whether the breaking of windows will assist in the overall operation or just cause unnecessary damage. Fear of causing excessive damage must be weighed against dangers encountered if ventilation is inadequate. Too many ventilation openings will supply large amounts of oxygen to a fire and may cause it to develop into a large fire that is past the control efforts of the firefighters.

Properly performed ventilation enables firefighters to quickly and efficiently extinguish a fire while minimizing heat and smoke damage in other areas of the structure. By opening above a stairwell, the stairs can be a viable means of entry and egress for occupants and firefighters. Without ventilation ahead of an advancing hose-line, heat and steam will prevent the firefighter operating the nozzle from being able to advance to the seat of the fire.

The Sound of Breaking Glass

Since time is a critical factor, window glass is often broken from the exterior, and then the opening is cleared of obstructions. Windows used by firefighters to enter the fire building should be trimmed completely prior to entering. This enables easy access and egress without injury or delay.

A firefighter with a hand tool can clear a window opening as quickly as he or she can gain access to it from either ground level or a ladder. He or she should ensure that no one who could be struck with the falling glass is beneath the window to be ventilated. The firefighter with full protective equipment should be positioned on the windward side of the opening, breaking the top panel first and then the lower panels, to prevent being exposed to the heat of the fire. Additional protection for the firefighter can be achieved by striking downward with the tool to break the glass. Striking the glass downward prevents the glass from sliding down the handle of the tool, injuring the firefighter performing ventilation. (See Figure 4-20.)

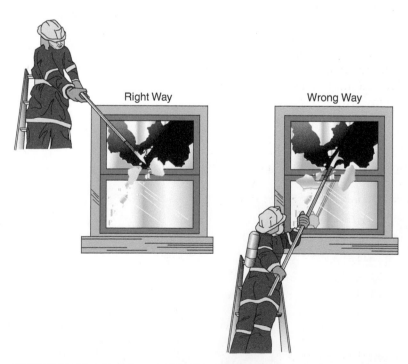

Right Way

Wrong Way

FIGURE 4-20 A firefighter should effect ventilation by striking downward to break a window. This will prevent injury to the firefighter from glass sliding down the tool. *Used with permission of Michael DeLuca.*

Windows may be broken from the interior of the structure without forewarning those on the exterior. Firefighters must be taught that when they hear the sound of breaking glass, they should slope their shoulders downward to allow any glass striking them to fall away from their body, reducing the risk of injury. They must not look toward the sound and should move away from the area if not involved in the ventilation operation.

On multistoried flat roof structures, one firefighter on the roof with the proper tool can reach over and quickly effect ventilation on a number of windows. (See Figure 4-21.) This up and over method can be used with town houses where restricted entry to the rear could cause a delay in rear ventilation. The first-arriving truck company or assigned engine company can raise a ladder

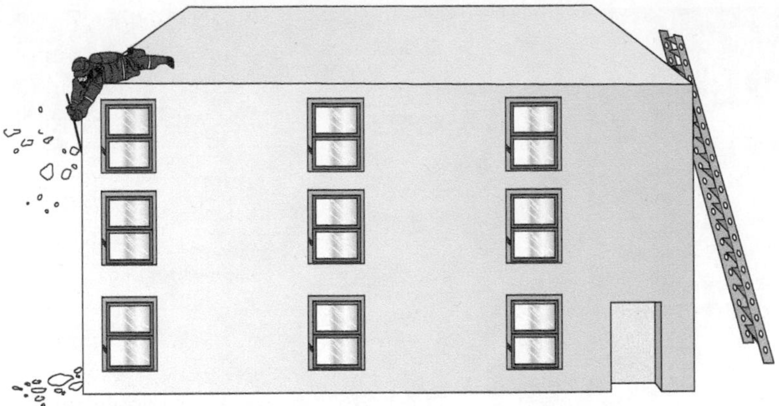

FIGURE 4-21 On multistoried flat roof structures, one firefighter on the roof with the proper tool can reach over and quickly effect ventilation on a number of windows. *Used with permission of Michael DeLuca.*

to the roof of the fire building or adjacent structure via the front of the building. The breaking of multiple windows should only be made in conjunction with units operating on the interior.

An unconventional method of window ventilation can be achieved by dropping a portable ladder into the window. The ladder is raised so that the head of the ladder is just below the top of the window. The ladder is then pushed or dropped into the window, pushing the glass into the building and allowing ventilation. If the head of the ladder is too low when dropped into the window, the higher glass can slide downward on the rails of the ladder, causing a potential injury to the firefighters. This method of ladder usage could damage the ladder and must be weighed against the immediate need for ventilation when a shortage of personnel exists or when there is no possibility of accomplishing the ventilation in another manner.

A main ladder can be used similarly when ventilation is required above the third floor. The head of the ladder can be extended through the glass, forcing it into the building. When utilizing this method above the third floor, the operator must ensure a safety zone beneath the windows being broken. Glass falling from these higher heights can cause serious injuries.

A drawback to these methods, in addition to possibly causing damage to the ladder, is that the quality of ventilation will depend on whether obstructions still exist that would require a firefighter to clean out the opening further. Caution must be exercised because glass can ride the beams of the main or ground ladder downward, striking the firefighters handling or operating the ladder.

Firefighters can vent the upper areas of a building by utilizing skylights or scuttles that are built into a roof structure above the staircase. Open stairways communicate heat and smoke throughout a structure, restricting evacuation of the occupants and entry by the firefighters.

Skylights should be removed intact since falling glass can injure firefighters operating below. Being struck by glass from a skylight that has fallen a few stories while operating on or climbing stairs can cause serious injury. If unable to remove the skylight completely, it should be broken and pulled back onto the roof. Its presence will act as a warning to other firefighters of openings in the roof.

Skylights vent only living spaces. If fire has entered the cockloft or attic, or if these locations need to be checked, it will be necessary to open the boxed-off areas around the skylight.

Opening the Roof

Roof ventilation should be in conjunction with horizontal ventilation on the upper floors. The opening of a roof directly over the fire area vents the fire to the exterior and, along with other flow paths, will increase the fire's intensity. The opening of the roof away from the fire will draw the fire to that opening. Improper ventilation can

FIGURE 4-22 Fire has caused the roof of the fire building to collapse. Ventilation holes have been placed in the adjoining properties for ventilation and to check their cocklofts for fire extension. *Used with permission of William J. Shouldis.*

endanger civilians and hose-line crews by pulling the fire onto them, as well as causing the fire to spread. (See Figures 4-22, 4-23, and 4-24.)

The experience of the firefighters assigned to the roof is important. They must draw on their knowledge and expertise to perform their duties. They must ascertain whether the roof is safe and be prepared to abandon roof operations if unsafe conditions develop. When a roof is no longer safe, it can affect the overall operation. The IC must decide whether the mode of operation should be changed from an offensive to a defensive attack.

New firefighters should not normally be the first to ascend to the roof. Their concentration may be totally consumed with venting the roof, causing them to miss danger signs. Constant size-up is needed to detect changing conditions. Monitoring radio communications of interior operations and providing the IC with information is also required.

Firefighters operating on roofs should work in pairs. They should sound out or probe the roof with a tool to ensure stability and have two remote ways of egress from the roof.

When ordered to open the roof, firefighters must determine an appropriate location. Observation of the building and the location of the fire prior to reaching the roof are important. Visual indicators on the roof can assist in determining where to place the vent hole. Fire burning through the roof, bubbling tar, melted snow on a snow-covered roof, or dry spots on a wet roof can assist in locating the fire below.

The wind should be at the firefighters' backs and they should make one large hole, not a few small holes. A ventilation hole should be 4 feet by 4 feet on a small roof and at least 8 feet by 12 feet on a large roof. Naturally, this is dependent on the size of the structure itself. Openings should end at the roof rafters. Rafters can be used as a guide for a straight cut and to ensure stability for the firefighters operating around the opening. After all the saw cuts are completed, the roofing material should be

FIGURE 4-23 Proper roof ventilation directly over a fire can pull smoke and flames away from trapped occupants. *Used with permission of Michael DeLuca.*

FIGURE 4-24 Improper ventilation can endanger civilians and hose-line crews by pulling the fire onto them, as well as causing the fire to spread. *Used with permission of Michael DeLuca.*

pulled back onto the roof, clearing the opening. A tool should then be used to probe the opening, pushing down any ceiling or obstructions that could restrict ventilation. (See Figure 4-25.)

Be prepared to encounter more than one ceiling. It is virtually impossible to see into the opening due to the heat and smoke being vented, thus your tool must be your eyes. If a second ceiling is encountered, push it down and probe for the possibility of a third ceiling. Experience teaches us when a vent hole is effective by the amount of smoke and

FIGURE 4-25 Roof openings should end at the roof rafters. After opening the roof, a tool should be used to probe the opening, pushing down any ceiling obstructions that could restrict ventilation. *Used with permission of Michael DeLuca.*

FIGURE 4-26 A ventilation cut in the peak of the roof allows the heat of the fire to be pulled to the exterior, permitting an interior attack to succeed. *Used with permission of Brian Feeney.*

gases being expelled. The units operating inside the structure should be notified when the opening is completed. They will be able to advance their lines into the area using the opening to vent their approach. (See Figure 4-26.)

Rooftop Solar Panels

The movement toward green energy includes the installation of thermal and photovoltaic (PV) rooftop systems for capturing sunlight. The thermal systems utilize sunlight for heating of liquids to use for hot tap water, hot water heat, or swimming pools, while PV systems turn it into electricity.

Firefighters need to identify the systems that are present on a structure. The greatest hazards are with PV systems, since they are always on during daylight and cannot be shut off. (See Figure 4-27.)

Tripping and slipping are the most common types of hazards for firefighters on rooftops with solar panels. Firefighters will need to consider the additional weight imposed on the roof by the solar panels, especially if fire has attacked the roof supports and weakened them. The presence of PV panels can impair rooftop ventilation tactics by preventing the roof from being opened for ventilation at the optimum location. Realize that the panels should never be damaged or compromised to perform vertical ventilation. Given the inherent electrical hazard, do not break, remove, or walk on PV modules.

 **ON SCENE**

A fire occurred in September 2013 in Burlington, New Jersey, in a 300,000-square-foot warehouse that had 7,000 solar panels on its roof. The presence of the solar panels caused the fire chief to keep his firefighters from going onto the roof of the burning building. The building was totally destroyed.

Another hazard is the potential flame spreading characteristics of the PV modules, for example in an exposed building fire or an approaching wildland fire. The panel materials do not necessarily have good fire-resistant characteristics, and they can produce toxic fumes when burning. Should PV panels become involved in fire, treat them as you would any energized electrical equipment: electric shock possibilities are constant. Though most other types of energized electrical equipment can be shut down, a solar panel exposed to sunlight is always on and energized. This can create problems for firefighters when shutting down the electricity to a structure; the panels can be a dangerous source of electric shock or may initiate a rekindle. Also, the possibility of residual electricity could cause electric shock. *Firefighters should at all times treat these systems as though they are energized.* This means that conduit or other components can also be energized.

PV modules, if covered with 100-percent light-blocking materials, will stop electrical generation. Yet, if using salvage covers, they may still permit some sunlight through and allow generation of electricity.

FIGURE 4-27 Firefighters should treat rooftop solar panels as though they are energized.

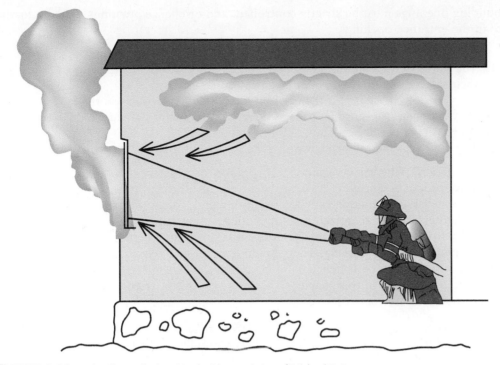

FIGURE 4-28 Hydraulic ventilation. *Used with permission of Michael DeLuca.*

Moonlight is reflected light and will not energize the panels, but lightning may generate a temporary surge. Incident scene-mounted lighting does not produce enough light to generate an electrical hazard in the PV system, but PV systems may contain an optional battery storage system for nighttime use, which could present an electrical shock hazard.

The IC should request assistance from the local electric utility at any incident where PV panels are present and the power needs to be shut down. See References at the end of this chapter to access UL's reports on fire service safety issues with PV systems.

Private Dwellings

Fires in private dwellings don't normally require the opening of roofs unless the fire has heavily involved the cockloft or attic. These areas can usually be checked from below by opening ceiling areas and extinguishing minor extension of fire with a hose-line. This minimizes structural damage and is easier for the firefighters.

Means of Ventilation

In addition to the conventional ventilation achieved by opening windows, doors, and roofs, mechanical means of ventilation can be utilized. Hydraulic ventilation is achieved by setting hose-lines with a nozzle on a fog pattern that covers about 90 percent of the window or door opening through which the smoke will be vented. The firefighter operating the nozzle should keep low to the floor and direct the stream to the outside through the opening. The action of the fog pattern will draw the smoke to the exterior. (See Figure 4-28.)

Negative pressure ventilation is accomplished by placing exhaust fans into windows or doorways and pulling the smoke or toxic fumes from a building. This is often used in the overhauling stages of an incident.

Positive pressure ventilation is a method that involves the placement of fans to pressurize the interior of the structure and remove the products of combustion by driving fresh air into a building and forcing out the smoke via openings made by the fire department. The size of the fans used must be in excess of 10,000 cubic feet per minute and often range to 20,000 cubic feet per minute. The fans can be stacked or placed one behind the other to increase the pressure. The air from the fan should encompass the entire door opening as it blows into the building. It should be set so that it does not interfere with entry and exit from the building. To be successful, building openings must be strictly controlled, and discharge openings must be smaller than the entrance openings.

Before operating a positive pressure fan, you should know:

1. The status of the building's occupants because operating the fan prior to ensuring their safety could place them in jeopardy.
2. The location of the fire. If the fire is in the cockloft or attic area, the fan can drastically accelerate the burning process there, causing major damage to the building.
3. Where the fans will be set up.
4. The door or window through which the smoke and fire will be vented.

Ventilation, whether conventional, hydraulic, or negative or positive pressure, is a necessary tool at fire scenes. Prior to the advent of positive pressure ventilation, many departments gave ventilation only a passing thought. Today they see the direct benefits of ventilation. Realize that each type of ventilation has benefits and drawbacks. Before a fire department implements any type of ventilation at an emergency scene, they should fully explore it during training. They should learn to recognize situations where a positive pressure fan can be beneficial and circumstances where it may be detrimental. See References at the end of this chapter to access a synopsis of research conducted by NIST on positive pressure ventilation.

Trench Cut Large structures with flat roofs that contain non–fire-stopped cocklofts can pose difficult control and extinguishment problems. These can include attached structures, such as those found in row house construction, town houses, strip malls, garden apartments, condominiums, or large apartment buildings. One tactic that can be employed to assist in the control effort is to contain the fire to one section or area of the cockloft. This can sometimes be achieved by using a trench cut.

Reports from roof sectors or radio reports from the interior can confirm heavy fire conditions in the cockloft as ceilings are opened from below. If this large body of fire cannot be controlled from the interior, then a trench cut can be utilized. The trench cut can cut off the spread and confine the fire to one part of a roof while protecting the unburned portions.

The number one rule when considering a trench cut is that it should not be attempted until an initial ventilation hole of sufficient size is placed directly above the fire area. (See Figure 4-29.) The trench should be located in the expected path of fire travel. This ventilation opening, along with the fire attack that is occurring on the interior, may be sufficient to control and extinguish the fire.

Locating the Trench The trench should be located in the expected path of fire travel. The location of the trench and the distance it is placed from the ventilation hole depend on a number of factors:

- The location of the fire.
- The expected speed of fire travel through the cockloft. There is a tendency to underestimate the speed of fire travel and place the trench too close to the vent hole, which then necessitates abandoning it as it is overrun by fire, and having to start another trench.
- The length the trench will cover. On some buildings, selecting a location where the roof narrows or where there are barriers already in place in the roof, such as elevator shafts, light shafts, or bulkheads, will reduce the overall size of the trench and shorten the length of time needed to cut the trench. On multistoried residential buildings, the configuration of the roof may contain some narrow areas that can be fully utilized as an excellent location for a trench cut. This is especially true on buildings in the shape of a T, E, F, or similar designs.
- The available staffing and saws needed to complete the task. This will be a labor- and equipment-intensive operation, in which the amount of time needed to accomplish the tasks is often underrated.
- The roof surface must be considered. An older building with many layers of roofing material will require a considerably longer time to cut the trench, thus slowing down the operation.

FIGURE 4-29 Before a trench cut is attempted, firefighters should first place a ventilation hole of sufficient size directly over the fire area. Openings or observation holes in the form of triangular cuts or kerf cuts should be placed between the ventilation hole and the location where the trench will be cut. *Used with permission of Michael DeLuca.*

These factors will also dictate the time it will take to cut the trench.

Multiple Trench Cuts   If the unburned roof area that is threatened is large, placing a trench cut on both sides of the fire should be considered. A major factor to consider is the time and availability of saws and personnel to complete both tasks. If more than one trench is to be cut, you must decide which trench takes priority, and sufficient resources should be given to accomplish that objective. Secondary consideration should be given to creating the second trench. A basic concern is placing the first trench where it will protect the greatest amount of property. This should take into account the location of fire walls, the size of the area threatened, and the potential for extension of fire to exposed areas.

Observation Holes   If a trench cut is being considered, openings or observation holes should be placed between the initial ventilation hole and the location near where the trench will be placed. If fire has already reached that location, you will have to decide whether there is sufficient time to cut the trench, or another location should be selected. The location of the observation openings should allow firefighters sufficient time to fully open the entire trench, even when fire is observed at these openings. Observation openings can be any shape. The most common are the *kerf cut*, which is a carpentry term for the width or thickness of a single cut of a saw blade, and the *triangular cut*, which typically is three cuts intended to remove a triangular piece of the roofing material.

Cutting the Trench   The cutting of a trench is the next step after the initial ventilation opening and the observation holes are completed. The trench consists of two parallel cuts approximately 30–36 inches apart. The distance must not be so narrow that the fire could jump it, nor so wide that a firefighter could not step over it. This is especially important; firefighters should not step onto the cut portions because they are unsupported. Perpendicular cuts made between the parallel lines approximately every three feet will facilitate the opening of the trench when ordered.

BASIC TRENCH CUT RULES

1. Place a ventilation hole of sufficient size directly over the fire area.
2. Decide on the location of the trench cut(s).
3. If multiple trenches are needed, prioritize which cut should be made first by deciding on the most critical side of the fire area to be protected.
4. Place observation holes in the direction of the anticipated fire travel to warn of impending extension of fire toward the trench.
5. Cut the trench 30 to 36 inches wide for the entire length with perpendicular cuts every 36 inches.
6. Once fire threatens to involve the area of the trench, have sufficient personnel available to quickly open the trench.
7. Once the trench is opened, push down the ceiling(s) from the roof with assistance from interior units to prevent fire from overrunning the trench.
8. Stretch hose-lines for use above and below the roof.

Opening the Trench After the trench is cut, the observation holes must be monitored. If the crews operating on the interior are able to extinguish the fire, it will not be necessary to open the trench. Once fire is showing through the observation openings, then the order to open the trench must be given. If fire involves the area where the trench is being cut prior to its completion, another location will need to be selected and another trench started.

Opening the trench will require a sufficient amount of personnel to accomplish the task quickly and ensure that the fire does not overrun the trench before it can be opened. Fire cannot be allowed to pass the trench opening. The opening should be cleaned out, and material removed to prevent ignition from the heat of the fire. As soon as the trench has been opened, the ceiling or ceilings below must be pushed down to create an upward draft and draw the products of combustion to the atmosphere above the roof through the original ventilation opening. Pushing down the ceiling also removes material that could feed the advancing cockloft fire. Units operating on the interior can assist in the removal of the ceiling.

Once the trench has been opened, the fire side of the roof must be written off as a defensive area. Firefighters must be restricted to the unburned side of the roof.

Communications between firefighters on the roof and in the interior are important. Opening the trench causes increased ventilation, which can escalate the fire. Lack of communication could place those operating on the interior in a dangerous location where the fire could be pulled onto them. Through good communication, interior firefighters can predict the path of fire travel and place units and hose-lines to confine the fire.

Once opened, the trench location must be monitored from above and below. A hose-line should be kept at both locations. The hose-line below the cut is usually more effective in preventing the fire from jumping across the trench.

Conditions will vary with regard to airflow. Ideally, the air from below will vent from the ceiling opening below the trench and into the cockloft, and then up through the initial vent hole. This airflow will deter the cockloft fire from spreading toward the trench. (See Figure 4-30.)

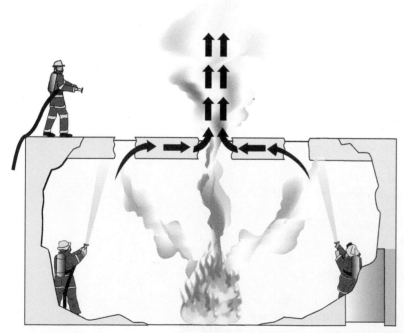

FIGURE 4-30 Ideally, the air from the fire area will vent from the ceiling opening below the trench and into the cockloft, and then up through the initial ventilation opening. *Used with permission of Michael DeLuca.*

Hose-Line Operations on a Roof

Ventilation openings in a roof are not placed for the use of hose-lines. Though a hose-line should be brought to the roof, its purpose is:

- To protect surrounding exposures that may be endangered by roof ventilation
- To protect the firefighters operating on the roof
- To protect firefighters' escape route
- To prevent flying embers
- To extinguish any roof fire that may occur during ventilation

Directing the line into any roof opening not only defeats the purpose of the vent hole, it will drive the heat of the fire downward, endangering the firefighters below.

Fire burning through a roof that is too weak to support firefighters should be allowed to burn through, creating an opening. The tendency to hit the fire with ladder pipes or other streams must be resisted. The self-venting that is created will be beneficial to the overall control efforts.

Ventilation is also defeated if hose-lines are placed into horizontal ventilation openings. An example would be a fire in a one-story store in a strip mall where the rear door is opened for ventilation. This ventilation would allow a hose-line crew to advance from the front door and through the store, extinguishing fire as it made its way to the rear. The heat and smoke would be pushed ahead of the hose-line and vented out the rear door. If a second hose-line entered the store through this rear door and attempted to extinguish the fire, it would negate the ventilation. Both crews would take unnecessary punishment by driving the fire at each other and would fail to be effective. As noted by the UL/NIST testing results (Chapter 3), the use of an offensive-exterior attack has proven beneficial in controlling dwelling fires.

Power Saw Usage   Before taking a saw to a roof, a firefighter should check that the proper blade is in the saw for cutting the roof. The saw should first be started when it is taken off the apparatus. After ensuring that it starts, it can then be turned off and taken to the roof. Never climb a ladder with a running saw. If the saw does not start, get another saw to do the job. Before starting the saw, make sure there is a safe operating area. The firefighter operating the saw must ensure his or her own personal safety and location on the roof. When not in use, the saw should be shut off and set in a safe location. Do not put the saw down while the blade is still rotating.

Know your equipment. Start the saw into the cut and work methodically. Don't attempt to rush the cut. It will only bind the saw. When a saw blade is bound in a cut, usually the cut has to be spread to allow the saw blade to be freed. Knowledge beforehand of the saw's limitations allows for a better fireground operation.

Know what kind of roof surface and roof supports you will be operating on. Is it solid beam or lightweight construction? The firefighter operating the saw must have a feel for the operation. He or she should not cut through roof supports. When a drag on the saw is felt, indicating a roof-supporting member, the saw should be allowed to ride over the supporting member, cutting only the roofing material and leaving the structural member intact.

Ventilation Safety   Firefighters should use a tool to sound out the roof prior to stepping off a ladder onto the roof. This allows the firefighter to find the roof surface and determine that it is sound. Finding the roof surface is usually routine, but in the case of a high parapet wall or an elevated cornice, the surface of the roof can at times be up to eight feet below the top of the cornice. A smoky nighttime fire with reduced visibility could injure an unsuspecting firefighter stepping from the cornice only to find no roof immediately underfoot. By sounding out or probing the roof with a tool, a firefighter can locate the roof surface and ensure stability. When encountering heavy smoke conditions while operating on a roof, probing with a tool can check for any weakened roof areas and any light or air shafts. Under heavy smoke conditions, it may be necessary to attach a guideline to the head of the ladder to allow a quick return to its location when evacuating the roof.

A dangerous situation can develop if skylights have been covered by tar paper. This is an economical way of repairing a leaking skylight; instead of placing wood over the opening, the area is tar-papered to seal it against water. This trick is also done in high crime areas to deter burglars. An unsuspecting firefighter stepping onto this area will fall through both the tar paper and the glass. There have been occurrences where elevator shafts have been similarly covered over. In one case, a firefighter fell four floors down a shaft onto the top of an elevator, causing career-ending injuries. Firefighters should avoid walking on any raised surface areas of the roof without sounding out those surfaces with a tool to ensure their stability. Openings in roofs should be marked to prevent firefighters from falling into them.

OVERHAUL

Overhaul involves checking a fire scene to determine that no fire remains. (See Figure 4-31.) A close examination should be made to ensure that every location where hidden fire could still be burning is thoroughly searched.

A rekindled fire is often attributed to poor overhaul practices. Many chiefs expect truck officers to check routinely for extension of fire in both the original fire building and exposed structures. This requires experience on the part of the firefighters, to recognize the signs of hidden fire and to open suspected areas.

Hidden Fire

After a fire is under control, suspected areas must be examined for hidden fire. The fire officer must determine:

1. Whether to open a specific area
2. Where to make the opening in a wall or ceiling
3. Whether a small opening or the entire wall or ceiling must be opened

FIGURE 4-31 Overhaul is the checking of a fire scene to determine that no fire remains. *Used with permission of Joseph Hoffman.*

If an officer suspects fire in a hidden location, a TIC, if available, should be used to check the area. If it is inconclusive whether fire exists in the hidden location, the order should be given to open the area. (See Figure 4-32.) Hose-lines should be stretched to locations that need to be opened. One hose-line can protect adjoining areas being investigated. Overhaul operations can occur in the fire building and exposed structures. Consideration must be given to the possibility of the spread of fire via horizontal and vertical openings. Signs of fire extending to concealed spaces can be detected through indicators:

- Discolored paint or wall coverings
- Smoke oozing or pushing from around baseboards or door/window moldings
- Continued smoke generation after all visible fire has been knocked down
- No abatement of heat conditions, though the fire appears to be extinguished

Areas of hidden fire can sometimes be detected by listening for crackling sounds of fire still burning or by touching walls or floors to feel for hot spots. If not found, a hidden fire can smolder for a long period, with the potential for developing into a serious problem. Firefighters could leave the scene, only to have to return to extinguish the rekindled fire. The primary causes of rekindles are incendiary fires and careless firefighting.

An excellent method to find hot spots and hidden fire is through the use of thermal imaging technology. This includes both handheld and helmet-mounted devices. This tool can reduce damage and create less work for the firefighters while ensuring that suspected areas are opened for closer examination.

Structural Stability

A concern of the IC during every phase of firefighting is whether the building is stable enough to continue interior operations. This ongoing size-up must continue during overhaul.

Structural damage caused by the fire must be thoroughly investigated to ensure that firefighters can operate safely in the building. Are collapse indicators present, such as wall cracks or walls out of alignment? Some floors in older frame buildings may have sagged.

Are they safe to operate on? Are there water tanks or air conditioner units on a fire-weakened roof that pose a threat to firefighters operating beneath them?

What are the building's contents? The weight of merchandise can affect the building's stability. Is there a heavy load of stock or machinery that could affect the structural integrity of the building?

A fire-weakened building should not be subjected to jarring impacts from axes. Saws should be used to open floors and roofs to minimize shocking jolts to the building that could create a collapse situation. (Chapter 6 discusses collapse indicators.)

Structural Overhauling

All buildings require thorough examination. The type of building construction can help determine what overhauling may be needed.

The fire-resistive building usually requires minimal overhaul. The investigation should concentrate on checking shafts and poke-through construction. The components of a noncombustible building will not add fuel to a fire, but an investigation must be made to determine if any structural damage has occurred.

Frame and ordinary building construction have components that will contribute fuel to the fire. Flooring may need to be opened to expose hidden fire. There can be many concealed spaces that need to be checked. The heavy timber-constructed building usually withstands the effects of fire or is totally destroyed.

A building struck by lightning may require extensive overhauling. The intense heat of the lightning can be conducted throughout the building by the piping and wiring, starting many smoldering fires in the inner cavities of the structure. It will require a time-consuming investigation that can be facilitated through the use of a TIC.

Parts of a building damaged by the fire may create a dangerous situation and need to be removed, such as a wall or roof sign, a window or roof air conditioner unit, or a hanging cornice.

FIGURE 4-32 If a fire officer suspects fire in hidden locations, the order should be given to open the suspected areas. This suspended ceiling was opened to reveal a tin ceiling above it. The tin ceiling was opened to reveal the hidden fire.

Minimizing Overhaul Damage After the emergency has stabilized, firefighters should consider how a location that has been opened to check for hidden fire will be repaired. An example would be how best to check an attic for fire extension. Ripping off roof shingles and placing a hole in the roof would provide access but would be expensive and time-consuming to repair. A better method would be to investigate the area from the interior. A trapdoor may exist through which a ladder can be raised into the attic, or the ceiling can be opened from below. When opening floors to check for fire extension, the cuts should be made close to floor joists, without cutting through the joists. This permits firefighters to inspect these areas, while allowing repairs to be made easily.

Removing Debris A window opening can be cleared out to allow firefighters to remove burnt debris to the exterior of the building. Windows can be removed undamaged and set aside for reinstallation at a later date. If damaged and not repairable, they can be removed and discarded.

Contents Overhaul

Care during overhaul operations can make a difference. Articles on the tops of desks or bureaus can be gently swept into the top drawers by a firefighter.

Smoldering furniture and burnt clothing and mattresses should be removed and thoroughly extinguished outside. A smoldering bureau should be checked to see if the contents are on fire. If not, the drawers should be removed and placed in a safe location so that the contents can be salvaged.

Do not make the mistake of thinking that a mattress or overstuffed furniture that has been damaged by fire can remain inside. It is of no use to the occupants and will have to be discarded at a later date, but most important, it can contain a smoldering fire that could erupt after firefighters have left the scene. Remove it to the exterior, pull apart all burnt areas, and wet it down thoroughly. Be careful if these items are taken through the building and out a doorway, rather than out a window. On many an occasion, they have lit up after being stuck in a doorway or when being carried down a staircase, seriously endangering firefighters.

Smoldering debris can be shoveled into debris cans and, together with fire-damaged items, deposited in one exterior location. The removal of smoldering items minimizes additional smoke damage and permits a safer working environment for the firefighters to complete their overhaul and salvage operations. These items can be thoroughly extinguished on the exterior, minimizing water damage to the fire building.

Exterior Overhauling

As debris is removed to the exterior, it should be placed where it will not:

1. Threaten the structure if the fire flares up
2. Damage shrubbery
3. Impede entry and egress from the fire building
4. Be handled again

When removing articles from upper floors via a window, the area below must be cordoned off to prevent firefighter injury. This can be accomplished by having a spotter on the ground to direct the firefighters operating above.

Water Damage

The best way to minimize water damage is strict control of all hose-lines. This is easily accomplished on a minor fire, but difficult when multiple streams are operating at different locations within a building.

The basic rule when operating a hose-line is to open the nozzle when fire is encountered and close it after the fire has been knocked down. The exception is a fog stream used to ventilate an area through a window or door opening. This ventilation can cause a smoldering fire to ignite. Firefighters can extinguish the fire with the backup hose-line, or they can redirect the nozzle being used for ventilation to quickly extinguish the fire.

Water Accumulation

Basements should be checked for any accumulation of water. If such an accumulation is found, pumps should be used to remove it. Stock on the floor can be elevated to minimize damage.

Water accumulated on floors or roofs can create a dangerous situation. Prompt action should be taken to relieve water buildup. An immediate size-up can determine the best method of removal. Are there floor drains or scuppers to remove the accumulated water? An inspection may reveal that the drains of these systems may only need to be cleared of debris to facilitate water removal.

We do not want a problem of water accumulation on the floor of a building to be solved by filling the basement with water. When removing water from a multistoried building, start on the first floor and work upward to prevent a buildup of water on the lower floors.

Overhaul Tools

A variety of tools can be utilized during overhauling. A pike pole or ceiling hook can be used for pulling lath and plaster walls and ceilings. There are tools that can effectively open large areas in plaster board or drywall to check behind or above the concealed spaces.

A prying tool can be used to open baseboards and moldings around doors and windows. The proven tools are the halligan, claw tool, and axe, but many new lightweight tools can facilitate these operations.

FIGURE 4-33 Grain elevator fires can be challenging. A major concern is the possibility of dust explosions. *Used with permission of Joseph Hoffman.*

ON SCENE

We normally associate dust explosions with grain elevators in rural areas. However, they can and have occurred in urban areas. A grain elevator exploded with the force of 1,100 pounds of dynamite in Philadelphia, killing three people and injuring 86 others.

Dust Fires

A particular concern of firefighters performing overhaul is dust. Dust can commonly be found in woodworking shops, grain elevators, bakeries, and other industrial occupancies. (See Figure 4-33.) Combustible dust lying on a heated surface is subject to ignition, due to carbonization of the dust. A cloud of combustible dust can be explosive. This occurs when the dust particles in the cloud are raised to their ignition temperature and come into contact with a source of ignition.

Once ignited, dust can smolder for long periods. Overhauling areas where there is a smoldering dust fire demands a close examination to ensure extinguishment. This must include examination of exposed beams, rafters, ledges, and other flat surfaces in industrial occupancies commonly associated with dust. A thermal imaging device is an excellent tool to detect smoldering dust fires.

Water will extinguish most dust fires, and moisture will raise the ignition temperature of the dust. When operating hose streams around dust, avoid creating dust clouds by using fog nozzles.

The most dangerous hazard associated with dust is explosion. Dust explosions usually occur in pairs. The initial explosion may not cause substantial damage, but it displaces dust that has rested on flat surfaces, creating a large dust cloud. If there is sufficient heat, a secondary explosion can occur. This secondary explosion is usually devastating. See References at the end of this chapter for access to the Occupational Safety and Health Administration (OSHA) publication on firefighting precautions at facilities with combustible dust.

ON SCENE

A fire that I worked at the United States Mint involved smoldering metal dust that had accumulated over many years on flat surfaces high above the machinery. The fire was not spectacular and received little press coverage, yet it took a prolonged overhauling and hard work to ensure complete extinguishment.

Basic Overhauling Concerns

- A fire officer should supervise the overhauling process, taking responsibility to ensure that all suspected areas are opened and that firefighters operate in a safe manner as a team. Teamwork will minimize damage.
- Fire officers must use their judgment when opening a wall or ceiling. If char or fire is found, a larger hole should be made. If cobwebs are discovered, it indicates that fire has not extended to that area; the heat of the fire would have destroyed them.
- When overhauling, be alert for signs of arson. If arson is suspected, minimize overhauling until the fire investigator can inspect the fire scene.
- Floors above a fire area can be checked for fire extension by removing the baseboards and checking the area behind them.

The goal of overhauling is to ensure that all fire has been completely extinguished, while minimizing further damage to the structure. It is not just the wetting down of a fire area but a thorough investigation.

 ON SCENE

During the final stages of overhaul at one particular fire, the engine company was washing down the area and a drum was knocked over. Its contents ignited and five additional alarms were needed to control the ensuing fire.

Incident Scene Preservation   Firefighters must realize that extinguishment, overhaul, and salvage are not the final steps at a fire scene. The scene must be fully analyzed by a trained fire investigator to find the origin and cause of the fire. A determination of whether a fire is accidental or was intentionally set will involve a careful and thorough investigation. A fire that is accidental in nature requires a diligent analysis to discover the exact cause. Did the fire start from faulty electrical wiring, or was a defective appliance the cause? Could careless use of smoking materials be involved? Is arson suspected?

The facts must support the conclusion. To assist the fire investigator, the scene must be reconstructed to how it was prior to the fire. This re-creation is highly dependent upon the information received from the firefighters who fought the fire. The data received from them, along with good investigative procedures, will assist in pinpointing the origin and cause.

As firefighters, we need to recognize the importance of preserving a fire scene for investigative purposes. Important points for suppression crews include:

- Whether the doors and windows were closed. Were they locked? Realize the difference between an open window or door versus an unlocked window or door. These may seem like minor details, but if the investigator has to testify in court, these seemingly minor factors can mean the difference between a conviction and allowing an arsonist to go free.
- If forcible entry was performed, make note of how and where.
- Did any occupants from the property offer any information on the fire?
- Was anyone seen leaving the scene, or was there anyone on the scene who stood out as peculiar?
- Did anything seem out of the ordinary? (A thief may set a fire to damage or destroy evidence.)
- Extinguish the fire without destroying the entire scene of potential evidence.
- The firefighters' ability to recall where each piece of furniture was located is critical.
- Remember what the scene/room looked like when they arrived, and once inside, where the fire was located.
- What color were the flames? Was it a quick-burning fire? Was there anything out of the ordinary with the way the fire was burning? Could any unusual odors, possibly an accelerant, be detected on the exterior?

- When overhauling, remember what was moved and to where it was moved.
- Tell the investigator whether they had set down any gas-powered tools within the fire area. This is important since an arson dog trained to smell flammable liquids may find a location where a power tool had been laid down.
- Cooperate with the investigator after the fire is extinguished and the smoke clears to allow the investigator to pinpoint the origin and determine the cause of the fire.
- Remember that firefighting and fire investigation take teamwork. An investigator who actively solicits information in a positive way at the fire scene tends to receive plenty of offers of assistance during the investigation with removing debris or digging out a scene. Firefighters instinctively want to help, and a good investigator can use this assistance.

If a fire investigator is on location at a fire scene as knockdown of the fire occurs, he or she can take a quick look at the scene prior to overhaul and can lend a hand in directing the overhauling process to assist in preserving any areas where further investigation may be needed.

Some progressive fire departments have initiated a program where after knockdown of the fire occurs and the smoke has lifted, yet prior to overhaul, the entire fire area is documented with pictures taken with a digital camera. These photos then become part of the investigative process.

Overhauling Safety There is a tendency to relax or let our guard down during overhauling operations. The demands of the firefight are past, and the overhauling process does not offer the challenge of the initial stages of a fire. Safety at this time is paramount. Firefighters are tired. They may be overheated and want to remove protective clothing to assist in cooling off. Rest and rehab of members must be a foremost consideration.

Many firefighters are injured in the overhauling stages due to fatigue and dangerous conditions. Common injuries include:

- Falls from stepping into holes
- Slipping on wet or icy areas
- Tripping over fire debris
- Being struck by building debris or tools swung by other firefighters
- Cuts and abrasions
- Foreign objects in eyes
- Strains and sprains from trying to lift or drag items that are too heavy
- Injuries after removing protective gear

Portable lighting should be utilized when overhauling building interiors during both day and night operations. Apparatus spotlights can illuminate a building's exterior in nighttime operations. Lighting enhances the efficiency of the operation while assisting in the overall safety effort.

It should not be assumed that since the smoke has lifted it is safe to remove self-contained breathing apparatus (SCBA). Carbon monoxide is colorless and odorless, and high levels may remain though visibility may seem normal. Air monitoring with testing equipment must be performed during overhauling, and SCBA should not be removed until the air reaches safe levels.

Interior doors can be removed and placed over holes burnt in the floor to prevent firefighter injuries during overhauling. Door removal also permits firefighters the freedom to move through door openings unimpeded.

Care must be exercised so that the overhauling process does not endanger firefighters. Knocking down bottles or throwing around cans could release flammable or dangerous chemicals, causing a chemical reaction or explosive fire. This is especially true in properties that use these materials in their processes or have them in stock. Containers may have been damaged or weakened by the effects of the fire or water used during extinguishment.

There may be a need to set up a fireground detail or watch to ensure that the fire has been fully extinguished. This is often necessary in structures that have experienced some collapse and smoldering debris may be buried. Remember, if a dangerous situation exists that could be a threat to the safety of the firefighters, it is best to step back and study the situation. The solution must consider the best and safest method of extinguishment and not how quickly it can be done.

There is a strong desire for occupants and owners to reenter the fire building. It may be to view the damage or collect valuable items, such as money, jewelry, or medicine. During active operations, no one should be permitted to reenter; he or she could be injured or impede operations. There is also the need to preserve the scene for the fire investigator. If you deem it necessary, a firefighter can be assigned to retrieve the valuables. A police officer should be involved to ensure that the articles are given to the rightful owners and that everything is intact (anyone can claim to be the occupant).

If people are observed removing items from a fire building or exposed building, a police officer should ensure that they are the owners of the items being removed. If the origin of the fire is suspicious, no one should be permitted to enter, and the police should secure the scene for the fire investigator when the fire department has completed its operations.

SALVAGE OPERATIONS

Salvage is the preservation of the structure and its contents from additional damage from fire, smoke, water, and firefighting activities. This can be accomplished by employing salvage operations as early as possible. One easy method is removing items from the path of heat and water. This can be done by:

- Taking down curtains or other window dressings in a nearby exposure
- Moving material from the path of water runoff
- Placing salvage covers over valuable items

Many fire departments have found that, in addition to conventional salvage covers, rolls of plastic sheeting make an excellent tool for salvage operations. The plastic can be left in place after firefighters have left the scene to guarantee that items are continuously protected. It can also be used to cover window, door, and roof openings.

In commercial properties, materials in direct contact with the flames should be removed. Other stock can be protected by salvage covers or sheet plastic. If there is considerable stock on shelving that has been damaged due to heat, it can be inspected, and if deemed safe, it can be left there. The items can be checked more than once to ascertain that no problem exists. If there is any doubt, the items should be removed to the exterior.

Forklift trucks can facilitate the quick removal of stock either to the exterior or to another part of the building. This includes baled material, rolls of paper, and pallets of material. Fire-damaged goods can be taken to the exterior for overhauling.

Though a fire may be contained to one room, damage and cleanup may be considerable. It can leave the building uninhabitable. In colder climates, the heating system may be needed to prevent further damage from subfreezing temperatures that can cause pipes to burst. After the fire investigator has cleared a building, local utility personnel can assist in restoring electricity and placing the heating plant back into operation. If this cannot be accomplished, the water supply to the building should be shut off to prevent further damage.

General Rules of Salvage

- Salvage must be a constant consideration. The goal is to restore the property to the state it was in prior to the fire.
- If an item is not damaged, place it aside to protect it. Many items may not have a significant dollar value but can be irreplaceable family heirlooms.
- After the utilities have been turned off, the fire department should not turn them back on. The local utility company should be called to restore them.
- Prevent unnecessary water damage by controlling water lines. Use fog streams to minimize water runoff.
- Avoid leaking couplings on hose-lines. During freezing weather, nozzles must be kept flowing to prevent hose-lines from freezing. They should be set on a window-sill and a slight stream should continue to flow to the exterior.
- Before leaving the scene, all roof, window, and door openings should be sealed to prevent the weather from causing additional damage.

Our goal is to bring the emergency to a rapid conclusion while preserving civilian life and protecting the firefighters operating at the scene. We should attempt to restore the property to the condition that it was in prior to the fire. A fire officer can accomplish this by supervising the salvage operation. Some forethought will assist in the overall implementation of salvage operations; it will better protect the building and contents and avoid duplication of effort while achieving better results.

Firefighters are looked to whenever problems arise. Our hard work has earned us a special place in our communities. We assist those in need during their worst times. This assistance does not take into account the income status of the recipients. In fact, the less affluent usually need our assistance the most. They often do not have sufficient insurance to cover the damage caused by their emergency. By performing salvage operations, we can assist them in restoring their home or place of business as quickly as possible. The best part is that the firefighter gets a feeling of accomplishment through a simple yet earnest thank you, or the smile of a grateful child.

Securing Utilities

Truck companies have many varied duties, one of which may be securing the utilities in the fire building. This could include:

- Shutting off gas valves to stop the flow of leaking gas, or ensuring that no leaks occur.
- Shutting down water supplies in the fire building to prevent broken pipes from flowing.
- Securing the electricity in the fire building, which may entail shutting off electrical circuit breakers to the affected area or pulling the electric meter to the property to completely shut down all electric supply.

Summary

Through a decision making process, the many problems found at an incident can be solved by teamwork. The implementation of engine and truck duties at an incident helps to fulfill the action plan by accomplishing the tasks assigned by the Incident Commander. The utilization of sprinkler and standpipe systems by firefighters can be beneficial. The importance of gaining knowledge about rooftop solar panels can pay dividends should a fire occur within a building so equipped.

Review Questions

1. Discuss the benefit of having standardized operational guidelines.
2. Discuss the benefits and drawbacks of 1½- or 1¾-inch hose-line versus 2½-inch hose-line.
3. How does water hammer occur and what problems does it create?
4. Discuss the considerations to determine the number of hose-lines needed and where to place them at an occupied apartment building.
5. Discuss the ability of your department to handle large operations involving numerous master streams.
6. Discuss the duties of a Logistics Section Chief in regards to water supply. When should this position be implemented?
7. What are the basic duties of an engine company?
8. List and describe the four types of sprinkler systems.
9. Describe the actions of firefighters when operating at a fire in a sprinklered building.
10. Define the three classes of standpipe systems.
11. What are the basic duties of a truck company?
12. Discuss the removal of civilians from a fire building via stairs and via ladders.
13. Discuss the many uses of portable/ground ladders.
14. What are the basic ladder positions?
15. What are the basic considerations of a trench cut?
16. What are the uses of a hose-line taken to a roof?
17. Discuss the different types of ventilation. Which types are in use in your department? What are the various benefits and drawbacks of each type?
18. What indicators would suggest a fire in a concealed space?
19. What concerns should firefighters have when confronted with combustible dusts?
20. List the common types of injuries that occur during overhauling.

Suggested Reading, References, or Standards for Additional Information

US Fire Administration (USFA)

Chubb, M. "Indianapolis Athletic Club Fire." Technical Report 063.

Routley, J. G., et al. "Aerial Ladder Collapse Incidents." Technical Report 081.

Routley, J. G. "Six Firefighter Fatalities in Construction Site Explosion, Kansas City, Missouri." Technical Report 024.

Routley, J. G. "Six Firefighter Fatalities in Construction Site Explosion, Kansas City, Missouri." Technical Report 024A, Appendices.

OSHA. "Firefighting Precautions at Facilities with Combustible Dust." Available at: https://www.osha.gov.

Underwriters Laboratories (UL)

"Achieving Increased Reliability in Photovoltaic Installations." 2014. Available at: http://library.ul.com/wp-content/uploads/sites/6/2014/07/UL_Final_Achieving-Increased-Reliability-in-Photovoltaic-Installations_v8-HR.pdf.

Backstrom, R., and D. A. Dini. 2011. "Fire Fighter Safety and Photovoltaic Installations Research Project." Listed in and linked to "UL Fire Safety Research: Recent and Current Studies." Available at: http://www.ul.com.

National Institute of Standards and Technology (NIST)

"Positive Pressure Ventilation." 2011. Available at: http://www.nist.gov/fire/ppv.cfm.

National Fire Protection Association (NFPA)

Standards
13: Standard for the Installation of Sprinkler Systems
14: Standard for the Installation of Standpipe and Hose Systems
1142: Standard on Water Supplies for Suburban and Rural Fire Fighting
1911: Standard for Inspection, Maintenance, Testing, and Retirement of In-Service Automotive Fire Apparatus

1932: Standard on Use, Maintenance, and Service Testing of In-Service Fire Department Ground Ladders

1961: Standard on Fire Hose

1962: Standard for the Care, Use, Inspection, Service Testing, and Replacement of Fire Hose, Couplings, Nozzles, and Fire Hose Appliances

1963: Standard for Fire Hose Connections

1964: Standard for Spray Nozzles

Related Courses Presented by the National Fire Academy, Emmitsburg, Maryland

Management Strategies for Success

Command and Control of Incident Operations

Residential Sprinkler Plan Review

5

Building Construction

“To accomplish an assignment does not necessarily mean that it was done properly. The decision on whether safety and success have both been accomplished must be made by the Incident Commander. His or her evaluation must be based strictly on what is right and proper. There is no room for attitudes or personalities to encroach in the review. A prime consideration is always firefighter safety.”

—James P. Smith

Wall collapse can occur with little or no warning. The reading of collapse signs will often be our only indication. *Used with permission of Joseph Hoffman.*

KEY TERMS

fire-resistive construction, *p. 173*
noncombustible/limited combustible construction, *p. 174*

ordinary construction, *p. 178*
heavy timber construction, *p. 183*
frame building, *p. 186*

timber truss, *p. 190*
lightweight building components, *p. 195*

OBJECTIVES

Upon completion of this chapter, the reader should be able to:

- Identify and discuss the five basic types of building construction.
- Discuss the benefits and safety concerns of fires involving timber trusses.
- Discuss the benefits and safety concerns of lightweight building components.

A fire department's ability to control and quickly extinguish a fire in a building depends on a variety of factors. It starts with early discovery of the fire and a quick and adequate response by the fire department. Receiving accurate information on the location and extent of the fire can save precious time for the first-arriving firefighters. They can immediately stretch hose-lines to the fire location and initiate an aggressive attack. Yet one of the most important factors is the building's type of construction. Chapter 5 reviews the different types of building construction that firefighters will have to confront.

The fire service has different methods of classifying buildings. One method used by the National Fire Protection Association (NFPA) categorizes buildings from Type 1 to Type 5. The following list shows the actual type of building and the NFPA's listing:

Fire-resistive	Type 1
Noncombustible/Limited combustible	Type 2
Ordinary	Type 3
Heavy timber	Type 4
Frame	Type 5

Each type of building has positive and negative features that can affect fire spread. Generally, firefighters refer to buildings according to the kind of construction. This chapter, in addition to examining the basic types of construction, will look specifically at the dangers of timber truss and lightweight building components. The timber truss has taken many firefighters' lives due to truss roof collapses. Likewise, the new lightweight components have made their mark by their deadly collapses.

Fire-Resistive Construction

In **fire-resistive construction**, the structural members are of noncombustible materials that meet or exceed requirements prescribed by the applicable code. A noncombustible material is defined in "NFPA 220: Standard on Types of Building Construction," as "a material which, in the form in which it is used and under the conditions anticipated, will not ignite, burn, support combustion, or release flammable vapors, when subjected to fire and heat."

The fire-resistive building contains steel, concrete, and other fire-rated materials.

fire-resistive construction ▪ Structural members are of noncombustible materials that meet or exceed requirements prescribed by the applicable code.

> A rule of thumb for firefighters when encountering steel in building construction is that steel expansion can be estimated at 1 inch for every 10 feet. Thus a steel beam 75 feet long could expand 7½ inches.

STEEL

The strength of steel makes it a common material in fire-resistive buildings. But steel is also a ready conductor of heat. When heated, it will expand about 9½ inches per 100 feet.

Due to its high conductivity of heat, steel must be protected. (See Figure 5-1.) The most common methods of protecting steel are:

▪ *Encasement.* This is accomplished with concrete, brick, terra cotta, plaster, or fire-rated drywall.
▪ *Sprayed-on protection.* Previously accomplished with asbestos, today a cementatious mixture, volcanic ash, and other coatings are used. The protection provided by this method is highly dependent on the quality of application and whether this barrier is negated by removal when other craftsmen install plumbing and heating or welders seek a suitable ground for their welding equipment.
▪ *Membrane protection.* A floor and ceiling assembly protects the steel members within the assembly. This system usually includes a suspended ceiling below the steel. Removal of any part of the ceiling will negate the protection of the steel.

CONCRETE

The strength of concrete depends upon how it is supported. Reinforced concrete is very strong if it is properly constructed and the steel reinforcement is well protected. Concrete will readily absorb the heat of a fire and retain that heat. If heated to a high temperature and cooled by a hose stream, the concrete can spall or crack. This can occur violently; the concrete facing can project pieces of concrete away from the wall or ceiling. Cracks can develop due to heat exposure that weakens the wall.

NEW CONCRETE

A typical question firefighters ask is, "How long does it take for new concrete to be cured to assure firefighter safety?" There are many variables when confronted with freshly poured concrete and it is not possible to give exact time frames. This is due to the many substances added to Portland cement in various parts of the country and the temperature the concrete

FIGURE 5-1 Steel beams supporting the 10th floor expanded due to fire impingement, pushing the granite wall in which the beams rested outward.

is subjected to during the curing process. Low temperatures slow the early-age strength of concrete substantially.

In general, structural concrete is supported with forms and shoring until sufficient strength has been developed to support its loads as well as the loads applied by construction activity. This is determined by strength testing to ensure that the concrete has achieved an appropriate strength.

A rule of thumb that firefighters could use: concrete that is still supported by forms and shoring should be suspect, while any self-supporting concrete has been deemed safe to support construction loads. (Concrete less than 48 hours old requires extreme caution.) Another factor is that fire impingement on newly poured concrete can affect its strength.

The person most qualified to assist firefighters in assessing the strength of concrete at a construction site is the design engineer. He or she may be able to provide the designed live and dead loads, the approximate load on the slab, and technical assistance.

> Spalling is a condition in which concrete will lose surface material.

FIREFIGHTING

Firefighters think of a fire-resistive building as one in which the structural components will resist the effects of fire for a period of time and allow them to make an aggressive attack. Though the building's protected structural members will withstand fire, extensive damage can occur, with the possibility of injuries and death due to the smoke and/or fire. A fire in a fire-resistive building can be a challenging fire to fight. If the fire is in a high-rise building, it will complicate the situation. The time delay in reaching the fire area and assuring the life safety of the occupants will compound the firefighters' problems. Firefighters can be faced with high temperatures while fighting these fires, especially in newer buildings that contain large amounts of drywall. The drywall does not absorb heat as masonry materials do, and a fire can spread at a much faster rate. Though the building is fire-resistant, flammable furnishings can allow a fire to spread rapidly in a short period of time.

Noncombustible/Limited Combustible Construction

noncombustible/limited combustible construction ■ Employs materials that do not contribute to the development or spread of fire.

Noncombustible/limited combustible construction employs materials that do not contribute to the development or spread of fire. Though the buildings themselves offer little fire resistance, they do not contribute fuel to a fire.

These structures can vary. They can be metal-framed or metal-clad buildings that are sometimes prefabricated and erected on site. Concrete or concrete block buildings with a metal deck roof also are classified as noncombustible/limited combustible construction.

The roof configuration of this kind of construction can be either peaked or flat. A peaked roof is often supported on a metal-framed truss consisting of small-dimension angle iron and a roof surface of corrugated metal. A flat roof is often constructed and supported on steel bar joists.

A noncombustible building can be negated by the use of combustible material during construction. This includes the installation of wood paneling in an office area or other types of flammable decorative wall coverings.

STEEL BAR JOIST

Steel bar joist is a lightweight component that is a flat parallel chord truss. The top and bottom chords are steel plates. The web member consists of a long, continuous, round steel rod, shaped and welded to the top and bottom chords, forming angles and distributing the weight of the roof to the bearing walls.

The steel bar joist allows large, wide-open floor areas. This design is attractive due to its low cost, but this type of construction is not typically protected for fire resistance or given any kind of fire protection. Although drywall ceilings may be installed in some occupancies, they may not be fire-rated. Exceptions are areas with strong fire and building codes that require sprinkler systems in these structures.

The trusses consist of relatively small pieces of steel that will not add fuel to a fire but will fail rapidly under fire conditions. The steel absorbs heat quickly and distributes it rapidly. As the fire increases, so does the temperature of the steel, causing elongation and eventual failure. When set in masonry walls, the expansion of steel may be seen from the exterior by cracks at the top of the wall. This can occur quite early in a fire.

BUILT-UP ROOF

A common method of roof installation on these structures is metal decking that sits on the steel bar joist. (See Figures 5-2, 5-3, and 5-4.) Insulation is typically placed on the decking to achieve the desired amount of protection. The roofing felt, paper, and tar adhesive are then installed over the insulation to waterproof the roof. (This is commonly referred to as a built-up roof.)

The tar adhesive, when subjected to high heat conditions, can break down into a flammable gas. These flammable vapors, unable to escape through the sealed roof, will

FIGURE 5-2 A typical bar joist assembly.

FIGURE 5-3 Expansion of the bar joist may be visible on the exterior as cracks develop in the masonry wall. *Used with permission of Joseph Hoffman.*

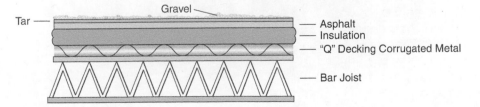

FIGURE 5-4 A cutout of a built-up roof sitting on steel bar joist. *Used with permission of Michael DeLuca.*

accumulate, ignite, and burn on the underside of the roof. The heat of this fire will continue to produce additional fuel by feeding on the vaporization of the remaining adhesive. This can sustain a roof fire independent of the original fire in the structure. In fact, the original fire may be extinguished and units may be unaware of the presence of the second fire. This can be referred to as a combustible metal deck roof fire. I have witnessed fires that ignited in stock located in a remote area of a structure, and after reaching the underside of the roof, it proceeded to destroy the entire building. The use of flammable tar adhesive on the steel decking has led to the destruction of many of these buildings.

OPERATING ON A BAR JOIST ROOF

When operating on an insulated bar joist roof, it is difficult to determine the location of a fire below. The usual visible signs (bubbling tar, dry spots on a wet roof, or soft, spongy wood) may not be present due to the insulation under the roof surface. Walking or working on a lightweight steel roof, especially the bar joist kind, is different than operating on a wood joist roof. Large roofs seem bouncy or unsound under normal conditions, let alone when a fire is occurring within the building.

When operating on a bar joist roof, firefighters must use caution since the bar joist can be spaced farther apart than the 16- or 24-inch centers usually found on solid wooden beam roof joists. The bar joist can be spaced on centers six–eight feet apart. A safety tip for operating on any roof that is suspect includes the use of a safety belt attached to a main ladder or aerial platform. Spreading a portable ladder to span the distance between the bar joists allows the firefighter on the roof to use it as a work platform. (See Figure 5-5.)

VENTILATION

The opening of bar joist roofs for ventilation must be a cautious operation. Care must be taken not to cut through the top chord of the roof truss. The roof opening should be made adjacent to the bar joist. If not, that area will be unsupported and a danger to firefighters operating there. Heated tar on the roof will make the roof slippery and dangerous.

Roof insulation will hide some of the telltale warning signs, and collapse can occur before the roof gives any indication of an impending failure. Because the bar joists are

FIGURE 5-5 Collapse of a bar joist roof can occur with no warning.

FIGURE 5-6 A–B As the steel bar joist is heated, it expands and can push walls outward and cause the roof to collapse. *Used with permission of Joseph Hoffman.*

set at a wider spacing, a firefighter falling through a roof opening cannot grab the sides of roof rafters as he or she could if the rafters were set on 16- or 24-inch centers. Since the joists are set on wider centers, failure of one joist will affect a larger area of the roof.

Operating on or opening a roof of unprotected steel during a building fire should be considered dangerous, and the Incident Commander (IC) must strongly consider other means of ventilation that can be accomplished with a lesser degree of risk.

COLLAPSE

A fire within a structure will quickly reach the temperature at which steel fails, commonly set at 1100 degrees Fahrenheit (F). When the steel reaches its failure temperature, collapse will occur. (See Figure 5-6.) The type of building and its contents will directly affect the temperature of the fire. Storage of flammable liquids, as well as other stock, can accelerate fire temperatures to more than 1500 degrees F quite rapidly.

If the overhead can be cooled to prevent the generation of flammable gases or to extinguish those already burning, the fire can be controlled. This requires large volumes of water with sufficient pressure to reach the underside of the roof. The difficulty lies in the ability to handle these streams. Firefighters will be in a position of danger. The reduced visibility accompanying interior firefighting will not permit hose-line crews to know where the fire is actually located. They will only know if their water streams are hitting the fire by feeling the temperature of the runoff water and by having the exterior walls monitored to determine if conditions have stabilized. (See Figures 5-7 to 5-10.)

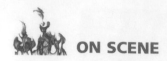

ON SCENE

In 1968, at an auto dealership in Wichita, Kansas, four firefighters were killed in the collapse of a steel bowstring truss roof. The building was occupied at the time. The fire was believed to have originated in an overhead heating unit and then spread between the ceiling and the roof, in the attic space. Though the workers smelled smoke, there was a delay of at least 45 minutes in calling the fire department.

Firefighters were on the scene approximately 16 minutes when the collapse, which trapped six firefighters, occurred. Two firefighters were able to escape; the chief of department and three firefighters perished.

Estimating time frames for the collapse potential of a building is difficult since the calculation would need to be computed from the time of ignition and not from the time of the fire department arrival.

NIOSH FIREFIGHTER FATALITY REPORT F2013-02

On January 22, 2013, a 34-year-old male volunteer captain (the victim) died when the floor collapsed and trapped him in the basement at a residential structure fire. Crews were attempting to locate and extinguish fire that might have spread from a nearby shed fire that they were dispatched to. While a crew was fighting fire that had spread into the basement, the victim and his partner were pulling the ceiling just inside the front door to extinguish the fire that had spread to the attic. The floor collapsed sending them both into the basement. The victim's partner was able to make it back up to the first floor doorway. He attempted to assist the victim out of the basement, but was unable to pull him up. The hole that the firefighter had fallen through became untenable due to the fire conditions and firefighters were unable to rescue him from the basement. The victim was recovered from the basement approximately 20 minutes later. Before he could be rescued, the victim's mask became thermally degraded and he was overcome by the products of combustion. He was transported to a local hospital where he was pronounced dead

Contributing factors included inadequate water supply, ineffective fireground communications, ineffective incident command, inadequate size-up, uncoordinated fire attack, lack of situational awareness, and deteriorated structural members. **See Appendix B: NIOSH Reports in www.pearsonhighered.com/bradyresources to read the complete report and recommendations.**

ordinary construction ▪ Constructed of wood, with exterior masonry walls.

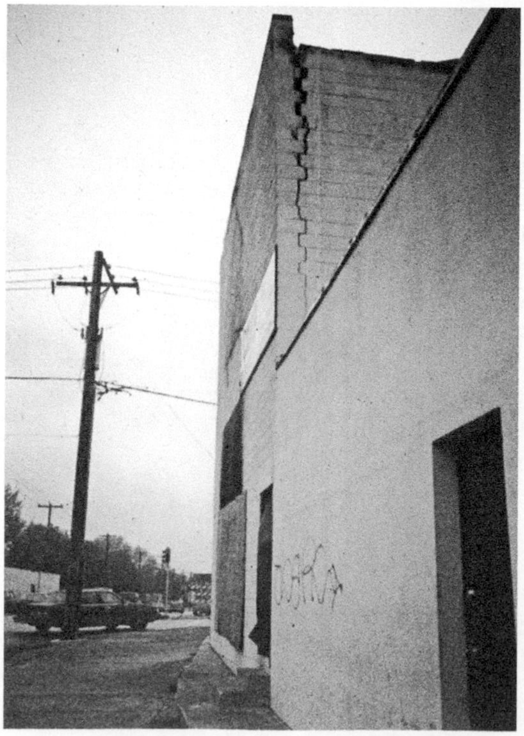

FIGURE 5-7 As bar joists expand, wall cracks can develop.

Ordinary Construction

Buildings of **ordinary construction** are used for offices, retail sales, commercial occupancies, mixed occupancies, and dwelling units. These buildings contain exterior masonry walls and interior floors and roofs constructed of wood. These buildings are commonly one to three stories tall. Taller buildings of ordinary construction can be found in some communities, and they occasionally reach 15 stories in height.

WALLS

The thickness of the exterior walls is dependent on the building's height. The minimum depth is generally 8 inches, though 6-inch walls can be found in some of these structures. Taller buildings can have walls up to 30 inches thick. Walls more than 8 inches thick will diminish in thickness as they rise. A larger-sized foundation is needed to support the building weight.

Walls can be constructed of concrete, stone, brick, or concrete masonry units that are colloquially known by many names, most predominately as concrete block, cinder block, CMU, or simply block. Today, it is common to find composite walls containing both brick and block. This economical method allows the beauty of the brick on the exterior, with the less costly block supporting it from behind. These walls are interconnected or tied together, either by metal trusses that are laid atop corresponding courses, or by common wall ties that are strips of metal laid between the courses. The finish on these walls can be brick, block, stucco, or parging.

FIGURE 5-10 Heavy fire conditions demand an aggressive interior attack to rescue occupants and control the fire.

A *course* of brick or block is one horizontal layer of the material. A brick would normally be approximately 2¾ -inches in height, while a typical CMU would be approximately 7⅞-inches in height. *Parging* is a thin layer of mortar covering a wall.

The building can contain bearing and nonbearing walls. Bearing walls will normally be the longest walls in length. This permits the floor and roof joist to sit in the bearing wall, spanning the shorter distance. Typically, lumber comes in standard 20-foot lengths. For longer spans, interior bearing walls, steel beams, lightweight building components, or supporting columns will need to be installed. (See Figure 5-11.)

Common bearing walls or party walls may be found in town houses, garden apartments, strip malls, row houses, and similar types of buildings. (See Figure 5-12.) The adjoining buildings, as the name implies, share the same bearing or party wall. (See Figure 5-13.) In older structures, the floor and roof joists from both buildings may sit in common wall sockets. This practice provides a ready means for fire extension via conduction through the wooden beams from one building to another.

FIGURE 5-11 An unprotected steel beam supporting a roof assembly was attacked by fire and failed, causing the roof to fail.

FIGURE 5-12 Common bearing or party walls may be found in town house construction.

FLOOR AND ROOF JOIST

The floor and roof are constructed of wood joist. In older construction, the floorboards consist of 1-inch-thick tongue-and-groove boards and the roof of 1- to 1¼-inch-thick planks. In newer construction, plywood or oriented strand board (OSB) is used for both areas. New construction may use lightweight trusses and wooden I-beams in both floor and roof assemblies. (See Figure 5-14.) The roof can be any assortment of typical roof coverings:

- Wood shingles
- Asphalt shingles
- Rolled asphalt
- Clay or slate tiles
- Tar and gravel
- Rubber roof covering

FIRE-CUT JOIST

With the exception of the older ordinary constructed buildings, the floor joists are commonly fire-cut on each end. The fire-cut is an approximately 30-degree cut on each end of the joist. The longer end of the cut is placed downward and rests in the bearing wall. The fire-cut allows the floor to collapse down into the building without pushing the masonry wall outward. The fire-cut concept is beneficial to the building owner because the exterior walls remain intact, but can be deadly to firefighters fighting a fire in the building. (See Figure 5-15.)

FIGURE 5-14 Ordinary construction may use lightweight floor or roof trusses.

FIGURE 5-15 Fire-cut beams can be seen where a bearing wall has been removed.

LINTELS AND ARCHES

To support the weight of the masonry wall above openings made for windows and doors, the load must be transferred to the sides of the opening. This is accomplished by installing a lintel or an arch. These can be made of wood, steel, or masonry. The strength of either the lintel or the arch depends upon the stability of the foundations that support them. A strong foundation ensures stability. Though often subjected to heavy fire conditions, both the lintel and the arch usually hold up quite well under sustained fire attack.

VOID SPACES

Void spaces in ordinary constructed buildings are common. The space under the roof in a peaked building is called an attic; in a flat-roofed building, it is referred to as a cockloft. In addition to the roof space, pipe chases can be found for plumbing and heating systems. Wall voids can be quite common and can be interconnected throughout the building. In older buildings, renovations can include the addition of chases or shaft ways, creating ready paths for fire to travel from lower to upper floors.

The Federal Emergency Management Agency's (FEMA) US Fire Administration (USFA) issued a special report examining the characteristics of attic fires in residential buildings. Developed by USFA's National Fire Data Center, the report is based on 2006 to 2008 data from the National Fire Incident Reporting System (NFIRS).

According to the report:

- An estimated 10,000 attic fires in residential buildings occur annually in the United States, resulting in an estimated average of 30 deaths, 125 injuries, and $477 million in property damage.
- The leading cause of all attic fires is electrical malfunction (43 percent).
- The most common heat source is electrical arcing (37 percent).
- Almost all residential building attic fires are nonconfined (99 percent), and a third of all residential building attic fires spread to involve the entire building.
- Ninety percent of residential attic fires occur in one- and two-family residential buildings.
- Residential building attic fires are most prevalent in December (12 percent) and January (11 percent) and peak between the hours of 4 and 8 p.m.

(Source: USFA, Attic Fires in Residential Buildings)

Wall voids are created by the framing of the walls. The depth of the upright wall studs creates a wall space between the outside and inside walls. In the past, that air space alone was used as a barrier from the outside temperatures. In modern construction, this space is filled with insulation.

Today the wall studs are typically finished on the interior with drywall (also referred to as wall board or plaster board). In the past, they were covered with rough-hewn wood strips that were approximately ¼ inch thick, 1¼ to 1½ inches wide, and horizontally spaced about ¼ to ⅜ inch apart. These wood strips (referred to as wood lath) covered the entire wall and ceiling surfaces (with the exception of door and window openings) and were then covered with plaster. They were susceptible to fires that penetrated the wall voids.

Wood lath are rough-hewn thin wood strips attached to the wall studs and ceiling joist to allow plaster to adhere to and form the finished plaster walls and ceilings.

FIREFIGHTING

Due to the building's overall stability, fires in ordinary constructed buildings can be extinguished with an offensive attack; an exception would be fire attacking lightweight

FIGURE 5-16 Fire attacking this truss roof caused an early failure.

structural components. Plaster or plasterboard walls will assist in containing a fire for a reasonable period of time and allow an offensive attack to be successful. Should there be holes in these walls, fire can enter the void spaces and create problems. Wood lath will supply fuel to a fire and can permit the fire to extend to areas above. A fire involving these void spaces will necessitate opening walls and ceilings to ensure complete extinguishment.

NIOSH FIREFIGHTER FATALITY REPORT F2010-38

On December 22, 2010, a 47-year-old male (Victim #1) and a 34-year old male (Victim #2), both career firefighters, died when the roof collapsed during suppression operations at a rubbish fire in an abandoned and unsecured commercial structure. The bowstring truss roof collapsed at the rear of the 84-year-old structure approximately 16 minutes after the initial companies arrived on-scene and within minutes after the IC reported that the fire was under control. The structure, the former site of a commercial laundry, had been abandoned for more than 5 years and city officials had previously cited the building owners for the deteriorated condition of the structure and ordered the owner to either repair or demolish the structure. The victims were members of the first alarm assignment and were working inside the structure. A total of 19 other firefighters were hurt during the collapse.

Contributing factors included lack of a vacant/hazardous building marking program within the city, vacant/hazardous building information not part of automatic dispatch system, dilapidated condition of the structure, dispatch occurring during shift change resulting in fragmented crews, weather conditions including snow accumulation on roof and frozen water hydrants, and not all firefighters equipped with radios. **See Appendix B: NIOSH Reports in www.pearsonhighered.com/bradyresources to read the complete report and recommendations.**

Heavy Timber Construction

Heavy timber buildings, or mill buildings as they are often called, are constructed with wooden timbers of large-dimension lumber. Through proper design, **heavy timber construction** provides an excellent degree of fire resistance by requiring a minimum dimension for all load-carrying wooden members. The quality construction features of these buildings afford firefighters time to make an aggressive attack on a fire.

The exterior walls are of masonry and can be up to eight stories in height. Larger structures contain masonry fire walls. A fire wall in heavy timber construction is customarily a bearing masonry wall that is a barrier to fire. Any openings in a fire wall must be protected with the same fire rating as the wall that is pierced. Door openings must contain self-closing fire doors.

heavy timber construction ■ Uses wooden timbers of large-dimension lumber. Also called *mill buildings*.

FIGURE 5-17 Tremendous heat from a fire below has caused the cross-laid wood flooring to expand and push upward.

If fire walls are located in a structure with a combustible roof, they must extend above the roof to prevent the spread of a roof fire. The roof can be supported by columns or fabricated of heavy timber truss. The common truss configurations are:

- Flat and constructed of parallel chords
- Triangular-shaped
- Bowstring-shaped

The floors are built to carry heavy loads. (See Figure 5-17.) This is accomplished through the installation of columns to support large floor timbers or parallel chord truss containing heavy timber components. The truss permits large open spaces with fewer columns.

Columns must be a minimum of eight inches thick. Girders that span the distance between the columns must be a minimum of six inches thick. The thickness of the floor must be a minimum of three inches. This is often installed using tongue-and-groove one-inch planks of lumber that are cross-laid.

The interior walls and ceilings are usually not finished. This leaves exposed masonry walls and creates few concealed spaces. This type of construction utilizes the underside of the exposed wood floors as the ceilings on the floor below.

BUILDING MODIFICATIONS

Modifications to the structure include makeshift offices that can create problems for firefighters:

- They may be built beneath the sprinkler system and not contain sprinklers.
- Portable heaters used in these makeshift offices can overload the electrical wiring.
- An accumulation of paperwork and improper storage may set the stage for a fire to be well-involved before activation of the building's alarm system.

FIGURE 5-18 Prolonged fire conditions combined with fire attacking multiple floors is a recipe for collapse.
Used with permission of Joseph Hoffman.

Many of these buildings are still used as factories and warehouses; others have undergone major changes and are occupied by retail and wholesale outlets. Others have been converted to apartments and condominiums. Some are multi-use structures that include offices and warehousing.

These modifications change some of the inherent qualities of the original construction. Where few concealed spaces existed, renovations may include many voids. Pipe chases are installed to provide the utilities for the upper floors. They also afford a ready means for fire to extend from the lower to the upper levels of the building. Heating, ventilation, and air conditioning systems are installed to provide centralized heating and cooling. Ceilings are suspended to reduce the living space needed for the changes in occupancy, creating more void spaces.

BUILDING DETERIORATION

The heavy timber building as built is not prone to collapse. The large timbers will withstand attack by fire and give firefighters time to control and extinguish a fire. Many of these buildings, though, are quite old and have been modified or have deteriorated.

A serious problem develops when a building sits vacant for many years. It is subjected to deterioration due to the lack of maintenance and the attack by the weather on the building. Leaks can develop in the roof and window areas. The deterioration of masonry walls can occur due to water infiltration, combined with freezing and thawing of the walls. Under fire conditions, these weakened walls can collapse, endangering firefighters and threatening nearby buildings with the extension of fire.

Rainwater from leaking roofs will rot the wooden roof planks and interior support beams, especially at the connection points. These deteriorating conditions may not be noticeable under normal circumstances but may cause rapid failure under fire conditions.

FIREFIGHTING IN MILL BUILDINGS

Fire in heavy timber or mill buildings can present unique challenges to a fire department. The fire service today does not have the number of large building fires as in the past. This is due to a number of factors:

1. Better building codes
2. Stronger code enforcement
3. Fully sprinklered buildings
4. Fewer buildings that present fire hazards

FIGURE 5-19 A tremendous explosion on the top floor of this building is caught on film. *Used with permission of Joseph Hoffman.*

The reduction in these fires is positive for the community. One problem, however, is that it makes it difficult for firefighters to keep the necessary skills sharp to fight these types of fires.

Mill buildings present a tremendous fire problem due to the contents, methods of stock storage, and large amount of exposed wood. Once ignited, they generate massive amounts of heat. These fires are difficult to control and can severely threaten exposed buildings.

Since there are few concealed spaces in unrenovated mills, firefighters are able to mount an offensive attack and control a fire in the early stages of development. The lack of concealed spaces permits hose streams to reach the fire. However, once a fire gets past the incipient stage, it can be quite difficult to control. These fires can and will spread rapidly.

FIRE ATTACK

Once the location of the fire is known, a hose-line should be stretched to attack the fire. This attack should not drive the fire into other areas of the building. When possible, the fire should be fought from the unburned side. Since these buildings can be of tremendous size, a decision about from which direction to fight the fire must consider access, fire wall locations, and minimizing building loss. The IC may decide to use a blitz attack in a defensive/offensive mode of attack (see attack modes in Chapter 3) by shutting the fire doors located in the fire walls to prevent spread of fire to the adjacent areas and then to knock down the fire from the exterior with master streams. This can be followed by an interior attack.

The interior attack on a fire above the first floor will initially be fought from the stairways. The IC should designate from which stairway to fight the fire. Should multiple stairways be used, close coordination is a must to prevent opposing hose-lines. If the fire threatens the floor above, a hose-line should be immediately stretched to that location.

In this case, additional hose-lines must be stretched to back up the initial line and the hose-line

FIGURE 5-20 Containing a fire to the original fire building must consider the surrounding exposures and implementation of numerous master streams. *Used with permission of Greg Masi.*

stretched to the floor above. Backup lines must be either the same size or larger than the initial hose-line. Realize that 2½-inch hose-lines should be strongly considered. Smaller hose-line, though deployed much faster, does not have the knockdown and cooling ability of the 2½-inch hose-line.

Floor openings that allow interconnection of floors will work against the containment efforts. These openings could have been created for the use of large tanks or for conveyor belts that service multiple floors. Hose-lines must be stretched to the floor above to check these areas for fire extension.

The large area of the building and the high ceilings permit an accumulation of heat at the ceiling level that may not be recognized by firefighters on an offensive attack. As the fire expands, it can flash over a large area, threatening the firefighters operating within the building.

COMPARTMENTATION

Compartmentation, accomplished through the use of fire walls and properly functioning fire doors, will assist in confinement efforts. This tactic also works in a large building where a defensive attack is underway. The IC should decide where the fire can be stopped. This consideration will depend heavily upon building construction features, resources, and water supply.

The defensive attack can be fought in a designated section of the fire building if it is separated from other areas by fire walls. Though the attack is defensive on the fire area, the large size of the building may allow companies to be simultaneously deployed in offensive mode on the building's interior, in areas adjacent to the defensive fire area. These companies should be located at the appropriate fire wall location. Recognize that their assignment is not to attack the fire that is being fought defensively. Their assignment is to ensure that the fire doors remain closed and that their hose-lines are in place to handle any minor extension of fire that may have spread past those fire door locations.

In multistoried buildings, this strategy requires coordination of companies from the ground floor up to and including the roof. The Division Supervisors must ensure that units on all levels are operating at the correct fire wall location.

The sturdiness of these buildings allows the fire department flexibility. A well-involved fire can initially be attacked with exterior streams. This blitz attack can knock down large amounts of fire while interior hose-lines are being stretched. After hand-lines are in place and the fire has been knocked down, the exterior lines can be shut down and an offensive attack can be initiated through the fire doors at designated locations selected by the Division Supervisors, in coordination with the IC.

The success of any offensive attack will depend heavily on proper ventilation. Creating ventilation openings in advance of the hose-lines is a must. Without adequate ventilation, the hose-lines will be unable to push into the fire area, and the attack will be useless.

Frame Buildings

A wood frame building is one in which all members are wooden or a similar material. **Frame buildings** contain features that can assist or impair firefighters' ability to contain and extinguish a fire, depending on the individual type. Frame buildings are used for dwellings and commercial buildings. Their size can be quite large. The components within these buildings can consist of solid beam lumber or lightweight building components.

TYPES OF FRAME BUILDINGS

The types of frame buildings are:

Balloon Frame
This system utilizes a continuous wall stud from the lowest level through the roof area. It interconnects the void space created by the floor joist at each level of the building with the wall void spaces. Fire-stopping is rare. In actuality, the entire void space within the building is one large interconnected area. A fire starting on a lower level and entering wall or floor voids can quickly extend to the attic area. Though this type of construction is no longer used, there are many balloon-frame–constructed buildings in communities throughout the United States.

frame building ■ Wood frame building in which all members are wooden or of similar material.

FIGURE 5-21 Balloon-frame buildings allow wall voids to interconnect with void spaces contained in the floors. This creates one large void space throughout the voids in the building allowing fire to spread throughout

Platform Frame

The most common type of new frame construction is the platform frame. It starts at the lowest level where the floor joist and flooring are installed on the foundation. The sidewalls are then erected on the floor. After the wall studs are in place, the floor joist and flooring for the floor above are set onto the studs. This reduces paths for fire to extend to areas above, forming a measure of compartmentation.

Post and Beam

This framing system uses posts as the vertical members and beams as the horizontal members. They are connected by rigid joints to support the structure. The posts support the walls in lieu of wall studs. Post and beam construction, though usually associated with commercial structures, is found today in buildings being renovated and converted to private homes. (See Figure 5-22.)

Log

Logs are used in the construction of log cabin homes and small commercial buildings. (See Figure 5-23.). The logs are interlocked by notching the ends. Modern log buildings use logs that are sawn and machined to a standard and uniform size. They can use a tongue-and-groove design on the top and bottom to interlock each log. The logs may be finished on the interior or covered with paneling or drywall. Under heavy fire conditions, the walls usually withstand the effects of the fire, though the roof may be destroyed.

FIGURE 5-22 A post and beam building under construction. *Used with permission of Jim Smith, Jr.*

FIGURE 5-23 Firefighters must study all types of building. A disciple of Abraham Lincoln tries his hand at constructing a log cabin in the inner city.

Plank and Beam

Similar to post and beam, this system uses beams of significant size covered with thick planks or laminated boards. This type of construction is popular for churches, skating rinks, and buildings requiring a large, clear span with no center supports. The beams are often boards laminated together to form large beams. This integrated structure typically uses tongue-and-groove planks that are a minimum of two inches thick. The interior of the plank can become a finished surface. These structures provide a massive fire load. The flammable interior finishes can allow a rapid fire spread.

BALLOON AND PLATFORM

Balloon frame and platform frame are the most common types of frame construction. The wall studs are normally 2-inch by 4-inch lumber set 16 inches apart. The bearing walls support the structure. The bearing walls can be any of the exterior walls, depending on the layout of the building. Similar to ordinary construction because standard framing lumber comes in a maximum of 20-foot lengths, the building is typically designed so that the floor and roof joists sit on or in the exterior bearing walls. The difference is that the bearing walls are constructed of wooden members. If spans over 20 feet in length are desired, then interior walls must act as bearing walls. Lightweight trusses and wooden I-beams may be used to span wider areas.

FRAME ROW DWELLINGS

Frame row dwellings as a rule utilize the adjoining sidewalls as bearing walls. Contiguous buildings have more support under fire conditions than a freestanding building due to their interconnection, but this attachment will also magnify the exposure problem. Common attics and basements will increase the potential for spread of fire and need to be addressed by an aggressive interior attack with adequate ventilation to release the pent-up gases and to prevent mushrooming and spread of the fire.

NOMINAL-SIZED LUMBER

The lumber industry today utilizes nominal-sized lumber. It is smaller than the named piece of material. For instance, a piece of wood referred to as a "two-by-eight" (meaning 2 inches wide by 8 inches deep) actually measures 1½ inches by 7½ inches. The difference is due to cutting and planing of the original-sized piece of lumber (2 inches by 8 inches) by the lumber mill. Because there is less wood contained in the finished piece of lumber, it contains less support than wood of full-dimensional size.

WOOD DETERIORATION

Wood can be attacked by a variety of causes. A major problem exists when wood-boring insects bore out the interior of a wooden beam. (See Figure 5-24.) The types of insects can vary throughout the country, and include termites, carpenter ants, and powder post beetles. Another complication occurs when wood is constantly exposed to dampness. It will develop rot or fungus that attacks and destroys the wooden fibers.

FIGURE 5-24 A building under attack from termites and other wood boring insects can be severely damaged if untreated. *Used with permission of William V. Emery.*

It is not uncommon to see older multistory frame buildings in which the floors sag. Overloading of the building or a prolonged roof or plumbing leak can cause this condition. The leaking water can weaken the wooden beams and cause serious structural problems. Firefighters anticipate that structural members will react a certain way under fire conditions. But wood that has been exposed to wood-boring insects or dry rot can fail quite readily and unexpectedly under attack by fire.

WOOD TREATMENTS

Wood can be treated to resist attack from fire, moisture, and insects. To retard attack by fire, wood is treated with phosphates and sulfates. In recent years, problems have arisen in some areas with the fire-retardant–treated plywood used in roofing assemblies. The plywood has deteriorated due to the high humidity and high temperatures that develop in attic spaces. This combination has resulted in a major weakening of the plywood, causing it to become so brittle that the roof may not support firefighters under normal conditions, let alone during a fire. To prevent rot, creosote has been used on piers and in other locations where wood is in contact with water. Firefighters must use care when fighting fires involving this material. Burning creosote gives off carcinogenic vapors. Exposure can occur by inhaling these vapors or absorbing them through the skin. Overhauling of pier areas or other locations that utilize creosote-covered wood can present a safety problem. Eye protection is needed to prevent inflammation or permanent eye damage.

Pressure-treated or green lumber is wood permeated with various chemicals to resist attack from moisture and insects. Common treatments have used phenol and arsenic, so the smoke and the remaining ash from pressure-treated wood may be toxic.

The weight of the treated lumber may lead one to believe that the wood is actually stronger. In reality, the impregnation of lumber with phosphates or mineral salts for fire protection, or with chemicals to prevent insect or moisture attack, results in sapping some of the strength of the lumber.

ATTACKING FIRES IN FRAME BUILDINGS

Since the type of construction has a direct impact on fire spread, our knowledge of construction features will assist in predicting likely routes of fire travel. A firefighter arriving at the scene of a fire in a building of balloon frame construction can be confronted with smoke pushing out from the attic vents. An inexperienced firefighter seeing the location of the smoke may incorrectly assume that the fire is confined to the attic. Actually, the fire could be burning at a lower level and extending to the attic via the void spaces in the walls.

EXTERIOR WALLS

The exterior walls of frame buildings can include materials that will supply fuel to a fire. They can be covered with shingles of wood or asphalt or a siding material of wood, asbestos, aluminum, or vinyl. Noncombustible finishes can include stucco, stone facing, or brick veneer.

Stucco (a masonry product) is applied over a wire mesh attached to an existing wall, bonding both surfaces together. It can also be purchased in ready-to-install four-foot by eight-foot sheets on which a thin layer of stucco is attached to a composition board. Radiant heat from an exterior fire can conduct heat through the stucco, igniting the material to which it is attached.

Stone facing can be installed in a way similar to stucco, in very thin layers. The protection it offers depends on its thickness and the method of attachment to the building.

Brick veneer is a single course or wythe of brick attached to a building for aesthetic purposes. (See Figure 5-25.) It offers no support for the structure;

FIGURE 5-25 Brick veneer offers no support for the structure; in fact, the brick adds to the building's dead load. *Used with permission of Jim Smith, Jr.*

in fact, the brick adds to the building's dead load. The brick does protect the wall to which it is attached from an exposure fire. The attachment to the wall studs is its weak point. If the brick veneer's attachment was attacked by fire and destroyed, it would be freestanding and could collapse.

> Wythe: the vertical section of a wall one masonry unit thick.

Communities can have a large number of frame buildings with combustible exterior walls and minimal spacing between them. A fire involving a building's exterior can quickly expose surrounding structures. Asphalt shingles, wood shingles, cedar shakes, and similar exterior finishes will burn readily, creating a hot, fast-moving fire.

Fire heavily involving the exterior of a building must be addressed with an aggressive attack using either a mobile 2½-inch hose-line or a master stream.

Care should be taken to leave window glass intact until all visible exterior fire has been knocked down. The exterior fire will find enough avenues to enter a building without the firefighting efforts assisting in providing paths through broken windows.

A quick knockdown of the fire will allow firefighters with handheld hose-lines to enter the structure and check for extension of fire. Areas around doors, windows, roof areas, overhangs, attics, and porch roofs should be considered priority locations where fire can enter a structure.

CHIMNEY FIRES

Chimney construction for frame structures can use traditional masonry or a metal prefabricated design within a framed enclosure. When incorporated into new construction, the framed chimney enclosure can create a void space from the lowest level through the roof, interconnecting other void areas.

A fire entering this void space will endanger the entire void space within the building. Likewise, a fire starting on the exterior and burning into the space can cause the same problem.

Timber Truss

timber truss ■ Truss construction using wood trusses at least four inches by six inches in size.

There are two types of wood truss construction: lightweight truss and **timber truss**. The parts in both of these trusses consist of a top and bottom *chord* and intervening members, referred to as *web members*. Trusses are distinguished by the size of the framing members (the top and bottom chords and any end pieces, if they are utilized). To be a timber truss, the minimum size of these members must be four inches by six inches. (See Figure 5-26.) The web members can be smaller in dimension. Trusses with framing members having smaller dimensions would be considered lightweight.

The timber truss comes in various designs. Triangular and bowstring trusses are the more familiar shapes. The parallel chord truss, which has gained notoriety utilizing lightweight components, is also used for the timber truss.

A timber truss can be built completely of wood. Wooden dowels and mortise and tenon joints can attach the intersecting members. However, the more common method of attaching chord to web members incorporates some form of steel into the truss. One method is the split ring connector. (See Figure 5-27.) The split ring spreads the pressure over a wide area, rather than just concentrating it on the

FIGURE 5-26 To be a timber truss, the size of the framing members must be a minimum of four by six inches. *Used with permission of William V. Emery.*

bolt, relieving some of the shear placed on the bolt. Steel rods can be used as integral parts of the truss. They may take the place of web members, or they may be used to stabilize the top and bottom chords. Turnbuckles may be attached to the steel components to adjust and align the truss.

The interconnection of adjoining trusses gives them lateral strength and is called *bridging*. Various methods of bridging are utilized in timber truss. It is usually accomplished by attaching wood or steel from the chord of one truss to an adjoining truss.

BENEFITS OF A TRUSS

The benefits of lightweight and timber trusses are similar. Since less material is used, the overall cost of the structure is less. Through the use of geometry and the triangle principle, smaller pieces of lumber properly constructed can support heavy loads and span wide areas. With the introduction of computer-aided designs, the possible shapes of a truss are endless.

IDENTIFICATION OF THE TIMBER TRUSS

The roof of a structure is the most important factor in determining whether a building will fail under fire conditions. (See Figures 5-28 and 5-29.) There are various methods

FIGURE 5-27 A split ring connector relieves the bolt of some shear and spreads it over a wide area. *Used with permission of William V. Emery.*

FIGURE 5-28 This photo shows damaged bridging and high-piled stock. A fire attacking the bridging could precipitate an early failure of the bridging and thus the truss. *Used with permission of William V. Emery.*

FIGURE 5-29 The bowstring-shaped truss is easily recognized in this bowling alley's roof to determine if a timber truss was utilized in roof construction. The most accurate and safest method is to inspect the building. An inspection and walk-through of the structure, followed up with a detailed preplan listing the building components, including the type of roof construction, is invaluable. If the roof consists of timber trusses, the actual size of the members and the method of attachment should be noted. A timber truss can consist of framing members that are 4 inches by 6 inches with steel attachments, or of members that are 12 inches by 16 inches containing no steel, which will better resist the effects of fire.

If no preplan exists and a determination is being made at the scene of an emergency, some indicators that a timber truss is supporting the roof include:

- Occupancy
- A large open area under the roof
- The shape of the roof

The timber truss is utilized in supermarkets, bowling alleys, theaters, large auto garages, churches, and manufacturing plants. Though a large unbroken area is an indicator of a timber truss, the problem with using this indicator is that interior walls can be erected to mask this feature.

Another indicator is the shape of the roof. The characteristic shape of the bowstring truss can warn of its presence. The problem is that modifications can hide the roof shape. The addition of a rain roof to facilitate the runoff of rainwater, or alterations to the external walls by adding a facade, can disguise the original roof shape.

REASONS FOR TRUSS FAILURE

Though timber trusses can be protected by covering them with fire-rated drywall or by installing sprinklers, most are not.

A weakness of all trusses and the major reason for truss failure lies in the connection points. The heat of a fire will attack the truss on many fronts. The exposed wood will ignite. The steel parts of the truss will also be attacked. Steel distributes heat very well. In doing so, it will conduct heat to the inner wood surfaces, destroying the wood fiber in contact with the steel. The attack of the fire on the truss causes web members to shift and accelerate the truss failure. As the steel absorbs the heat of the fire, it will elongate. At temperatures in excess of 1100 degrees F, it will fail. (See Figure 5-30.)

Truss members connected together with steel gusset plates may not have wood-to-wood contact, allowing the heat of the fire quickly to distort the steel, causing the truss to shift and fail. (See Figure 5-31.) The severity and duration that a fire has been burning and exposing the truss members are important.

The major problem associated with the use of steel in any truss is that it fails readily under fire conditions. A truss consisting of all wooden members will better resist the effects of fire than a similar-sized truss that employs steel in its construction. (See Figure 5-33.)

A

B

FIGURE 5-30 A–B Fire has attacked these gusset plates. Heat is conducted to the interior of the wood through the bolt and the split ring connector, contributing to the collapse of the truss. *Used with permission of William V. Emery.*

A

B

FIGURE 5-31 A–B Timber truss may not have wood-to-wood contact; when under attack by fire, this will cause the truss to fail. Fire attacking this truss will cause the steel to expand, twist, and cause the truss to fail, as can be seen in the gusset plate from a failed truss. *Used with permission of William V. Emery.*

FIGURE 5-32 Renovations and additions to this building, which was previously a supermarket, make identifying the bowstring difficult if the IC is too close to the building. Heavy fire destroyed the building and caused early failure of the truss roof.

FAILURE OF TIMBER TRUSS ROOFS

Timber truss roofs fail, as do all types of roofs. A heavy body of fire in the truss area must act as a warning sign. Supporting members may be overloaded, whether by storage of material or other loads. The installation of loads could be atop the roof, such as air conditioning units, or beneath the roof, such as stationary heaters, shelving units, or light fixtures. These modifications place an additional load on the truss and can accelerate a collapse under fire conditions.

Timber truss roofs that contain sloping hip rafters (sloping hip rafters are considered framing members) have four bearing walls. The sidewalls usually carry the weight of the truss, and the front and rear walls carry the sloping hip rafters. When truss failure occurs, there can be a violent collapse of all four walls. As the truss fails, it can push out the sidewalls it rests upon and shift its weight onto the hip rafters. This action exerts outward force on either the front or rear walls, or both, setting the stage for a violent collapse.

Timber trusses can span areas of 100 feet and be spaced 20 feet on center. Failure of one truss with this spacing can create a roof opening of 40 feet by 100 feet.

A timber truss may have a ceiling attached beneath the truss, or the truss area can be one open void together with the top floor of the building. During a fire, the void can be fully charged with smoke and be venting to the outside. It may be difficult to estimate the seriousness of the fire. This necessitates constant monitoring from both the interior and exterior.

We often judge interior conditions by our ability to advance a line into a fire area. If we are unable to push forward into a building, it indicates unfavorable conditions and increased damage. Interior divisions may have no visible fire, yet have high heat conditions. Checking whether runoff water is hot to the touch can assist in determining if water lines are hitting the fire. With high ceilings, the heat of the fire may be contained in the roof area and may not be felt by firefighters on entering the building. This may lead firefighters to the conclusion that the fire is not that serious. Because visibility on the interior is usually impaired, one of the best indicators is the volume, intensity, and color of the smoke being monitored from the exterior.

The tendency of units operating on the interior is to say, "Chief, just give us five more minutes, and we'll have this fire under control!" This is a great attitude, often guided by a desire for completion of their assignment without full knowledge of the situation.

Progress reports from both interior and exterior divisions and groups will keep the IC abreast of all developments. If control isn't accomplished in a reasonable period of time, withdrawal from the structure and the initiation of a defensive attack must be considered. After the fire has heavily involved the truss area, it is a lost cause to continue an interior attack.

How soon will a timber truss fail under fire conditions? By utilizing our experience and having accurate building information, we can make safe decisions under critical fire scene conditions. (See Figure 5-34.)

FIGURE 5-34 A bowstring truss roof has collapsed, causing truss members and roofing material to fall onto the top floor of the building.

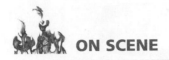

ON SCENE

The fire service has a less than distinguished record with fires in wood bowstring truss roofs. In 1978, in Brooklyn, New York, firefighters were operating on the roof of a supermarket when one of the wooden bowstring roof trusses collapsed into the building, sending twelve firefighters into the burning inferno, six of them to their death. In 1988, in Hackensack, New Jersey, five firefighters perished while operating on an interior attack under a collapsing wooden bowstring truss roof of an auto dealership. In 2010, in Chicago, Illinois, two firefighters were killed in a building collapse while fighting a fire in an abandoned building with a bowstring truss roof.

Lightweight Building Components

The introduction of **lightweight building components** has changed the way buildings are constructed. Previously, the use of solid beam lumber led to well-built, sturdy structures. Lightweight building components have come to replace full-dimensional lumber in both frame and ordinary constructed buildings. Their popularity is based primarily on economics: There is a lower cost per unit, coupled with reduced labor due to the lighter weight. They allow for faster assembly and require fewer skilled workers. Today, more than 90 percent of all housing units use some form of lightweight construction. Preferred components include triangular and parallel chord trusses, wooden I-beams, and steel bar joist. (See Figure 5-36.)

lightweight building components ■ Have replaced full-dimensional lumber in both frame and ordinary construction. Preferred components include triangular and parallel chord trusses, wooden I-beams, and steel bar joist.

Lightweight construction uses fewer raw materials and has the advantage of allowing wide-open areas without the need for columns or interior walls. Because lightweight construction uses less material, the structure weighs less. Well-built lightweight components are dependent on proper installation and bracing. Trusses are built with set bearing points. If installation and bracing specifications are not followed, local or general collapse may result. Because they are composed of less material, these trusses fail more readily than conventional construction when under attack by fire. (See Figure 5-37.)

SHEET METAL SURFACE FASTENERS

A common type of lightweight truss used in residential construction is the triangular and parallel chord truss with sheet metal surface fasteners. (See Figures 5-38 and 5-39.) These fasteners are referred to as gusset plates, gang nailers, or staple plates. The plate is of galvanized steel, 16 to 22 gauge, and has 90-degree V-shaped points punched into it that take the place of nails (the higher the gauge number, the thinner the steel). The points are from ¼ inch to ½ inch long. They attach the various members of a truss together. To build a truss that covers spaces up to 70 feet, smaller pieces of lumber must be connected

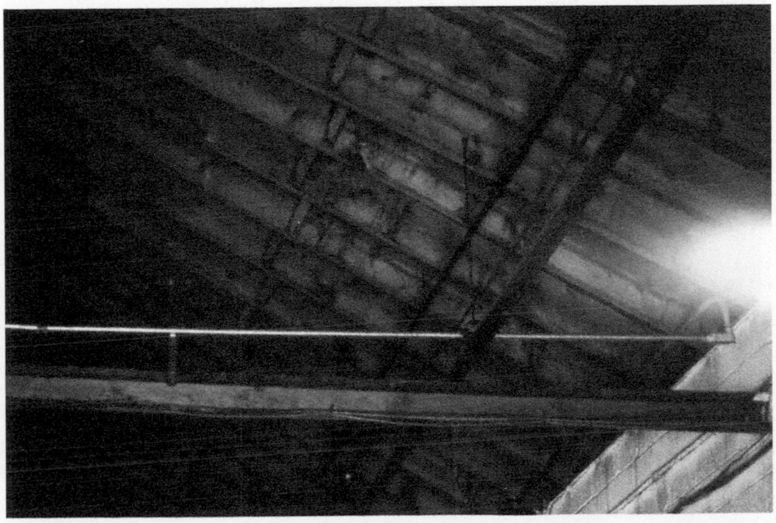

FIGURE 5-35 Sloping hip rafters in a bowstring truss roof. *Used with permission of William V. Emery.*

FIGURE 5-36 A–B Today more than 90 percent of all housing units use some form of lightweight construction. This roof is constructed with one-inch by three-inch wood truss components attached to a wooden I-beam that is being utilized as a ridgeboard. This roof is meant to keep the weather out of a building and not as a platform for firefighters to operate on. Roof collapse will be quick and without warning. *Used with permission of John Christmas.*

FIGURE 5-37 Lightweight truss allows wide-open areas with few if any fire-stops.

together. This is done by butting the ends of the wood together and attaching them with sheet metal surface fasteners. The connection point is an area where failure can and does occur under fire conditions.

PLYWOOD AND ORIENTED STRAND BOARD

Common materials used in frame buildings and other types of construction include plywood and oriented strand board. These materials are used for roof decks, flooring, and exterior walls. Thicknesses for these materials are commonly from ⅛ inch to 1⅛ inches. The common size of these sheets is four feet by eight feet, but larger sizes are available.

Plywood is manufactured from thin sheets of cross-laminated veneer and bonded under heat and pressure with strong adhesives. The more layers used, the greater the strength of the plywood. Wood is strongest along its grain, and shrinks and swells most across the grain. By alternating grain direction between layers, strength and stiffness in both directions are maximized, and shrinking and swelling are minimized in each direction.

OSB is widely used in construction sheathing as the web materials for wooden I-beams and other applications. It is composed of compressed wood strands arranged in layers to form a mat. The individual strands are typically three to six inches long. These mats are then oriented at right angles to one another. The orientation of layers achieves the same advantages of cross-laminated veneers in plywood. Like the veneer in plywood, these mats are layered and oriented for maximum strength, stiffness, and stability.

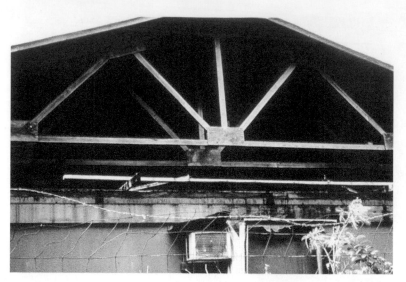

FIGURE 5-40 One of the first types of lightweight wood truss was constructed utilizing wooden gusset plates. Many of these trusses still exist in buildings today.

WOODEN GUSSET PLATES

Trusses also have been constructed with plywood gusset plates. (See Figure 5-40.) Under fire conditions, plywood has a tendency to delaminate, exposing additional fuel to the fire, and thus accelerating collapse. Conversely, when lumber chars, it has been found that the char offers some protection to the wood in the form of insulation.

FAILURE UNDER FIRE CONDITIONS

The main problem we find with either method of truss member connection is that failure under fire conditions can occur quite readily. The plywood gusset plate will delaminate, and the sheet metal surface fastener will fall out when the wood is charred to a minimum depth of ¼ inch. Either action will cause failure of the truss.

When steel and wood are combined to form a truss, the wood is used for the top and bottom chords and the steel is used for the web members. (See Figure 5-41.) One method uses a corrugated steel web member. Each end of the web member is formed into a sheet metal surface fastener that attaches to the sides of the top and bottom chords. Another method uses web members with steel tubing flattened on each end and a hole drilled into the flattened part. A slot is cut in the chord members, and the tubing is held in place by a bolt or steel dowel pin through the chord.

A problem associated with steel is that it conducts heat to the interior of the wood. The heat destroys the fibers of the wood on the interior, while the fire simultaneously attacks the exterior. This two-prong attack causes a quick failure of the truss.

FINGER-JOINTED TRUSSES

Gusset-less trusses are produced and used primarily to support floors. The parallel chord trusses use a finger-joint process to connect two wooden members (chords to the web members or to extend the length of a chord) in lieu of sheet metal surface fasteners or wooden plates. They are connected using the adhesive phenol resorcinol. A concern is how these trusses will hold up under fire conditions. Phenol resorcinol adhesive's auto-ignition temperature is approximately 1130 degrees F compared to wood's 520 to 880 degrees F. It has been reported that the glue does not soften, lose bonding capabilities, or break down chemically at temperatures below the wood's charring temperature.

Tests conducted at the Forintek Canada Corporation showed that the finger-joints survived fire exposure until the wood members themselves were consumed; and when a web or chord failed, it did not cause the others to dislocate.

(For additional information on fire performance of finger-jointed trusses, visit:

- Carbeck Structural Components Institute training at www.fire. carbeck.org and click on "Fire performance of finger-joints"
- Structural Building Components Association (SBCA) at www.sbcindustry.com/firepro.php
- Open Joist Triforce at http://www.openjoisttriforce.com/

WOODEN I-BEAMS

Another lightweight component in use is the wooden I-beam. (See Figure 5-42.) It is also referred to as the I-joist, and manufacturers also use trade names to identify their brand of I-joist. As with the truss, the top and bottom members are called chords, and they are connected by a web consisting of plywood or OSB that is also referred to as a spline or stem. The chord is slotted to accept the web, which is glued into place.

FIGURE 5-41 When wood and steel are combined to form a truss, the top and bottom chords are usually wood and the web is metal. *Used with permission of William V. Emery.*

FIGURE 5-42 This wooden I-beam consists of chords that measure 1½ by 1¾ inches and has a ⅜-inch plywood web. *Used with permission of William V. Emery.*

Common stock sizes for the wooden I-beam contain chords as small as 1½ inches by 1¾ inches, with a web of ⅜-inch plywood or OSB (longer spans can increase the chord size to 1½ inch by 3½ inch and the web size to ⁷/₁₆ inch).

Wooden I-beams can span wide distances when used in floor systems and roof assemblies. When properly engineered and installed, the I-beam has proven to be an economical building component.

Improper installation and alterations will affect an I-beam's strength. When purchased, wooden I-beams are accompanied by detailed instructions on proper installation. Field experience has shown that required steps are ignored by many installers, leaving buildings that contain minor and major flaws. (See Figure 5-43.)

This is compounded by the fact that plumbers, electricians, and other workers run their piping, ductwork, and wiring through the joist. (See Figure 5-44.) They often fail to adhere to directions of where to place their openings in the web. Openings too large, too close together, or placed in the chord can drastically reduce the carrying capacity of a wooden I-beam.

Temporary bracing must be in place during the erection of wooden I-beams. Field modifications and shortcuts during construction can result in collapse of the I-beams. Response to such a collapse may find workers trapped, and rescuers must anticipate the possibility of secondary collapse.

When burning, the plywood web of an I-beam can delaminate, rapidly reducing its strength.

FIGURE 5-43 A wooden I-beam set in this masonry wall has been fire-cut and will fail readily if attacked by fire. *Used with permission of William V. Emery.*

FIGURE 5-44 Electricians, plumbers, and other tradespeople routinely cut holes in wooden I-beams to install wiring, piping, ductwork, and other utilities. If not properly sized and spaced, the holes can drastically weaken the joists.

FIGURE 5-45 Wooden I-beams installed in this assembly are showing some deflection. They may hold up under normal circumstances but can fail readily under attack by fire. *Used with permission of William V. Emery.*

CONSTRUCTION SITE DANGERS

Lightweight building components that are stored outside on a construction site prior to installation are prone to damage. (See Figure 5-46.) Because extra components are not manufactured, damaged or hastily repaired components may be installed. These damaged units may hold up in day-to-day use but can fail quite readily under fire conditions.

FIGURE 5-46 Trusses are typically stored outside at a construction site and may be damaged before being installed. *Used with permission of William V. Emery.*

During construction, these components are not a completed unit and lack strength. Demolition can occur when a building is being destroyed by fire.

Failure of a connector or any part of the truss causes web members to shift and can cause the entire truss to fail. Because trusses are interconnected, multiple truss failures can occur.

INCREASING THE CARRYING CAPACITY

There are various methods to increase the carrying capacity of a structural member. If the width of a beam or truss is doubled without a change in the depth (2 by 6 inches to 4 by 6 inches), the carrying capacity of that member is doubled. However, if the depth is doubled without an increase in the width (2 by 6 inches to 2 by 12 inches), the carrying capacity is increased four times. This formula heavily favors lightweight construction because the components are often only 2-by-4-inch pieces of lumber. Increasing the depth of the web members drastically increases the overall carrying capacity without a corresponding increase in the size of the individual members.

There are other methods to increase the carrying capacity of a truss:

1. By increasing the number of web members, the amount of top chord support increases.
2. Trusses can be placed side by side to support a critical point.
3. The size of truss members can be increased from two by four inches to two by six inches.

TRUSS VOID AREAS

When solid joist lumber is used in a floor or roof system, a void area is created between joists. This area consists of the length and depth of the beam and the width between adjacent joists. A length of up to 20 feet long and spacing of 16–24 inches wide is common. A fire starting in this void area has a limited supply of air to sustain it.

With parallel chord trusses, one large void area is created. (See Figure 5-47.) This is often referred to as a *truss void* or *truss loft*. A parallel chord floor truss assembly creates a situation similar to balloon frame construction. This large space allows an entire floor area to be interconnected, and it contains a sufficient amount of air to sustain a fire.

Many buildings have interconnected void areas. This occurs when horizontal truss voids combine with vertical voids where utilities pass through the various floor levels. This interconnection increases the size of the void space that is vulnerable to attack by fire. (A floor void space covering an area 20 feet by 45 feet and 1 foot deep contains 900 square feet. This is equivalent to a 9-foot by 12-foot room with a 7-foot ceiling. Combine this with attic spaces and other interconnecting void spaces, and our problem is compounded.)

To minimize the potential problem of a fire being able to spread throughout large void spaces, most codes have limits on the size of these spaces. This is a form of compartmentation in the truss area and is called *draft-stopping*. Some codes require compartmentation of floor areas over 1,000 square feet and attic areas over 2,500 square feet depending on the building's use.

FIGURE 5-47 A truss void space or truss loft is created by the parallel chord truss in this building. Many times, these void spaces are interconnected with the void spaces on other floors due to chases created by tradespeople for utilities to pass through the floors.

The main problem is that the draft-stopping is too easily negated. This happens due to the use of poor materials, or poor installation or piercing by workers or maintenance crews. Draft-stopping is not a priority area for either the builder or the occupant and can easily be circumvented by creating *poke-throughs* (holes created during postconstruction renovations), which negate the draft-stopping by again creating one large area.

VOID SPACE FIRES

Probably no other factor indicates a more dangerous condition for a firefighter than a fire that is attacking lightweight building components. There is no surplus lumber in the individual unit. Each member is critical to its overall strength. (See Figure 5-48.) Failure will occur quite early, shifting the load previously carried by the collapsing member onto the adjacent members and leading to multiple failures.

Void space fires attack lightweight components that are especially vulnerable to fire, causing them to fail quickly and easily. The existence of a void space fire may not be recognized until the collapse of lightweight components occurs. Fire can spread from floor to floor in multistoried structures. The wall voids often interconnect with floor voids.

Fire can attack the void in a variety of ways. It can start in the void from electrical wiring. It can burn into the void from below or above.

A mattress fire can project enough heat downward to allow fire to enter the void. Firefighters must be careful, checking whether a fire that has burned for a period of time has damaged the flooring.

When a truss is damaged, all firefighters operating in the area must be told. This is especially important for those operating above the damaged areas. Weakened trusses may not be recognizable by the firefighters operating above. Firefighters can easily fall through flooring damaged by fire. The collapse may not occur until the weight of a firefighter is placed on the floor. If fire is suspected in any concealed space, it should be opened quickly and investigated.

A critical cue and probably the most dangerous situation from the standpoint of firefighter safety is a fire burning in a truss void area. It is difficult to place time frames on anticipation of the collapse. Though there have been tests to determine how long it takes for collapse under direct flame impingement, it is usually not known when the flame contact first occurred.

SITE INSPECTIONS

Knowing whether lightweight components have been used in the construction of a property must be accomplished through preplanning and documentation during the construction of the building. (An exception is if buildings that have used lightweight materials in their construction are visibly marked with placards on the exterior of the building, indicating the presence of truss construction or other lightweight components used as structural components.)

REACTING UNDER FIRE CONDITIONS

How should firefighters react when confronted with a fire in a structure containing lightweight

FIGURE 5-48 Truss defects can include cracked wooden members. *Used with permission of William V. Emery.*

components? Should firefighters change their method of fire attack? We must still consider our incident priorities. Due to inherent weaknesses in these structures, it may be difficult for firefighters to protect both the occupants and ourselves. Lightweight constructed buildings fail rapidly and with little warning. This complicates rescue efforts. The potential for firefighters to become trapped or involved in a collapse is increased. As always, firefighters should err on the side of safety by protecting themselves, while affording the best chance of survival to a civilian trapped or overcome in a building (this balance can be extremely difficult).

It is easy to say that buildings made of lightweight materials are disposable buildings and that we should fight the fire from the exterior after the building is evacuated. In a one-story single-family dwelling with fire through the roof, this would probably be considered a sound decision. But what about a large commercial building or a strip mall containing many stores, one of which is on fire? We must attempt to contain the fire to as small an area as possible while protecting our firefighters. This requires fighting the fire from the interior to achieve containment and control. Continuous monitoring of the building's stability will be needed.

FIGURE 5-49 Truss defects can include chord members of smaller dimension. *Used with permission of William V. Emery.*

FIREFIGHTERS' ACTIONS

There are a number of actions firefighters can take to protect themselves when confronted with buildings using lightweight building components as structural members.

1. Let those in power in your community (the authority having jurisdiction) realize the tremendous danger to both citizens and firefighters when buildings containing lightweight components are involved in fire.
2. As firefighters, get involved. Push for laws requiring sprinklers in living areas and in void spaces containing lightweight components. Argue against tradeoffs allowed by our generous codes, architects, and engineers that would eliminate sprinklers in these locations.
3. Read a building for indicators that could denote the presence of a truss (age of building, occupancy, type of building, large unsupported spans).
4. Find a way of identifying these buildings. We always speak of preplanning, but marking the front of these structures to indicate the presence of lightweight components is a better method that will be immediately visible to firefighters on their arrival.
5. Study known weaknesses found in lightweight components and problems related to improper installation. These include sheet metal surface fasteners not properly set, knots located at joint locations, use of split or smaller dimension of lumber at joint locations, defective or damaged lumber, and improper installation. (See Figure 5-49.)
6. Visit construction sites and observe the workmanship in assembling buildings with lightweight components. Be prepared to point out potential flaws to the construction crew that could affect building stability under fire conditions. (See Figures 5-50 through 5-54.)
7. Realize that some tests conducted to determine how lightweight components will react under direct fire attack have found them to fail in less than two minutes. This should be interpreted by firefighters as meaning that *if fire is burning in a void or a location containing these components, an immediate failure should be anticipated.*
8. Understand that draft-stopping and fire-stopping is nonexistent in older structures. Though required in new construction, these newer assemblies are often violated and should be considered one undivided

FIGURE 5-50 Truss defects can include a web member not full-sized. *Used with permission of William V. Emery.*

FIGURE 5-51 Truss defects can include a damaged sheet metal surface fastener not embedded in the wood. *Used with permission of William V. Emery.*

FIGURE 5-52 Truss defects can include a sheet metal surface fastener that is offset. *Used with permission of William V. Emery.*

structure. Draft-stopping is frequently violated by using the head of a hammer to make an opening in drywall. Though this may seem like a minor problem, testing on poke-throughs in other types of construction has found that the heat of a fire will be quickly transmitted through openings as small as a pencil and equated on the other side.

9. Realize that draft-stopping can be a material (such as plywood or OSB) that will add fuel to a fire.

10. Recognize that triangular truss roof spaces are often used for storage. This can have a drastic effect on the load-carrying capacity of the truss because the live load can be much greater than anticipated, and the potential for a rapid collapse can exist. A cue would be the presence of a drop-down collapsible ladder placed in the ceiling area.

FIGURE 5-53 Truss defects can include a damaged chord. *Used with permission of William V. Emery.*

11. Educate firefighters to the signs of failure, collapse indicators, and the weaknesses associated with lightweight components. Encourage them to be vigilant for these signs.

12. Get feedback. Alert the division or group supervisors to communicate collapse indicators to the IC immediately.

13. Be concerned when operating in an altered or renovated building. It may contain deadly surprises. Full-dimensional lumber is often replaced with lightweight components. This can change how a fire will travel within the building and increase the collapse potential.

14. Anticipate the interconnection of void spaces in a building, allowing rapid spread of fire throughout the structure. (Conditions

will be similar to those found in balloon-frame-constructed buildings in which void spaces throughout the structure are interconnected. The major difference is the smaller components will fail much faster than in the older balloon-frame buildings.)

15. When forced to operate above potentially weakened areas that contain lightweight components (search and rescue), try to stay near the exterior walls or spread a portable ladder across the area to distribute the weight of the firefighters over a greater area. If possible, select a smaller firefighter for this assignment to reduce the load placed on the structure. Make greater use of thermal imaging devices to both quicken and enhance the search.

16. When pulling or opening ceilings, attempt to do so from doorways. This will utilize the safety of the doorway itself and permit a ready path of egress if an emergency evacuation is necessary.

FIGURE 5-54 Frame buildings provide ample fuel and can create a fast-moving and dangerous fire situation. This residential structure collapsed due to the heavy fire conditions. *Used with permission of Deputy Chief Thomas Lyons.*

17. If opening ceilings in a large room, utilize only those personnel who are needed to complete the task, and have them operate from near the exterior walls for additional safety.

18. If pulling or opening ceilings and a part of the truss is hooked by the tool, do not attempt to pull down the ceiling. If the truss is weakened, this action could cause the collapse of a large portion of the ceiling.

19. Be alert to firefighters who may fall through weakened floors from the load imposed by their own body weight. Wall-to-wall carpeting may keep them from falling completely through an opening because the carpeting can restrict their fall. They may also be able to hold onto a hose-line and use it to climb out of the opening or be aided by other firefighters who can pull them to safety.

20. Realize that the large void spaces created by lightweight trusses can supply a fire with a large amount of fuel and oxygen. This combination can create a free-burning fire that produces little smoke. This lack of heavy smoke showing on the exterior can lead an IC to underestimate the amount of fire involvement and the seriousness of the situation (again, feedback is needed).

21. Be aware that abnormal amounts of smoke with little visible fire present may indicate the presence of a partition or void space fire. These concealed spaces are often interconnected throughout the building. Interior crews can experience:
    - Continued generation of heat after all visible fire has been knocked down
    - Blistering paint and smoke pushing from around door, window, and baseboard moldings and from electrical switches or outlets

    Exterior crews can experience:
    - Continued generation of heavy smoke from the structure
    - What appears to be lack of progress by interior lines, despite reports to the contrary

22. Anticipate rapid failure of lightweight assemblies. Recognize that theoretical time frames are practically useless. There is no way for firefighters to know when a fire started, only when the fire department was called. Tests on lightweight assemblies are performed in sterile laboratories, with perfectly constructed samples installed to exact specifications, and with properly set nails and every possible penetration sealed. Real-world assemblies are often much different, due to shortcuts, poor workmanship, alterations, or areas that may inadvertently be overlooked. Collapse within 10 minutes after arrival of the first-due units has occurred in a number of incidents, resulting in firefighter deaths (Memphis, Tennessee, seven minutes after arrival; Branford, Connecticut, nine minutes after arrival).

23. Realize that sheet metal surface fasteners (metal gusset plates used in lightweight truss construction) will be attacked by the fire, which will destroy their connection

FIGURE 5-55 Systems installed on the roof will add to the dead load of a building.

to the wood. The breaking of this adhesion will accelerate the failure of the truss. Should firefighters encounter sheet metal surface fasteners lying on a floor when entering a fire area, they should immediately retreat to a safe area. These sheet metal surface fasteners would have fallen out from overhead trusses in an unfinished ceiling and would indicate the potential of an imminent collapse.

24. When ventilating a roof or cutting through a floor, firefighters must not cut through the top chord or any part of the truss. This action could cause an immediate failure of the assembly, or of a large portion of the roof or floor.

25. Recognize that operations on roofs containing lightweight components are different than operations on a roof with solid wooden roof beams. Roof failure can occur without the warning signs apparent on a solid wooden-beam roof, such as creaking noises, sponginess of the roof, or roof sagging.

26. Be cognizant of dead loads on roofs (heating, ventilation, and air conditioner units, etc.). Firefighters sent to the roof should report these findings to the IC. (See Figure 5-55.)

27. Understand that a buildup of ice, snow, or water adds to the live load a roof must support and could precipitate an early collapse.

28. Realize that lightweight beams may be set farther apart than solid wooden beams. Failure of one beam will create a weakness over a larger area.

29. Recognize that the bridging that ties one truss to another truss interconnects lightweight assemblies. A collapse can be localized or, due to the interconnection or bridging of an adjoining truss, a collapse can cover a large area or can cause a general collapse of a roof or floor area.

When considering stability of truss assemblies, there is very little experience to draw upon. Trusses will fail quite readily under direct attack by fire. Realize that the truss is strong and, when built to specifications and installed properly, it will respond well if protected from the heat and flames of the fire. This assumes that the fire does not start in the truss area. The problems facing firefighters are shortcuts, poor workmanship, and changes that have occurred in the building since the time of initial construction. These changes can affect how a building will react when involved in fire. (See Box 5-1.)

 BOX 5-1: BUILDING REVIEW

How a building reacts under attack by fire depends on the type of construction. The following list summarizes strengths, weaknesses, fire resistance, and collapse potential of the basic types of construction.

Fire-Resistive

Strengths:	A well-constructed building in which no structural steel is exposed and all vertical openings are protected.
Weaknesses:	Sprayed-on fire protection on steel may be removed, exposing the steel. Spalling of concrete is possible under prolonged attack by fire.
Fire resistance:	Structural members generally receive two- to four-hour fire protection.
Collapse potential:	Only under sustained attack by fire does failure occur, and in those cases, it is usually localized collapse.

Noncombustible

Strengths: Structural elements will not contribute fuel to a fire. This type of building is often recognizable from the exterior.

Weaknesses: The early collapse potential of the unprotected steel members.

Fire resistance: The structural elements rarely receive fire protection and are exposed to the heat of a fire.

Collapse potential: Because the structural members are unprotected and exposed to the fire, they will fail rapidly, and an early collapse should be anticipated. Steel expansion can cause collapse of exterior walls.

Ordinary

Strengths: Full-dimensional lumber. Fire-cut joist. Masonry exterior walls are noncombustible.

Weaknesses: Common party walls. Joist may sit in same wall socket. Common cocklofts or attics may exist. Renovations should be expected. Void spaces exist. Lightweight construction under attack by fire can fail readily.

Fire resistance: Structural members are usually protected by plaster or drywall construction. Exterior walls are noncombustible.

Collapse potential: Older buildings of ordinary construction contain structural members of substantial size. These larger structural members hold up well under attack by fire but can and will fail, causing collapse. Lightweight members subjected to attack by fire will fail readily. Fire-cut joist will allow the interior of the building to collapse while the masonry walls remain intact. This feature can affect firefighters operating within the building.

Heavy Timber

Strengths: Large structural members will support a structure for an offensive attack. Load-bearing walls are noncombustible. There may be floor drains to drain the water used in firefighting. Normally, there are no void spaces.

Weaknesses: Floors may be oil-soaked. Unprotected openings may exist between floors. There may be an excessive fire load of stock, manufacturing process, or storage of finished goods. Alterations can create concealed spaces.

Fire resistance: Structural members are of substantial size and will contribute a large fuel load to a fire. After a fire is past the initial stages, it is very difficult to control and can burn for a prolonged period of time.

Collapse potential: Though constructed of substantially sized pieces of lumber and not prone to collapse, under prolonged attack, these buildings will fail.

Wood Frame

Strengths: Platform construction provides some barriers for vertical extension of fire. Interior wall coverings can be of plaster or drywall. Log, post and beam, and plank and beam buildings have substantially sized structural members.

Weaknesses: Wood will burn. Older buildings may have dry rot or damage from wood-boring insects. Void spaces are common and in balloon-frame can be extensive. Renovations are commonplace in older structures.

Fire resistance: Plaster or drywall can offer some protection for structural members. Exposed wooden members will contribute fuel to a fire.

Collapse potential: Frame structures do pose a collapse concern because they lose their load-carrying capacity as they burn. The type of frame construction will determine collapse potential. Log cabins are substantial and usually sustain only roof collapse. Other frame structures are prone to localized and general collapse.

Summary

The knowledge of how a particular type of building will react when it is being attacked by fire will allow a firefighter to predict how a fire can spread through a structure and whether stability will need to be a consideration.

The heavy timber truss and the lightweight building components have each caused firefighters problems when attacked by fire. There have been incidents with both types that have resulted in firefighter deaths. It is incumbent upon every firefighter to recognize when these types of construction are contained in a fire building and to realize the potential dangers that each type presents.

Review Questions

1. List the five types of building classifications used by the fire service.
2. What are the characteristics of a fire wall?
3. What are the problems associated with fires in heavy timber buildings?
4. Discuss some common reasons that vertical openings exist in heavy timber buildings.
5. Discuss the problems associated with fires in buildings of balloon-frame construction.
6. How do fire-cut floor joists exposed to heavy fire conditions affect firefighting operations?
7. List the five types of frame buildings and identify which types are found in your response district. Give a street address for each type found.
8. What are the indicators of a timber truss roof?
9. Discuss collapse considerations of timber truss roofs.
10. Discuss methods to strengthen the truss to increase its carrying capacity.
11. List the common types of defects found in lightweight trusses.
12. Describe how you would start a public relations campaign to emphasize the importance of sprinklers in residential buildings, especially those containing lightweight building components. What are the most critical points needed for a successful program?
13. Review the basic strengths and weaknesses of the five types of construction and discuss additional strengths or weaknesses for each type.
14. How should firefighters, when arriving on the scene of a working fire, use the exterior marking on a building that indicates the presence of lightweight structural components within?

Suggested Readings, References, and Standards for Additional Information

Smith, M. 2012. *Building Construction: Methods and Materials for the Fire Service*. 2nd ed. Upper Saddle River, NJ: Pearson.

US Fire Administration (USFA)

Routley, J. G. "Floor Collapse Claims Two Firefighters, Pittston, Pennsylvania." Technical Report 073.
Routley, J. G. "Wood Truss Roof Collapse Claims Two Firefighters, Memphis, Tennessee." Technical Report 069.
Schaenman, P. "Apartment Complex Fire, 66 Units Destroyed, Seattle, Washington." Technical Report 059.
"Attic Fires in Residential Buildings." *Topical Fire Report Series* 11, no. 6. January 2011. Available at: https://www.usfa.fema.gov/downloads/pdf/statistics/v11i6.pdf.

National Fire Protection Agency (NFPA)

Standard 220: Standard on Types of Building Construction

Related Courses Presented by the National Fire Academy, Emmitsburg, Maryland

Command and Control of Fire Department Operations at Natural and Man-Made Disasters
Fire Protection Systems for Emergency Operations
Fire Inspection Principles
Evaluating Performance-Based Designs

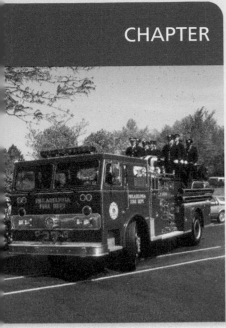

CHAPTER

6

Building Collapse and Scene Safety

66 Leadership is doing what is right, realizing your duty, accepting responsibility. 99

—US Army General
Norman Schwarzkopf

A firefighter's worst fear: losing a brother or sister firefighter in the line of duty. *Used with permission of James H. Bampfield, Jr.*

KEY TERMS

building collapse, *p. 210*

wall collapse, *p. 210*

collapse search, *p. 229*

safety, *p. 234*

OBJECTIVES

Upon completion of this chapter, the reader should be able to:

- Describe building collapse indicators.
- Describe a plan to address search at a building collapse.
- Discuss the role and responsibilities of an incident scene Safety Officer.

Chapter 6 covers three areas: building collapse, collapse search, and firefighter safety. Being trapped in a collapsing building is one of a firefighter's greatest fears. There are many reasons for building collapse during firefighting. Yet it is difficult to predict when a structure will collapse. By analyzing the building and monitoring it during a fire, some degree of predictability can be achieved.

Various types of collapses can occur. Interior collapse can involve floors, interior walls, ceilings, or the roof. Exterior collapse can include exterior walls, cornices, or overhangs. Every collapse is different, yet the common bond that exists is the potential for firefighter injuries or deaths.

This chapter looks at building collapse, with special emphasis on collapse indicators. If we can predict the potential for collapse, we can protect firefighters by removing them from the collapse area.

After a collapse has occurred, firefighters will be called on to find and extricate those who are trapped. To accomplish this in a safe manner will require control, coordination, and having a plan in place to ensure that safety procedures are followed.

This chapter will also review safety. Safety must be an attitude that starts with the fire chief and includes everyone in the department. The implementation of and adherence to all safety measures should be our goal. Safety begins with training as a new cadet, when safety rules must be reinforced in every training evolution. It must continue to be an important process in daily activities and incident scene operations. The appointment of a Safety Officer at an incident scene will assist in accomplishing a department's goal of reducing firefighter death and injuries.

Building Collapse

building collapse ■ The partial or major collapse of a building.

Most **building collapses** are not spectacular; they occur on the interior of a structure and usually are not noticeable from the exterior. There may be a collapsing plaster ceiling, a burnt-through floor, or a falling staircase. Suspended ceilings, or dropped ceilings as they are referred to in some communities, are installed below an existing ceiling. Tiles (two foot × two foot, two foot × four foot, or one foot × four foot in size) may be inserted into metal grid work along with lighting systems, and are typically supported by lightweight wires strung from the original ceiling. In situations where heavy fire attacks these ceiling assemblies, there is the possibility of a partial or total collapse of the suspended ceiling. Generally there is not a heavy dead load involved, but when ceilings covering large areas collapse and strike firefighters, it can cause disorientation and injury. It can also make exiting the area difficult by trapping a firefighter in a tangle of wires and ceiling material. Most firefighters are injured by these smaller and less spectacular occurrences. For this reason, firefighters and company officers must be alert to such hazards during operations.

Though most problems occur with partial collapses, preparation for a major collapse must be made when the indicators are present. A falling wall is an awesome sight to witness. The seemingly slow-motion demise of a structure collapsing, along with the massive destruction it wreaks, is breathtaking when viewed from a safe distance.

AGGRESSIVENESS

Aggressiveness will be counterproductive when dealing with buildings that have a high potential for collapse. It is a foolhardy decision to enter and fight fires in structures prone to collapse where no life hazard exists. The life safety of firefighters dictates that the safest method be employed, which could result in a longer period of time for total extinguishment, but an overall safer operation. If any doubt exists, the first-arriving officer must step back and consider whether firefighters should be committed to an offensive attack.

Wall Collapse

wall collapse ■ The failure of a wall.

Collapsing walls can trigger the devastating failure of a building. (See Figure 6-1.) A bearing wall will release the supporting connections of floors and roof members, resulting in their failure.

There are three basic types of **wall collapse:** (See Figures 6-2 through 6-4.)

1. 90-degree collapse
2. Inward-outward collapse
3. Curtain collapse

The 90-degree wall collapse assumes a wall will fall outward its entire height, encompassing a 90-degree angle. In the inward-outward collapse, the top of the wall falls into the building and the lower part of the wall outward and away from the building. The curtain collapse resembles a curtain falling straight down as if dropped from a curtain rod.

The problem with wall collapse is that we are not able to predict the type of collapse that may occur. There are many so-called rules of thumb on predicting how a wall will fall. One rule states that a solid wall can fall 100 percent of its height, but if the wall contains windows or doorways, it will fall only one-third its height. Many older texts stated that a masonry wall would fall only one-third of its height. This theory is still contained in the books of many fire department libraries. There is the possibility of a wall falling less than its full height, but we can never be 100 percent certain in predicting this. With

A

B

FIGURE 6-1 A–B A frame wall starts to fail, as indicated by the wall cracks and the missing shingles. Failure results in a 100 percent collapse of the frame wall. *Used with permission of Robert T. Burns.*

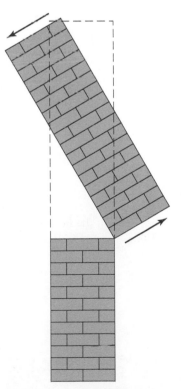

FIGURE 6-2 An inward/outward collapse of a wall causes the top of the wall to fall inward and the bottom to fall outward. *Used with permission of Pearson Education.*

one exception, every rule of thumb has been proven wrong. The only rule I have found to be consistent in ensuring firefighter safety is, "the minimum safe distance in predicting a wall collapse must anticipate that the wall can fall 100 percent of its height and that parts of the wall can project farther and increase the area of concern."

When dealing with structural stability, the building must be monitored constantly for any indicators or combination of indicators that could signal a problem.

If part of a wall fails, the reason for the failure must be determined.

a. Did the wall lose its carrying strength?
b. Did another part of the failing structure cause it to fall by the sudden overloading caused by an impact load?
c. Is additional collapse imminent?
d. Is there sufficient time to complete a search within the building?
e. Is immediate and total withdrawal necessary?

COLLAPSE ZONES

The collapse zone must be recognized as a safety zone. (See Figures 6-5 through 6-7.) After a collapse or safety zone has been established, the area should be cordoned off by stretching banner tape or ropes to isolate the area and deny entry. Safety zones are exactly that. There should be no excuse to violate the zone. The comment "I'll be only a second" is not valid. This pertains not only to firefighters but to chief officers as well. Too often, a firefighter will be chastised for walking past the banner tape and through a collapse zone, but a chief officer who does the same will not. Chief officers should lead

FIGURE 6-3 A curtain wall collapse causes the wall to fall straight downward.

FIGURE 6-4 A–B A 90-degree wall collapse causes the wall to fall a minimum of the height of the wall, and parts can be thrown farther, as indicated in the photo. Notice the mark on the collapsed wall where it met the roof and how bricks have fallen farther than 100 percent of the height of the wall.

A

B

FIGURE 6-5 Though the masonry wall is collapsing, the corner of the building maintains its strength.

A

B

FIGURE 6-6 A–D A fireground detail was still on the scene at a cold storage plant two days after the fire. One engine company was wetting down buried hot spots. A collapse zone had been established the night of the fire and was still being maintained. The row of two-story dwellings in the rear had been evacuated the night of the fire and the residents relocated until demolition was completed. The rear wall lost its stability and failed. It demolished the structures beneath it. Remember, once a collapse zone is established, entry must be denied to everyone. *Used with permission of Robert T. Burns.*

FIGURE 6-6 A–D
(Continued)

C

D

FIGURE 6-7 The collapse zone should be considered a safety zone. *Used with permission of Pearson Education.*

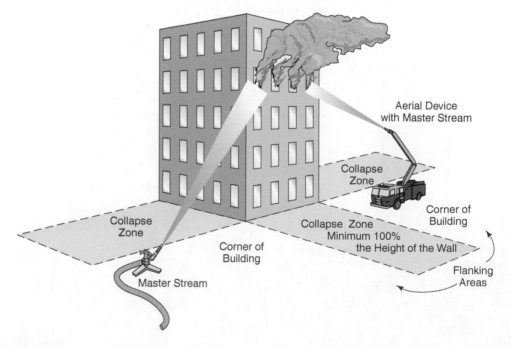

by example, not by exception. The longer walk around the area can be an inconvenience but is a safety necessity. The zone should cover an area at least equal to the height of the wall in question. Many fire departments mandate a minimum of one-and-one-half times the height of the wall. If master streams are operating, they should be placed in a flanking position or to the side of the danger area. The corner of a building is its strongest point. Too often, a master stream that is set up in the collapse zone needs readjustment, and so firefighters approach the stream to move it. This places them in a danger zone. The master stream should be moved to a safe location or shut down.

Firefighters in an adjoining building, an exposed building, on the roof of another building, or those operating on main ladders or elevating platforms must ensure that they remain outside of the collapse zone.

> The only accurate rule of thumb in predicting the distance that an exterior wall can collapse is that you must anticipate that the wall can fall 100 percent of its height and that part of the wall can project farther.

NIOSH FIREFIGHTER FATALITY REPORT F2012-13

On April 9, 2012, a 60-year-old male career lieutenant (Victim #1) and a 25-year-old male career firefighter assigned as Tiller (Victim #2), both assigned to Ladder 10 (L10), died when a wall collapsed during firefighting operations at a commercial structure fire. One engine company (Engine 2) was initially dispatched. Upon arrival, Engine 2 radioed that the fire was spreading throughout the structure. The vacant and abandoned warehouse covered more than half a block. This incident would eventually grow to five alarms. The fire would extend to an occupied furniture store. L10 was dispatched on the second alarm and assigned to deploy an elevated master stream. The fire originated in Building 1 on Side C and rapidly extended to the other structures within the vicinity. Building 2, located east of the building of origin and situated on Side A, sustained the structural and wall collapse that resulted in the firefighter fatalities and injuries. L10 set up a ladder pipe operation on Side A of the fire building. The collapse occurred after the lieutenant and three firefighters from L10 were sent inside the furniture store to operate a hand line to stop the fire extension. Two firefighters were trapped by the collapse and were injured. The lieutenant and another firefighter were buried by the collapse and died as a result.

Contributing factors include a multi-alarm fire in a vacant/abandoned structure; dilapidated building conditions; high winds; collapse zone maintenance, control, and compliance; fireground communications; personnel accountability; training on fireground operations; and situational awareness. **See Appendix B: NIOSH Reports in www.pearsonhighered.com/bradyresources to read the complete report and recommendations.**

BUILDING COLLAPSE INDICATORS

Critiques of various collapses have shown that only after a collapse occurred was it realized that seemingly unrelated occurrences (or indicators) contributed to the collapse. The probability of only one indicator or defect causing the collapse was likely remote, but combined with other defects or occurrences, it caused a devastating problem.

Each scene has indicators that provide varying degrees of information. The ability to manage an incident scene will be enhanced when firefighters utilize every piece of information at their disposal. The prompt reporting of collapse indicators to the Incident Commander (IC) will allow a clearer picture of the building's stability. The following are some indicators of building collapse.

Fire Conditions Fire attacking a structure creates constantly changing conditions. A continuous size-up can detect dangerous situations as they occur.

Collapse Indicators Involving Fire Conditions

- Two or more floors fully involved
- Continued or heavy fire
- High heat and heavy smoke conditions coupled with inadequate ventilation

Two or More Floors Fully Involved A large commercial and industrial building with two or more floors fully involved in fire creates a sustained high heat condition.

Commercial and industrial buildings often have exposed structural members. There may be interconnection between floors for an industrial process or a conveyor belt. These construction features allow a fire quickly to spread from floor to floor.

Large open areas favor a fast-moving fire that is difficult to control. The sustained fire will attack structural members, leading to their failure. Conversely, fire on two or more floors in a residential structure or small commercial building is often controllable by an interior attack. This is because these buildings are smaller in size, are compartmentalized, and have well-protected structural members, which permits an aggressive attack to be successful.

Continued or Heavy Fire for 15–20 Minutes

The amount of fire and where it is burning can be used as a collapse indicator. Continued or heavy fire for 15–20 minutes is a reference point to indicate collapse potential. This time frame considers that the fire is attacking solid wood structural members. Continual attack by fire can destroy the structural integrity of these members, creating a collapse situation. This rule of thumb cannot be applied to buildings using lightweight components.

Structural stability must be a prime consideration in a fire involving large areas for a sustained period of time. The time required to effect knockdown on a large volume of fire may be a sufficient amount of time for structural failure to occur.

High Heat and Heavy Smoke Conditions with Inadequate Ventilation

These conditions set the stage for a backdraft. Though backdrafts happen infrequently, they can have devastating results. Similar to any explosion, its force can knock down walls and be deadly to firefighters. Firefighters within the fire building will be severely endangered should a backdraft occur. The signs of a backdraft situation consist of smoke being forced from the fire building around windows and doors under pressure. Window glass can be blackened because the fire is in a smoldering state and starving for oxygen. Heavy smoke conditions can occur, with little or no visible fire in evidence and air being pulled or sucked into the building. (Backdraft is discussed in Chapter 1.)

Type of Construction

The type of building construction is important. A building is the sum of its parts. How sturdy it is and how it will react under fire conditions are based upon the structural components. Certain components can withstand a great deal of attack by fire, while other components fail readily, with ensuing collapse.

FIGURE 6-8 An unprotected steel column supporting a heavy load and subjected to high temperatures shows signs of failure. *Used with permission of Jim Smith, Jr.*

Construction Features That Should Be Considered Collapse Indicators

- Unprotected steel columns and beams exposed to heavy fire
- Unprotected lightweight steel and steel bar joist roofs subjected to heavy fire conditions
- Fire burning in an area containing lightweight wooden building components

Unprotected Steel Columns and Beams Exposed to Heavy Fire

Unprotected steel columns and beams exposed to heavy fire and high heat conditions can precipitate an early collapse. (See Figures 6-8 through 6-11.) Steel beams attacked by the heat of the fire expand. As steel is subjected to high temperatures, it will absorb the heat until its failure temperature is reached. It will then fail, often pulling or pushing down other structural members with it.

As a horizontal steel beam expands, it can push against an exterior wall, knocking the wall out of plumb and causing it to lean outward. When the steel reaches its failure temperature, it will sag and can slide down the inside of the leaning wall, exerting a tremendous pressure and violently pushing the wall outward. This causes a 100 percent collapse of the wall and often projects parts of the wall even farther. The function of the failing steel structural member will determine the severity of the collapse. Failure of a beam can affect a localized area. A column failure can trigger a substantial building collapse.

Unprotected Lightweight Steel and Steel Bar Joist Roof Assemblies
Lightweight steel truss-constructed roof assemblies are susceptible to attack by fire. The bar joist is a form of lightweight steel truss. These roof assemblies are commonly found in noncombustible buildings. This is an inexpensive method of construction and, though easily protected against the effects of fire, this protection is rarely provided because it is not economically feasible. These roof assemblies are often left exposed to fire with no protection provided. (See Figure 6-12.)

These assemblies have all of the inherent qualities and faults of steel. They will quickly absorb the heat of a fire below, and the bar joist and the lightweight steel truss will fail. Due to the rapid failure of these roofs, firefighters should not be operating on them if the building is heavily involved in fire.

Fire Burning in an Area Containing Lightweight Wooden Building Components
Lightweight components consisting, for example, of parallel chord truss, triangular truss, and wooden I-beams that utilize wood or wood and steel tubing are prone to early collapse under direct fire attack. Fire burning in a truss area under a roof or between floors will be directly attacking these structural members. These components will not withstand direct flame impingement.

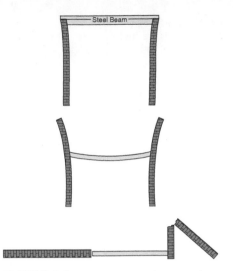

FIGURE 6-9 As steel is heated, it expands and exerts a tremendous amount of pressure on the exterior walls, causing them to bow outward. As the walls are forced outward, the steel and the roof or floor it was supporting falls downward, pushing on the walls until they collapse. *Used with permission of Michael DeLuca.*

FIGURE 6-10 A horizontal steel beam has expanded, pushing the masonry wall outward. The beam was cooled by water streams and frozen in place. Had the heat of the steel beam reached its failure point before being cooled, it would have sagged in the center and slid down against the interior of the wall, forcibly pushing it outward.

FIGURE 6-11 The destruction wrought on the steel of a building after a disastrous fire.

FIGURE 6-12 Lightweight steel bar joist roofs are susceptible to attack by fire, causing roof collapse.

Steel components and sheet metal surface fasteners (gusset plates) will act as heat sinks and conduct the heat of the fire through them, destroying the wood fibers adhering to the steel. This will loosen the connections, causing plates to fall out and the truss to fail. (Chapter 5, Building Construction, contains more information on this subject.)

Exterior Walls As fire attacks a building, it will destroy the building's integrity. The exterior walls of a building, more than any other component, can contain a multitude of collapse indicators. There are many indicators that can be seen and many that will not be visible. As the exterior wall is being attacked, it can react in a variety of ways.

Typical Exterior Wall Collapse Indicators

1. Smoke showing through walls
2. Old wall cracks enlarging
3. New wall cracks (See Figure 6-14.)
4. Fire showing through wall cracks
   (See Figure 6-15.)

FIGURE 6-13 A wood truss supported by steel beams is only as strong as the steel that supports it. Failure of the steel will cause the truss to collapse. *Used with permission of William V. Emery.*

FIGURE 6-14 When a new wall crack is discovered, the building is experiencing movement at that time. The steel beams supporting the wooden roof in this photo are expanding and pushing out the wall.

5. Bulging walls
6. Leaning walls
7. Failure of part of a wall
8. Visible spalling of a brick wall
9. Spalling of concrete and exposure of steel
10. A wall breaking down under a hose stream
11. The presence of wall spreaders

Smoke Showing through Walls   The pressure of a fire can force or push smoke through any available openings. If the interior walls are not tight or have been breached by the attack of the fire, smoke will be pushed into the wall, ceiling, and floor cavities. Once in these void spaces, it will seek the path of least resistance. Smoke may be seen pushing from around door and window openings in an exterior wall. Loose-fitting siding on a frame building or missing mortar in a masonry wall will also be avenues of escape for the smoke. The movement and pressure of the smoke may indicate that conditions are present for a backdraft explosion.

FIGURE 6-15 The expanding steel has reached a point where fire is now visible through the wall cracks and collapse is imminent.

Old or New Wall Cracks   As building components fail, they can directly impact the exterior walls. The collapse of interior floors can push against masonry exterior walls, causing wall cracks.

When a new wall crack is discovered, it indicates that the building is experiencing movement or failure at that time. In either case, old and new wall cracks must be monitored. If a wall crack increases in size, this may be a sufficient indicator for the IC to abandon an interior attack.

Fire Showing through Wall Cracks   As the intensity of the fire increases, flames may become visible through cracks in the exterior wall. This can indicate that the fire is increasing in size and is attacking the structural members in the void spaces. A burn-through of a frame exterior wall can provide fresh air to feed the fire.

Bulging, Leaning, or Partial Wall Failure   As failure occurs on the interior of a building, the failing structural members can exert an outward pressure on the exterior wall. This may be indicated by walls bulging or leaning outward. (See Figure 6-16.) Movement or failure of a wall may occur slowly or quite rapidly, with seemingly little or no warning. Through a combination of interior collapses and movement of structural members, or due to a major interior collapse, the failure of part of an exterior wall can occur.

FIGURE 6-16 Water freezing in a masonry wall can cause a separation between the face brick and the backup brick. Once this bond is broken, the wall is weakened and collapse can occur.

FIGURE 6-17 Wall spreaders are visible in the wall as fire attacks the top floor. The amount of fire in this photo would not ordinarily be a problem if no other collapse signs were present.

FIGURE 6-18 The front wall was removed, showing the exposed wall spreader and method of attachment.

Spalling of a Brick or Concrete Wall   Sustained heat can cause visible spalling of a brick or concrete wall. The wall is weakened when spalling occurs. If concrete is protecting structural steel or steel rods, spalling can expose the steel to the heat of the fire. Intense heat concentrated on the steel can raise the temperature of the steel, causing expansion and the possibility of collapse.

A Wall Breaking Down under the Pressure of a Hose Stream   Firefighters must ensure that their actions do not accelerate building collapse. Hose streams can cause a breakdown in a masonry wall. Water can wash out mortar that is holding bricks in place, weakening the wall and reducing its load-carrying capacity.

Wall Spreaders   A building feature that can be utilized as a potential collapse indicator in a masonry wall is a wall spreader. (See Figure 6-18.) Wall spreaders can be installed when the wall is initially constructed, as additional support for the wall, or when a wall has developed structural problems to act as stabilizers to stop the problem from becoming worse.

Wall spreaders come in a variety of shapes. They can be in the form of a star, an S, steel I-beams, or flat steel plates of diverse shapes. A common installation to strengthen a weakened masonry wall starts on the exterior. A one-inch hole is drilled through the exterior wall and a minimum of three adjoining wooden floor joists. The spreader is held in place by a large bolt on the end of a solid steel rod of all-thread. The all-thread is then inserted through a hole in the front wall and the adjacent joist. At each joist, it is attached with washers and bolts. This spreads the load of the front wall to all three joist members. Fire attacking these supporting joists will directly affect the wall's load-carrying capacity. Their failure can impact the stability of the wall. (See Figures 6-19 through 6-23.)

It is common to find the shortcuts used when wall spreaders were installed. A common fault is that the spreader is attached to only one joist. Should a problem develop in the supporting joist, the wall would lose its support and could be freestanding.

Today, prior to installing a wall spreader in a wall with a structural problem, most jurisdictions require that an engineer design and submit drawings for approval. This system alleviates many shortcuts that have occurred in the past.

Firefighters finding wall spreaders in a structure involved in fire usually do not know why the spreaders were installed. They must assume the spreaders were installed to rectify a problem with the wall and be guided accordingly.

FIGURE 6-19 Angle iron was used as a wall spreader to support this weakened and failing wall.

Present State and History of the Building   As fire attacks a building, failure can start with localized interior collapse. As fire-stops break down, the fire area will enlarge and structural members can become exposed. The destruction of fire-stops will cause a much larger area to be affected and increase the danger of building collapse.

Present State and Building History Collapse Indicators

- Previous fire damage
- Windows, doors, floors, and stairways out of level
- Sagging wooden floors
- Excessive snow or water on a roof
- Cracking noises coming from a building
- Interior collapse
- Plaster sliding off walls in large sheets

Previous Fire Damage   Previous fire damage in a building is cumulative in sapping a building's strength. It creates a

FIGURE 6-20 Wall spreaders come in a large variety of shapes. A fancy plate can prove as deadly as a piece of scrap angle iron.

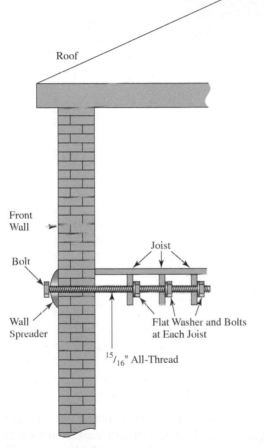

Roof

Front
Wall

Bolt

Joist

Wall
Spreader

Flat Washer and Bolts
at Each Joist

$^{15}/_{16}$" All-Thread

FIGURE 6-21 A wall spreader is placed in a wall that has been weakened. A $^{15}/_{16}$-inch all-thread is placed into a hole drilled in the wall. It passes through three joists and is attached to each joist. This spreads the weight of the wall across these joists. *Used with permission of Pearson Education.*

FIGURE 6-22 A building owner has installed multiple wall spreaders in an attempt to stabilize a wall. Firefighters in their size-up should recognize these signs and anticipate the possibility of wall collapse and that the roof may be unsupported where the wall has pushed outward.

FIGURE 6-23 A building under renovation in a flood zone is raised with shoring.

dangerous situation for firefighters attacking a fire. The building may be vacant and may not have been repaired from the previous fire, so the damage could be easily recognizable. However, fire damage may be covered over and remain in walls, floors, or ceilings of occupied buildings.

Windows, Doors, Floors, and Stairways Out of Level Observation of windows and doors prior to entering a building should include not only their location for finding a secondary means of egress, but also whether there are other telltale indicators, such as doors or windows out of plumb, which could indicate a structural problem. After entering a building, firefighters may find interior doors stuck or jammed because a building has shifted. Staircases that are out of level also can indicate that a building has shifted. If the staircase shifts enough, it may not support the weight of a firefighter. Firefighters finding this condition should keep close to the wall when ascending or descending the stairs. This permits their weight to be supported on the stair support. A safer method is to place a portable ladder up the staircase.

Sagging Wooden Floors Fire conditions can damage floor supports. A wall leaning outward may not be supporting a floor joist that was once set in the wall. A sagging wooden floor must be checked to determine whether it is safe to operate on. If possible, a visual inspection of the supports that the floor rests on will assist in the determination.

By checking the wall sockets in a wall of ordinary construction, or the wooden wall connection in a frame building, the floor's stability can often be ascertained. This can be accomplished by opening the ceiling beneath the floor and checking the ends of the floorboards. Recent movement of the floor joist can be indicated by a swatch of clean lumber at the connection point, or the fire cut on the end of the joist may be visible, indicating that the joist is pulling away from the wall's support. The floor joist should be checked at each bearing point to determine whether firefighters can operate safely above.

Excessive Snow or Water on a Roof Rainwater or water from firefighting operations can be retained on flat roofs that have clogged or frozen drains. Though some water on the roofs of exposed buildings can be beneficial in preventing fires from flying embers, a large buildup can be dangerous. Water buildup can reach the height of the parapet walls before overflowing. Firefighters may be able to utilize a tool to unclog the roof drains of debris or ice, or the dislodging or removal of downspouts may alleviate the problem.

Roofs are built to hold the average snow load for the specific region of the country. During winter operations, a buildup of snow and ice on a roof must be recognized. This condition will impair operations. Snow and ice can mask fire conditions normally observed on a roof. For example:

FIGURE 6-24 An owner of a building heavily damaged by fire is attempting to salvage the front wall. Should another fire occur while the temporary shoring is in place, an early collapse must be anticipated.

1. Dry spots indicate where a fire burning below may be attacking roof supports.
2. No sponginess will be felt.
3. Smoke conditions may be masked.
4. The ice itself can act as a platform over a weakened roof area, failing to reveal the hidden dangers below. As firefighters attack the roof to open it, they will break through the ice platform. This action can add a sudden impact load to the roof from the weight of the ice and the firefighters. The sudden weight shift may be sufficient force to cause roof collapse, dropping the firefighters into a raging fire below.

Cracking Noises Coming from a Building As fire attacks the building components, they can shift. This movement can cause sounds that firefighters can hear. Cracking noises could indicate that parts of the building are failing. Upon hearing building noises, a visual inspection should be attempted to see if any distortion is noticed or additional indicators are present. Loud and repetitive noises should be a warning to vacate the area immediately.

Interior Collapse Though the most sensational type of collapse is the exterior wall collapse that is captured on video or camera, the type of collapse that causes most firefighter injuries is the localized interior collapse. As portions of the interior fail, load-bearing walls can be affected. When interior collapse is noticed, the firefighter observing the collapse should try to ascertain whether additional dangers exist to the firefighters on the interior and exterior of the building.

Plaster Sliding Off Walls A severe condition that can be observed is plaster attached to wooden lath sliding off the walls in large sheets. The twisting movement of the building breaks the attachment of the plaster from the wooden lath, causing this dangerous situation. This occurrence must act as an indicator for immediate withdrawal from the building by all personnel.

Buildings under Construction, Renovation, or Demolition Buildings go through various phases, from their initial erection, to a variety of renovations over the years, to the final stage of demolition. Each stage can create dangerous situations for firefighters responding to an emergency within the building. For this reason, firefighters should consider a building going through any of these stages as a potential collapse or as an indicator of collapse itself.

Buildings under construction, renovation, or demolition can present different problems for responding firefighters. New construction will not have the safety features required of a completed building. Sprinkler and standpipe systems may not be functional, delaying an attack on the fire. Unprotected structural members will be prone to attack by fire. Alarm systems may not be installed, delaying fire department notification.

Buildings under construction and under demolition often have welding operations or what is referred to as *hot work*. These welding operations are a source of numerous fires and should require a permit to ensure that the welding is being done in a safe manner. Precautions can include a 35-foot area around the welding operation:

- Clear of flammable liquids, dust, and the removal of lint or oily deposits
- Eliminate any explosive atmosphere in the area
- Floors swept clean
- Combustible floors wet down, covered with wet sand, or fire resistive sheets
- Remove combustible materials where possible or protect them with welding pads, blankets, curtains, or other approved materials
- Cover all wall and floor openings
- Install approved welding pads, blankets, and curtains under and around all work
- Post a fire watch during welding and for 60 minutes after work is completed

The issuance of a permit allows for evaluation of the hazards in the immediate work area where the hot work will be done, specifications of requirements for fire suppression, elimination of hazardous activities, and notification of the security service of the location of the hot work.

Buildings undergoing renovation can have varying stages of fire protection. A nearly completed structure would probably be well-protected, whereas a building in the early stages could have much of its protection removed. Weekly inspections of a building being renovated, either by the first-due company or the fire prevention or inspection bureau, will keep the fire department abreast of changes that could affect the building should a fire occur. (See Figure 6-25.)

The building undergoing demolition will have many of the same faults as a building under construction or a building being renovated. Safety features will be removed, and fire-stops may be nonexistent. Structural supports may be replaced with temporary shoring, which could lead to an early collapse. Trash accumulation can be a constant problem and can become a source of fire and a cause for fire spread. Good housekeeping must be enforced at these sites.

FIGURE 6-25 A–B Regentri-
fication is the restoration of older
properties to their original state by
preserving as much of the old work
as possible. Too often shortcuts are
taken. The shoring on this property
is questionable. The rear wall is
braced, and a fire will create the
conditions necessary for an early
collapse.

A

B

Buildings being prepared for demolition by explosives fall into this category. The demolition experts will weaken critical areas of the structure to ensure that the building will collapse in an exact and specific manner. Should a serious fire occur prior to the actual demolition, an early collapse must be anticipated.

> A building being attacked by fire must be considered a building under demolition.

Water and Building Loads
Water is the principal tool used to combat fire. For all its positive aspects, water can have a negative effect on a structure. It is important to monitor water usage.

Collapse Indicators Associated with Water Usage and Other Building Loads

- Excessive water in a building
- Water not coming out of a building as fast as it is going in
- Water runoff from between bricks
- Bales of absorbent material in a building
- Large machinery or heavy contents in a building
- Excessive or unusual roof loads

Excessive Water in a Building Firefighting operations can drastically increase the live load on the fire building. This can be due to the weight of:

1. The firefighters with their protective equipment and tools.
2. The hose-line brought into the fire building.

3. The water used to attack the fire. A 1¾-inch hose-line can deliver approximately 175 gallons of water per minute. This adds about 1,500 pounds per minute into the fire building. If multiple hose-lines are operating, the weight of the water can be tremendous.

When operating in an offensive mode, a buildup of water within a building requires that immediate action be taken to alleviate these conditions. The remedy may be as simple as controlling the excess flow from the hose-line or moving fire debris that is restricting runoff. Realize that at the same time this additional weight is being introduced into the fire building, the fire is weakening the structure.

Water Not Coming out of a Building as Fast as It Is Going In
Defensive attacks are handled by master streams from the exterior. These devices deliver large quantities of water to control the fire. This effort places tons of water per minute into the fire building. The IC must monitor the amount of water going into a building in relation to the amount of runoff coming from the building. If water is being retained within the structure, the building must support more weight. A point will be reached when the weight of the firefighting efforts places such a strain on the building that collapse will occur. The question is sometimes asked, "How do we know if the water runoff is sufficient?" The best answer is that it is a judgment call best guided by experience. If in doubt, consider the worst and act accordingly.

Water Runoff from between Bricks
The water being used to attack a fire can present a problem if the water runoff washes out mortar joints. The water flowing through the wall indicates a weak point in the wall. If this is allowed to continue, additional mortar joints will be destroyed, along with the supporting strength of the wall.

Bales of Absorbent Material in a Building
Another way that water is retained in a building is through absorption. The building will absorb some water. A danger exists if the building contains stock or material that can absorb large quantities of water. This is especially true with bales of absorbent material. Through absorption of water, their weight can drastically increase, placing extremely high loads on supporting members.

Another serious problem with absorbent bales is that they can expand and push out load-bearing walls or supporting columns. Bales piled one upon another can shift when expanding. These piles can become unstable, and if they fall, they can create a tremendous impact load to the floor of the building. Falling bales have seriously injured and killed firefighters in the past.

Besides forklifts, overhead cranes are often used to move these bales. Heat attacking the crane's cables can cause the crane's contents to fall on firefighters operating below.

Large Machinery or Heavy Contents in a Building
The building contents or live load can impact the building's stability. Large machinery or heavy stock must be considered. Auto parts stores and hardware stores maintain a heavy stock. Industrial buildings can contain many materials that will place tremendous loads on a building.

Excessive or Unusual Roof Loads
Roof loads must be monitored during a building fire. Excessive or unusual roof loads can cause early roof failure. Large roof air conditioning units, water tanks, billboards, and signs can place a heavy dead load on a building. It should be determined if their support is dependent upon the roof or the bearing walls. Trucks and other vehicles that have outlived their usefulness are sometimes painted and placed on the roof of a building for advertising purposes. This practice can prove dangerous to firefighters fighting a fire below. (See Figures 6-26 through 6-28.)

FIGURE 6-26 Undesigned roof load: A truck on the corner of a masonry wall. How is it attached to the building? Is it secured to the wall? To the roof? Or is it just balanced there? How will it react if there is heavy fire in the building? These questions must be answered before a fire occurs.

FIGURE 6-27 Undesigned roof load: A truck is placed on the roof of a building near an interstate highway for advertisement. This is a heavy load to be carried by this truss, which is in the shape of a sawtooth roof.

FIGURE 6-28 Undesigned roof load: A bus on a roof. This is a reinforced concrete structure that can easily support this weight. Knowing a building's strengths and weaknesses is necessary to predict collapse potential of a building.

FIGURE 6-29 The presence of billboards on a building places a dead load on the structure that in most cases it was not originally designed to carry.

Live Loads, Dead Loads, Eccentric Loads, Impact Loads, and Undesigned Loads

Firefighters should be aware of how different loads affect a building. *Loads* are weights applied to a building that can take different forms. The firefighter, in performing size-up, must consider the type of construction, occupancy, stability, and distinguishing features of a building. Distinguishing features include eccentric loads or dead loads that may be visible to responding firefighters. A load placed on a building can affect its stability.

Live Load   The live load of a building is the weight of the material in a building that is not permanent. This load can constantly change. It includes:

- Desks
- Furniture
- Machinery
- Kitchen utensils
- Tools
- Merchandise

The amount of weight can vary greatly. Merchandise can add a tremendous load to a commercial building, such as a hardware store, plumbing supply store, or other commercial establishment. (A fire that took the lives of five firefighters at Hackensack Ford in Hackensack, New Jersey, had auto parts stored in the overhead area of a heavy timber truss. This storage added a tremendous amount of weight to the live load while placing combustible goods on the truss.)

Live loads can also be weather-related, such as snow or water accumulating on a roof. Also, firefighting can introduce many live loads into and onto a building.

Dead Load   The dead load is the total weight of the building components. This weight is substantial. It includes every part of the building that is permanent. The dead load of a building includes:

- Structural members of steel, masonry, or wood
- Air-handling systems
- Plaster or wallboard
- Paint and wallpaper
- Plumbing fixtures and piping
- Electrical components

The presence of billboards on a building's roof places a dead load on the structure that, in most cases, it was not originally designed to carry. (See Figure 6-29.) A fire attacking the structure will weaken it and reduce its load-carrying capacity. Since billboards are commonly constructed of unprotected steel, fire can attack the exposed steel of the signs. If sufficient heat is present, this fire impingement can cause collapse of the sign much sooner than the structure.

Eccentric Load   Eccentric loads essentially create a pulling action on a wall in a downward thrust. Examples include:

- Wall signs (See Figure 6-30A.)
- Large ornate cornices (See Figure 6-30B.)

- Corbelled brick (See Figure 6-30C.)
- Marquees (See Figure 6-31.)

Marquees are an excellent example of an eccentric load. They are usually installed with wooden beams cantilevered through the front wall of a structure. They may

FIGURE 6-30 A–C Wall signs, decorative corbelled brick, and large ornate cornices place an eccentric load on a wall. These walls should be monitored for stability if heavy fire conditions exist in a building.

FIGURE 6-31 A marquee consists of beams cantilevered through a wall, and as can be seen in the photo, it may also receive additional support from cables attached to the front wall. Sagging of a marquee could precipitate a wall collapse.

receive additional support from vertical posts supporting the end of the marquee or from cables or rods attached to the upper part of the wall from the end of the marquee farthest from the supporting wall.

An eccentric load acts like a downward thrust on a wall. A wall sign places a downward weight on the wall to which it is attached. This pulling action that is created by an eccentric load on the wall of a building must be offset by distributing the weight. It can be dispersed to interior structural members by using cables or additional supports.

Should fire attack and weaken the connection points supporting an eccentric load, collapse can occur.

If engaged in an interior attack in a building with a marquee, the wall must be monitored to ensure that fire doesn't attack the supports of the marquee. Failure of the marquee could precipitate the collapse of the front wall. Placing a wooden ladder as a brace under the marquee will allow any slight movement of the marquee to be reflected as a strain on the ladder. This will act as a warning to the firefighters to withdraw from the building. Wooden ladders often can be obtained from the building and will show any compression more readily than an aluminum ladder.

Corbelled brick is built out upon itself to form an ornate cornice.

Impact Load The impact load can be a critical factor in a building's stability during firefighting operations. An impact load is a weight forced upon a building. Under firefighting conditions this can come in the form of the fire department's actions: by a main ladder being dropped upon a cornice or roof, a firefighter chopping with an axe to open a roof or floor, or a firefighter jumping onto a roof surface from an adjoining roof or ladder.

A location or area that under normal conditions would not be affected by an impact load may already be weakened by the damage caused by the fire. The force of an impact load may be the "straw that breaks the camel's back" and become the impetus for a collapse. A structural failure may plunge a firefighter using a tool or jumping onto a roof into a collapsing area.

The impact load can cause failure of a portion of a building or initiate total building collapse, such as seen in a pancake collapse of multiple floors of a building.

Undesigned Load A designed load is a load that a building was designed to support. Undesigned loads on a building are loads that a building was not originally designed to carry. Though these loads are typically supported by the structure under normal conditions, failure can readily occur when a building is under attack by fire.

Auto garages may suspend hoists from roof trusses built of lightweight wood, or steel bar joists for removing vehicle engines or transmissions.

Undesigned loads commonly come in the form of alterations or renovations to a structure. The stability of a building that contains undesigned loads becomes unpredictable for firefighters under fire conditions.

Firefighters must take care in how they apply force to a fire building. It is usually not apparent how susceptible the building is to collapse because many components that may be affected by a fire cannot be viewed by the firefighters.

An example of an undesigned load is the deadly fire in Chesapeake, Virginia, that took the lives of two Chesapeake firefighters. The roof contained three air-conditioner units that weighed more than 3,000 pounds and collapsed under fire conditions. The roof was not designed to carry these units.

Effect of Building Collapse on a Fire

A building collapse can affect a fire in different ways. Collapse might result in near extinguishment of the fire caused by the smothering effect of the resulting debris on the burning area. The fire can be snuffed out or reduced to a manageable situation, instead of remaining an open, free-burning fire.

Another effect can be the involvement of a much larger area and a spreading fire. This occurs because the fire now is getting a sufficient supply of air to sustain free burning.

Removal of the collapsing wall may allow the extension of fire to nearby exposures that previously had been protected by the collapsed wall. The endangered property would then have to be covered by hose-lines for protection.

COLLAPSE CONSIDERATIONS

The previous list of collapse indicators is by no means complete. Certain situations are dangerous in themselves. If you are the IC and notice one or more of the signs mentioned previously, you should take immediate action. The proper action will depend on whether any civilian life is in jeopardy. If all civilians are safe, then firefighter safety is your Number One consideration. Waiting for a variety of warnings and not acting immediately could result in your final warning or indicator being a collapsing building. (See Figure 6-32.)

Collapse Search

Possibly no other fire department operation needs more coordination than **collapse search** rescue. Victims can be found trapped within a building or buried on the exterior of a structure trapped by parts of the falling building.

The ability to rescue victims is dependent on many factors: the type of collapse and the extent of the area involved, as well as how many floors of a structure have collapsed. The physical health of victims and their ability to withstand such an ordeal is also important.

Rushing into an area and pulling a person from a would-be grave is a firefighter's first instinct. This is especially true if fellow firefighters are involved in a collapse. However, the IC must establish a rescue plan, and it must be followed to the letter. The plan should be adjusted according to the reports received from Division and Group Supervisors. All rescue efforts must be coordinated, and the IC must ensure that none are in conflict with other operations being performed. If firefighters rush into a collapse area with no prior thought or size-up on the collapse, they often become part of the problem instead of the solution.

The company officer must take control of his or her unit's efforts and coordinate its operations at the scene through the IC. An incident management system is imperative. Use of a command structure will minimize and handle problems that are apt to arise when confronted with a time-consuming search.

SAFETY OFFICER

The Safety Officer (SO)'s responsibilities will be heavily taxed in assisting the IC. It is important for the SO to prevent or minimize injuries by a variety of causes, as well as being able to recognize warning signs of further collapse. Reconnaissance and monitoring of firefighters operating at the scene enables the SO to recommend when firefighters should be rotated to avoid fatigue that can result in rescuers taking unwarranted risks or

collapse search ■ Occurs when firefighters are called on to find and extricate those trapped. Requires skill and the implementation of a plan.

FIGURE 6-32 A wall collapse can permit a fire to spread to a larger area because the wall is removed. *Used with permission of Joseph Hoffman.*

making mistakes in judgment. Responsibilities also include monitoring members for signs of critical incident stress.

STAGING

Assigning a Staging Area Manager and establishing a staging area accommodates the movement of personnel and equipment (especially ambulances) to and from the scene without delay. If the collapse involves many injuries, a Medical Group Supervisor may be required to set up triage, treatment, and transport of victims. The first-aid station can tend to minor injuries while also monitoring the fatigue level of the rescuers during rehabilitation.

SIZE-UP

Response to a collapsed building involving a rescue operation must contain key elements. The first consideration is size-up. The site must be surveyed and information gathered.

a. Was anyone in the building at the time of the collapse?
b. How extensive is the collapse?
c. Are other sections of the building in danger of collapsing?
d. Have adjoining or nearby buildings been damaged, posing a danger to rescuers?
e. Is fire involved?

An attempt must be made to establish the number of people in the building at the time of the collapse. This can be a time-consuming and frustrating assignment. These data must include any occupants who were rescued prior to the fire department's arrival.

Surrounding structures may need to be checked for possible interior damage and people trapped or injured. Evacuation may be necessary if the potential for additional collapse or the accumulation of gases may endanger these buildings. Determining the kind of structure that has collapsed allows the anticipation of some normal hazards that could be encountered. Residences would be expected to contain heavy objects such as a refrigerator, freezer, or large bureau. An office building would contain heavy file cabinets, safes, and desks. Large objects can provide protection for trapped victims by creating voids in which they can seek refuge. These items can also be quite unstable due to their weight, and movement can cause further collapse.

DETERMINING THE CAUSE OF THE COLLAPSE

Try to determine the cause of the collapse. Eyewitnesses to the actual event can assist, not only in giving information on those in the building but also on what actually happened. Someone may have either witnessed the collapse or heard something that will assist the fire officer. Whether the sound of an explosion or a call for help, responders must collect all available information and use it in their search and extrication efforts.

Talking to people who live or work in a building can provide valuable information on building layout, floor layout, and where occupant activity would normally take place within the structure. As with any incident, information gathering is initiated by first-arriving firefighters. When dealing with a collapse involving a prolonged search, a relocation area should be established. Check with area hospitals to see if anyone reported to be in the building may have been taken to an emergency room. Individuals with information on the collapse should remain at the scene, if possible, to update and provide additional information if needed.

If firefighting is in progress when a collapse occurs and the possibility exists that firefighters are trapped, a personnel accountability report or roll call must be taken immediately. When the data are gathered on who is suspected to be in the building, their possible location should be investigated. Searching the most likely places first has the greatest probability of producing positive results.

FIGURE 6-33 Temporary shoring placed under a frame bay should act as a warning sign of instability.

CALLING FOR HELP

Assistance by using the appropriate equipment and personnel is a must. Is a heavy rescue unit available either within your department or on a mutual-aid call? Contracts must be in place to call for heavy equipment, such as cranes or backhoes. Many kinds of fire department hydraulic extrication equipment have been found to be ineffective in collapses of buildings involving heavy concrete slabs.

Heavy rescue units have the necessary tools and equipment required for a complex rescue operation. Shoring, air bags, hydraulic tools, and special cutting equipment are but a few of the items needed. If a heavy rescue is unavailable, some locales utilize their truck companies to carry these tools, or utility companies often have the necessary shoring and tools used in their everyday operations. The incident scene, however, is not the place to find out what is available. A book or list of what special equipment and expertise can be supplied on request or under contract to the fire department should be maintained both in response vehicles and in the dispatch center.

FIGURE 6-34 A facade wall 10 feet high and 50 feet long collapsed onto a sidewalk below, killing three pedestrians and injuring 10 others.

The need for medical personnel must be established and requested early. This could include basic and advanced life support units or hospital surgical teams to assist at the scene.

Major collapses involving many injuries may necessitate the institution of a mass casualty procedure. (Mass casualty is discussed in Chapter 8.) (See Figure 6-34.)

UTILITIES

Control of the utilities supplying the collapsed structure will be an immediate concern. The IC should check with the utility representatives to establish if natural gas, electric, and water service is provided to the involved property. Liquid petroleum gas (LPG) or natural gas escaping from broken pipes and accumulating in pockets within the building can cause asphyxiation, as well as an explosion. Pilot lights on gas-burning equipment are a ready source of ignition, as are arcing electrical wires damaged in the collapse.

Water from broken pipes, fog nozzles used to absorb or disperse gas or toxic fumes, or water from firefighting operations can flood basements or pool in lower areas. A person trapped at a lower level could drown in this water. Therefore, it is imperative that minimal amounts of water be used to control a fire. Water runoff must be monitored and the structure's water supplies shut down.

Electric service must be shut off. Firefighters or the utility company can accomplish this. A building with live wires buried in the debris can electrocute those trapped, as well as rescuers, by either direct contact or by energizing parts of the structure.

LOCATING VICTIMS

Surface victims are first to be rescued. (See Figure 6-35.) A close inspection must be conducted because a surface victim may have only one arm or leg visible to rescuers. Walking

FIGURE 6-35 Surface victims are the first to be rescued. A close inspection must be conducted because a surface victim may have only an arm or leg visible to rescuers. *Used with permission of Michael DeLuca.*

on debris can cause further injury to those trapped just beneath the surface. If possible, those rescued should be interviewed for additional information. What was their location in the building at the time of the collapse? Was anyone else near them? If so, who were they, and was there any large piece of furniture or machinery near them? This can help establish a focal point that rescuers can zero in on in the overall search process.

If the large piece of furniture or machinery is found, the person may be nearby. Realize that people reported trapped may have been rescued prior to the fire department's arrival. It has happened that people thought to be trapped had already left the building and did not realize they were considered missing. A commercial building may have a delivery person or salesperson in the building at the time of the collapse. There could be service people or employees that either no one noticed or were forgotten due to the confusion. Someone should be assigned to interview bystanders for information. This assignment is best handled by police officers who can then advise the IC of their findings.

Operations must proceed slowly. The possibility of a secondary collapse or shifting debris dictates caution on the part of the rescuers. After the removal of surface victims, voids can be checked using the data collected from eyewitnesses. Searchers must not give up too early on the hope of finding survivors. Victims not visible can sometimes be located by a thermal imaging camera (TIC) or by stationing rescuers around the collapse site and having each rescuer take a turn calling out or tapping the debris and then listening for a response. These attempts require a high degree of quiet to be successful. (Heavy rescue units carry sophisticated listening devices that can be set at various locations to listen for sounds of people who are trapped.)

Once contact is established, verify the sound to get as exact a location as possible. It is very important to maintain contact. The person trapped can assist rescuers in establishing his or her physical condition, location, and possible pitfalls the rescuers may encounter in their rescue attempt. Communication will also assist the person trapped by keeping his or her morale high, knowing help is on the way.

 ON SCENE

On June 5, 2013, in Philadelphia, a collapse during the demolition of a building caused a four-story brick wall to fall onto an adjacent ordinary constructed one-story Salvation Army Thrift Store. The store, which was open for business, measured 15 feet wide by 100 feet deep. The collapsing wall caused a pancake collapse of the roof and injured 14, while killing 6. The collapse initiated a long and arduous task of searching for those trapped. Thirteen hours after the incident started, a 61-year-old woman was located alive in the rubble and transported to the trauma center.

A building that is open to the public can make it very difficult for responders to determine who may still be in the collapsed building. The collapse and ensuing mass casualties required the response of three alarms and 66 pieces of fire and EMS apparatus to handle the incident.

FIREFIGHTERS INVOLVED IN COLLAPSE

If firefighters are involved in the collapse, and they are equipped with a personal alert safety system (PASS) device, it can be manually sounded, or it will automatically sound after a preset period of inactivity. If the firefighters have no PASS device but have walkie-talkies, shut down other radios in the immediate area, key another portable, and aim the radio at the missing firefighters' suspected location to cause a loud noise or feedback through the missing firefighters' portable radio. Another method is to cause audible feedback by holding two portable radios close together while pressing the transmitter of one of the radios. This will cause a loud noise or feedback on the radio band and can possibly assist in locating the trapped firefighters by hearing this feedback on their radio.

If there are sufficient radio frequencies available, the IC should remove all units from the radio frequency that the trapped firefighter(s) are operating on to use it to communicate with those trapped. Another radio band should then be assigned for the rescue efforts. (See Figure 6-36.)

FLOOR COLLAPSE

The various kinds of floor collapse include:

1. The pancake collapse occurs when one or more floors collapse on top of each other.
2. The V-type collapse occurs near the middle of the floor and creates voids on the perimeters.
3. The lean-to collapse is supported or hinged on one side, with the opposite side resting or supported and hanging freely.
4. Unsupported floor collapse occurs when a bearing wall has failed and the floor has sagged due to the missing wall, yet still remains somewhat intact.

Remember, a collapse may be any combination of the aforementioned types. (See Figure 6-37.)

At a collapse incident, vibrations from all sources must be eliminated because they can trigger further collapse. Vibrations can originate in the rescue operation itself or from a nearby source. Highway traffic or nearby worksites may have to be stopped until after rescues have been accomplished. This will resolve two problems: vibration and noise.

Hose-lines should be stretched as a precautionary measure even if no fire is present. If fire is present and not supported by leaking gas, it should be extinguished with a minimum amount of water. If a water buildup occurs, it will be necessary to use pumps in low-lying areas. This can sometimes be accomplished remotely through a basement window or an outside basement door.

If leaks of natural gas are suspected and the gas cannot be immediately shut off or a pocket of gas has been formed, an attempt to eliminate all sources of ignition must be made. Naturally, a fire fed by a natural gas leak should not be extinguished, but surrounding areas can be cooled down until the supply can be shut down. If natural gas is escaping with no fire present, hose-lines can be used to dissipate any buildup of gas.

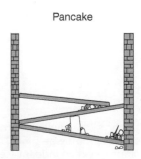

FIGURE 6-36 A front wall collapse struck and injured seven firefighters, including two buried under the wall who sustained serious injuries.

SHORING

Freestanding walls or unsupported floors can be temporarily stabilized by the use of shoring. If the necessary equipment is not available at the scene, it should be requested for immediate delivery. As members enter a collapsed structure, they should shore as they move along. Basically, the shoring is placed either against an unstable wall or beneath an unsupported floor or object. This is meant to prevent further movement and should not be an attempt to restore the area to its original position. When shoring a floor or object, the shoring should not be forced into place but situated so the weight of the floor or object will secure it at its present position.

A larger area can be searched if firefighters can connect the voids that occur in the collapse. This interconnection of voids allows an escape route from more than one direction should additional collapse cut off the primary means of egress. Lower-level areas or voids can sometimes be entered through basement windows or exterior basement doors.

The search and rescue teams must work in pairs under the supervision of their company officer or Division or Group Supervisor. Too many teams can

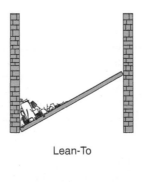
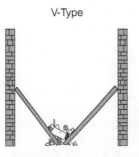

Lean-To

V-Type

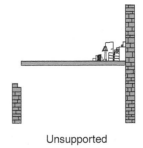
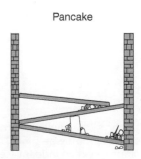

Unsupported

Pancake

FIGURE 6-37 Various types of floor collapse can occur: lean-to, V-type, unsupported, or pancake. *Used with permission of Michael DeLuca.*

cause coordination problems. The teams must have communication capability and keep their Division and Group Supervisors informed of their approximate whereabouts.

The search process can be time-consuming. The risk taken by firefighters must be weighed against the benefits to be achieved. Body retrieval should not be performed at the risk of firefighter injury.

SAFETY AT COLLAPSE SCENES

Firefighter injuries at this kind of incident can vary widely. Serious injuries can occur from secondary collapse of the structure. Firefighters can be operating in areas deficient in oxygen or receive puncture wounds from nails or jagged parts of the structure. Proper protective equipment should be worn, and members must remain cautious and alert.

Storage containers of gasoline or kerosene may be damaged and leaking. The storage of chemicals and pesticides in an average household can be dangerous if containers are breached. Cleaning solvents containing ammonia and chlorine stored in the same area can be damaged, and then mix together, creating a deadly gas that can asphyxiate trapped victims or rescuers. Sewer gas from broken sewer or sanitary lines can be both an explosive and an asphyxiant. For this reason, air monitoring is important.

EMERGENCY EVACUATION

When the IC determines that a structure should be evacuated of all firefighters, it can be an orderly withdrawal or an emergency evacuation. The decision to abandon interior operations can occur due to the fire department being unable to control a fire from spreading or the potential for the structure to be compromised. If the IC decides on an orderly withdrawal, this order may be given face-to-face or via portable radio. It allows hose-lines to be backed out and equipment removed as all firefighters leave the structure. Once an order is received to evacuate the building, it must be immediately acknowledged and complied with.

Conversely, an emergency evacuation is ordered when some type of emergency has occurred (such as an explosion), or is about to occur (structural failure or collapse). During an emergency evacuation, equipment may need to be left behind to ensure an expedient exit.

To expedite an emergency evacuation, a signal is needed to ensure that all members involved are notified that conditions have drastically changed and immediate withdrawal from the area is required. Some departments use a continuous blast of either an air horn or a siren for at least one full minute. Other departments use sequences of either three or five blasts on the air horn or siren. These warning signals are then followed by radio notification to evacuate the structure immediately. Relying strictly on radio messages for emergency evacuation signals can be disastrous if a message is not received for a variety of reasons, including problems with a radio. A secondary system is a must.

Training in an emergency evacuation of a building should be part of an annual training program. We should do it routinely, similar to the US Navy practicing a man-overboard or an abandon-ship drill.

DEBRIS REMOVAL

When debris removal begins, it should be done selectively. Selected debris removal can start as a hands-on operation. This part of the operation is a dangerous time in which secondary collapse can occur. At some point, it becomes necessary to use overhead cranes and other special equipment to remove debris. All debris must be checked for victims, even if everyone has been accounted for.

Safety

The **safety** of the operating personnel at an incident scene is the prime responsibility of the IC. The frequency of firefighter deaths, coupled with the many injuries sustained, demands that ICs try to find ways to reduce these numbers.

Since ICs must prioritize their time, one way to prevent becoming overwhelmed is by delegating. Appointing a SO should be considered on incidents that pose a significant hazard or could adversely affect the well-being of those operating at the scene. Much time and effort may be required at an incident scene to establish a safe environment. The appointment of a qualified individual to perform the function of SO will help to reduce the number of injuries and lessen the severity of those that do occur. The delegation of safety must occur when the complexities of an incident could prevent the IC from handling safety on the priority basis that is demanded. It ensures continuity of this important task and allows the IC time for other important decisions. Some departments address this problem by the response of the departmental SO.

Appointing a SO helps ensure that safe procedures are followed. However, it is not prudent to assign the duties of safety to just anyone at the incident scene. If a SO is to be appointed before the arrival of the departmental SO, that individual must have received training in performing the required tasks. To use someone without specific training often results in confusion, and little is accomplished in terms of safety enhancement.

SAFETY OFFICER'S RESPONSIBILITY

The SO has core responsibilities to ensure the safety, welfare, and accountability of all incident personnel. This includes the need to recognize potentially hazardous situations. The SO has to inform personnel of any possible hazards. This should be accomplished by communicating the necessary information both up and down the chain of command. The SO needs to make certain that special precautions are taken that will protect all responders when both ordinary and extraordinary hazards exist. The SO must ensure that adequate rest and rehabilitation is provided at proper intervals, to protect the individual responder and collectively all responders at an incident.

SELECTING A DEPARTMENTAL SAFETY OFFICER

The position of SO should be entrusted to an officer of sufficient rank. Some fire departments utilize captains in this role. This can suffice on smaller incidents, but larger emergencies often require a chief officer as SO. With the realization that command systems refer to function and not rank, this would seem to be a contradiction. Reality dictates that a higher-ranking officer commands more respect and that his or her suggestions and orders carry more weight. A SO witnessing an unsafe act cannot hesitate to correct the problem, but problems might occur if a lower-ranking officer has to tell one of higher rank to change the operation to a safer one.

A departmental SO is someone who not only fills this position at an emergency incident but also reviews all injuries and determines their cause and methods to prevent similar occurrences. The process used to select a departmental SO must consider the many aspects of this important position. This individual should be highly motivated, have good managerial skills, and have sufficient exposure in emergency situations to determine safe or unsafe acts. He or she must command the respect of supervisors, peers, and subordinates alike. The SO must have a good working knowledge in many areas, such as building construction, strategy, tactics, human behavior, and fire science.

The departmental SO must develop programs to reduce injuries. To initiate this program, the number and types of injuries must be identified. How are they occurring? What must be done to reduce or eliminate the causes? A close analysis often reveals that injuries occur when a department's operational guidelines are violated. Firefighters involved in an operation can enthusiastically perform their assignment and, in doing so, may overlook correct procedures. Unsafe acts must be corrected immediately, and the person must be shown the correct methods. Compliance with safe operating guidelines is a primary responsibility of all fire officers and firefighters. Rules and regulations must be enforced or they are worthless. This is especially true with safety issues. The SO should review violations of standard operating guidelines. Strong enforcement at all levels of the department is necessary.

Some departments have divided the roles of safety by assigning the administrative roles to a Health and Safety Officer and the incident scene duties and responsibilities to an Incident Safety Officer (ISO). This permits a sharing of these essential roles and ensures a greater emphasis by the department on safety issues.

Health and Safety Officer

The Health and Safety Officer must consider all types of injuries. Unlike most workplaces, incident scene conditions are unpredictable and firefighters cannot be assured of a safe workplace by government standards. Firefighters are often placed in circumstances involving life-and-death situations and these incidents can lead to stress and exposures that can have long-term detrimental consequences.

Stress and heart attack is a major cause of injury that occurs to firefighters and it is the leading cause of firefighter deaths. The best method for a firefighter to protect against stress is to participate in a personal physical fitness program and maintain a proper diet. These positive actions keep firefighters in good physical condition and prepare them for the rigors of firefighting.

Illnesses can be caused by exposure to toxic chemicals or communicable diseases while operating at incident scenes. They include hepatitis and cancer and occur without the knowledge of those exposed. These exposures may cause firefighters to acquire a disease or be the basis for developing cancer.

Hepatitis

Hepatitis A and hepatitis B have been known viruses for many years, while hepatitis C was previously denoted as hepatitis non A, non B. Though all three types cause inflammation of the liver, they are genetically different. Hepatitis C was discovered in 1988 and has been in the spotlight in the fire service due to the high number of firefighters and paramedics who have contracted it. There are also hepatitis D and hepatitis E, but both of those infections are rare in the United States.

The spread of hepatitis A is primarily transmitted by the fecal-oral route, by either person-to-person contact or consumption of contaminated food or water. Bloodborne transmission is uncommon and transmission by saliva has not been demonstrated. Hepatitis B and C are spread through contact of blood from an infected person. Firefighters can be exposed during medical calls or accident responses. Another method of spreading the disease is through infected needles. Firefighters can be exposed when fighting fires in places where high drug activity occurs, or vacant buildings where drug users have discarded infected needles. A common problem is needles penetrating firefighter's personal protective equipment (PPE) when crawling into these structures to fight a fire. Paramedics can be exposed due to needle sticks from infected needles carried by drug users.

 ON SCENE

In Philadelphia, hundreds of firefighters and paramedics tested positive for hepatitis C. These incidents came to light when Local 22 of the International Association of Firefighters (IAFF) offered testing to its members. Once firefighters and paramedics discovered they had hepatitis C, they filed workmen's compensation claims. The city administration took the blatant step to automatically deny line-of-duty status to anyone with hepatitis C. This put the onus of proof that the disease was caused by on-duty contacts on the firefighters and paramedics. Judicial hearings became witch hunts as city-hired lawyers attempted to destroy the character of the infected individuals. These heroes were penalized twice: first by being infected with this cruel disease and then from attacks by lawyers that they were guilty until proven innocent. The only saving grace in Pennsylvania was the state legislature. It stepped forward and, despite the strong dissent by Philadelphia officials, introduced and passed legislation to protect emergency responders.

Cancer

Cancer poses another problem. The latency period from exposure to a carcinogen until development of cancer can span from a few years to more than 50 years. With the numerous chemicals used in the workplace and the daily development of new ones, it becomes impossible to know what a firefighter is exposed to. This worsens when a fire involving unknown material burns and creates smoke that can be inhaled or absorbed by firefighters. No one knows exactly what occurs when a fire involves plastics, solvents, PCBs, pesticides, and asbestos and routinely emits known or suspected carcinogens. But exposure to these materials over a career puts firefighters at an increased risk for developing cancer.

The following chart shows the latency period for some carcinogens:

LATENCY PERIOD

CARCINOGEN	TIME ELAPSED FROM INITIAL EXPOSURE
Arsenic	30–50 years
Asbestos	20–40 years
Chromium	10–20 years
Nickel	15–30 years

Cancer-Causing Chemicals

Certain chemicals are known to cause certain types of cancer. It is difficult, if not impossible, to determine an exact incident when an exposure occurred that caused cancer in a firefighter. Even if numerous exposures have been documented over the career of a firefighter, the long latency periods allow many years to elapse from the time of exposure until the development of cancer.

Benzene   Benzene may cause leukemia in as little as six years. It is one of the most widely used chemicals in the United States, and firefighters most likely inhale or absorb a small amount through the skin at each fire. Absorption is probably one of the most serious ways that it enters the body.

Carbon Tetrachloride   Carbon tetrachloride, previously used in fire extinguishers, is a known carcinogen. These extinguishers were often found mounted in heater rooms. They operated similar to a sprinkler head with a fusible link, which melted and released its contents into the confined space below. (Carbon tetrachloride was also manufactured in a glass bulb for throwing at a fire.) It was present in many fires in which firefighters responded. If a carbon tetrachloride extinguisher released while firefighters were fighting a fire, it excluded the oxygen from the air in the immediate area, and firefighters couldn't get a breath; thus, it rapidly caused unconsciousness.

Creosote   Creosote is a known carcinogen that causes cancer of the face, lung, neck, penis, scrotum, and skin. It is found in piers to protect lumber from rotting when in contact with water.

 ON SCENE

As a captain, my company was assigned to a fireground detail the morning after a multi-alarm fire at a pier. It was a hot summer day and the detail lasted the entire 10-hour shift. To facilitate the movement of hose-lines and counteract the heat of the hot summer day, bunker coats were removed. As we maneuvered around the creosote-soaked pillars, our unprotected arms came into contact with the creosote. In the course of the day this occurred numerous times. On returning to the firehouse that evening, we found that we had received burns to our arms consisting of reddening of the skin and blisters. What long-term effect this poses for us can only be conjecture.

Asbestos   Asbestos is a carcinogen and was used for years in many facets of building construction. It was contained in insulation, acoustical tile, joint compounds, wallboard, floor tile, and more. It can be found on or in walls, ceilings, floors, exposed structural steel, air ducts, plenums, and return air spaces. It can be found in asbestos cement shingles on a building's exterior. In the past it was used in firefighter turnout coats, and asbestos blankets were carried on firefighting apparatus for smothering a fire. Covers for brooms used for fighting grass fires were made of asbestos.

In a solid state, asbestos presents little hazard, but as a dust it can be inhaled or ingested. Firefighting, especially in the overhauling stages when opening ceilings, walls, and floors, is typically accompanied by airborne dust. Should firefighters remove their self-contained breathing apparatus (SCBA) due to absence of smoke and not utilize dust masks, these dusts can be inhaled. As a dust, asbestos can be found on bunker gear, which must be decontaminated prior to leaving the incident scene.

 **ON SCENE**

A voluntary study conducted in Philadelphia by Local 22 of the IAFF tested approximately 1,200 active and retired firefighters for asbestosis. The tests showed that more than 24 percent tested positive, well above the norm for the general population.

Personal Protective Equipment

Firefighters' exposures can occur due to absorption of chemicals or fumes through eyes, ears, or exposed skin. It can occur by inhalation of smoke or fumes from direct contact with a chemical or from off-gassing that occurs when PPE is impregnated with chemicals or smoke-laden carcinogens. This includes immediate exposure from off-gassing after firefighters remove their SCBA prior to decontamination, and future exposures if the PPE is not properly decontaminated. Many firehouses keep their PPE on the apparatus floor or in an adjacent area. I have seen gym areas where firefighters perform physical fitness training in close proximity to the PPE area. If all firefighters are not diligent about decontaminating their gear, it will expose others to contamination by off-gassing.

How many firefighters keep their PPE in their personal vehicle? Volunteer firefighters do because it enables them to proceed directly to an incident scene, as do career firefighters who may be detailed or temporarily assigned to another station for a tour of duty. Do all firefighters routinely clean their gear after every fire? In many fire departments, it is routine for PPE to appear to be darkened or dirty in order to represent the number of working fires attended by the firefighter. This mindset permits contamination to remain and allows the off-gassing from the PPE to affect anyone who enters the area where the PPE is kept. It will directly affect the firefighter wearing the gear in the future.

Diesel Fumes in Firehouses

Diesel exhaust fumes from fire apparatus are a known carcinogen. To protect firefighters, there needs to be a system in place that captures the diesel exhaust from the apparatus and directly expels those fumes to the exterior. There are numerous commercial systems. They function by attaching to the exhaust pipe of the apparatus and are equipped with a quick release, allowing an automatic disconnect that is activated when the apparatus pulls away. Though some firehouses use large fans on their apparatus floors to exhaust the fumes, their operation is not as reliable and depends upon someone to operate them. In my experience, fans are inefficient. They steal heat from the apparatus floor when in operation during cold weather. If the fans are turned on when an apparatus gets a response, they are typically not turned off once the apparatus leaves the station. In too many cases they are not used, and diesel fumes permeate the station, leaving a coating on all surfaces including any PPE located in the area. I have seen many firehouses where the walls and ceilings had a coating of diesel residue.

Safeguards

With the many dangers firefighters are exposed to in regards to disease and exposure to carcinogens they should:

- Implement the safeguards provided by their department.
- Protect themselves at all times: on medical calls wear all the protection necessary, and on fire scenes fully utilize their PPE.
- Not operate in smoke or dust-borne atmospheres without breathing protection; and use a SCBA or a dust mask if no smoke conditions are present.
- Decontaminate PPE at the scene of all hazardous materials incidents and working fires.
- Ensure that PPE in personal vehicles is kept in a sealed bag where it will not be in contact with household items, or a child's toy that may be reclaimed by children and invariably put in their mouths.

- Ensure that brother and sister firefighters are doing the same since their disregard for protection can contaminate the apparatus, equipment, and possibly the firehouse.
- Read the fire service trade journals and compare the level of protection that is available in other fire departments to what is available within your department.
- If a Health and Safety Officer is available, discuss safety concerns with that person.
- In addition to annual physicals conducted by the fire department, firefighters should have an annual physical with their physician (or more frequently if recommended).
- Discuss with their personal physician the fact that they are a firefighter and are concerned with possible exposures to communicable diseases, toxic smoke, and chemicals found at incident scenes.
- Have blood work performed at least annually.

 ON SCENE

Fire injuries and deaths from smoke inhalation caused by exposure to carbon monoxide (CO) have been well-documented. A potentially greater hazard to firefighters was uncovered in Providence, Rhode Island. There it was found that firefighters had been exposed to dangerous levels of hydrogen cyanide (HCN) and the serious effects it can have on the body. Twenty-seven firefighters were tested for HCN levels after three separate structure fires. Eight of the 27 firefighters had elevated levels of HCN and required treatment. One firefighter collapsed at the scene and was treated for HCN with an antidote. He retired on medical disability two years later.

Hydrogen Cyanide

Hydrogen cyanide (HCN) is 35 times more toxic than CO and is generated by the burning of ordinary materials contained in insulation, carpets, clothing, synthetics, and manmade plastics, which release cyanide if they catch fire. Naturally, these basic materials are found in almost all buildings. High temperatures and low oxygen levels favor the formation of cyanide gas.

HCN can enter the body by absorption, inhalation, or ingestion and targets the heart and brain. It can incapacitate a victim in a short period of time. It has a half-life in the blood of one hour, but its effects can last much longer. It is highly flammable and most HCN will burn away during combustion. Workers exposed to low concentrations of HCN have been found to develop enlarged thyroid glands.

The pre-hospital treatment of acute cyanide poisoning entails removing the patient from the source of cyanide, administering 100 percent oxygen, and providing cardiopulmonary resuscitation if necessary. The antidote, hydroxocobalamin, can be administered for known or suspected cyanide-poisoned patients.

Cyanide exposure is an expected outcome of smoke inhalation in closed-space fires. Research has established that cyanide poisoning can be a crucial cause of incapacitation and death to victims as well as chronic health complications to firefighters. The full utilization by firefighters of their PPE and SCBAs is critical until air monitoring shows that levels are safe.

SAFETY OFFICER RESPONSE

Ideally, a SO position should be staffed around the clock. This is costly and rarely done. Instead, the position is usually staffed through the use of on-duty or on-call personnel. Career fire departments may utilize on-duty battalion or district chiefs to fulfill the role of SO. The on-duty chief may be the second-due chief on an assignment, or may be automatically assigned once a working fire has been reported by the IC. This timely response allows chief officers to routinely work with their assigned personnel. It establishes credibility for the SO and enhances safe operations. It also ensures activation of the ISO function in the early stages of an incident when the potential for serious injuries exists. For this program to be successful, all chief officers and captains who could fill the SO role must be properly trained. Realize that training is not a one-time program. There needs to be annual refresher training to keep everyone current on safe operations and to recognize the commitment to safety required of each individual.

To ensure availability, fire departments utilizing on-call personnel during emergencies should have a minimum of three trained staff personnel who rotate on call. Volunteer departments can utilize a safety staff position. This arrangement provides a group of officers who rotate on call as the SO. The nonemergency functions can be assigned to ensure that training, proper documentation of injuries, and performance of other SO functions are provided.

The SO is part of the command staff and should work through the IC. The duties and responsibilities of the SO are an extension of the IC's. When problems arise, the SO can recommend to the IC possible solutions and be guided by his or her direction to rectify these problems. The SO has the authority to bypass the chain of command to correct unsafe acts and remove firefighters from imminent danger. If this power is used, the IC must be immediately informed. (Realize that the SO works for the IC, who has the authority and final say in safety matters.)

The utilization of a SO strengthens and supports the Incident Command organization by providing specialized knowledge. Monitoring of the incident by the SO maintains a systematic safety analysis of the scene. This ensures that recommended safety procedures are being followed.

Recognize that the diversity of calls and emergencies that fire departments respond to makes it very difficult for a SO to be an expert in all areas. During complex incidents with multiple areas of concern, the SO may need technical assistance. This can be especially true when dealing with specialized types of incidents such as high angle rescue, trench rescue, swift water rescue, hazardous materials incidents, and so on. In dealing with these types of events, an Assistant Safety Officer (ASO) should be assigned to work for the SO. This individual must be a qualified SO who is also well-versed in the safe operation of the specific activities that need to be performed. It usually involves a person who has received the same level of training as those performing the assignments. The level of training is especially important when addressing hazardous materials incidents. A qualified ASO must be assigned to work with the hazardous materials team. Competencies for this position include training as a responder at the awareness, operational, and technician levels, in addition to the required training as a SO. Experience helps the SO learn, but utilizing ASOs for technical safety support can augment that knowledge. Large complex incidents may require the use of ASOs to ensure that all areas are being monitored for safe operations. To ensure that the position of safety is fully addressed, the IC can request additional personnel to fill the roles of ASOs. Some incident management teams assign an ASO to each division or group on complex assignments, or to those covering multiple operational periods.

At the incident scene, the SO should be highly visible. In addition to full protective gear, the SO should wear a vest with "Safety" or "Safety Officer" across the back in bold letters. (See Figure 6-38.) For communications purposes, the radio call sign of "Safety" or "Safety Officer" should be used.

INCIDENT SCENE CONDITIONS

Firefighters often operate in a noncontrolled environment. They perform hazardous activities that can jeopardize their safety. Since firefighters are routinely exposed to these high-risk situations, the distinct possibility of injury and death exists.

Though risk cannot be totally eliminated, it can be managed. By examining dangerous situations, safe operating guidelines can be established. This process of risk analysis can help minimize the number of surprises to which firefighters will be subjected. Information about potential risks can be received by listening to the firefighters express their concern of certain activities or situations. (See Figures 6-40 and 6-41.)

FIGURE 6-38 The Safety Officer should wear a vest with "Safety" or "Safety Officer" across the back in bold letters. *Used with permission of Joseph Hoffman.*

FIGURE 6-39 A truss can collapse with no prior warning. *Used with permission of William V. Emery.*

FIGURE 6-40 A common bearing wall has failed, causing the roof and floors that are supported by the wall to collapse.

FIGURE 6-41 A serious wall crack was found in this wall. The wall is eight inches thick, and it is now leaning outward more than seven inches. A collapse zone should be established.

A role of the incident scene SO is to observe operations and act as a risk manager. The effectiveness of the SO will be based on his or her ability to monitor an incident scene. As the SO does a 360-degree walk-around of the incident and reviews the actions of the firefighters, he or she is performing a risk analysis of the potential dangers.

Hazards and Risks

Hazards are things that can cause harm to people or equipment. Hazards can be found pretty much everywhere. Several hazards can exist at an incident scene. Once a hazard is identified, it can be controlled or mitigated.

Risks are potentially unsafe actions taken by personnel when responding to hazards at an incident. A hazard does not put a person at risk; it is the actions taken that can put a person at risk. Risks are the chances that people take in relationship to hazards. For example, a downed electric wire on a highway is a hazard. The electrified wire will not threaten anyone as long as no one approaches it. However, if the downed wire on the highway is crossing a person's body, then a responder may deem it necessary to assist the injured person by attempting to remove the wire. The actions of the responder in this example are the risk being taken.

A SO must monitor an incident scene for both hazards and any risks taken that are associated with those hazards.

RISK ANALYSIS

Risk analysis starts with identifying various types of responses where a high degree of risk could exist. These could involve:

- Fire situations (structures, high-rise, refinery, wildland fires)
- Medical calls (exposure to blood-borne pathogens, contagious diseases)
- Special operations (hazardous materials responses, high angle rescue, swift water rescue, trench rescue)

All aspects of an operation should be analyzed for risk. Research can include departmental records, the experience of fire department members, national journals, and statistics from the US Fire Administration (USFA) and the National Fire Protection Association (NFPA).

Many fire departments require near-miss reports of events where a firefighter received minor injuries or was able to escape injury due to circumstances. Filing a near-miss report brings to everyone's attention the potential hazardous situations that need to be avoided in the future.

The International Association of Fire Chiefs (IAFC) maintains the Fire Near Miss website, which defines a near miss as "an unintentional, unsafe occurrence that could have resulted in an injury, fatality, or property damage if not for a fortunate break in the chain of events." The use of a near-miss reporting system by individual fire departments and the monitoring of national reporting systems will assist in safer operations. The IAFC's website can be found at http://fire.nationalnearmiss.org/.

After risks are identified, the frequency and anticipated threat to firefighters from those risks can be examined for the individual jurisdiction. Some risks can be present at many incident scenes. Other types of risk may occur occasionally.

FIRE DEPARTMENT HISTORY

A major consideration when examining risk is past occurrences of firefighter injuries or deaths in your fire department. What types of firefighter injuries or deaths were involved? What caused the injuries? If a firefighter is injured, a thorough examination of how and why it happened must be done. This investigation should look for ways to prevent future occurrences. The investigation must be an objective analysis to discover if any safety rules were violated or whether operational guidelines need to be rewritten to specify a safer method of handling similar situations in the future. In retrospect, what can be done differently to assure a positive outcome? When a solution is found, it is important that this data be shared with all departmental members.

Managing risk for emergency responders is not something that is performed once, has guidelines written about it, and then is forgotten. It must be a continuous process.

Guidelines must be reviewed to ensure that they continue to meet the safety needs of the department. Situations may arise that require the guidelines to be changed. Firefighter injuries can indicate areas that require additional analysis. The person responsible for reviewing and updating the guidelines is often the departmental SO. How well he or she accomplishes the job will be seen through the departmental safety record.

SAFETY MONITORING

Monitoring is the constant surveillance of an incident scene by visual observation, listening to communications, and through instrumentation to detect or identify problems or dangerous situations. This monitoring process starts when the SO is dispatched or one is assigned at the scene. The first step is to identify the concerns of the IC through a briefing. The SO must find out what has already occurred and what is currently underway. What are the IC's immediate and long-term safety concerns?

The next step is a systematic 360-degree walk-around of the scene. What is observed? What is heard? What problems exist? Are assignments being carried out in a safe manner? Has the SO received accurate information? If this is not the case, the SO must find out why. Have the orders been misinterpreted? Have conditions changed that caused the difference? Is the IC aware of the changes? Are the current actions creating an unsafe operation? Are there unnecessary risks being taken?

In his or her initial walk-around of the incident scene, the SO should make a note of areas that could pose safety risks. By talking to Division and Group Supervisors, the SO can get a feel for specific problems that have been recognized in a division or group and determine what impact they can have on the overall operation. Discussing any observed unsafe situations with the Division and Group Supervisors could increase the effectiveness of the SO. This method ensures immediate compliance while allowing the Division and Group Supervisors to maintain control of his or her specific area. It also establishes respect for both the SO and Division and Group Supervisors.

The SO must prioritize hazards at an incident scene. One method is to control or mitigate the most serious hazards first (downed electric wires that could electrocute a firefighter) versus minor hazards that are found.

In addition to the SO's monitoring, all personnel operating at an incident need to keep the SO informed of any hazards they encounter. This information will allow the SO to inform everyone at the scene of the known hazards. It should be department policy that when a hazard is discovered it must immediately be brought to the attention of the IC and the SO.

MONITORING REHAB

The SO will monitor the rehab area. As firefighters are sent to rehab, emergency medical services (EMS) personnel must evaluate them and take their vital signs. Elevated blood pressure or other problems with vital signs will necessitate that EMS personnel keep track of whether those vital signs return to normal. A firefighter must not return to firefighting until receiving clearance by medical personnel. It cannot be a decision on the individual's part. Too often, firefighters, in their desire to assist in the firefighting effort, ignore sound medical advice and return to firefighting only to be stricken and require additional medical treatment. If EMS is a part of the fire department, those personnel must be given the authority to prevent a firefighter from leaving rehab until he or she has been cleared to return to firefighting. If EMS is not part of the fire department, then a fire officer should be assigned as liaison to ensure that medical advice is heeded.

FIRST IN, FIRST OUT

In many areas, the first-arriving or first-due unit is also the last unit to leave the scene. This procedure can keep a unit at an incident well past the time where fatigue can set in. When dealing with a fire or incident where a company has exhausted themselves, a much better method is the first-in, first-out system. The first-arriving unit will be the first unit to leave the scene of a working fire or incident. In congested inner-city areas, it sometimes warrants the temporary swapping of apparatus. This first-in, first-out method will ensure an evenly distributed workload. It also enables the unit that may be fatigued to recover and be prepared should another incident occur.

CHECKLISTS

Safety issues can be reviewed by using an incident safety checkoff sheet and note pad. The development of a checklist by the SO should consider problems encountered during past incidents. For instance, if back injuries are occurring at a frequent rate, the SO should be checking that a sufficient number of personnel are being utilized for stretching hose-lines and raising portable ladders. The checklist allows follow-up on problem areas and assists in preparation of the final safety report.

Monitoring will assist the SO in the goal of preventing or reducing injuries. A checklist will assist in monitoring the scene, but a good common sense approach must prevail. You must not be limited only to areas noted on a checkoff sheet. An unlit loading dock filled with water could be a pitfall for an unsuspecting firefighter, yet wouldn't be on a checklist unless the list was so long as to be impractical.

 BOX 6-1: WHAT MUST BE MONITORED

THE BUILDING

- Access and egress points—doors, adjoining roofs, ladders, and any obstacles that may hinder egress (e.g., bars on windows or security gates on doorways)
- Danger to firefighters operating on roofs (e.g., avoiding raised roof surfaces, which could indicate the presence of covered-over skylights or shaftways. Are hose-lines stretched to the roof for protection of the firefighters?)
- How various types of construction present different problems (e.g., quick collapse of some types of roofs and floors of lightweight components)
- The size and stability of the structure
- The presence of collapse indicators
- Sounds of building movement (i.e., groans, cracking of wood, grinding of masonry)
- Suspended loads that could place an eccentric load on a building (i.e., wall signs, marquees)
- Age of a building (may indicate the use of full-dimensional structural members or lightweight components)
- Building density (closely built structures can reduce the SO's ability to monitor locations where firefighters are operating)
- Sprinkler and standpipes being utilized (Have they been pressurized by the fire department?)

CONTENTS

- Contents in the building (hazardous materials, sophisticated electronics, overstuffed furniture, etc.)
- Housekeeping (Well-kept or cluttered?)

TIME

- Length of time the fire has been burning (the longer the fire is burning, the more potential damage to a structure)
- Length of time firefighters have been operating at the scene

WEATHER

- High winds driving into window openings (can create a wind-driven fire causing dangerous interior conditions for firefighters)
- High winds driving a wildland fire, making a hazardous situation worse
- High winds at lumberyards and other establishments with outside storage of combustible materials (may require firefighters to flank a fire)
- Extreme heat (can cause fatigue and dehydration of firefighters; more rest periods with monitoring of vital signs will be needed) (See Figure 6-42.)
- Extreme cold (can slow operations; frostbite will be a concern, and icy conditions can affect footing) (See Figure 6-43.)

FIGURE 6-42 Extreme heat can cause firefighter fatigue and dehydration. *Used with permission of Joseph Hoffman.*

FIGURE 6-43 Extreme cold will slow down operations, and frostbite and falls on icy surfaces can occur. *Used with permission of James H. Bampfield, Jr.*

SURROUNDINGS AND STAFFING

- Rehabilitation of responders
- Life safety (Has the life safety of the occupants already been established? Are fire department efforts now being directed toward incident stabilization and property conservation, while minimizing risk to the responders?)
- Personnel (firefighters, police, media personnel, and other emergency workers; there should be a sufficient number of responders to achieve the goals of the IC without overtaxing them)
- Progress reports and any communications indicating problems
- Fire behavior (flame spread, rollover, potential for backdraft or flashover. Do conditions warrant any changes in the modes of attack?)
- Exposed interior electrical wiring (See Figure 6-44.) [safety concerns with exterior overhead electrical wiring (for operation of ladders and observing any downed wiring)]
- Modes of attack (ensure that offensive and defensive operations are not employed simultaneously)
- Defensive operations (once initiated, ensure that everyone is in compliance, not in the interior, and operating a safe distance from the fire building)

FIGURE 6-44 Exposed electrical wiring in large attic areas or other unfinished spaces can shock firefighters coming into contact with them. *Used with permission of William V. Emery.*

- Collapse/safety zones defined, properly marked to deny entry, and limits adhered to by personnel
- Safe positioning of firefighters operating at the incident scene
- Positioning of ground and main ladders on all sides of a structure to allow secondary means of egress
- An accountability system in place; firefighters working in teams; rapid intervention crew/s (RIC) in place
- Time limits on firefighters operating SCBA
- Proper protective gear being utilized (PASS activated, adherence to guidelines for interior operations in large structures)
- Coordination of incident scene activities to ensure a safe operation (See Figures 6-45 and 6-46.)
- The environment [smoke, toxic releases, areas that are immediately dangerous to life and health (IDLH)]
- Operating areas must be monitored for changing situations
- Equipment [using the correct tools, proper placement of apparatus, proper use of equipment, sufficient lighting to illuminate areas of concern (e.g., flooded loading docks, unprotected swimming pools, dangerous conditions)]
- Ensuring that vehicular traffic does not interfere with operations nor injure firefighters
- Securing the safety of crowds of curious civilians who could be endangered and impact operations

FIGURE 6-45 Firefighters and other emergency workers operating at a trench collapse must be monitored. Their actions can make a bad situation worse. Here, firefighters are standing too close to the open trench and could cause additional collapse, further endangering those already trapped and the rescuers in the trench below.

FIGURE 6-46 A shipboard fire in Wilmington, Delaware, tested the Wilmington Fire Department. Technical experts from the Navy and Coast Guard may be of assistance when confronted with these unusual types of incidents. *Used with permission of William J. Shouldis.*

MEDICAL RESPONSES

- Infection control procedures in place
- Proper protective equipment being utilized
- Proper disposal of medical waste
- Mass casualty procedures being properly implemented

SPECIAL OPERATIONS

- Safety lines at a water rescue
- Proper shoring at a trench rescue (Are firefighters and other emergency workers operating in a safe manner? Is the shoring properly installed or is it bowing or moving, threatening rescuers?)
- Approved lifelines at a high-angle rescue
- Properly qualified firefighters for the activities in which they are participating
- Proper use of technical experts
- Hazardous materials incidents (Does the spill, leak, or fire threaten personnel? Are zones set up and is there restriction of personnel within those zones?)
- Confined space operations
- Recommending critical incident stress debriefing for units

TUNNEL VISION

Tunnel vision can occur when you become so engrossed in a particular phase of an operation that you fail to see the big picture. An example is a vehicle striking a building, resulting in injuries to the driver, passengers, and building occupants. In an effort to give immediate assistance to the seriously injured, firefighters and emergency medical personnel may disregard their personal safety. It is foolish to ignore preliminary safety precautions in stabilizing an incident.

People trapped and seriously injured may require the removal of a vehicle from the structure to extricate them. The IC and/or SO must assess the scene thoroughly with regard to scene safety. An assessment of the structural stability must be made. A careful survey must ensure that the vehicle's removal will not cause the building to collapse. In striking the building, the vehicle may have displaced structural supports and could be holding the building in place. If time does not permit the arrival of a building engineer (which it seldom does) and the IC feels the vehicle must be removed immediately, all personnel should be removed from the potential collapse zone to allow the careful removal of the vehicle.

A

B

FIGURE 6-47 A–B A firefighter gets into trouble while performing a primary search. Flashover of the second floor is occurring. He sounds a Mayday. The only means of escape is hanging from a window sill awaiting the arrival of the rapid intervention crew. Fortunately he was protected by his gear and was not injured. *Used with permission of Terrance Grady.*

Most situations warrant the restriction of unnecessary personnel from specific areas. Medical personnel may not be protected properly in some situations until the incident can be stabilized. There is a difference between aggressiveness and foolishness. Firefighters, by their nature, are action-oriented people. They are risk takers. These traits are commendable when a person is trapped and in need of rescue. However, after rescues have been made, chances should not be taken to save property if it presents a substantial risk to the firefighters.

The same inner drive that compels firefighters as risk takers and doers often prevents them from admitting to themselves that they are tired, and they often push themselves past a safe working limit. This overexertion makes them susceptible to injuries.

APPARATUS PLACEMENT

Apparatus placement must consider the use of the apparatus. It should not be placed where it will impede movement of other apparatus. Fire hydrants should not be used if they are directly in front of or alongside a burning building because their use can affect the safety of the operator (assuming that the fire building is located close to the street/s).

A good practice is to reserve the front of the fire building for the truck company, allowing use of its aerial equipment to effect rescues. If rescues are not required, the truck company may be used for master streams and can set up on a corner of the structure. This places the apparatus at the strongest part of the structure and allows flexibility to attack the fire or protect exposures on two sides. Other apparatus must not park near the rear of a ladder truck where it would limit the removal of portable ladders.

NIOSH FIREFIGHTER FATALITY REPORT F2013-19

On June 13, 2013, a 36-year-old male volunteer firefighter (the victim) died after being electrocuted as he investigated the source of a small structure fire. The fire department was dispatched to a vehicle fire caused by an energized power line that was downed after severe weather passed through the area. The energized power line fell across the roof of a small metal storage shed, causing the wooden support structure to arc and catch fire. The victim, dressed in street clothing, walked down a rain-soaked gravel and dirt driveway and knelt down to look underneath the building where fire and smoke were emitting, and immediately fell to the ground unconscious. An eyewitness reported the victim did not touch the building before he fell to the ground. Rescuers dragged the victim approximately 30 feet away and began performing cardio-pulmonary resuscitation (CPR) and attempted to use an automated external defibrillator (AED). The victim was transported by ambulance to the local hospital where he was pronounced dead.

Contributing factors include an energized power line in contact with a metal building; pooling water and runoff from recent rain storm; victim not wearing any personal protective clothing or non-conductive boots; lack of situational awareness; unrecognized electrical hazards—especially on ground gradients and step potential hazards; no other first responders in the immediate area at the time of the incident. **See Appendix B: NIOSH Reports in www.pearsonhighered.com/bradyresources to read the complete report and recommendations.**

Electricity and Firefighting

A tactical and safety consideration when placing apparatus is the proximity of overhead wires. Engine companies should consider flame impingement on the overhead wires and the possibility of their failure and falling onto their apparatus. Additionally, truck companies need to decide whether the overhead wires will limit the ability of a main ladder or platform to operate. Apparatus placement must consider the possibility of falling overhead wires due to flame impingement and ensure that the apparatus is not parked under threatened overhead wires.

Where overhead electrical wiring is present, the SO needs to ensure that a safe distance is kept when raising or operating with ground ladders, main ladders, and platforms.

Electrical service wires that are coming from electric poles and attached to the upper part of a structure, commonly called a house drop, need to be avoided by portable and aerial devices. This is especially important in case the wires are concealed or hidden when they pass through tree branches.

In some locales, high voltage electric wires may be attached to the front or rear of properties and extend over the length of the collective buildings. This situation can be found in some row houses, town houses, and strip mall construction. Fire lapping out of windows or doors can attack the electrical service lines and will require a quick response from the electric company to alleviate the potential of electric shock from downed wires or electrified portions of the building.

DOWNED ELECTRIC OVERHEAD WIRES

If downed overhead electric wires are encountered, all firefighters must realize that direct contact with the wire can cause serious injury or death. On learning of a downed wire, the SO must immediately broadcast a safety alert to all personnel operating on the incident scene and should immediately cordon off the area. Use whatever means are necessary, including banner tape, ropes, barricades, and even apparatus. If deemed necessary, assign an ASO or other personnel to deny entry and ensure strict compliance with the restricted area around the downed wire.

The IC should immediately request an emergency response from the utility company, stating that a downed overhead electrical wiring exists at the scene. This will alert the electric company of the importance and high priority of their response.

If a live electric wire comes into contact with a vehicle or apparatus, it will become energized. Anyone trying to step out or off of the vehicle while in contact with the vehicle and the ground will allow the electricity to pass through them, causing serious injury or death. Personnel in a vehicle should remain inside the vehicle until the wire can be removed, or the electric company can cut the power to that wire, usually by cutting the wire on the electric pole. If confronted with someone physically in contact with a live electric wire, firefighters may be able to move the electric wire by using a dry rope or pike pole. This should only be attempted when a person's life is at stake; otherwise, the best avenue is avoid the risk and wait for the electric company to arrive and mitigate the problem.

If an apparatus is electrified, everyone should be kept away. Should a firefighter be on an apparatus, he or she could safely jump away from the vehicle if necessary, as long as they can clear themselves from the apparatus without making contact with the ground and apparatus at the same time.

 **ON SCENE**

At the scene of a multi-alarm fire in Philadelphia, the extended main ladder on an aerial apparatus that was operating in front of the fire building made contact with an overhead electric wire, which electrically charged the entire apparatus. The RIC was positioned alongside the ladder truck. All five members of the RIC received electric shock in varying degrees. Some members standing alongside the apparatus had their body in physical contact while others received a shock by their close proximity. Fortunately the injuries were not life-threatening.

FIGURE 6-48 What may appear as a vacant building may be occupied. Firefighters found a family in need of rescue from this structure with a major structural problem. Rescues were effected and a defensive attack initiated.

LADDER PIPE OPERATION

Too often on a defensive attack, firefighters are at the head of the main ladder manually operating the ladder pipe. There are a few times when directing the stream from the head of the main ladder allows for better control and direction; however, this must be weighed against the risk involved. Often, the only reason that firefighters are positioned at the head of a main ladder is their desire to be actively involved. This attitude is commendable, but the safety of the practice is questionable. Malfunctioning ladder pipes cause many serious injuries and deaths. The use of ropes attached to the ladder pipe and operated from a safe location on the ground should be the accepted practice.

RISK VERSUS GAIN

Too often, the enthusiasm of firefighters leads them to do anything to get the job done. This attitude is noteworthy and from a fire officer's point of view leads to the accomplishment of many difficult tasks and lifesaving rescues. The drawback is that this mind-set also leads to injuries that many times can be prevented while still accomplishing the task. There is a time and a place for risk. The question that must be considered is, "Is the risk worth the possible gain?"

Risks to firefighters can be classified as low, medium, or high. A high-risk situation would be fighting a fire in a vacant building that is heavily involved. A low-risk situation could involve the same structure with a minor fire on the exterior impinging on the building.

The gain or benefit to be realized by firefighters can also be classified as low, medium, or high. A building fire involving trapped civilians would be classified as a high-gain situation. (See Figure 6-48.) A fully involved occupied structure in which the occupants would be considered past helping or a vacant building devoid of occupants would be a low-gain situation.

There are also ways that risk can be avoided. A hazardous materials spill in a remote location might best be handled by withdrawing all firefighters and allowing the product to evaporate into the atmosphere. A fire impinging on an LPG tank where a boiling liquid expanding vapor explosion (BLEVE) is an imminent possibility might require evacuating an area and allowing the fire to burn out. These nonintervening actions are a way of protecting responders.

The classification of risk versus gain seems to place everything into clear-cut categories. The problem is that many factors are unknown at an incident scene. Vacant/abandoned buildings can be occupied. Information received from Dispatch or bystanders can be faulty. Conditions can change rapidly from a relatively stable situation to one that is out of control. Ideally, we would like to be confronted with low-risk, high-gain situations. Yet there will be times where a high-risk, high-gain situation will be encountered, and difficult decisions will need to be made. High-risk situations still must be made as safe as possible. The proper utilization of all protective equipment must be enforced. The seconds necessary to ensure safe entry and exit can be the difference between life and death for the rescuers. For this reason, the IC and SO must continuously evaluate the risk to firefighters. By weighing the dangers, a determination can be made as to the degree of risk to which firefighters will be subjected.

A typical safety policy could state:

a. Firefighters may risk their lives to save a life.
b. Firefighters may place themselves in moderate risk to save property.
c. Firefighters will risk nothing to save life or property that is already lost or destroyed.

SELF-CONTAINED BREATHING APPARATUS USAGE

During an offensive attack in a building, firefighters must monitor their SCBA to ensure an adequate air supply. Most 30-minute air cylinders actually have only 15 minutes of usable air. This means that if a company is relieving another on a hose-line and it takes approximately three minutes to enter the fire area, they must allow another three minutes of air to ensure safe exit. This leaves approximately nine minutes of usable air for firefighting. This time restriction means the IC must have sufficient forces for rotation of hose-line crews. Some departments now use 45-minute or 60-minute air cylinders to increase working time. Although the air supply is increased, the weight and/or the size of the cylinder is also increased.

When using air cylinders, there is a tendency for firefighters not to monitor their time. They are attuned to listening for the low-air warning signal to notify them when they are low on air. In large-area operations, this might not leave enough time to exit the fire area safely. The SO or an assistant can keep track of the length of time units are on interior attack and their physical condition when exiting the building. Firefighters operating in large structures must also be cognizant of the possibility of becoming lost due to the large area, darkness, and heavy smoke conditions encountered in such fires. In large buildings, they must be in constant contact with a hose-line or utilize guideline ropes to ensure that they can safely exit the building.

"NFPA 1404: Standard for Fire Service Respiratory Protection Training" mandates that firefighters need to have individual air management. The standard states that the reserve air supply (the air remaining in the cylinder after the low air alarm has sounded) should be intact when a firefighter leaves the IDLH area. This standard specifies that:

- Firefighters should exit the IDLH atmosphere before consumption of the reserve air supply begins.
- The low-air alarm is notification that the individual is consuming the reserve air supply.
- Activation of the low-air warning alarm is an immediate action item for the individual and the team.

ACCOUNTABILITY

Firefighter safety must be a main concern of every IC. One method of ensuring firefighter safety during the course of an incident is an ongoing accounting of all firefighters operating at the incident scene. An accountability system must be in place to track personnel and allow for the oversight of firefighters' safety. The accountability system will assist in knowing what everyone is doing and where they are operating. Should a dangerous situation develop, those endangered can be notified immediately. An accountability system can be simple or complex, depending on the needs of the individual department. The key point is that the system must be practiced, and all personnel must be familiar with it. This key factor means that mutual aid companies need to use the same system. Training with the accountability system will hone the firefighters' skills to ensure that it will work properly if a firefighter actually becomes lost, trapped, or in trouble.

Accountability can be assisted by using tactical work sheets as well as a Planning Section Chief to track personnel through the use of situation status and resource status sheets. Some fire departments make accountability a function or dual responsibility of the SO. To accomplish accountability, there is a need to know:

- The number of personnel operating at the scene
- Their approximate location
- The task or function they are performing

Types of Accountability Systems

There are a variety of accountability systems. Many departments use a riding list, which is a sheet that lists the names and assignments of each member. The list can be given to the IC, Operations Section Chief, SO, or Planning Section Chief upon arrival. Another method is a tag system. Members have a removable tag attached to their turnout coat or helmet. Those operating at an incident place their tag on a board attached to the apparatus or located at the entry point into the fire building. Off-duty firefighters who respond directly to the scene can attach their tag on arrival. The firefighters are then assigned under the command of an officer to form a company or a crew. The tags are grouped on the board to show each unit formed. This information can also be used to determine which units will need relief or rotation. There are also more sophisticated systems utilizing bar codes and on-scene computers.

Global positioning is currently being developed for firefighter accountability. Funding for some development and testing has been provided by the Department of Homeland Security. Global position tracking of firefighters has been accomplished by each one wearing a unit that transmits their location. Hopefully, future systems will allow tracking of everyone operating at the incident scene by placing a computer chip in each firefighter's PPE. Should a Mayday be initiated, immediate knowledge of that firefighter's location will be known.

Accountability ensures control at an incident. Firefighters should operate in teams, not alone. No one can be permitted to freelance. Units must be part of a known plan, and company officers must know where their firefighters are operating at all times. Though this is not an easy task, it is an essential one.

Personnel Accountability Report

The implementation of a personnel accountability report (PAR) must be accomplished quickly and accurately. A PAR is a head count or roll call of all personnel operating at an incident. The IC must contact divisions and groups at timely intervals throughout an incident to ensure accountability. It is not acceptable for a Division or Group Supervisor to just verbalize that they have PAR. Each Division or Group Supervisor must verbalize to the IC the exact units operating in their division or group, which allows the IC to check his or her tactical worksheet for verification. Though some fire departments utilize their dispatchers to conduct a PAR, it should be done by the IC or his or her designate at the scene. Dispatch will not know what divisions and groups are operating at the incident or what companies or units are assigned to those divisions and groups. Additionally, when the IC conducts a PAR, he or she should contact the most endangered units first. Should the need arise to communicate with other units, the PAR can be delayed until those important communications can be made. A PAR should be taken routinely to ensure accuracy of where personnel are operating, and during drastic incident changes, such as a building collapse, a shift from an offensive to a defensive attack, or after an emergency evacuation of a building has been sounded. When doing a PAR, there can be no assumption that a firefighter who is not physically present is accounted for. Each firefighter must be seen or personally report via radio to their company officer.

TWO-IN-TWO-OUT RULE

Fireground staffing varies from fire department to fire department. Large-city fire departments may have six apparatus with four or more firefighters per apparatus responding on an initial assignment. Rural areas may have the response of a single apparatus with two firefighters to a structure fire. To ensure firefighter safety, rules and standards have been adopted by the Occupational Safety and Health Administration (OSHA) and the NFPA to protect firefighters operating at an incident scene.

OSHA adopted the Respiratory Protection Final Rule (01/08/98, 29CFR 1910.134). This legislation mandates the use of respirators in IDLH atmospheres, including structural firefighting. OSHA describes structure fires beyond the incipient stage as IDLH atmospheres. In these atmospheres, OSHA requires that personnel use SCBA. OSHA also mandates that a minimum of two firefighters work as a team inside the IDLH environment and that a minimum of two firefighters be on standby outside of the area to provide assistance or perform rescue. This legislation is referred to as the Two-In-Two-Out regulation. OSHA states that "once firefighters begin the attack on an interior structural fire, the atmosphere is assumed to be IDLH and the two-in-two-out applies." It is further stipulated that firefighters operating on the interior must operate in a buddy system and maintain direct voice or visual contact with each other at all times; radio contact cannot take the place of voice or visual contact. Prior to a team of firefighters entering the IDLH area to fight a fire, a team of two firefighters must be positioned outside of the IDLH atmosphere. This team must at all times account for and be available to assist or rescue members of the interior team.

The location of the team outside of the IDLH atmosphere can be on the exterior of the fire building or within the building outside of the IDLH atmosphere. An example would involve a room and contents fire on the second floor of a two-story dwelling. Everyone has evacuated the building prior to the arrival of the fire department, so life safety of civilians is not a concern. Fire is venting from a second-floor window. The initial-arriving firefighters would stretch hose-line to the interior stairs because there is no smoke at that location. When a sufficient number of firefighters are on the scene, the two initial firefighters (inside team) would be able to climb the stairs and enter the IDLH area. The other two firefighters (outside team) could be inside the dwelling on the first floor. One firefighter inside the front door can be stretching the hose-line to the other member of the outside team, who can be positioned on the stairs. This firefighter will ensure that a sufficient amount of hose-line is available for the inside team. Both of these members are still outside of the IDLH and in direct contact with each other. The firefighter on the stairs is in contact with the inside team operating on the second floor. The IC can be within sight of the outside team should communication with them become necessary.

OSHA allows those designated as the outside team to be engaged in other activities, though these tasks cannot be critical to the health and safety of other firefighters working at the incident. Fire departments should establish specific guidelines to comply with this rule.

OSHA does allow an exception to the two-out rule where immediate action is necessary to save a life. Where firefighters find that immediate action could prevent the loss of life, deviation from the two-out rule may be permitted. These occurrences must be exceptions to the fire department's standard operational guidelines and not standard practices.

Realize that with heavy smoke conditions at a fire scene, an IDLH location could exist on the exterior of a fire building, including on the roof of the fire building, or around an exposed building. When entering IDLH locations, the SO must utilize an SCBA and must be accompanied, not operating alone.

In addition to OSHA, the NFPA has written "NFPA 1500: Standard on Fire Department Occupational Safety and Health Program" and "NFPA 1561: Standard on Emergency Services Incident Management System and Command Safety." These standards cover incident scene safety, among other things. Both standards recommend that a RIC be established. This crew should consist of a minimum of two firefighters who are trained in performing this function.

RAPID INTERVENTION CREW

RICs, rapid intervention teams (RIT), and firefighter assist and search teams (FAST) are standby crews that are assigned for rescue of firefighters that become lost, incapacitated, or trapped in a building due to a flashover, a backdraft, a collapse, an injury, an SCBA malfunction, or a similar event. The purpose of a dedicated rescue crew is to permit an immediate response to the firefighter in distress and effect a quick removal to the exterior. (See Figure 6-49.)

The assignment of the RIC must occur in the early stages of the fire. The early implementation of the RIC is needed because many firefighter deaths and serious injuries occur during the period prior to a fire being placed under control. Ideally, a company (engine, truck, or rescue) should be assigned this duty, but a crew of two or more firefighters can perform this function. The RIC reports to the IC. If an Operations Section is established, then the RIC reports to the Operations Section Chief.

Rapid Intervention Crew Training

There must be training of personnel before they are assigned as part of a RIC. This training should be in individual skills,

FIGURE 6-49 A rapid intervention crew should be located near the Command Post. They should monitor radio communications and be ready for immediate deployment. *Used with permission of William J. Shouldis.*

such as using a TIC or utilizing buddy breathing, and crew drills, such as implementing a search of a large area or removing a downed firefighter.

Training for crew members must include basic extrication procedures. The duties of the RIC start on their arrival at the incident. They should be observing all activities and listening to radio communications. By monitoring the radio at all times, they can visualize the conditions on the interior of the building and the approximate locations of where units are operating.

The RIC should make note of all means of entry and egress from the building. Special emphasis should be placed on the location of raised ground and aerial ladders, the location of windows that access roof areas, and any impediments to exit from a window, such as bars on windows or window air conditioners. They should note the location of aerial ladders or platforms that could be utilized to assist in the rescue of a trapped firefighter. If there are sufficient personnel in the RIC, two members can stand by the Command Post while the other members of the RIC raise ground ladders to ensure additional means of egress from every level of the structure.

Many factors need to be considered in rescuing a firefighter, including the type of building, the location of the missing firefighter, the fire and smoke conditions, the building's stability, the time needed to locate him/her, debris hindering the operation, and the number of firefighters available to effect the rescue.

 ON SCENE

Fire conditions in a supermarket in Phoenix caused a firefighter to declare a Mayday. The firefighter was disoriented and lost inside the building. It took two initial RICs and additional rescue crews 53 minutes to retrieve him in a search area of about 2,000 square feet.

In the aftermath of this death, the Phoenix Fire Department did comprehensive studies to determine how long it takes to effect the rescue of a firefighter who declares a Mayday. Their research found that the fire service has underestimated both the number of rescuers and the time required for RICs to enter the building, locate, and move a firefighter to safety. They found that on average it took 12 firefighters 21 minutes to enter, locate, and retrieve a downed firefighter.

Location of RIC

Each situation is different. If a minimum of personnel is available, the positioning of the RIC must be flexible. In multiple company operations, the RIC should be located near the Command Post and visually review the resource sheets that track the location of all units. If an Accountability Officer has been designated, this individual will be an excellent resource for determining the approximate location of a missing or downed firefighter.

Once Operations is established, the RIC will be positioned in the vicinity of the Operations Section Chief. During a high-rise fire, the RIC will typically be located on the floor below the fire floor in the vicinity of the fire stairs designated for firefighting. If multiple stairs are being utilized for firefighting, multiple RICs could be assigned.

Tools and Equipment

RICs should gather and make ready the tools that will be needed on an assignment. They include:

- Thermal imaging camera (if available)
- Hand light
- Portable radio
- Hand tools for prying and forcible entry
- Search rope
- Spare air cylinders and buddy breathing, if available. Buddy breathing allows two firefighters to share the air supply from one air cylinder.

Other tools may be helpful, including saws and bolt cutters. However, the more tools that need to be carried, the slower the crew, and time is a critical factor.

UPGRADING A RAPID INTERVENTION CREW TO A TASK FORCE

Missing or downed firefighters trapped under debris in a collapse will need to be extricated. Testing has shown that a four-person RIC will find it difficult to remove a downed firefighter(s) in an open area near an exit. Rescue becomes compounded when a firefighter is trapped or located in an area where removal is complicated, such as in a basement or on upper floors. A downed firefighter(s) may be unable to assist in their removal and may require a supplemental air supply for their SCBA.

Certain observations by the IC could indicate the need to expand the RIC from a single company to a task force. The exact size of the task force would be determined by the available resources. If possible, a minimum of two companies should be assigned, along with a chief officer as the task force leader. Optimally, the inclusion of multiple companies, such as an engine company, two truck companies (or a heavy rescue unit if available in lieu of a truck company), and a chief officer as the task force leader would be an adequate response to a Mayday to assist a firefighter(s) in need of assistance.

Factors or cues that could indicate the need for upgrading the RIC from a single company to a task force are:

- A working fire in a commercial building
- A working fire in a high-rise building
- A working fire in a large residential structure
- A significant or unusual fire situation
- Difficulty in performing effective ventilation
- A report of firefighters missing or in trouble
- Personnel not reporting back after two PAR attempts
- If RIC advises Command that they cannot accomplish an assignment

Realize that some of the above indicators could indicate the need for a shift from an offensive attack to a defensive attack. The decision to remain on an offensive attack could be due to the report of occupants still in need of rescue. The deployment of a RIC should immediately signal the demand for assigning at least one other company to assume the standby position of a RIC.

CREW INTEGRITY

Safety means working together as a team. Teamwork creates a synergy in which a crew or company of firefighters working together can accomplish the tasks needed to mitigate the emergency in a safe manner. Safe fireground operations demand crew and company integrity at all times. Units should enter and leave a fire area together or in teams of two or more. When climbing stairs in multistoried buildings, the crew or company is as fast as the slowest member. A firefighter who is not in top shape can slow down the entire company. The benefit of the team concept is that it permits a firefighter in trouble to be assisted to safety by the other team members.

MAYDAY

The term *Mayday* is the internationally recognized distress call that is used by ships and aircraft. The fire service has adapted it to indicate a firefighter who is faced with a life-threatening emergency, such as being trapped, disoriented, low on breathable air, or endangered by threats of physical harm from civilians.

The word Mayday sends a chill through even the toughest veteran firefighters. The sounding of a Mayday means one of our own is in trouble. It is not a call for help that is taken lightly by the firefighter calling the Mayday, nor by the other responders at the incident scene. Once it is heard, the focus of the entire incident needs to be adjusted to ensure that whoever is in trouble is given top priority. On more than one occasion, the IC has turned over the handling of the overall incident to another fire officer and has personally taken over supervising the operation of locating and removing the firefighter(s) in trouble.

Many types of portable radios contain an emergency activation button (EAB). A common feature is that once the EAB is depressed to sound a Mayday, that radio frequency is captured for a period of time; usually from 10 to 20 seconds, allowing an open microphone without the sender having to depress the talk button. This open microphone feature allows firefighters to call the Mayday and give important information, which should include:

- His/her unit number
- Their exact location, as much as possible (e.g., second floor, middle room)
- What is happening to them (e.g., trapped in the room with heavy fire in the hallway)
- Conditions at their location (e.g., heavy smoke, conditions worsening, and no egress except through windows)
- What is needed to assist/remove them from their critical situation (e.g., place portable ladder to second floor window for rescue)

Another piece of information required by some fire departments is for those who transmit a Mayday to state their name. The reasoning is that it allows units that find a firefighter in distress to realize that the person found is indeed the one who sounded the Mayday. Other fire departments do not allow the names of anyone trapped or injured to be transmitted via radio since it can be monitored by non-fire department personnel.

RESCUE EFFORTS FOR MISSING FIREFIGHTERS

After sounding the Mayday, firefighters should manually activate their PASS device to assist the RIC in locating them. Depending upon the radio system, Dispatch may be able to identify (through an identifier system) the radio user and/or company number who is sending the Mayday and notify the IC. If simplex channels are being utilized, it may restrict the notification of the dispatcher when a Mayday is sounded. Only those units working on that specific talk group channel would know that a Mayday was declared, and they would be the only ones who would be able to hear the Mayday transmission.

Simplex channels are also referred to as talkaround channels. They transmit from radio to radio without the use of towers. They have limited range, usually of one-quarter mile or less, and are not typically monitored by Dispatch.

When a Mayday is called at an incident scene, immediate measures need to be taken to assist the firefighter(s) in distress. The IC needs to keep personnel on the scene focused on their specific role, especially if it does not involve the assistance to the missing firefighter. The IC needs to ensure that the ongoing emergency is handled properly. The word *properly* means that the emergency that brought the fire department to that location still has to be resolved. The missing firefighter(s) will certainly be the major priority, but the other problems at the incident cannot be forgotten with everyone dedicated only to the rescue of the missing firefighter(s). To ignore the other ongoing problems could drastically affect rescue operations. For instance, shutting down the hose-lines that are fighting the fire or stopping ventilation operations could worsen conditions surrounding the trapped firefighter(s). Like many fireground operations, the correct actions to be taken will be incumbent on the amount of resources available and what needs to be accomplished.

A system with high-frequency radio waves is now being utilized to locate downed firefighters. It employs an activated transmitter that a firefighter can carry on his or her person or have integrated into their SCBA. Once activated by a firefighter in trouble, the system gives out a signal that can be tracked and located through the use of handheld receivers. These receivers are programmed for individual transmitters and can be used to identify the user. The signal strength from the downed firefighter's transmitter acts as a homing device. As the searchers near an activated transmitter on the downed firefighter, the receiver shows the strength of that transmitter. Should the searchers make a wrong turn and start going away from the transmitter, it will be indicated by a weaker signal until the correct path is taken. The monitor is equipped with both visual and audible indicators to show signal strength. These systems are designed to be activated by a short period of the wearer being motionless or by manual activation of the transmitter.

INCIDENT COMMANDER'S RESPONSIBILITIES
WHEN A MAYDAY IS SOUNDED

Once a Mayday is sounded, the IC needs to do the following:

- If the IC did not hear the Mayday report, either have Dispatch repeat it or speak to someone on the scene who heard the report. This will allow the IC to understand the situation.
- The reason that the firefighter is missing must be discovered. If there is an interior collapse or other dangerous condition, will the RIC be endangered? Could other firefighters be missing?
- Contact the member's officer or other companies working in the same approximate area where the firefighter is located, to determine the current conditions, actions being taken to extricate or remove the firefighter, and what additional resources they require.
- If there are sufficient radio channels, dedicate the current fireground channel to the rescue efforts, leaving those units involved in the rescue of the missing firefighter(s) on the same frequency.
- Move all other units to another available fireground channel, if possible. If there is only one channel available, radio discipline becomes paramount, and there should be no unnecessary transmissions.
- Perform a PAR to determine exactly who is missing.
- Have Dispatch initiate an announcement to alert all units on the incident scene that a Mayday request has occurred.
- The incident objectives will need to address the firefighter(s) in trouble.
- Have all radio channels monitored at the Command Post and in Dispatch.
- Immediately assign the RIC or a RIC task force to search and rescue operations in the known area or last known area where the firefighter(s) need assistance. If there are units in staging, immediately assign at least one additional company to assist the RIC.
- Assign the ISO to work with the RIC and request additional ASOs.
- Assign a Supervisor (chief officer if available) to the RIC.
- Assign a Supervisor (chief officer if available) to strictly control staffing for the RIC and provide/rotate resources as requested.
- Should a firefighter(s) be trapped and need to be extricated from an IDLH, have sufficient units on the scene to ensure short working intervals of a maximum eight to ten minutes. This has to be strictly enforced with no exceptions. Firefighters may not want to leave the immediate area, but they must. Rotation of firefighters will reduce congestion and fatigue. Tired firefighters will be less productive and increase the risk to those trapped.
- Assign another RIC.
- Immediately request from Dispatch:
  - Additional resources to ensure sufficient personnel, including a Staging Area Manager
  - Additional chief officers for supervisor positions
  - Any specialized equipment that is needed
  - EMS personnel to treat injured personnel
  - EMS Supervisor to set up and supervise the Medical Unit
  - ASOs
- Initiate and/or maintain firefighting locations and reinforce those positions with extra alarm or mutual aid companies as needed.
- Expand the command organization as needed.
- Maintain strong supervision in all work areas.
- Maintain ALS units standing by to treat the trapped firefighter(s).
- Request police to assist in scene control.
- Provide scene lighting.
- Monitor the building if stability and additional collapse is an issue.
- Consider requesting a MEDEVAC helicopter to stand by.
- Consider family notifications and providing of transportation to the hospital for the family of any injured firefighters.
- Anticipate media requests and have a PIO dispatched.

PASS DEVICES

PASS devices can assist in locating a missing or trapped firefighter. They will activate due to a lack of movement when a firefighter is overcome or incapacitated. When a firefighter becomes trapped or missing, the RIC members and other firefighters often place themselves in extreme danger to locate and rescue the missing crew member. The quicker they find the firefighter, the faster they can remove him or her to safety, and they can leave the dangerous area.

Current standards require a dedicated PASS device in all SCBA. When the SCBA cylinder is opened, it automatically activates the PASS.

The PASS device can be manually triggered if firefighters find themselves in trouble. Firefighters should be urged to utilize this tool when needed. A firefighter's pride may prevent him or her from activating the device until there is serious endangerment. This is foolish pride. If lost or in trouble, the immediate activation will permit the RIC time to enter, locate, and remove the firefighter from the peril. Likewise, any firefighter who does call a Mayday should never be chastised by anyone if the individual is either able to safely reach safety on their own or is quickly found and removed by others.

Locating a Firefighter with an Activated PASS Device

Training is needed to locate a firefighter who is down with an activated PASS device. Trying to go directly to the location of the activated PASS can be difficult. The sound of the device echoing from walls, ceilings, and floors can be distracting. The sounding of smoke detectors, fire alarm bells, and SCBA low-air warning devices can cause confusion. Building components or a firefighter buried under a collapsed portion of the building can muffle or absorb the sound of the PASS. Accepted methods of search must be employed after a room is entered where a firefighter is reported down.

UNSUCCESSFUL RESCUE

Every effort must be expended to rescue the trapped or missing firefighter. If a firefighter cannot be found, the IC will be faced with the task of making difficult decisions. Each situation is different. There could be structural failure or increased fire conditions that make it untenable for firefighters to continue to operate in the location where the firefighter(s) was reported lost or last seen. In these extreme cases, the IC may have to stop the search to prevent the loss of other firefighters. After this determination has been made, the firefighters must be withdrawn. This will not be an easy decision. No one, especially the IC, ever wants to give up trying to save the life of a brother or sister firefighter. The firefighters will want to continue searching with no regard for their own safety. But a realization that body recovery is too dangerous must take place, and the safety of the RIC and search crews has to be a prime consideration.

DOCUMENTATION

The SO must investigate any injuries that occur on the incident scene. This includes interviewing witnesses to discover how the injury occurred. Major incidents should be documented with a report covering all safety aspects of the incident and suggestions on methods to improve future operations. To ensure a complete and thorough investigation, the SO should request ASOs at the incident scene to assist in interviewing everyone who had involvement at the incident. Once the investigation is completed, a report should be written and later distributed describing the injuries, how they occurred, and how they can be avoided at future incidents.

Mayday activations should be thoroughly documented and analyzed to discover the methods deployed, actions that worked well, and actions that were not as efficient or successful. This analysis will be beneficial to future Mayday operations.

PUBLIC SAFETY OFFICERS' BENEFIT PROGRAM

The Public Safety Officers' Benefit (PSOB) Act (42 U.S.C. 3796, et seq.) was enacted in 1976 to assist in the recruitment and retention of law enforcement officers and firefighters.

Specifically, Congress was concerned that the hazards inherent in law enforcement and fire suppression and the low level of state and local death benefits might discourage qualified individuals from seeking careers in these fields, thus hampering the ability of communities to provide for public safety.

The PSOB Program provides a onetime financial benefit to the eligible survivors of public safety officers whose deaths are the direct and proximate result of a traumatic injury suffered in the line of duty. The benefit was increased from $50,000 to $100,000 for deaths occurring on or after June 1, 1988. Since October 15, 1988, the benefit has been adjusted each year on October 1 to reflect the percentage of change in the Consumer Price Index. For eligible deaths occurring on or after October 1, 2017, the benefit is $350,079.

Additional information on the PSOB Program can be found at https://www.psob.gov/.

Safety Officer Actions to Be Taken When a Line-of-Duty Death Occurs

When a firefighter death occurs in the line of duty, the SO must see that proper steps are taken to ensure compliance with the PSOB Program.

The SO should understand that:

- It is necessary to obtain a blood sample and test for blood alcohol level (not merely the presence of alcohol in the blood).
- In cases of nontraumatic injury, such as heart attack, it is also necessary to measure the level of carbon monoxide saturation in the blood (not merely to test for the presence of carbon monoxide in the blood).
- It is advisable to impound and secure any equipment involved in a firefighter fatality incident (such as protective gear, SCBA, fire apparatus), communications, and other records (tapes, dispatch report, incident reports, casualty report).

Local Assistance State Team

The National Fallen Firefighters Foundation (NFFF) has founded the Local Assistance State Team (LAST). This is a line-of-duty death state response team that operates in every state. Their primary mission is to bring expertise to the surviving family in filing for Department of Justice-Public Safety Officers' Benefits (DOJ-PSOB). The team can also provide:

- Honor guard and ceremonial support
- Chaplain services for department members
- A fire service survivor to work with the family
- A behavioral specialist to work with the department members and the family
- Information access to various federal, state, and local benefits that may be available to the family
- Other information and resources to assist the family and department

These teams will only respond if requested and will be on the ground within six hours. Their priority is the family and the department at all times. They will ask first before taking any actions while maintaining transparency by working behind the scenes. They will provide timely and accurate information while practicing personal and team integrity. They are trained to ensure quality over expediency, and to show empathy at all times while always thinking in terms of honor, dignity, and respect. Telephone numbers for state teams and additional information can be found at http://www.firehero.org/resources/department-resources/programs/local-assistance-state-team/

Summary

Safety is the duty of not only the IC and the SO—it is everyone's responsibility. Safety procedures must be exercised throughout the entire operation. Though the pressure of controlling a situation is much greater initially, many injuries occur after an incident is under control because firefighters are fatigued. In an effort to protect firefighters, a strong emphasis on safety must be employed at all times.

Review Questions

1. List the three types of wall collapse.
2. What is the minimum distance a wall can be expected to fall outward?
3. Discuss the collapse indicator involving fire conditions that states "continued or heavy fire for 15–20 minutes."
4. List how different structural steel elements will react under attack of heat and fire.
5. What is the difference between live load and dead load? Give five examples of each in a single-family residence that are not listed in the preceding text.
6. List and discuss exterior wall collapse indicators.
7. What is a wall spreader? Do any buildings in your response area contain wall spreaders?
8. How can a buildup of water, snow, or ice on a roof impact fire department operations?
9. Discuss how eccentric and impact loads on a building can affect firefighting efforts.
10. Discuss what measures are in place in your department to secure special equipment to assist in collapse rescue situations. Are there contracts already in place? How long would it take for this special equipment to arrive on the scene?
11. What emergency evacuation signal does your department use? How often does your department test it in drills?
12. List the important traits necessary for a Safety Officer to be effective.
13. List items that should be contained on a Safety Officer's checklist.
14. Discuss your department's accountability procedures. Do they accomplish the intended goal? Can they be improved? How?

Suggested Reading, References, or Standards for Additional Information

Smith, M. 2011. *Building Construction: Methods and Materials for the Fire Service*. 2nd ed. Brady Fire Series. Upper Saddle River, NJ: Pearson Education.

Occupational Safety and Health Administration, Department of Labor. 1988. "Respiratory Protection, Final Rule (29CFR 1910.134)." Washington, DC: US Government Printing Office.

US Fire Administration (USFA)

"Firefighter Fatalities in the United States [by year]." USFA Publication.

"Search and Rescue Operations Following the Northridge Earthquake, Los Angeles." USFA Publication.

Routley, J. G. "Three Firefighters Die in Pittsburgh House Fire, Pittsburgh, Pennsylvania." Technical Report 078.

Routley, J. G. "Four Firefighters Killed, Trapped by Floor Collapse, Brackenridge, Pennsylvania." Technical Report 061.

Thiel, A. K. "The Aftermath of Firefighter Fatality Incidents, Preparing for the Worst, Special Report." Technical Report 089.

National Fire Protection Agency (NFPA)

Standards

1500: Standard on Fire Department Occupational Safety and Health Program

1521: Standard for Fire Department Safety Officer

1581: Standard on Fire Department Infection Control Program

1583: Standard on Health-Related Fitness Programs for Firefighters

1851: Standard on Selection, Care, and Maintenance of Structural Firefighting and Proximity Firefighting Protective Ensemble

1971: Standard on Protective Ensemble for Structural Firefighting and Proximity Firefighting

1975: Standard on Station/Work Uniforms for Emergency
   Services
1982: Standard on Personal Alert Safety Systems (PASS)

Related Courses Presented by the National Fire Academy, Emmitsburg, Maryland

Incident Scene Safety Officer
Safety Program Operations

Health and Safety Officer
Incident Command for Structural Collapse Incidents
Methods of Enhancing Safety Education
Community Risk Issues and Prevention Interventions
Initial Fire Investigation for First Responders
Fire/Arson Origin-and-Cause Investigations
Fire Cause Determination for Company Officers

7

Special Situations and Occupancies

> "Tell me and I forget. Teach me and I remember. Involve me and I learn."
>
> —Benjamin Franklin

> "Without continual growth and progress, such words as improvement, achievement, and success have no meaning."
>
> —Benjamin Franklin

Row house fires can quickly spread to adjoining properties.

KEY TERMS

basement and cellar fires, *p. 263*
garden apartments, *p. 268*
row house, *p. 274*

town house, *p. 274*
renovated buildings, *p. 282*
hotels and motels, *p. 289*

vacant buildings, *p. 297*
wildland urban interface (WUI), *p. 303*

OBJECTIVES

Upon completion of this chapter, the reader should be able to:

- Understand the special occupancies discussed and the types of fires that may confront firefighters.
- Identify pertinent characteristics of these occupancies.
- Recognize the 13 points of size-up that pertain to these special occupancies.
- Recognize the strategic considerations for these special occupancies.
- Understand the incident management considerations of special occupancies.

Chapter 7 discusses specific types of structures, fires, or operations with which a firefighter may be confronted. These occupancies, situations, or types of fires have specific dangers that have had deadly results.

At the end of each section are size-up factors as well as strategic goals and tactical priorities that need to be taken into account when confronted with fires in these occupancies. These lists are not meant to be all-inclusive. Other factors can and will exist that also must be considered.

The strategic goals and tactical priorities are for both offensive and defensive attacks, with suggested considerations for implementing an incident management system to meet the potential problems. The strategic goals and tactical priorities noted are in response to some of the potential problems that may occur. The suggested incident management positions do not mean that they must be used or that using them is the only correct method.

These are initial positions that should be considered. Depending on the problems presented by each individual situation, the IC can implement these or other incident management positions to deal with the problems.

Basement and Cellar Fires

A basement or cellar is described as a level of a building either partially or fully below grade. One definition of the difference between a basement and cellar is that a cellar is more than 50 percent below grade or ground level and a basement is less than 50 percent below grade. Another definition is based solely upon the use of the area. If unfinished, it is referred to as a cellar. If a finished room, it is called a basement. In either case, the basement or cellar requires some type of descent to reach, whether entering from an exterior or interior door. In the following discussion, the terms will be used interchangeably. A fire occurring below the ground level will present problems.

Fighting **basement and cellar fires**, if you are forced to enter the basement via the interior stairs, can be like crawling down a chimney with a fire burning in the hearth below. The heat and smoke slam into you if you open the basement or cellar door. The fresh air introduced by opening the door feeds the fire below, increasing its intensity. You can only hope that the pent-up heat will be released and make it bearable to enter the basement to extinguish the fire.

basement and cellar fires ■ Fires occurring below ground level in a building.

Responding firefighters may be confronted with subcellars, which present additional dangers. Usually, there is only one way into the subcellar, which makes entry during a fire very hazardous. There may be drainage problems if the subcellar is below the sewer lines. The presence of a French drain (a drain that allows water to leach into the soil) may be suitable for minor water runoff, but it will allow a water buildup during a cellar or subcellar fire. This water buildup could catch an unsuspecting firefighter off guard when he or she steps into an area of deep water.

Common cellars are similar to common attics, in which adjoining properties use the same area with no fire walls in between. They can be found in older mercantile areas. Wooden partitions or chicken wire can separate the individual cellar areas. Poor housekeeping by one store owner will jeopardize not only the contents of the common cellars but the businesses above. Locating the seat of the fire can be extremely difficult since smoke may be showing from many properties. The common cellar is not discernible from the exterior.

PREPLAN CONSIDERATIONS

If unprotected steel is supporting the first floor, a well-involved cellar fire can cause an early failure. Similarly, the increased use of lightweight building components in residential and commercial properties will be a serious concern. With no sprinkler systems, failure of the first floor will occur quite early in the fire. A decision must be made on the stability of the building and whether an interior attack is feasible.

A danger in basements and cellars is the storage of unwanted hazardous items, including paint, paint products, and pesticides.

The cellar can be one large, open area, or it can be partitioned into several rooms. One large, open area will allow a fire to spread to a greater area, but can be more conducive to firefighting. The hose-line can reach most areas from the base of the stairs to knock down any large body of fire, and then it can be moved in to overhaul the area. Having a large open area also makes ventilating the area easier. The reach of a stream will be reduced if the height of storage reaches near the ceiling. In this case, hose-line advancement will be difficult and additional lines may be required.

DOORWAYS

The kind and location of cellar entrances will vary but can be interior, exterior, or a combination of the two.

Interior cellar doorways can be found in different locations within buildings. In residential, multistoried buildings, they usually will be located beneath the stairs to the upper floors. In one-story residential properties, interior cellar doorways often will be located in the kitchen. The location of cellar doorways in commercial properties varies widely.

Exterior entrances can be either in the front of a building with outside stairs and a door located at the bottom of the stairs, or a door opening onto stairs leading into the basement. A door may be at ground level if the building has the basement exposed in the front, side, or rear of the property. A commercial property might have a metal door flush with the pavement. It can contain stairs, a ladder, or a hoistway. If the property is closed, forcible entry will have to be made. If an exposed padlock is present, it can be pried open or cut off. If locked from the inside, the easiest way to gain entry is to break out the concrete in each corner where the door is anchored. The entire door can then be lifted or slid aside. Residential and commercial properties also may have a lift-type metal door located on the side or rear of the property.

ONE- AND TWO-FAMILY RESIDENTIAL BUILDING BASEMENT FIRES (2010–2012)

According to data collected by the US Fire Administration's (USFA) National Fire Incident Reporting System:

- An estimated 6,500 one- and two-family residential building basement fires were reported to fire departments within the United States each year and caused an estimated 65 deaths, 400 injuries, and $278 million in property loss.
- One- and two-family residential building basement fires are considered part of the residential fire problem, and they comprised about 3 percent of all one- and two-family residential building fires.
- The leading reported causes of one- and two-family residential building basement fires were "electrical malfunction" (19 percent); "heating" (14 percent); "appliances" (12 percent); and "other unintentional, careless" actions (12 percent).
- January (12 percent) was the peak month for one- and two-family residential building basement fires, followed by December (10 percent) and February (10 percent).
- Electrical arcing was the most common heat source in residential building basement fires (19 percent).

The full report can be found at: http://www.usfa.fema.gov/downloads/pdf/statistics/v15i10.pdf

---

RECOGNIZING A CELLAR FIRE

The presence of heat and smoke at the first-floor level or on all floors and the absence of visible fire can indicate a cellar fire. Another indicator is smoke emitting from the baseboards on lower floors and banking down on the top floor. Delayed discovery of a cellar fire will allow a fire to gain a foothold and is a common occurrence.

STRATEGY AND TACTICS

The best protection afforded any property is automatic sprinklers. Cellars and basements are no exception. From a strategic point of view, if sprinklers are present, one of the first considerations is to pressurize the system to guarantee a continuous and adequate supply of water.

Entrance to the basement area may be required to shut down the building's utilities. Care must be taken to avoid extinguishing a fire involving a natural gas meter if doing so prevents shutting down the meter. If the meter is involved, the hose crew should protect the surrounding area from fire extension until the gas supply can be shut off. (Many areas still have natural gas meters located in basements.)

From a life safety standpoint, protection of the interior stairs is paramount. The simple act of closing the interior door to the cellar will provide some protection to the first floor. It allows people on upper floors to exit the building, enables firefighters to enter the upper floors to assist with rescues and extinguishment, and if necessary, permits interior access to the cellar.

A staircase that is located against an exterior wall is much easier to protect than a staircase that terminates in the middle of a cellar. A hose-line may be required just for that purpose.

It may be necessary to place a line outside an exterior cellar door that is being used for ventilation. This precaution prevents fire extension up the outside of the building.

A basement fire can initially be hit from the exterior via a window or door, to get an initial knockdown of the fire. If an outside doorway exists, then the hose-line should be taken through that opening for final extinguishment of the fire. This entrance is the best method of entering the building to attack the fire. The firefighters are not subjected to the heat and smoke that accompanies descending interior stairs. As the fire attack is underway in the basement, an interior hose-line can be utilized to prevent the upward spread of the fire via pipe chases and the interior stairs. This can be a punishing and difficult assignment. Not only must the cellar door be closed and monitored for fire burning through, but if the stairs to the upper floors are stacked above the cellar stairs, it is possible for fire to burn through at the stair treads or risers to the upper floors. Studies by UL and NIST at Governors Island found that the fire is more likely to extend directly upwards through pipe chases and other shafts leading from the basement, than via the stairs. This demands that the entire first floor be continuously monitored for any upward fire extension.

If no exterior entrance exists, entry will need to be made from the interior stairs. When descending cellar steps, there are some basic rules:

- A charged hose-line must be ready to descend the interior stairs immediately after the fire is knocked down through a basement window.
- The hose-line(s) must be able to control the fire.
- After deciding to descend the stairs, there must be sufficient hose-line so the firefighter operating the nozzle can reach the cellar floor without stopping.
- After the hose-line(s) are on the cellar floor, an aggressive attack on the fire must be made. Retreating back up the steps is difficult, to say the least.
- Additional ventilation openings can be made at the direction of the interior crews.

The proper use of resources will facilitate fighting a cellar fire. Placing a firefighter at the door to the cellar near the top of the stairs and another at the bottom of the stairs will assist hose-line advancement.

In fires above grade, a firefighter who becomes disoriented often can locate a window to gain his or her bearings. This is not feasible in a cellar fire. Closer contact must be maintained among personnel operating in below-grade locations. Contact with the hose-line or the placement of guidelines for search personnel is necessary to ensure the location of exits.

An associated problem with cellar fires is that water buildup on the floor, storage strewn about by firefighters operating in confined areas, and the disarray of the contents caused by hose streams can bury the hose-line and make it impossible to follow the hose-line to locate an exit. In any case, accountability of all personnel minimizes the possibility of someone becoming lost.

The presence of what appears to be only light smoke in a basement or cellar still requires the continued use of self-contained breathing apparatus (SCBA). The Safety Officer (SO), before allowing removal of breathing apparatus, must test for carbon monoxide levels.

It is important with cellar fires to check all floors above for extension of fire. This is necessary due to the presence of pipe chases and other connections between the basement and the upper floors through which fire can travel. It may be necessary to open walls and ceilings and stretch hose-lines to these locations.

CELLAR PIPES AND DISTRIBUTORS

If it is impossible or impractical to enter a cellar, the Incident Commander (IC) may decide to use cellar pipes or distributors. This involves feeling the floor for the hottest spot to discern the location of the fire and cutting a hole between the joists. Water must be at the shutoff of the cellar pipe or distributor and at the tip of a backup line before cutting the hole. If using a cellar pipe, it then is inserted into the hole, and the tip is opened.

A distributor is basically a large sprinkler head that is attached to a hose-line (usually 2½ inch) and lowered through the opening created into the cellar until it touches the floor or an obstruction. Then it is withdrawn half the distance and raised and lowered to cover

FIGURE 7-1 An alternative method of ventilating a basement is by cutting a hole in the floor beneath a window on the first floor and using a hose-line on fog to hydraulically push the smoke to the exterior. *Used with permission of Michael DeLuca.*

the maximum area. A hose-line is stretched for the protection of the firefighters operating cellar pipes or distributors.

Cellar pipes are not a practical option if the floor joist is of lightweight construction, since a fire burning for any amount of time would attack and weaken the lightweight joist and it would not support firefighting activities.

HIGH-EXPANSION FOAM

There has been some success in extinguishing cellar fires with high-expansion foam. Ventilation must be provided in front of the foam for proper distribution. The major drawback is that the foam is restricted greatly by walls and closed doors, and hose-lines must be shut down to prevent breaking down the foam.

VENTILATION

If adequate ventilation is not provided by an outside cellar door or sufficient windows, an alternative method is cutting a hole in the floor beneath a window on the first floor. (See Figure 7-1.) Knowing the approximate location of the fire will assist in choosing the location of the window or windows to be used. If at all possible, the location of this hole should be directly over or just past the fire. If incorrectly placed, it can endanger the hose-line crews because the fire will be drawn to the ventilation hole.

The opening should be the width of the window and extend out from the wall approximately one foot. Ventilation can be enhanced by installing a fan in the window (negative pressure ventilation) or stretching a hose-line, placing it on fog, and directing it out the window (hydraulic ventilation). This will pull the smoke and heat from the basement and allow the crews operating in the basement to advance on the fire. A hose-line should be stretched to monitor for any extension of fire onto the first floor. This protective line must not be directed into the cellar while firefighters are operating there; doing so would drive the heat back onto them, seriously jeopardizing their lives.

A cellar fire in a store containing showcase windows offers two additional ways of effecting ventilation. The showcase is an elevated floor. There is normally no additional flooring between it and the cellar. The windows can be broken out and the floor opened. An easier method of ventilation is to break out the material on the exterior under the windows, allowing immediate access to the cellar.

The material under the showcase windows can be masonry, marble, or wood. After removing this material, probe the opening to see if any ceiling exists in the basement. If present, it will have to be pushed down. Care must be taken if the front door is recessed with showcase windows in the recess. Breaking out under the windows alongside the door entrance can cut off any exit from the front door if the cellar fire extends to that location.

Older commercial buildings often have basements that extend out past the front of the building and under the sidewalk. Deadlights were installed in the sidewalks to provide natural lighting. These deadlights are made of round, thick glass, embedded in the sidewalk. They can be broken out to effect ventilation.

Recessed window wells allow for natural lighting where a cellar is wholly beneath grade. These wells often have a metal grate cemented into the sidewalk. Ventilation can be effected by reaching a tool down through the openings between the grate or by removing the grate. The grate can be removed by breaking the concrete at each corner where the metal is attached. If maintenance of these wells is ignored, they become catchalls for trash. Many times, the smell of smoke in the basement can be traced to a carelessly discarded cigarette starting a fire in accumulated trash in a window well.

SIZE-UP FACTORS FOR CELLAR FIRES

Water
- Stubborn cellar fires may require large quantities of water.
- Water can accumulate in cellars and especially subcellars.

Area
- The basement may be one open area or partitioned into numerous small rooms.

Life Hazard
- A cellar fire will threaten occupants on all floors above.
- Cellar fires are difficult to control and dangerous for firefighters.
- Thermal imaging cameras (TICs) can assist in the primary and secondary searches.

Location, Extent
- Locating a fire in a cellar can be time-consuming.
- Entry into the basement may be from the interior, exterior, or both.
- Entry through trapdoors into cellars is too dangerous an entry point if confronted with a well-involved cellar fire.
- For a below-grade fire, it will be difficult to effect ventilation and to gain access to the area.
- TICs can assist in checking walls and ceilings for hidden fire.

Apparatus, Personnel
- A sufficient number of personnel will be needed to attack the fire and perform other necessary functions.
- High-expansion foam may be successful in extinguishing a cellar fire.
- Cellar pipes or distributors may be successful in extinguishing a cellar fire if entry into the cellar is not possible.

Construction/Collapse
- A portion of the cellar will be below grade.
- There may be subcellars.
- Common basements can be found.
- If an unprotected steel or lightweight wood truss supporting the first floor is attacked by a cellar fire, it can fail rapidly.
- There may be no cellar windows that can be used for ventilation.
- Deadlights may exist in the sidewalk for ventilation.

Exposures
- The major exposures will be the floors above the cellar and adjoining buildings if a common cellar exists.

Weather
- High humidity can retard ventilation.

Auxiliary Appliances
- Basement sprinklers may be present.

Special Matters
- High-piled stock, narrow aisles, and low ceilings will impact on operations.

Height
- There may be subbasements.

Occupancy
- Basements can contain paint, paint products, pesticides, and other hazardous materials.

Time
- Delayed discovery can occur, especially when the building is unoccupied.

CONSIDERATIONS FOR CELLAR FIRES

Strategic Goals and Tactical Priorities for an Offensive Attack
- Perform search and rescue for occupants.
- Call for a sufficient amount of resources.

- Confine the fire to the area of origin, if possible. Immediately check for upward extension of fire into walls, ceilings, or shaft ways.
- Protect the interior stairs from upward extension of fire.
- Stretch 1½-inch, 1¾-inch, or 2½-inch hose-lines.
- Ventilation must be in coordination with the interior crews. Open doors and windows to release pent-up heat and smoke. It may be necessary to vent through the flooring of the first floor.

Incident Management System Considerations/Solutions for an Offensive Attack
- Incident Commander
- Safety Officer
- Rapid Intervention Crew(s)
- Operations (if needed)
- Staging
- Ventilation Group
- Search and Rescue Group
- Basement or Cellar Division
- Division 1, 2, 3, etc. (a division can be implemented for any floor where deemed necessary)

Strategic Goals and Tactical Priorities for a Defensive Attack
- If changing from an offensive to a defensive attack, ensure that a personnel accountability report (PAR) is taken.
- Set up and maintain a collapse zone, if necessary.
- Utilize master streams.
- Protect exposures, if necessary.

Incident Management System Considerations/Solutions for a Defensive Attack
- Incident Commander
- Safety Officer
- Operations (if needed)
- Staging
- Logistics (for water supply, if needed)
- Divisions on the exterior on Sides A, B, C, D
- Exposure Divisions (if needed)

Garden Apartments

garden apartments ■ Buildings that are set back from the roadway.

Garden apartments are common throughout most of the United States. The reference to "garden" infers that the building is set back from the roadway. This category can include town houses and row houses. Depending on the area of the country, they have many similar features.

CONSTRUCTION

Garden apartments can be built of wood frame or ordinary construction. They can be three stories or greater in height. Frame buildings are of platform construction. A typical construction feature includes a brick veneer covering the front of the frame buildings.

Garden apartments are often built in sections. A common arrangement is to have four apartments per floor in a section, though larger sections can be found. Sections usually are constructed adjacent to other sections. This interconnection of sections can create a rather large building.

The interior partition walls are usually constructed of wood studs that are set either 16 or 24 inches on center and covered with drywall. The walls between the adjoining sections can be of masonry construction or other fire-rated material.

ROOF ASSEMBLIES

Roof assemblies can be pitched or flat. They can be constructed of solid wooden beams, lightweight truss, wooden I-beams, precast concrete, or steel bar joist.

The roof can overhang the sidewalls, creating eaves that protect these walls from the weather. The eaves can have openings that permit air circulation in the roof space. This vents any heat buildup in the summer and prevents condensation from forming in the winter. The overhang may not be fire-stopped and can circumvent fire walls. An overhang of 36 inches is not uncommon. The eaves can be finished with plywood or with only aluminum or vinyl soffit products. These materials will rapidly break down from the intense heat of a fire.

Within each section, there may be partitioning with fire-stopping between each apartment, or a section may share a common roof area. A fire starting in one apartment that extends to an open, common roof area will endanger other apartments sharing that roof area.

A problem encountered is that the eaves can be in close proximity to the windows on the top floor. Fire coming out of these windows can easily extend into the roof space. Draft-stopping is used to compartmentalize an attic or cockloft into smaller areas. Draft-stopping may be found there, depending upon when the building was built, the size of the roof area, and the local code.

Even when draft-stopping is required and installed during the construction of the building, it can be removed or breached many times during building renovations. This creates *poke-throughs*, which are holes or openings in the draft-stopping material that negate the compartmentation and create a large non-fire-stopped area in the concealed space.

ACCESS

The terrain can hamper access to the building. Illegal parking can prevent apparatus from entering parking areas. Apparatus may not be able to reach the sides or rears of some garden apartments. Parked vehicles in front of the buildings may restrict the proper placement of aerial equipment and ground ladders.

The front doors of garden apartments typically enter onto a common hallway containing the stairs. Fire reaching these stairways will threaten the safety of civilians trying to evacuate the building and firefighters attempting to control and extinguish the fire. Many garden apartments have the first floor partially below ground level. The building's terrain may be landscaped in the rear to permit sliding glass doors opening onto a rear patio. This would provide a ready means of escape.

Windows may be sliding glass placed high in the wall to allow wall space for the placement of furniture. The height of this secondary means of egress can hinder rescues. It will be difficult for firefighters and occupants to escape via these windows. They will have to stand on chairs or other furniture in the apartment to mount ladders raised to these windows from the exterior. Firefighters can gain access to apartments that are partially below grade by lowering ladders downward through window openings.

PROBLEMS

The problems associated with garden apartments are the same as with any multiple occupancy dwelling. The possibility of someone being in the building at any time is a reality. The greatest challenge for firefighters will be between midnight and 8 a.m., the time when most residential fire deaths occur. Delayed alarms are commonplace. People assume that the fire department has already been called. Many times, occupants attempt to extinguish the fire themselves. A small fire can quickly get out of

FIGURE 7-2 This garden apartment had no fire wall in the attic area, allowing the entire roof to be destroyed.

FIGURE 7-3 Fires lapping out of windows can enter the eaves and the attic area.

control, and valuable time can be lost. To compound the problem, tenants fleeing the building may leave the doors to their apartments open.

There may be interconnected void spaces throughout the building. This can include soffits above kitchen cabinets, in wall spaces, and floor or roof areas.

FIRE CONSIDERATIONS

The location and extent of the fire must be determined. Has it entered the floor or roof assemblies? Has it spread to adjacent apartments?

Through information received from occupants and by observing conditions, a determination of the location and the approximate extent of the fire can be made. Conditions should be monitored to see the amount and density of the smoke in the various apartments in that section.

Observing fire through the windows of more than one apartment indicates that fire has breached common areas and the entire section could be under attack by fire. A heavily involved apartment will severely threaten surrounding apartments. Each section should be considered one building.

FIRE TRAVEL

We must predict the travel of the fire and have units checking for extension well ahead of the fire in suspected areas.

1. Is smoke or fire coming from roof vents?
2. How serious do conditions appear?
3. Is fire already through the roof?
4. Are there additional signs that the fire is spreading? To where?

Fire lapping out of windows will threaten the floors above by entering windows directly above them or by extending into the roof space via the eaves. (See Figures 7-3 and 7-4.) Checking for spread of fire to the floors above the fire floor can be accomplished by having hose-lines stretched to those locations.

When sufficient hose-lines have been positioned on the interior of the building to fight the fire and effect rescues, additional hose-lines on the exterior may be successful in preventing the fire from entering the roof space. Firefighters on the exterior need to coordinate any fire attack with the firefighters on the interior attack.

FIGURE 7-4 The eaves on this garden apartment are located close to the windows on the top floor. The eaves are not fire-stopped and circumvent the attic fire wall—a prescription for disaster.

Fire Burning out of Control

A fire can burn out of control due to:

a. Insufficient resources
b. Lack of an adequate water supply
c. Late discovery of the fire
d. An incendiary fire
e. A fire involving flammable liquids
f. A poor plan of attack
g. Poor construction
h. Improper ventilation

If the initial number of resources is limited, an immediate rescue effort by the first-arriving units could delay an attack on the fire, permitting it to gain control of the building. If only one stairway exists, fire and smoke conditions may prohibit occupants from exiting the building by their normal route. Firefighters can be confronted with rescues from multiple locations. Aerial and portable ladders will need to be placed quickly.

Another problem can occur if these buildings are set back from the street. The engine companies must be prepared for long hose-line stretches to reach the seat of the fire. If preconnected hose-lines that contain only 150–200 feet of hose are normally used, additional lengths of hose will need to be added; this can delay the operation.

STORAGE AREAS

A common location for fires to start is in the storage areas, normally located on the lowest level. These rooms are often rather large and contain a variety of combustible materials. There is often an individual storage space for each apartment. A closet or storage area consists of a wooden framework with wood planking and/or chicken wire. The storage can include flammable liquids, pesticides, poisons, furniture, automobile tires, and a variety of household articles. Housekeeping is often lax. A fire in these storage areas demands an aggressive interior attack on the fire. The immediate goal is to contain the fire to the room of origin.

The gas and electric meters for the building and facilities for washing and drying clothes can be located in the same room as the storage area. To facilitate the ease of entering and exiting this area, the door is often propped open. Should a fire start, the open door allows the fire to spread quickly to the common hallway. An immediate action is to close the door to gain door control. Fire can then be attacked through a window for an immediate knockdown and then by entry with an interior hose-line to complete extinguishment.

A fire originating in the storage area can spread upward through ceiling openings. Holes in the ceiling are created for cable television wiring, piping, electric wiring, and renovations. Damage to the drywall ceiling can occur by tenants jamming large items into storage bins.

The immediate threat is to the units directly above the storage area. There is also the possibility of a fire following pipe chases upward to the underside of the roof, attacking the roof, and mushrooming at that location.

ROOF FIRE

A roof constructed of lightweight wooden components that is under attack by flames may be in a weakened state prior to the fire department's arrival. It may not support the weight of a firefighter attempting to open it for ventilation.

When fire has control of a roof, an exterior attack may be able to knock down the fire. This depends upon whether there are sufficient openings in the roof either from ventilation holes or from openings where the fire has burned through. The indiscriminate use of water, however, can be as destructive as the fire in causing damage to the building contents. (See Figure 7-5.)

OPERATIONS

The initial strategies must be based upon the best method of protecting the occupants and firefighters. Close coordination is needed between suppression, search and rescue, and ventilation crews. Spread of fire to common areas and other apartments must be anticipated. The assigning of divisions must be made to attack the fire in these exposed areas.

FIRE ATTACK

The dangers must be weighed and the IC must decide on the correct plan of attack.

Coordinated ventilation with the interior crews is critical. If hose-lines are in place, a consideration could be roof-top ventilation. All ventilation must be approved by the interior crew or the IC. Opening the roof may assist in reducing horizontal fire spread in roof spaces. Opening skylights

FIGURE 7-5 Once fire has control of the attic space, the water used for extinguishment can be as destructive as the fire to the building's contents.

directly over stairways can allow firefighters a safer entry via the stairs and a means of egress for the occupants.

The best method to control garden apartment fires is through an aggressive offensive attack. A quick knockdown can solve many problems before they start. Determining whether an offensive attack is the correct mode of operation should be based on the following factors:

1. The need for rescue of occupants
2. Accessibility to the building
3. The amount and location of the fire
4. The available resources on the scene
5. The resources responding
6. The available water supply

Garden apartment fires will be challenging. The key to successful operations will be preplanning by the fire department, good housekeeping, and strong code enforcement. The most successful strategy is fire prevention.

SIZE-UP FACTORS FOR GARDEN APARTMENTS

Water
- Hydrants may be a considerable distance away.
- When buildings are set back from the street and surrounded by parking areas, additional lengths may have to be added to preconnected hose-lines.
- Illegally parked cars may interfere with stretching hose-lines and overall operations.

Area
- The buildings are often set back from the street.
- Parked cars may interfere with firefighting efforts.
- Large areas may exist under one roof without fire-stopping.
- With the location of the storage room on the lowest level, a fire would threaten all apartments located above.

Life Hazard
- High: there are people at home at all hours.
- There may only be one exit from the building.
- Windows may be located high on the wall and restrict occupants from exiting from them.
- Multiple family units under one roof can increase the danger to others.
- Lack of self-closing doors can expose apartments to fire if fleeing occupants do not close their doors behind them.
- Ventilation may be required to save lives and prevent mushrooming of the fire on the upper floors.
- TICs can assist in the primary and secondary searches.

Location, Extent
- Has the fire entered void spaces?
- Where is the fire located? One apartment? Multiple apartments? Is the fire contained to one section or has it spread further?
- Depending on the location and extent of the fire, a decision must be made whether an interior attack can be accomplished in a safe manner, or whether a primary search should be made and an exterior attack implemented.
- Where can this fire be stopped?
- Are roof vents clear, or is smoke or fire emanating from them?
- Is fire through the roof?
- TICs can assist in checking walls, ceilings, and floors for hidden fire.

Apparatus, Personnel
- There may be difficulty in maneuvering due to the building being set back from the street and the presence of parked cars and sloping grounds around the building.
- Limited resources can impact on fire control. Initial personnel may be required to rescue trapped occupants and not fight the fire.

Construction/Collapse

- These buildings are usually wood frame, wood frame with brick veneer, or ordinary construction.
- The interconnection of multiple sections of apartments can create rather large buildings.
- Common attics or cocklofts will permit the spread of fire.
- Lightweight building components are often used, especially in roof and floor assemblies.
- Interior partition walls can consist of gypsum wallboard, fire-rated drywall, or masonry construction.
- Required draft-stopping for large void spaces may be violated by poke-throughs.
- There is a possibility of interconnected void spaces.

Exposures

- The fire apartment can expose other apartments in the same section.
- Common roof and floor voids can severely affect adjoining apartments.
- Draft-stopping must be in place between sections to prevent fire extension from one section to another.

Weather

- Rain, snow, and ice can affect evacuation and response time. Wind can have a negative effect on ventilation and threaten exposures.

Auxiliary Appliances

- Heat and smoke detectors should be present.
- With the exception of new construction, it is rare to find sprinklers or standpipes.

Special Matters

- Delayed alarms can occur because occupants often think others have already called the fire department.
- Arson fires often start in the storage area.

Height

- These buildings are typically three stories or greater in height.

Occupancy

- Classification is multiple occupancy/residential.
- Storage room can contain an assortment of junk, automobile parts, flammable liquids, and household goods.

Time

- As with any residential property, midnight to 8 a.m. is when the greatest loss of life occurs.

CONSIDERATIONS FOR FIRES IN GARDEN APARTMENTS

Strategic Goals and Tactical Priorities for an Offensive Attack

- Perform search and rescue for occupants.
- Call for a sufficient amount of resources.
- Confine the fire to the apartment of origin, if possible. Immediately check ceiling area for extension of fire into common cockloft or attic areas.
- Hose-lines of 1½ or 1¾ inches should be sufficient for apartments. Storage room fires may require 2½-inch hose-lines.
- Ventilation must be in coordination with crews operating on the interior. Consideration should be given to horizontal ventilation, and if fire is suspected to have a hold on the attic or cockloft, vertical ventilation may be required. There may be skylights in the roof that will vent living spaces.

Incident Management System Considerations/Solutions for an Offensive Attack

- Incident Commander
- Safety Officer
- Rapid Intervention Crew(s)
- Operations (if needed)

- Staging
- Ventilation Group
- Search and Rescue Group
- Division 1, 2, 3, etc. (a division can be implemented for any floor where deemed necessary)
- Medical Group

Strategic Goals and Tactical Priorities for a Defensive Attack
- If changing from an offensive to a defensive attack, ensure that a PAR is taken.
- Set up and maintain a collapse zone.
- Utilize master streams.
- Protect exposures, if necessary.

Incident Management System Considerations/Solutions for a Defensive Attack
- Incident Commander
- Safety Officer
- Operations (if needed)
- Staging
- Logistics (for water supply, if needed)
- Divisions on the exterior on Sides A, B, C, D
- Exposure Divisions (if needed)

Row Houses and Town Houses

row house ■ Vary in size; up to four stories high.

town house ■ Usually larger than row houses and not typically set in the same contiguous style.

There are many cities and towns in the United States with block after block of **row house** or **town house** dwellings. They were built to meet the demand for economical housing. Because they share common party walls, they were inexpensive to build. A block containing more than 60 contiguous dwellings is commonplace. (See Figure 7-6.) The buildings can be either ordinary or frame construction.

Row houses and town houses vary in size. They can be up to four stories high and range from 13–20 feet wide by 30–65 feet deep, though depths of more than 100 feet are not uncommon. The average row house is 14 feet wide by 40 feet deep and is two to three stories high; town houses are typically somewhat larger and average 18 feet wide and 40 to 50 feet deep. Town houses are not typically set in the same contiguous style as row houses. Though long blocks can be found, more often they are in smaller groupings with some open space between sets of town houses.

The upper floors can be the same depth as the first floor, or they may be set back from the rear of the property. This means that if the first floor is 65 feet deep, the second floor can be set back from the rear by one room, or 50 feet deep, and the third floor can be set back from the rear by another room, or 35 feet deep.

Town houses may be large complexes of buildings. Though commonly built as single-family occupancies, some locales will have multifamily units. They may be built as rental units or condominiums with each floor containing a separate living space. Row and town houses can contain flat or peaked roofs. The flat roof has a cockloft formed between the roof and the ceiling of the top floor. The cockloft in the front of the structure is normally higher and the roof pitches to the rear where the roof rafters meet the ceiling joist. This angle or pitch allows rainwater to drain to the rear of the property. The peaked roof may be attic space or contain living quarters. Town houses may have very large attic or cockloft areas constructed of lightweight construction. These areas will supply ample fuel to a fire that extends to that location.

FIGURE 7-6 A block containing more than 60 contiguous dwellings is commonplace. *Used with permission of Joseph Hoffman.*

Buildings of ordinary construction have a common party, or bearing, wall that contains a double (eight-inch) wythe of brick or eight-inch concrete or composition block that is shared with the adjoining property. The front and rear walls are normally nonbearing walls. The town house may contain fire-rated drywall in lieu of the masonry party wall. (See Figure 7-7.)

The older row houses were built with full-sized lumber used for floor joists, roof rafters, flooring, and roof planks. The floor joists and roof rafters range from full 2-by-10-inch to 3-by-10-inch wooden beams. Flooring is full-sized 1-by-4-inch tongue-and-groove boards, and the roof planks are an inch thick.

Newer row houses and town houses mimic typical ordinary construction but include the use of lightweight components. Town houses are often of frame construction. They may contain brick veneer, which could give the appearance of ordinary construction. There is usually a light and air shaft between the buildings. This allows windows to open onto the shaft in rooms that ordinarily would not have outside light. There are different types of light and air shafts. In large buildings, the windows open onto an enclosed shaft that is open at the top. Shafts can also be formed by having the width of the building at the rear of the structure narrower on one side, leaving the shaft open on the rear. (See Figure 7-8.)

The rear of the row of houses on one street backs up to the rear of those on the opposing street. There may be narrow alleyways between the rears of the row and town houses. In some areas, the small rear yards are lined with fences that abut the rear fence of the yard on the opposing street, limiting access to the rear yard. This may necessitate that units attempting to gain access to the rear go through an adjoining property or climb over many fences with ladders and tools to reach the rear, a time-consuming and sometimes dangerous chore. The firefighters can be faced with a variety of fences. They can be wood, masonry, or metal fabric. They are often to a height in excess of five feet for privacy and may be topped with barbed wire or glass embedded in the tops of walls to deter the criminal element. The presence of attack dogs is another potential concern. Newer town houses may have ample back yards, which are located a distance from the rear of those on the opposing street.

There will be a problem if these structures are situated on a waterfront where there is access to only one side of the structure, limiting the fire department's approach.

Some newer row houses and typically all town houses have garages located at the basement or first-floor level of the dwelling. Garages located in the rear have driveways that access the rear of the properties, facilitating the approach for apparatus and firefighters. It is common to find the first floor at the ground level in the front and the basement and garage at ground level in the rear.

Because the basement is at ground level in the rear, it raises the level of the living quarters another story above grade, requiring longer ladders and causing severe injuries if occupants jump from these upper-floor windows to escape a fire prior to

FIGURE 7-7 This frame town house has a masonry party/bearing wall that protected the adjacent property from the fire. *Used with permission of William J. Shouldis.*

FIGURE 7-8 The rear of these row dwellings contains a light shaft. Note the electric wires attached to the buildings that provide electric service to the properties.

FIGURE 7-9 A–B Heavy fire conditions will threaten adjoining properties. This will require an aggressive attack on the fire building and assignment of companies to the exposed buildings. *Used with permission of Brian Feeney.*

arrival of the fire department. These multiple levels can deceive responders and require a 360-degree walk-around to ensure a proper size-up.

The garage can also be found in the front of the house on the first-floor level and the rear of the property is situated at the same level.

I have seen three-story town houses constructed with only a front door and no other means of exiting the building, except from front windows. The rear of the property is built against a similar town house located on the opposing street. Life safety concerns in these buildings will be a challenge for firefighters if confronted with a well involved first-floor fire and open stairways connecting all floors.

INTERIOR WALL CONSTRUCTION

In many older row houses, the bearing walls constructed of brick have wood studs (typically two-inch-wide by one-inch-deep studs) attached to them, with the floor butted up against the studs. This creates continuous voids from the basement or cellar to the cockloft or attic, similar to balloon-frame construction. A fire occurring in a cellar or basement with voids leading to the attic or cockloft area demands that firefighters check the void areas above for fire extension. Fire in concealed spaces can burn unnoticed by firefighters. If discoloration of the plaster is seen or if any doubt exists as to whether fire has entered these voids, the area should be inspected with a TIC. If any doubt exists, a tool should be used to open the suspected areas to check for spread of fire.

In row houses the stairs are often steep, hallways can be long, and both are often narrow. These tight quarters restrict firefighting operations. Transoms still exist in many of these structures. These are operable glass windows above the doorways into rooms. They were installed for air circulation from the hallway. Under fire conditions, fire and smoke can extend from the fire room to adjoining hallways and into other rooms through open transoms.

RESIDENTIAL LIFE SAFETY

We know from statistics that approximately 80 percent of all civilian fire fatalities occur in residences and:

- Most deadly residential fires start between midnight and 0800 hours.
- The most common rooms of origin are the bedroom, lounge areas, or kitchen.

In relation to life safety, firefighters want to give a trapped occupant the benefit of the doubt. If there is any chance they are alive, a concerted effort to rescue them will be made.

Information is critical—we must find out where the people are located, who are the most threatened, and the quickest and safest way of reaching them:

- Through the interior stairs?
- From portable or main ladders?
- From platforms?

Other decisions that need to be made are: What is the safest way of protecting occupants? Can those trapped be protected in place? This is a better option in a fire-resistive building than in a two-story dwelling, but it may be a consideration.

FIREFIGHTING PRACTICES

Fires have claimed many lives in these structures. Their small size confines the smoke and heat of a fire and limits the number of firefighters who can operate within these buildings.

During initial size-up, it can be difficult to observe conditions in the rear. It is usually not practical, due to time constraints, to physically encircle the full row of dwellings. One way to gain access is through an adjacent property to see what problems exist in the rear.

Fireground strategy must address the many possible areas of fire extension. With long blocks of dwellings and entry to the rear limited, a standard operational guideline can designate specific areas of assignment for each unit on the initial dispatch. One method mandates the first-due engine and the first-due truck to the front of the dwelling. The second-due engine and the second-due truck would respond to the rear. This type of sectoring ensures adequate coverage. Assignments can and must be changed by the IC to address incident priorities. (See Figure 7-10.)

Coordination and communication are essential. Firefighting in these structures is punishing to the firefighters. If possible, the fire should be fought offensively. The life factors involved demand a quick and coordinated attack on the fire while rescues are being made. Information gleaned from the rear can be given to the units responding to that area. They can then bring the necessary hose-lines, ladders, and tools needed to perform their tasks. Narrow streets restrict entry of apparatus and necessitate the stretching of hose-lines by hand from the intersecting streets.

Though these structures may be four stories in height, they are not equipped with standpipes. This necessitates the manual stretching of hose-lines to the upper floors. Long stretches mean that preconnected hose-lines may not reach the fire, and extra hose-line must be added to the stretch.

The hose-line of choice is 1½ inch or 1¾ inch. A good practice is to stretch a 2½-inch or 3-inch hose-line and attach a gated wye to break down to 1½- or 1¾-inch hose-lines.

Interior partition walls separate the floor area into many small rooms and reduce the effective reach of hose-lines. Smaller properties require only

FIGURE 7-10 A response to the rear or Division C side of a town house will ensure personnel for rescue and ventilation if required. *Used with permission of Brian Feeney.*

FIGURE 7-11 Heavy fire conditions demand a defensive attack on the fire building, while simultaneously mounting an offensive attack in the exposed attached structure.

one length of hose-line per floor. The many partition walls and doorways tend to hang up the hose-line as you attempt to advance, and firefighters must be stationed to assist in feeding the hose-line forward. As the structures get bigger, the problems increase almost in proportion to the building's size. The higher and deeper the property, the more challenging the fire problem. (See Figure 7-11.)

Interior spread of fire can be difficult to contain. Many of the older structures initially had hot-air coal heat in the basement. Heated air was carried by gravity through ducts to the upper floors. The duct was often sheet metal lining a wall void with a hot-air register placed in each room. The cold-air return was accomplished by placing open grates in the flooring of the first floor. A fire in the cellar or basement of a building with such a system has a ready avenue to extend quickly to the floors above.

BUILDING FRONTS

The fronts of the older row houses may have a continuous wooden porch roof extending the length of the block. Fire starting on a porch or lapping out porch front windows can quickly spread to many properties via combustible porches. A coordinated attack must be made on arrival in such a situation. (See Figure 7-12.) A hose-line must be deployed from each end to stop the lateral spread. This may encompass master streams in a blitz attack or handheld 2½-inch or 1¾-inch hose-lines, depending on the amount of fire and the access available. It is not unusual to have 10 or more properties involved. A quick response can knock down the fire and minimize the problem on the adjacent properties. After the fire extension under the porch roofs is addressed, hose-lines can be stretched into the exposed structures and original fire building, and the fire can be brought under control. The overhauling required can be quite involved. Each property must be checked for fire extension. There can be interconnection between the porch roof and void spaces within the building. Fire can easily travel into the walls of the upper floors, necessitating that firefighters open these areas to check for fire spread.

Front cornices can be constructed of sheet metal. They are often non–fire-stopped and may interconnect from one property to the next. On arrival at a working fire, it is

FIGURE 7-12 A fire burning under front porch roofs involving multiple properties must be dealt with by a coordinated attack to stop the lateral spread. *Used with permission of Joseph Hoffman.*

FIGURE 7-13 Loss of life and damage to more than 10 dwellings resulted from a fast-moving row house fire. *Used with permission of William J. Shouldis.*

not unusual to have smoke pushing from the cornices of many properties. The immediate need for roof-top ventilation is indicated, and a close watch must be maintained to see if the smoke subsides after ventilation has been performed or continues to be produced. If it continues, it could indicate a common cockloft or that the fire has entered the adjacent properties.

Many row houses of ordinary construction have combustible frame rears attached. This leads to fire originating in one property spreading to properties alongside and the possibility of jumping across rear yards and extending to the rear of the exposed properties on the opposing street. A fast-moving fire involving many properties needs to be addressed with a minimum of 2½-inch hose-lines working from each side to contain and knock down the fire, preventing further escalation. (See Figure 7-13.)

Another problem with fire extension confronting firefighters in a building heavily involved in fire is that joists laid in the common party wall may abut the joists in the adjoining buildings. This allows conduction of fire from one building to the next. Overhauling must ascertain that fire has not extended to another building. This can be checked in the overhauling stages through the use of a TIC or by pulling ceilings in the fire building, and if fire is found, playing a stream into the affected areas. It may necessitate opening walls and ceilings in exposed buildings. (See Figure 7-14.)

TRUCK COMPANY OPERATIONS

Due to narrow streets restricting aerial apparatus and limited rear yard access, portable ladders are the mainstay of row house and town house firefighting operations. When the upper floors are set back from the rear, shorter ladders can be used. It is a common practice to place one ladder that will reach the roof and top floor windows from the ground in the open light and air shafts, along with two 16-foot portable ladders to cover the rear. (See Figure 7-15.) With multiple roofs, the 16-foot ladders can be placed for each roof, or they can be pulled up as each level is reached to gain access to the next roof level. Full-depth properties require portable ladders that allow access from the ground level to the roof and upper floor windows.

If adjacent buildings are the same height, roof access can be gained from another building and firefighters can walk down roofs to the fire building. Adjoining roofs allow firefighters a secondary means of egress when performing roof operations. If driveways exist in the rear of the properties, aerial ladder use may be possible.

FIGURE 7-14 Each property must be thoroughly checked for hidden fire. *Used with permission of Joseph Hoffman.*

If attic or cockloft areas are heavily involved in fire, and no fire walls exist between the houses, the IC should be notified to recommend that the roof be immediately opened. This action can prevent both the mushrooming of fire under the roof and the spread of fire to adjoining properties. If the fire has already spread to the other properties, additional roof openings may be needed.

LIGHT AND AIR SHAFTS

Light and air shafts present a danger to firefighters operating on roofs. (See Figure 7-16.) During nighttime or smoky operations, firefighters accidentally stepping into these shafts can fall a number of stories to the ground. These same shafts can provide a means of exterior ventilation on flat-roofed buildings. Ventilation can be achieved from the roof by firefighters utilizing tools and reaching down to break windows on the upper floors of the fire building. Realize that fire may extend via these windows to the next building.

OVERHEAD ELECTRIC WIRES

Electrical service supplied from overhead wiring can enter the property from the front or the rear. The electric wiring may be attached to the rear of the properties and extend the length of the block.

Fire lapping out of windows can cause a problem with the electrical service lines as they enter or are attached to the property. This will require a quick response from the local electric company to alleviate the potential of electric shock from downed wires or electrified portions of the building, allowing firefighters to ladder the building, ventilate, make rescues, and extinguish the fire.

FIGURE 7-15 Carrying and raising portable ladders in the rear of a row house can be challenging yet is a necessary assignment. *Used with permission of Joseph Hoffman.*

SIZE-UP FACTORS FOR ROW HOUSE AND TOWN HOUSE FIRES

Water
- Small, narrow streets can restrict entry by pumpers, and additional lengths of hose will have to be hand-stretched, causing a delay.
- In the larger properties, additional hose-line may need to be added to preconnected hose-lines to reach the fire location.

Area
- These attached structures can stretch the length of a city block and include more than 60 separate contiguous dwellings.
- They may be 13–20 feet wide and reach depths of 30 to more than 100 feet.

Life Hazard
- Limited rear access may cause a delay for firefighters making rescues.
- Portable ladders will normally be required in the rear.
- Stairs are often steep, and hallways can be long and narrow.
- Overhead electrical services can impede operations. Firefighter safety and the potential for downed wires due to fire exposure can be a concern.
- Rear driveways will allow access for apparatus.
- TICs can assist in the primary and secondary searches.

Location, Extent
- It is not uncommon to have restricted access to the rear of the property.
- Fences can restrict rear access.

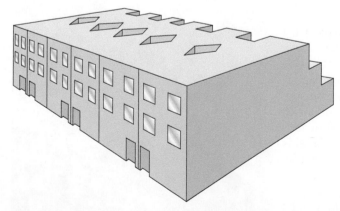

FIGURE 7-16 Light and air shafts present a danger to firefighters operating on roofs. Heavy smoke conditions will obscure these openings and become pitfalls for firefighters. *Used with permission of Michael DeLuca.*

- Newer construction may include rear driveways and garages under the home.
- TICs can assist in checking walls, ceilings, and floors for hidden fire.

Apparatus, Personnel
- Narrow streets will limit the access of apparatus.
- Hose-lines may need to be stretched from the intersecting streets. Hose-lines of 2½-inch or 3-inch diameter with gated wyes may be needed to the front door, and then 1½- or 1¾-inch hose-line can be attached to the gated wye.
- Light and air shafts can be dangerous to firefighters operating on roofs.
- Hose-lines may be needed on each level inside the structure.

Construction/Collapse
- Construction is typically ordinary or frame.
- In ordinary constructed row and town houses, the front and rear walls are usually nonbearing.
- Light and air shafts can commonly be found.
- Continuous void spaces may be found from the lowest to the highest level.
- Transoms may exist that allow the spread of smoke and fire.
- Floor and roof beams of adjoining properties may be set in the same wall socket in a masonry wall and therefore abut. A fire involving one of the beams can extend into the next property via conduction.

Exposures
- Continuous front porches of frame construction enable a fire to involve multiple properties by burning on the underside of the attached porch roofs.
- Attached properties of frame construction and ordinary constructed buildings with rooms of wood frame in the rear will be severely exposed.
- Closely built row or town houses on the next street will be exposed to the rear of burning properties.
- Common attics or cocklofts may be present.
- Fire can extend to the adjacent building via light and air shafts.

Weather
- Normal weather considerations apply.

Auxiliary Appliances
- Smoke detectors may exist in private dwellings. Hardwired heat and smoke detectors often exist in multiple-occupancy dwellings.

Special Matters
- Nothing unusual to be concerned with.

Height
- Row and town houses are typically two to four stories in height.

Occupancy
- Residential single-family and multifamily dwelling units are typical.
- It is common to have stores on the first floor and living quarters on the floors above.

Time
- As with any residential property, midnight to 0800 hours is when the greatest loss of life occurs.

CONSIDERATIONS FOR FIRES IN ROW HOUSES AND TOWN HOUSES
Strategic Goals and Tactical Priorities for an Offensive Attack
- Evacuate and perform search and rescue for occupants.
- Call for a sufficient amount of resources.
- Check exposures on Sides B and D for extension of fire.
- Hose-lines of 1½-inch or 1¾-inch diameter are usually effective.

FIGURE 7-17 A supermarket with the characteristic bowstring truss roof is being converted to offices. The common attic space under the roof remains an open area above the offices.

- Horizontal ventilation will be needed in coordination with interior crews. Roof ventilation may be required if fire heavily involves the area under the roof or if a common attic or cockloft exists, and fire has entered these areas.
- Positive pressure ventilation may be effective.

Incident Management System Considerations/Solutions for an Offensive Attack
- Incident Commander
- Safety Officer
- Rapid Intervention Crew(s)
- Staging
- Ventilation Group
- Search and Rescue/Evacuation Group
- Division 1, 2, 3, etc. (a division can be implemented for any floor where deemed necessary)
- Medical Group

Strategic Goals and Tactical Priorities for a Defensive Attack
- If changing from an offensive to a defensive attack, ensure that a PAR is taken.
- Set up and maintain a collapse zone.
- Utilize master streams.
- Protect exposures, if necessary. This may involve interior operations within the exposed buildings.
- Multiple dwellings may be involved in a fire extending down the porch fronts or the rears of the properties. Initially, a defensive attack with 2½-inch hose-lines or master streams can knock down the fire, and then an offensive attack can commence.

renovated buildings ■ Buildings in which changes have been made that can alter how a fire affects them structurally.

Incident Management System Considerations/Solutions for a Defensive Attack
- Incident Commander
- Safety Officer

- Rapid Intervention Crew (for exposures if needed)
- Operations (if needed)
- Staging
- Logistics (for water supply if needed)
- Divisions on the exterior on Sides A, C
- Divisions in Exposures B, C, D

Renovated Buildings

Fires in buildings undergoing renovations can cause problems during the renovation stage and after they are completed. **Renovated buildings** should be viewed as altered buildings. (See Figure 7-17.) Renovations may be undertaken without obtaining proper permits. Plans will not be reviewed. This can lead to the use of material not in compliance with the applicable codes. Workmanship can be substandard. Shortcuts in the construction process become magnified under fire conditions. Good housekeeping is not usually a priority on construction sites. A fire starting in accumulated rubbish can feed on the stored supplies.

FIGURE 7-18 Multiple renovations are exposed. The old wood clapboard is visible at the A/B corner. Asphalt shingles cover most of the house, and remnants of another wall covering remain on the upper part of the wall on Side B. Anticipate many changes when confronted with a fire in older structures due to numerous renovations over the years.

A major problem with renovated buildings is that a higher number of firefighters are injured and killed in these structures than in most other types of occupancies. This can occur due to changes in the buildings' layout, the stage of renovations at the time of the fire, or the materials used in the renovation project.

Renovations can affect the firefighters' abilities to control and extinguish a fire. During renovations, firefighters can be confronted with:

1. Automatic detection and extinguishing equipment removed
2. Standpipe systems inoperable
3. Walls, ceilings, stairways, elevators, and other vertical and horizontal fire-stops removed
4. Exposed electrical wiring
5. Welding operations without the required fire watch
6. Storage of combustible building supplies within the building
7. Small rooms where walls have been removed to create one large room
8. Doorways sealed off, changing normal means of entry and egress
9. Confusing interior layouts that could cause firefighters to become disoriented under heavy smoke conditions

Building renovations can entail a minor change, such as the relocation of office walls or the addition of a recreation room to a dwelling. Larger operations can involve the conversion of a large single-family home to apartments or the conversion of a factory or loft building to condominiums. (See Figure 7-19.)

The material installed in a building can increase the fire load. Vinyl wall coverings give off large amounts of deadly smoke when ignited. Decorative thin wood paneling, once ignited, burns fiercely and can resemble a natural gas fire in its intensity. Large amounts of heat are generated, resulting in a fast-spreading and potentially deadly fire.

Concrete fire-resistive buildings or mill buildings with 20-foot ceilings are often converted to apartments, with two stories made out of each original 20-foot story. This is accomplished with large quantities of wood framing. These changes will not be recognizable from the exterior or from the elevator lobbies. The second floor of each unit has to be accessed from the interior of each apartment. Because the exterior of the building is left intact for aesthetic purposes, central heating and ventilation systems are installed, making ventilation and fire attack very difficult.

In some buildings of ordinary construction, the interior is stripped, leaving only the exterior walls standing. Then the interior is completely rebuilt utilizing lightweight building components. If the responding firefighters are not aware of the reconstruction, they could be completely surprised to find new components in the structure.

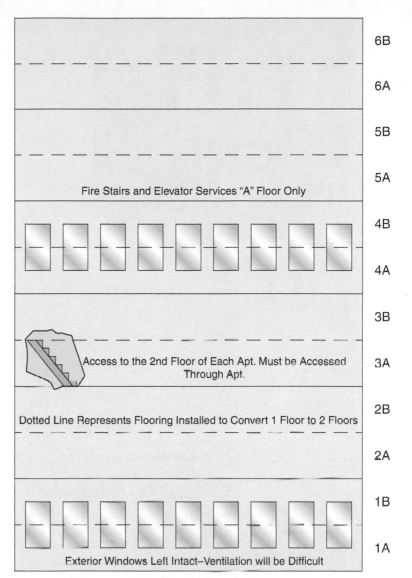

FIGURE 7-19 Commercial or loft buildings with 20-foot ceilings and higher are often converted to apartments, with two stories made from each original floor level. These changes will not be recognizable from the exterior or the elevator lobbies. *Used with permission of Michael DeLuca.*

TEMPORARY SHORING

Temporary shoring is often utilized to stop a structure from failing until a permanent correction can be accomplished. All too often, building owners allow these stopgap measures

FIGURE 7-20 Too often, temporary shoring becomes a permanent situation, as has occurred with this failing bay.

FIGURE 7-21 This temporary shoring is supporting a heavy roof load. Should fire attack the shoring, an early roof collapse would occur.

to become permanent. Though stable under normal conditions, temporary shoring fails quite readily under fire conditions. A fire originating around the shoring can cause a fast and deadly collapse. (See Figures 7-20 and 7-21.)

FIGURE 7-22 Scaffolding is supporting the interior floors while a masonry wall is being replaced. A fire occurring within this building would require a defensive attack.

FIREFIGHTING

Because renovations can involve minor or drastic changes in varying stages of construction, the problems encountered will differ. (See Figure 7-22.) The responding firefighter may see that renovations are occurring by the construction activity underway. During a daytime fire, information can be readily available. After working hours, it is not unusual to find little or no information to assist the responding firefighters.

Large-scale building changes will be a serious challenge to a fire officer arriving at a well-involved structure. The challenge can differ, depending on the stage of the renovation. If occurring after the building has been stripped of its protection and fire-stops, fire will be difficult to suppress. After fire has control of the structure, an exterior attack is the only option left. Sprinkler and standpipe systems that are out of service will allow fires to get past the incipient stage. Hose-lines will have to be stretched longer distances and will require a larger commitment of resources from the fire department. If the building is nearing the completion stages of renovation, it is possible that protective systems in the building can assist the firefighters.

When renovation projects are completed, sprinkler and standpipe valves that were turned off during the construction phases must be checked to ensure that they have been turned on and that the systems are serviceable and meet code requirements.

VENTILATION

Changes and modifications to window openings can impede ventilation and delay extinguishment. Brick or glass block can seal basement or cellar windows. Windows on upper floors can be covered on the interior with drywall or paneling, leaving the window intact, rather than detracting from the exterior appearance. This occurs in residential and commercial buildings.

ROOFS

The roof of a structure is another area where renovation can be cause for concern. Sawtooth roofs were constructed to allow natural light for industrial and manufacturing plants. (See Figures 7-23 and 7-24.) Today, these roofs are expensive to maintain and are not necessary due to the quality of electric lighting available. The greater need for security in these buildings has led to many modifications. One alteration involves tar-papering over the window openings without sealing the area with plywood. This deters burglars but allows quick acceleration of fire reaching the tar paper. Firefighters attempting to ventilate these modified roofs can become trapped on the roof by a fast-moving fire.

Most builders renovating large buildings with a timber truss roof find it too costly to rebuild the roof using the original size lumber. Older roofs were constructed with full 1-inch to 1¼-inch roofing planks. In many buildings being renovated today, builders install plywood or oriented strand board roof coverings as thin as ½-inch.

Another method of renovating large timber truss roofs is by constructing a lightweight roof directly over the old roof using parallel chord truss or wooden I-beams. (See Figure 7-25.) These changes could lead an IC to fail to recognize that a timber truss roof exists on the structure. This roof assembly will compound firefighting problems.

Firefighters look for signs on where to ventilate the roof, such as roof tar bubbling, melted snow, or a very dry spot on a wet roof, indicating that the fire area is directly beneath. However, when a new roof is set above the old roof, this creates a space between the two. Ventilating by opening the new roof will do little to relieve the heat and smoke from the fire building. Access to the old roof can be accomplished only by firefighters lowering themselves into the newly created opening. This is too dangerous a task and *is not* an alternative. Fire entering the void between the roofs is not accessible from below because the old roof structure is intact, nor from above, and will usually result in the complete destruction of both roof structures and often the entire building.

PARTITION FIRES

Openings in walls, floors, and ceilings are created during renovations and are rarely sealed. (See Figure 7-26.) Ceilings are lowered for economic or decorative reasons. High ceilings mean a larger living space that must be heated in the winter and cooled in the summer. Older ceiling styles may not fit the changing décor, so suspending a ceiling below the existing one can be quicker and cheaper than updating the old ceiling. Multiple ceilings create voids through which fire can travel and burn undetected.

Unprotected vertical shafts starting at the lowest level and extending to the attic or cockloft are created to install ducts, piping, and wiring. They will usually be found in, or adjacent to, kitchens and bathrooms. The shafts, interconnected on all levels, allow a ready avenue for transmission of fire throughout the building. These shafts act like a

FIGURE 7-23 A sawtooth roof under attack by fire. These roofs permitted natural light for manufacturing plants. They were designed for a northern exposure to take advantage of the longer daylight, with less heat and glare. *Used with permission of Joseph Hoffman.*

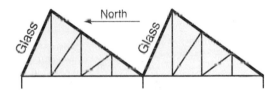

FIGURE 7-24 A small section of a sawtooth roof. As the name implies, the roof has a profile similar to the teeth in a saw. *Used with permission of Pearson Education.*

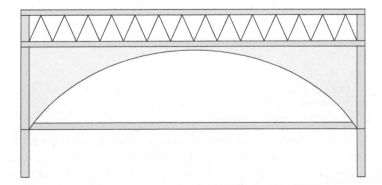

FIGURE 7-25 A method of renovating buildings with old timber trusses is achieved by constructing a lightweight roof directly over the old roof using parallel chord truss or wooden I-beams. Firefighting problems will be compounded when confronted with this roof assembly. *Used with permission of Michael DeLuca.*

FIGURE 7-26 Openings in walls are created during renovations and rarely sealed. Workmanship can be substandard. Old siding is being covered with a new layer of siding. Void spaces can be avenues for fire to spread throughout the building.

flue and draw the fire to them. When reaching the highest level, the smoke and heat will bank down in the attic or top floor, quickly extending the fire to that location.

The original fire may be extinguished, leaving a partition fire that may not be readily recognizable. Indicators of a partition fire include:

- Continued heat and smoke generation after all visible fire has been knocked down. This smoke generation is more visible from the exterior, leading to conflicting reports from the interior and roof sectors. Firefighters operating on the interior will see that the heat is not decreasing after all visible fire has been knocked down. This is due to the continued generation of heat in the void space from the unabated fire.
- Smoke pushing from baseboard and window moldings and from around electrical outlets and switches
- Discoloration of plaster or wallboard due to the heat behind it

Walls and ceilings will have to be opened to ensure that all fire has been exposed and extinguished. Firefighters must look for signs of a partition fire. They can feel walls for heat and observe whether smoke is pushing from around baseboards or door and window moldings. TICs can detect heat in partitions, allowing firefighters to quickly find and open the necessary areas to extinguish the hidden fire.

Fire-stopping is a rare occurrence in renovated buildings that did not get permits for the construction. In instances where fire-stops or draft stopping have been installed, later changes often negate them.

Where unprotected vertical voids transmit fire to multiple levels in a building, the amount of fire damage and overhauling can be extensive.

ASBESTOS ABATEMENT

Renovations of buildings may include the abatement or removal of asbestos. This operation can be performed in different ways. One method often used is to seal the area involved with sheet plastic to prevent the escape of airborne asbestos.

Workers enter the enclosed area with respirators and fully encapsulated suits while they perform the asbestos abatement. They are then decontaminated as they leave the contaminated area. Firefighters who enter these areas must continue utilizing their SCBA until they leave the contaminated area and have their PPE decontaminated.

Ignition sources must be controlled; this could involve smoking materials and portable heaters. A fire starting in an area sealed with plastic will quickly fill with smoke and seriously jeopardize the safety of anyone working within. The fire could ignite the plastic, which will burn furiously, quickly spreading the fire and releasing the airborne particles to the surrounding locations. (See Chapter 6 for more on asbestos dangers.)

SIZE-UP FACTORS FOR RENOVATED BUILDINGS
Water
- If larger areas are involved in fire, a greater amount of water will be required to contain and extinguish the fire.

Area
- The area may differ from before the renovation. There may be interconnection of areas and void spaces, negating compartmentation. These alterations can increase the area of fire spread.

Life Hazard
- There may be exposed electrical wiring that poses the threat of electric shock for firefighters.
- TICs can assist in the primary and secondary searches.

Location, Extent
- During renovation operations, there may be storage of combustible supplies within the building.
- Interior renovations can add material to fuel a fire, such as combustible wood paneling or vinyl wall coverings.
- TICs can assist in checking walls, ceilings, and floors for hidden fire.

Apparatus, Personnel
- Whether the building is undergoing renovations at the time of the fire or whether firefighters are dealing with a completed renovation will impact on the response. Buildings undergoing renovation can be more dangerous than a new building under construction. Fire-stops may be removed.
- Specialized apparatus may be needed for a defensive attack.
- Additional firefighters may be needed to open void spaces to check for fire.

Construction/Collapse
- Horizontal and vertical fire-stops may be nonexistent during renovations.
- Poke-throughs may be created to allow utilities to pass through fire-stops.
- Multiple false ceilings may be found, lowering the ceiling height and creating numerous voids.
- The use of lightweight structural members will increase the potential of collapse under attack by fire.
- Temporary shoring may be used during the renovation process. Fire attacking this shoring could precipitate an early collapse.
- Multiple roofs may be found.

Exposures
- Interior exposures could be at risk if fire-stops and protective systems are removed or not functional.

Weather
- The practice of wrapping a building in sheet plastic to keep out the weather or for asbestos abatement can contribute to a fast-moving fire.

Auxiliary Appliances
- Automatic detection systems may be out of service during renovations.
- Sprinklers or standpipes may not be operating during renovations.

Special Matters
- Housekeeping is often overlooked during renovations, and rubbish can accumulate.
- Many fires are started by improper welding operations. Oxygen and acetylene tanks must be removed or protected from the heat of a fire.

Height
- Renovated buildings can range from one story to a high-rise building.

Occupancy
- Any type of building can be involved.

Time
- If construction workers are on the scene, they can assist in an early discovery of the fire and provide information of the ongoing progress of the renovations.

CONSIDERATIONS FOR FIRES IN RENOVATED BUILDINGS
Strategic Goals and Tactical Priorities for an Offensive Attack
- Perform search and rescue for occupants.
- Call for a sufficient amount of resources.
- Check exposures if they are threatened for extension of fire.

- Check for voids in walls and ceilings through which fire can travel.
- Depending on the size of the building and the amount of fire, 1½-inch or 1¾-inch hose-lines may be effective. Larger buildings may require the use of 2½-inch hose-lines.
- Horizontal ventilation will be needed in coordination with interior crews. Roof ventilation may be required if fire heavily involves the area under the roof or if a common attic or cockloft exists with adjacent buildings, and fire has entered these areas.
- Positive pressure ventilation may be effective unless renovations are currently underway and numerous openings exist in the building.
- Anticipate the use of lightweight building components in the renovation and an early failure if they are attacked by fire.

Incident Management System Considerations/Solutions for an Offensive Attack
- Incident Commander
- Safety Officer
- Rapid Intervention Crew
- Staging
- Ventilation Group
- Search and Rescue/Evacuation Group
- Division 1, 2, 3, etc. (a division can be implemented for any floor where deemed necessary)
- Medical Group (if needed)

Strategic Goals and Tactical Priorities for a Defensive Attack
- If changing from an offensive to a defensive attack, ensure that a PAR is taken.
- Set up and maintain a collapse zone.
- Utilize master streams.
- Protect exposures, if necessary. This may involve interior operations within the exposed buildings.

Incident Management System Considerations/Solutions for a Defensive Attack
- Incident Commander
- Safety Officer
- Rapid Intervention Crew (if needed for exposures)
- Operations (if needed)
- Staging
- Logistics (for water supply, if needed)
- Divisions on the exterior on Sides A, B, C, D

Hotels and Motels

Hotel and motel fires have proven deadly over the years, from the sensational fires causing multiple deaths and creating front-page headlines, to the seemingly nondescript fires that injure or kill a single resident and receive only a brief notice in a newspaper article. Yet each fire must be fought by the firefighters responding to the alarm, and they will find themselves challenged with life safety considerations for the occupants and themselves. There has been an influx of hotels and motels in many areas that are sometimes referred to as "express" hotels or motels.

According to the USFA, there are an estimated 3,900 hotel and motel fires reported to US fire departments each year that cause an estimated 15 deaths, 150 injuries, and $76 million in property loss.

- Hotel and motel fires are considered part of the residential fire problem. However, they comprise only approximately 1 percent of residential building fires.
- Half of hotel and motel fires are small, confined fires.
- Cooking is the leading cause of hotel and motel fires (46 percent). Almost all hotel and motel cooking fires are small, confined fires (97 percent).

- Eighteen percent of nonconfined hotel and motel fires extend beyond the room of origin. The leading causes of these larger fires are electrical malfunctions (24 percent), intentionally set fires (15 percent), and fires caused by open flames (12 percent). In contrast, 42 percent of all nonconfined residential building fires extend beyond the room of origin.
- While bedrooms are the primary origin of nonconfined fires (23 percent), when confined cooking fires are considered, the kitchen or other cooking area is the most prevalent area of fire origin.
- Hotel and motel fires are more prevalent in the cooler months due to increases in heating fires and peak in February (9 percent).

SIZES AND TYPES OF HOTELS AND MOTELS

There are many different sizes and types of **hotels and motels.** A hotel or motel being built today and complying with the current codes will ensure the life safety of the occupants. The foremost reason is the installation of a sprinkler system, which is the best measure of protecting life in any structure. In addition to sprinklers, there are adequate exits and early fire detection and warning systems.

hotels and motels ■ Classified as quarters that have more than 16 sleeping accommodations used by transients for less than 30 days. The quality of the hotel or motel can vary from upscale to rundown.

The large hotel will complicate fire department operations due to the vast size and number of occupants that can be found at any one time. Yet what is commonly found in many communities is what can be referred to as small hotels built in different eras. The small hotel is typically less than eight stories and 75 feet in height, and in many cases three stories or less. In some communities these buildings will not exceed 35 feet in height to adhere to height-limiting ordinances. Typically these structures do not contain the level of safety measures found in new construction and large hotels. The best that can be hoped for is that the authority having jurisdiction has not continually grandfathered these buildings when stronger codes were enacted that required the installation of more stringent fire safety measures.

The quality of the hotel can differ from upscale to rundown. The fire service has typically had more fire incidents in the latter category. The clientele in any hotel may show little or no regard for the other occupants, and many times the carelessness of one occupant has seriously jeopardized others.

The motel is a single building of connected rooms whose doors typically face a parking lot and were originally situated near highways, versus the hotel, which favored the urban setting. The motel is typically one to three stories and commonly is an I-, L-, or U-shaped design. Some motels may contain a center corridor and have rooms located on both sides of the building. They may contain a basement under a small part of the structure, though normally they are erected on a concrete slab.

The length of stay of the hotel and motel occupants can vary; they may be long-term tenants or guests on a short-term basis as normally associated with these occupancies.

It should be recognized that problems found in motels are similar to condominiums. In many areas of the country, motels are being converted to condominiums. Depending upon many factors, this can become a problem for responding firefighters. Access to the individual units may not be available and forcible entry may be needed, sometimes causing unnecessary damage.

 ON SCENE

At a fire that I had in a high-rise hotel, an occupant compounded her careless smoking when, after the mattress ignited, she grabbed her belongings and exited the room, leaving her door open to the hallway. She proceeded to leave the building without notifying anyone of the fire. The open door and an open window provided the fire with an adequate supply of air. The lack of a self-closing door allowed the fire to spread to the hallway, threatening other guests in the hotel. On their arrival, firefighters were faced with a fire being driven down the hallway by the strong wind blowing into the open window. A fire that should have been contained to the room and contents turned into a four-alarm incident.

PROTECTIVE SYSTEMS

The building's fire protective systems can range from smoke detection to fully sprinklered buildings. Some buildings may be partially sprinklered providing protection for basements, public assembly areas, or hallways, but not guest rooms.

There may be a tendency by the staff to delay reporting alarm system activations to the fire department until building security or maintenance personnel investigate the alarm. The staff may not want to disturb their guests should it turn out to be a false or unnecessary alarm, or the staff may want to avoid being penalized by the fire department. In some locales there is a response charge by the fire department if constant false alarms are received. Though all fire departments reinforce the need for immediate notification, this practice by the staff will create a delayed alarm, and the valuable minutes lost can be deadly.

STANDPIPES

The presence of a standpipe can be advantageous to firefighters. Strong consideration should be given to their use. They require less hose-line and allow firefighters to attack the fire in less time.

Many hotels and motels do not contain standpipes. This will complicate the fire department's efforts to quickly stretch hose-lines to the fire location. Hotels of three stories or fewer in height may allow firefighters to use their preconnected hose-line from an apparatus. In many cases this may not be feasible or could require additional lengths of hose to reach the fire.

Motels usually have ready access for hose-line stretches from the parking area for all rooms, unless a center corridor exists and then standpipes may be available in those buildings.

LIFE SAFETY

A fire in these occupancies has the potential for large loss of life. The fire department should ensure that its actions assist and not deter the evacuation and protection of the occupants. How well the firefighters can accomplish this will depend upon the location of the fire, the building layout, the floor area, the number of exit stairs, and the number of stories in the building.

The benefit of many motels is that the door to their room leads directly to the exterior. Open stairways that are found in some hotels and motels will allow smoke and fire to spread to the floors above. Evacuation under heavy smoke conditions through these stairs or hallways will not be possible and will impact upon the rescuing of the trapped occupants. Occupants will need to be:

- Protected in place until the fire is controlled or extinguished
- Rescued via fire escapes
- Rescued via portable or main ladders
- Rescued via platforms

The transient nature of these occupants and past fires in hotels and motels have identified that:

- There is often a lack of knowledge of the exit routes by the occupants.
- Though evacuation plans are posted on the door leading to the common hallway, most guests will have little or no knowledge of them.
- Fire alarms are often ignored or guests fail to realize that the fire alarm is sounding.
- There is the possibility of guests exhibiting irrational behavior or panic. These factors can set the stage for disastrous results.

Though it is usually required to provide guest rooms with information on what to do in case of a fire, the average person rarely reads the instructions. In reality many occupants ignore activated alarm systems if there are no signs of smoke or fire visible to them. This lack of initial action on their part will compound and delay the building evacuation.

Once the smoke and fire permeates other areas of the building, there will be many calls to the dispatcher or the front desk asking for help. This will be especially true if there are smoke-filled hallways. Telephone calls will need to be made by the staff at the front desk to those occupants still in their rooms. They can be given advice on how to protect themselves if they stay in their room. It can be as basic as opening windows to ventilate the smoke or placing towels under the door to prevent smoke from entering the room. These phone calls can assist in identifying:

■ Who is still in the building
■ Who can be protected in place
■ Who needs assistance

MULTIPLE OR MASS CASUALTY INCIDENTS

Hotel and motel fires can easily become multiple or mass casualty situations. Difficulty breathing and chest pains are common. If numerous injuries are present, the necessary procedures must be put in place to handle the problems found. (Mass casualty incidents are discussed in Chapter 8.)

The immediate request for ambulances and medical personnel must be made. The impact on resources to control the fire; perform search, rescue, and ventilation; and tend to the injured occupants will overwhelm most fire departments. A large mutual-aid response will be required.

ACCESS

Interior access to upper floors may be limited to only one stairway. If multiple stairways are present, firefighters should designate one stairway for firefighting and others for evacuation.

Though open stairways can be found, stairways should contain doors to prevent smoke or fire from traveling from floor to floor. In some hotels and motels it is common practice for these doors to be secured in an open position, negating their value.

Hotels that are more than eight stories in height should be handled as any other high-rise building. Since many of these buildings are less than eight stories in height, access to all floors via the stairs should not cause a prolonged delay and elevators should not initially be used by firefighters until their safe use can be assured.

Exterior access may be restricted by:

■ Narrow alleyways or driveways
■ Parked vehicles
■ Large trash containers
■ Adjoining buildings
■ Sloping terrain
■ Lack of access for apparatus on all sides of the building
■ Overhead electrical wires
■ Trees

These restrictions can affect apparatus placement, stretching of hose-lines, and ladder placement.

STRATEGIC AND TACTICAL CONSIDERATIONS

Time, height, and the size of the building will be factors in controlling fires and assuring the safety of the occupants. The taller and larger the structure, the longer it will take for units to perform firefighting and rescue operations.

Size-up will assist in determining the initial concerns and in predicting future problems. The exterior 360-degree walk-around can provide a wealth of data. Through information received from occupants and by observing conditions, an initial determination of the location and the approximate extent of the fire can be estimated. An initial interior assessment from the first units inside the building can inform the IC of:

■ The conditions found
■ The exact location of the fire
■ If fire has spread to adjacent rooms or hallways

The first units to arrive at the fire area should report not only the problems found, but also recommendations on solving those problems.

The initial report from the first unit to arrive at the location of the fire will be brief. It should be followed up with an ongoing assessment. This is meant to determine what fire department actions will do the most good.

If the fire can be quickly controlled and extinguished, many problems will be solved. Yet once a fire has spread past the room of origin, it must be decided how control of the fire can be accomplished. The following questions must be answered:

- Where is the fire located and what is burning? Guest rooms, ballrooms, restaurants, meeting rooms, or storage rooms containing bedding, linens, towels?
- Who is endangered and how can they be protected or removed?
- What are the smoke and fire conditions in the hallways?
- Where and how will the fire most likely spread?
- What additional problems can develop?
- What additional resources will be needed?
- Which stairway should be used for fire attack? For evacuation?
- Can the fire be attacked from the unburned side? Should it be?
- Will the attack on the fire push the fire and smoke in a direction threatening other lives?
- Are there sprinklers operating, and are they controlling the fire?
- Does the sprinkler system need to be supplemented by fire department apparatus?
- Can the standpipe be utilized?
- What size hose-line should be used?
- What type of ventilation is needed?
- Has fire entered walls, floors, or roof assemblies?
- Can a stair shaft be pressurized by positive pressure fans?

Typically, hotel and motel fires are contained to one guest room. An aggressive attack on the fire should be the aim of the first-arriving units on the fire floor. If the door between the room and the hallway is closed, the firefighters can usually attack and quickly knock down the fire. The attack line of choice will often be 1¾-inch hose-line. If the door to the hallway is left open or the fire burns through it, the firefighters' problems will be compounded. A fire that extends from the original room into the hallway or upward through stairways, or fires involving larger areas (i.e., ballrooms, restaurants, linen storage closets, etc.) should be attacked with 2½-inch hose-lines. This will permit a faster knockdown and cooling of the fire area. It will also require more firefighters to operate those hose-lines.

A fire located on a lower floor can endanger all the floors above. The smoke and fire will threaten occupants who access the hallway on the same level as the fire area and if open stairways exist (or doors to the stairways are propped open), it will jeopardize occupants above.

There should be an immediate assignment of a company to check the room directly over the fire room for search and rescue and fire extension. Additional arriving units should be given the tasks of search and rescue of other areas and ventilation.

FIRE SPREAD

The type of building construction can consist of ordinary, frame, fire-resistive, and mixed construction. It is common to find numerous renovations in older buildings. These altered buildings can contain many void spaces. Lowered ceilings and shaft ways can be found that were installed to repair piping, or to upgrade electrical wiring or cable television.

If a pipe chase or void space extends from the basement through to the underside of the roof, fire entering the chase can reach the attic or cockloft and then spread horizontally. Fire-stopping should be used in pipe chases to prevent the upward spread of fire, yet it is seldom found, with the exception of new construction that requires the approval of an inspector. Draft-stopping is utilized to reduce the size of a large space into smaller areas. An example would be a roof area of substantial size. Though draft-stopping may be required and installed in a building, it is removed or breached many times. These building changes will be ready paths for a fire to spread through the interconnected void spaces.

Fire may spread to the floor above by burning through the floor (a ceiling light fixture can become a ready avenue). Fire can also extend upward by lapping out of windows (auto exposure) and enter the windows on the floor(s) above.

 **ON SCENE**

A fire response to an older and often renovated 10-story hotel found smoke on a number of floors in the building. An orderly evacuation was accomplished, yet the search for the fire was proving futile. The building had numerous shops and stores on the first four floors and the mezzanine. As these areas were checked, some suspected walls and ceilings were opened to check for the location of the fire. In some cases, fire-damaged areas were found that initially were suspected of causing the problem. However, it was quickly realized that this was old fire damage that had occurred in the past. The fire was eventually found in an isolated closet, where an auxiliary heating unit had been installed to provide supplemental heat to handball courts located on the mezzanine level. The smoke had spread through the additional ductwork that had been tied into the main ductwork for the heating, ventilation, and air-conditioning (HVAC) system in the building. This interconnection spread the smoke throughout the building until the HVAC was shut down.

VENTILATION

Entry or egress may be limited by smoke and fire conditions. Horizontal ventilation can be an effective life-saving tactic. In conjunction with interior hose-lines attacking the fire, ventilation will remove smoke on the fire floor, which will permit advancement of the hose-line by the engine companies and removal of guests from the fire floor. When smoke travels vertically, it creates additional problems. Smoke can travel throughout a building via open stairs, open chases, dumbwaiters, and elevator shafts. Ventilation should be considered early to prevent mushrooming of fire and smoke on the top floors and under the roof. Since heavy smoke conditions will impact on the firefighting and evacuation efforts, ventilation can be used to clear the smoke from the hallways and stairs, but hose lines or door control must be in place to control the fire. If the stairs have a rooftop exit, the firefighters assigned to the roof may be able to open the doors from the roof. If no exit exists at the roof, it may contain a skylight over the stairs that can be opened or removed, or a hole can be placed in the roof over the stairs. The roof team must recognize if fire is in the attic or cockloft spaces. Opening the roof will assist in reducing horizontal fire spread in the roof spaces. After a hole is cut in a roof, firefighters can push down ceilings through the roof opening to relieve the smoke conditions in the living spaces. Likewise, opening skylights can vent living spaces.

Ventilation on the floors above the fire floor may be addressed by opening windows on both sides of the building (if the windows are operable) and opening guest room doors to achieve cross-ventilation. Hallways that have doors to the exterior could be used to provide cross-ventilation to clear them of smoke.

The coordination of ventilation with the attacking hose-lines can be the most important factors in saving lives. Likewise, lack of ventilation, delayed ventilation, or venting without coordinating with the fire attack and search and rescue crews can prove disastrous, since all types of ventilation create flow paths. Poor ventilation practices can cut off escape routes in hallways or on stairways. Creating ventilation openings at the wrong location can pull a fire toward trapped occupants or firefighting crews, or spread smoke and fire into other areas, endangering occupants who otherwise would not be threatened.

OVERHAUL AND SALVAGE

The indiscriminate use of water can be as destructive as the fire in causing damage. After the fire is knocked down, the use of water should be controlled. Consideration should be given to the spreading of salvage covers or plastic sheeting to safeguard the building and its contents.

To restore the building to the condition that it was previous to the fire, it is necessary that:

- Alarm systems are reset.
- Elevators are removed from firefighter service and restored to normal operation.
- Dry standpipes are drained in freezing weather to assure that the system will be serviceable.
- Sprinkler heads are replaced and any sectional valves that were shut down are reopened.
- Any evidence that is suspected of causing or contributing to the fire is preserved for the fire investigators.
- No fire remains by doing a thorough overhaul of the fire area.
- All burnt debris is removed to the exterior.

EVACUATION

Evacuation of a hotel during a fire can be compounded after the fire is out if the building is uninhabitable. This can occur due to:

- The damage caused by the fire and smoke
- The lack of heat in wintertime
- The inability to restore the building's electricity or other utilities

If this occurs, it will be necessary for most occupants to reenter the building to secure their personal items. There will be a need to retrieve medication, clothing, credit cards, and money. This operation will require assistance from the police and fire departments. In parts of the building that have sustained fire damage, items should be retrieved by fire department members after the police have found that they are the rightful owners. Police can assist the occupants in relatively undamaged areas of the building. Police should also secure the building to preserve the scene until completion of the investigation by the fire investigators. Once the fire investigators release the building, the building owners will handle building security.

Securing the building's utilities will vary with each building. There may be shutoffs on each floor or main shutoffs located on a lower level of the building. Restoring utilities that have been shut off by the fire department is the responsibility of the utility companies or the building owner and not the fire department.

SIZE-UP FACTORS FOR HOTELS AND MOTELS

Water

- Water supply can be critical.
- Hydrants may be a considerable distance.
- Larger buildings may require additional lengths added to preconnected hose-lines.
- Dry standpipes will need to be pressurized by fire department pumpers.
- Wet standpipes may need to be supplemented by fire department pumpers.
- Sprinkler systems should be pressurized by the fire department to ensure an adequate supply of water.

Area

- Parts of the building may be a considerable distance from the street.
- Large open areas may be contained under one roof without draft-stopping.
- Ballrooms or large reception areas may exist.
- Dead-end hallways can be found.

Life Hazard

- High: there are people in these buildings at all hours.
- Fire escapes may be used for rescue, evacuation, and roof access.
- Secure master keys to use for primary and secondary searches.
- Alarm system activations are often ignored by the occupants.
- The occupants are probably unfamiliar with the building.

- Diverse population of transients under one roof can increase the danger to others.
- The location of some endangered guests may be identified by phone calls to the front desk or Dispatch.
- Lack of self-closing doors can expose other rooms to fire if fleeing occupants do not close the door to the hallway behind them.
- Ventilation may be required in coordination with interior crews to save lives and prevent mushrooming of the smoke and fire on the upper floors.
- TICs can assist in the primary and secondary searches.

Location, Extent
- Determine where the fire is located. One room? Multiple rooms? Is the fire contained, or has it spread?
- Determine if the fire has entered void spaces.
- Depending upon the location and extent of the fire, a decision must be made how to best protect the occupants and control the fire.
- Utilize the hotel's phone system to receive phone calls in the lobby from guests.
- Determine if the roof vents are clear or if smoke or fire is emanating from them.
- TICs can assist in checking walls, ceilings, and floors for hidden fire.

Apparatus, Personnel
- There may be difficulty in maneuvering ladders or platforms due to marquees, overhead wiring, trees, parked vehicles, or inaccessibility to parts of the building.
- Limited resources can impact on fire control. Initial personnel may be required to rescue trapped occupants and not fight the fire.

Construction/Collapse
- Buildings are wood frame, brick veneer, ordinary construction, fire-resistive, or mixed construction.
- Attics or cocklofts may be quite large in size, which will allow the spread of fire.
- There may be lightweight building components used in roof and floor assemblies.
- Interior partition walls can consist of wood lath and plaster, metal lath and plaster, gypsum wallboard, fire-rated drywall, or masonry construction.
- Required fire-stopping may be violated by poke-throughs.
- There is a possibility of interconnected void spaces.

Exposures
- The fire room can expose other rooms that share a common hallway.
- Common roof and floor voids can severely impact on adjoining rooms if fire enters these spaces.
- Fire-stops must be in place in attics or cocklofts to prevent extension of fire throughout the roof area.

Weather
- Rain, snow, and ice can impact on evacuation and response time.
- Wind can have a negative effect on ventilation and threaten exposures.
- Ice and snow can render fire escapes too dangerous to use.

Auxiliary Appliances
- Heat and smoke detectors should be present.
- Standpipes may be present in some structures.
- Sprinklers may protect only hallways, basements, or assembly areas.
- Specialized extinguishing systems may be found in hotel kitchens or restaurants.

Special Matters
- Delayed alarms can occur due to occupants thinking others have already called the fire department.
- Irresponsible or careless actions by one occupant will threaten other occupants.

Height
- Hotels that are less than eight stories in height are well within the reach of aerial devices.
- Motels and hotels up to four stories in height are within the reach of ground ladders in most fire departments.
- Fires above the eighth floor will require dependency on the building's systems.

Occupancy
- Classification is multiple occupancy/residential.
- There may be occupants in the hotel restaurants.

Time
- As with any residential property, midnight to 0800 hours is when the greatest loss of life occurs.
- Guests should be considered occupying their room until a primary search has proven otherwise.

CONSIDERATIONS FOR FIRES IN HOTELS AND MOTELS

Strategic Goals and Tactical Priorities for an Offensive Attack
- Perform search and rescue for occupants.
- Secure a list of occupied guest rooms to assist in search and rescue.
- If practical, ladders should be placed at each level for entry and rescue.
- Call for a sufficient amount of resources.
- Confine the fire to the room of origin. Immediately check the room above for extension of fire.
- 1½-inch or 1¾-inch hose-lines should be sufficient for fires in guest rooms.
- 2½-inch hose-line may be required if fire has entered common areas, ballrooms, storerooms, restaurants, etc.
- Ventilation will be needed in coordination with interior crews. Vertical ventilation can be used if smoke and/or fire is banking down in stairways or is suspected to have a hold on the attic or cockloft. There may be skylights in the roof that will vent above stairs or living spaces.

Incident Management System Considerations/Solutions for an Offensive Attack
- Incident Commander
- Safety Officer
- Rapid Intervention Crew(s)
- Operations (if needed)
- Staging
- Ventilation Group
- Search and Rescue Group
- Division 1, 2, 3, etc. (a division can be implemented for any floor where deemed necessary)
- Medical Group

Strategic Goals and Tactical Priorities for a Defensive Attack
- If changing from an offensive to a defensive attack, ensure that a PAR is taken for accountability purposes.
- Set up and maintain a collapse zone.
- Utilize master streams.
- Protect exposures if necessary.

Incident Management System Considerations/Solutions for a Defensive Attack
- Incident Commander
- Safety Officer
- Operations (if needed)
- Staging
- Logistics
- Divisions on the exterior on Sides A, B, C, D
- Exposure Divisions (if needed)

NIOSH FIREFIGHTER FATALITY REPORT F2011-05

On February 16, 2011, at 2320 hours, the victim's department and a mutual aid department were dispatched to a structure fire at a three-story 12,500 square foot, single-family dwelling located on a hillside. Fire was observed on an exterior wall upon arrival. Additional fire was discovered within an interior wall that extended into a drop ceiling void space and into an attic. At 0003 hours (February 17, 2011), a 61-year-old male career firefighter/paramedic (the victim) and several other career firefighters were injured when a large section of the first floor interior ceiling suddenly collapsed onto them while they were attempting to gain access to the fire above them. Emergency traffic over the radio was immediately transmitted and the firefighters and officers were quickly rescued from under the debris and treated. The victim succumbed to his injuries on February 18, 2011. The other injured firefighters and officers were treated for non-life-threatening injuries.

Contributing factors include interior gas-burning fireplace not installed and constructed to applicable building and fire codes; unique ceiling construction with large void space allowed fire to burn freely and undetected for unknown period of time, deteriorating ceiling support members; sprinkler system unable to control the fire; difficulty in getting water on the seat of the fire; and unexpected ceiling collapse. **See Appendix B: NIOSH Reports in www.pearsonhighered.com/bradyresources to read the complete report and recommendations.**

Vacant Buildings

Every building—inhabited or vacant—goes through various stages of natural deterioration. Proper maintenance will minimize most breakdowns, whereas little or no maintenance, even in an inhabited structure, will result in degradation. Experience has shown that people tend to fix not only what they can afford to fix, but also that which inconveniences them.

Any fireground problem normally encountered will be accentuated in a vacant structure. Fires in vacant buildings cause more firefighter injuries and deaths than other structural fires. The many dangerous conditions found at vacant building fires will test firefighters. They need to remain alert to the task at hand. When operating at a vacant building fire, the IC must control the natural aggressiveness of the firefighters to ensure a safe operation.

vacant buildings ■ Can be categorized into two distinct types: vacated and awaiting resale, or vacant for some time and stripped of all contents.

DEFINITIONS OF VACANT BUILDINGS

Vacant buildings can be categorized into two distinct types. The first type is a structure that has been vacated and is awaiting resale. The structure itself is basically sound. The second type is a building that has been vacant for a period of time. These buildings have been stripped of any contents that would have a resale or scrap value, including all piping, toilet and plumbing fixtures, and kitchen cabinets. Though these buildings are empty of furnishings, they can become receptacles for trash. They are in varying states of decay, and they can be death traps for firefighters responding to fires in these virtually unusable structures.

These buildings are frequently abandoned by owners who do not want to be found. The cost of demolition far outweighs the value of the property. The expense to rid the community of the blight created by these structures is often borne by the local government.

A concern of fire departments is the possibility of arson for profit in declining areas. Unscrupulous owners, realizing that they are unable to sell the property, may attempt to take the easy way out of an unprofitable situation. These fires leave behind shells of buildings in various degrees of decay, awaiting insurance settlements. The vacant buildings become an invitation for trespassers to vandalize the remaining contents and start additional fires. (See Figure 7-27.)

Some cities with large numbers of vacant/abandoned properties, which have no occupants and are neglected, with no efforts made to preserve their value or condition, are now reassessing the practicality of offensive attacks in these structures.

FIGURE 7-27 This vacant structure has been abandoned by the owner and is classified as an unusable structure. *Used with permission of Joseph Hoffman.*

EFFECTS OF WEATHER

Further damage to these already deteriorated buildings is caused by the weather. Leaking roofs and lack of paint on windows and cornices allows water to attack structural members, weakening the structure. (See Figure 7-28.) Rainwater entering masonry walls can erode the mortar holding the wall together. It can freeze between the layers of brick in older buildings, causing the walls to separate, diminishing their strength.

The problem is compounded if the vacant structure is attached to other buildings. Whether a commercial or residential building, the adjoining properties will be affected. The party or bearing wall that is a fire-stop between structures can break down, and a fire entering the cockloft or attic can spread to adjacent properties. This breakdown may affect the structure's load-carrying capabilities, causing an early collapse.

Due to their diminished strength and exposure to the elements, vacant buildings can be in a very weakened state. The additional weight of firefighters and their equipment, combined with the weight of tons of water used in the extinguishment process, may be too much for the building to bear. Wood rotted due to constant exposure to the weather will fail rapidly under fire conditions. The potential for an early collapse in vacant buildings must be anticipated due to their dilapidated condition. Localized collapse is most common, yet major building collapse can also occur. It must be remembered that once a building enters into this condition, deterioration is continual and is usually followed by additional collapse.

MARKING OF BUILDINGS

The outward appearance of a building may give immediate signs of neglect and potentially dangerous conditions, or it may indicate structural stability. It must be recognized that the stability of exterior walls doesn't guarantee the strength of the roof or interior areas. An ongoing size-up or a closer look on the interior must be performed. At the incident scene there is typically not enough time to perform a thorough investigation of the building's stability. Size-up will discover some obvious collapse indicators, but smoke conditions and darkness can hide some collapse signs that could be found with a thorough inspection accomplished prior to an incident.

FIGURE 7-28 Rainwater entering masonry walls can erode the mortar, weakening the walls, and setting the stage for wall, floor, and roof collapse. This roofline photo shows the erosion of a brick wall taken after a pancake collapse of the roof that trapped firefighters.

A

B

FIGURE 7-29 A–B This vacant structure could be marked "danger exists; enter only if life is threatened." The brick bay has twisted. Cracks have developed. A later photo shows that it has collapsed.

One method of identifying vacant building problems that some fire departments employ is a routine inspection of buildings and then marking the exterior walls of a vacant structure. Markings can entail painting symbols on the front wall or attaching a wooden placard at a specific location to identify and categorize buildings as to their structural stability. Categories can include:

- Too dangerous to enter
- Dangers exist; enter only if life is threatened
- Minor interior damage; entry can be made depending on fire conditions

The buildings are inspected at timed intervals, and if conditions have changed, the markings on the buildings are updated to reflect the observed hazards.

Responding firefighters can utilize building placards as part of their size-up. Markings are weighed together with information available at the scene. Firefighters must realize that a placard on a vacant building is part of the preplanning stage and just one component of the size-up process. It is certainly not a fail-safe method of predicting structural failure during fires. Having a building marked as "minor interior damage; entry can be made depending on fire conditions" does not mean that an offensive attack must be made. Changes may have occurred between the time the building was last marked and the time of the fire, which may affect the building's stability. (See Figures 7-29 and 7-30.)

FIGURE 7-30 This vacant structure would be marked "too dangerous to enter."

OCCUPIED VACANT BUILDINGS

Runaway children or homeless people often occupy so-called vacant buildings. They may select structures in the poorest of conditions. These buildings can be structurally unsound. They are selected because the building owners, who would otherwise chase away these unwanted squatters, are not around. The collapse dangers contained within these buildings also keep out anyone who might otherwise hassle them.

Most fires in these structures are incendiary and meant to further vandalize the building. Fires may also be started to cook food or provide warmth to the occupants.

Though a building appears vacant, it still must be given a primary search to ensure that it is unoccupied. Fires that establish a foothold in these buildings can severely test the responding firefighters. Firefighters may be informed on their arrival at a fire that squatters are living in the fire building. Information may not be available as to whether these homeless occupants have escaped the fire. These individuals rarely remain at the scene for fear of being charged with trespassing and arson. A sad commentary is that all too often these unwanted occupants become victims and succumb to the very fires that they started to keep themselves alive.

Fires in unusable structures that "house" homeless people will challenge responding firefighters. These responders must consider a multitude of hazards with which they will be confronted, and rescues must be made in the safest manner. Smoke can hide many danger signs.

Multistory structures may have stairways that are not structurally sound. Rescue from upper floors via portable or main ladders may be a safer operation. Fire escapes installed to allow the escape of trapped occupants can become death traps. Deterioration due to a lack of maintenance can result in loose treads, rusted railings, and lack of support where railings are anchored into the walls of the structure. Firefighters or occupants can be injured. Portable ladders, main ladders, or aerial platforms should be used if a fire escape is in questionable condition.

FIGURE 7-31 Attempts by local, state, and federal agencies to seal vacant buildings does not deter illegal entry, but they make a time-consuming challenge for firefighters to effect ventilation and create entry and egress points during firefighting operations.

FIREFIGHTING CONSIDERATIONS

Vacant building fires must be handled in the safest manner possible. Some experts state that firefighters should never fight a vacant building fire in an offensive mode. Their point is that no vacant building is worth the life of a firefighter.

No building, vacant or otherwise, is worth the life of or injury to a firefighter. However, all buildings must be sized up and the factors weighed as to the best method of protecting life while controlling and extinguishing the fire. A vacant building with known or suspected occupants must be treated the same as any other occupied structure. It is not realistic to automatically fight every vacant building fire in a defensive mode. The IC must apply common sense while controlling the aggressiveness of the firefighters to ensure a safe operation. The lessons learned in fighting dwelling fires with an offensive-exterior attack can be used at a vacant building fire, adding a level of safety for the firefighters. Numerous openings that cannot be controlled may exist, creating multiple flow paths that will supply oxygen to the fire. Conditions will dictate whether the offensive-exterior attack can be followed by an offensive-interior attack. If deemed safe by the SO, crews can enter the building to perform a primary search and complete the extinguishing process. If interior conditions are unsafe, then a defensive attack should be used to extinguish the fire.

Attempts to seal vacant buildings by local, state, and federal agencies involve closing door and window openings with wood, tin, or masonry. These efforts are often foiled, and illegal entry continues. (See Figure 7-31.)

Under fire conditions, these doors and windows will need to be opened by the firefighters to ensure adequate ventilation for firefighting efforts. The removal of sealing material from window and door openings can be time-consuming and challenging.

EXTERIOR OPERATIONS

An exterior attack on fires past the initial stages, or in buildings that contain no occupants, can be a successful tactic. Unlike occupied structures that contain valuable furnishings, water damage is not a concern in vacant properties. The only consideration is the weight that the water adds to the live load in the building.

Illegal Electric Hookups

It is common to find varying methods of illegal electric hookups in vacant buildings. Since the building's electric supply has been disconnected, one method that squatters use is to reattach the electric wires to the previously cut overhead service wires supplying electricity to the building. In place of the electric meter that would normally be needed to complete the circuit, they install jumper wires in the box that once housed the electric meter. Once this has been accomplished, they electrify the building's wiring system.

Another commonly used method is to attach wiring to the electric service and then to attach the wire to a haphazard setup electrical outlet to plug in the electrical appliance they wish to utilize. This is often accomplished by stretching wiring along the floor, usually entering a window directly from where the illegal hookup to the service wires occurs. The wiring used for this installation has often been removed from within the structure and shorter pieces of wire are connected together so the wire can reach the desired location. As the wire stretches from room to room and floor to floor, it may be draped over doors or railings. Twisting the bared ends of the wire together secures one length of wire to the next wire. The neutral wire and the hot wire are separated from touching each other. In this type of installation rarely will you find electrical tape or wire nuts utilized. The possibility of electric shock is high since there is no protection by fuses or circuit breakers. The occupants avoid the locations where the wiring connections are located. A problem exists when firefighters enter a structure with an illegal electric hookup to fight a fire. A firefighter advancing a hose-line under smoky conditions can come into contact with the unprotected electrical wiring, or the water used in the firefighting effort can come into contact with these bare wires and electrify a portion of the structure.

 ON SCENE

At one incident, an illegal electric hookup was being employed in a vacant building. As firefighters advanced a hose-line up an interior staircase, the entire area was electrified. As the walls, railing, and some flooring were charged with electricity, it began to shock the firefighters advancing the hose-line. As other firefighters came to their assistance, they too became victims as they made contact with the firefighters initially injured. Nine firefighters received varying degrees of electrical shock before a quick-thinking firefighter utilized a tool to remove the illegal hookup and alleviate the situation.

The Abandoned Building Toolbox program contains background materials, slides, and lesson plans for dealing with abandoned buildings. Look for it on the International Association of Arson Investigators' website: https://www.firearson.com/Publications-Resources/Abandoned-Building-Toolbox.aspx

SIZE-UP FACTORS FOR VACANT BUILDINGS

Water
- If a defensive attack is utilized, master streams will be required, with larger flows needed.
- Water damage should not be a concern; only the weight the water adds to the structure itself is a concern.

Area
- Fire-stops (fire doors, sprinklers, doors to hallways and rooms) may be nonexistent, allowing a fire to spread quickly.
- Abandoned stock in a commercial building will fuel a fire.

Life Hazard
- There is a severe hazard to firefighters fighting fires in vacant buildings. A large number of firefighter injuries and firefighter deaths occur in these structures.
- Homeless people often inhabit these buildings, including families with children.
- Runaway children may seek refuge in these buildings.
- Establishment of a collapse zone may be needed.
- TICs can assist in the primary and secondary searches.

Location, Extent
- There may be sealed windows and doors that will hide the severity of the fire and delay the entry of firefighters.
- A blitz attack from the exterior may achieve a quick knockdown of the fire.
- TICs can assist in checking walls, ceilings, and floors for hidden fire.

Apparatus, Personnel
- If a defensive attack is initiated, additional personnel may be needed.
- A defensive attack will often mean an extended operation employing master stream devices.

Construction/Collapse
- Vacant buildings can encompass all types of construction, from fire-resistive to frame.
- Deteriorated conditions may exist, which could accelerate a collapse.
- Previous fire damage can weaken a structure.
- Stable exterior walls do not guarantee the stability of the roof and floors.
- The weight of firefighting operations and firefighters can overload an already weakened building.
- The building may be placarded to indicate structural defects.
- Holes in floors and roofs can cause injury to firefighters.
- These buildings may accumulate discarded furniture and large amounts of trash.

Exposures
- Interior exposures can involve a much larger area due to removal of fire-stops.
- Exposures can be vacant or occupied.

Weather
- The long-term effects of weather can cause a breakdown in the building's stability. Structural members are subjected to the elements. Wooden members will rot. Water can freeze and then thaw in masonry walls, or mortar can wash out of the walls, severely weakening them.

Auxiliary Appliances
- There is the possibility of the presence of sprinklers and standpipes. The problem is that there is minimal or no upkeep on these systems, and they may be unusable.

Special Matters
- There is the possibility of an arson fire started with flammable liquids, causing a fast-spreading fire.
- The fire may have been started by homeless people for warmth or for cooking, and they may still be in the building.

Height
- Vacant buildings can be from one story to a high-rise.

Occupancy
- In addition to homeless people, there is the possibility of stock being left in a commercial building that could add significant fuel load to a fire.

Time
- Delayed discovery may occur due to the building being vacant.

CONSIDERATIONS FOR FIRES IN VACANT BUILDINGS
Strategic Goals and Tactical Priorities for an Offensive Attack
- Perform search and rescue for occupants.
- Call for a sufficient amount of resources.

- Check exposures if they are threatened for extension of fire.
- Depending on the size of the building and the amount of fire, 1½-inch or 1¾-inch hose-lines may be effective. Large vacant buildings will require the use of 2½-inch hose-lines.
- Horizontal ventilation may be needed. Roof ventilation may be required if fire heavily involves the area under the roof or if a common attic or cockloft exists with adjacent buildings, and fire has entered these areas.
- Window and door openings may be sealed, and entry will be restricted until forcible entry is accomplished.
- Positive pressure ventilation is usually ineffective because of numerous openings in the building.
- Constantly monitor the building and conditions and do not hesitate to initiate a defensive attack after a primary search or should conditions warrant it.

Incident Management System Considerations/Solutions for an Offensive Attack

- Incident Commander
- Safety Officer
- Rapid Intervention Crew(s)
- Staging
- Ventilation Group
- Search and Rescue/Evacuation Group
- Division 1, 2, 3, etc. (a division can be implemented for any floor where deemed necessary)
- Medical Group (if needed)

Strategic Goals and Tactical Priorities for a Defensive Attack

- If changing from an offensive to a defensive attack, ensure that a PAR is taken.
- Set up and maintain a collapse zone.
- Utilize master streams.
- Protect exposures, if necessary. This may involve interior operations within the exposed buildings.

Incident Management System Considerations/Solutions for a Defensive Attack

- Incident Commander
- Safety Officer
- Rapid Intervention Crew (if needed for exposures)
- Operations (if needed)
- Staging
- Logistics (for water supply, if needed)
- Divisions on the exterior on Sides A, B, C, D

Wildland Urban Interface

Wildland urban interface (WUI) is becoming a common term in many fire departments. *Wildland* is an area with essentially no development except for roads, railroads, power lines, and similar transportation facilities. The term *urban* refers to the characteristics of, or an actual city. *Interface* refers to a place where two independent systems meet and interact with each other. The mixing of the distinctly different characteristics of wildland and urban areas can create varied challenges for firefighters.

wildland urban interface (WUI)
- A mix of wildland with urban areas.

Though a high percentage of wildland fires are ignited naturally as a consequence of lightning, the impact of people has changed the occurrence and extent of these fires in several ways. In addition to arson and accidental fires, natural fires are now being suppressed to protect lives and property. While some wildfires burn in remote forested regions, they can cause extensive destruction of homes and other property located in the WUI: a zone of transition between developed areas and undeveloped wilderness.

FIGURE 7-32 Housing developments built in wildland areas can be endangered by wildland fires. *Used with permission of Don Forsyth, Batt. Chief, Orange County Fire, CA.*

In a strictly wildland setting, structures, if any, are widely scattered. The WUI creates a different situation because of the conversion of vacation homes into year-round residences or due to the construction of new housing that extends into what was previously wildland or forested areas. Development in many communities has encroached into wildland areas and often involves an integration of structures into large areas of green space. Many who live in these communities, or isolated locations, strive to return to nature and enjoy the closeness of the wooded areas to their homes. The perception of fire is usually not a consideration.

HOUSING DEVELOPMENTS IN WILDLAND URBAN INTERFACE AREAS

In addition to isolated single occupancies, there is the danger that exists in large housing developments that are nestled into a wooded area. It could be a small development of 10 to 15 homes or hundreds of homes situated adjacent to, or surrounded by, wooded areas. The houses can be situated on a relatively level plain or contoured into hillsides completely enveloped by trees and vegetation.

This isolation of structures embedded into wooded areas will challenge firefighters to assure the life safety of the residents when confronted with a fire. In addition to protecting dwellings, firefighters need to consider cars, trucks, snow mobiles, boats, garages, barns, and outbuildings. The cost of these items can quickly add up to sizable losses if destroyed by fire.

There should also be a serious concern for the infrastructure that would be threatened by a wildland fire, which could include communication towers, electric service wires, high tension wires delivering electricity to distant locations, and water treatment facilities. Occasionally there are above-ground flammable liquid pipelines that run through the wildland for miles.

FIREFIGHTING IN WILDLAND URBAN INTERFACE AREAS

The reality is that when wildland fires occur near these WUI areas they create difficult and unique firefighting problems for firefighters. The intermingling of structural and wildland areas introduces a totally new set of considerations into the firefighting equation.

The major difference between wildland firefighting and fires that threaten WUI locales is that wildland firefighting poses a life safety threat mainly to the firefighting forces since these are strictly wooded areas.

SIZE-UP FACTORS IN WILDLAND URBAN INTERFACE AREAS

The normal size-up factors that are contained in the mnemonic WALLACE WAS HOT still pertain to the structural firefighting involved when dealing with a WUI fire, but additional concerns will need to be considered to address the wildland fire problems.

LOCATING STRUCTURES IN WILDLAND URBAN INTERFACE AREAS

The ability of a fire department to respond to a reported structure fire starts with an accurate address. In remote areas the structure may not have a formal address and though local people may recognize the location by using terms like "it's the old Mitchells' cabin," dispatchers or outside resources will need better information to locate the dwelling. Units that have responded from outside the area may be confused by the road layouts and lack of street signs.

When assessing areas in the wildland fire environment, trying to determine if/or how many homes may be up some remote road or driveway, counting mail boxes at the roadway's edge provides a good estimate.

Another problem that can occur is that calls for help by the property owner to the fire department may embellish the situation and be made for the sole purpose of having the fire department respond as a precautionary measure for a distant fire. These types of responses will tax potentially limited resources.

ACCESS TO WILDLAND URBAN INTERFACE AREAS

The fire-front can deny access by closing off roadways from the base of the hills or mountains where there may be only one road to access the property and fire may limit its use due to the imminent danger of firefighters and apparatus being trapped with no escape route. This may require scouting before committing large apparatus to areas where they may not be able to turn around.

There may be poor access roads or the only access may require the use of narrow roads that can easily be blocked by fire or fallen trees and deny entry. Steep grades, undercarriage problems, stream crossings, and terrain can restrict apparatus movement. The composition of the roadway also needs to be considered; is it paved, gravel, or dirt? Can a full-sized apparatus maneuver along the roadway, or will a smaller 4 × 4 unit be needed? Bridge load limits can restrict access of fire apparatus to some locales, preventing a timely response, and can be life threatening should firefighters need to utilize those roadways or bridges as a secondary means of egress. Before committing apparatus and personnel on a WUI fire, a secondary means of egress must be determined. This should include a visual check of these secondary roadways to ensure that they are passable. In many areas secondary roadways are not maintained and fallen trees, damaged bridges, or heavy vegetation may restrict their use. In many cases the narrowness of the roadway may make it impossible for vehicles to pass each other and will force units to drive in reverse until a location can be found where vehicles can pass one another or make a U-turn. Smoke conditions reducing visibility can further compound an already dangerous situation.

EVACUATION OF WILDLAND URBAN INTERFACE AREAS

Life safety will be the main consideration of firefighters. Occupants may self-evacuate or they may refuse to leave their homes. Accountability of family members may cause confusion and delay evacuations.

It may be decided that a forced evacuation of an area is necessary for life safety of the occupants and the firefighters. It must be realized that forced evacuation of residents is a police duty. The important consideration is the time necessary to effectively deploy, notify, and evacuate the community before the fire arrives or cuts off their egress. Considerations of traffic flow, direction, volume of traffic, etc. must be considered. Firefighters can strongly recommend that occupants leave their homes, but only the police have the power to enforce a mandatory evacuation. In some states, even this is not supported by law. Naturally this is dependent on local or state regulations.

The unwillingness of some residents to leave their homes can place undue pressure on the field forces attempting to further assist in fighting a fire in a location in which conditions can be deteriorating rapidly. The incidence of residents waiting until conditions get severe and then attempting to leave has proven fatal, due to panic, low visibility, being cut off or outrun by the fire, and can complicate the ability for firefighters to do their job. At this point, sheltering in place may be the best option. In addition to firefighters having a plan for a secondary means of exit, they also must recognize that under rapidly changing conditions that become life threatening they should consider having a tertiary plan. This could include seeking protection within the threatened structure they are trying to protect until the fire front passes, and exiting the building safely even though portions of the building may be burning.

UTILITIES

There may be a loss of electrical service to areas due to downed wires from the fire, or from electrical substations being knocked out of service. It may be decided to shut down electric power to areas to prevent electrocutions from downed wires that could otherwise injure firefighters and civilians or close roadways. Keep in mind that heavy smoke traveling through high-voltage power lines may cause them to arc significantly, causing a severe safety issue in that immediate area.

The impact of the loss of electrical power will affect water pumping stations and private wells due to their being shut down. Firefighters will need to consider securing water from swimming pools, storage tanks, lakes, streams, and ponds, along with water tenders. Air drops from helicopters and planes can be helpful in containing a fire and protecting structures.

Communications via fire department radios or cell phones may be limited or nonexistent in rural areas, and stationary phones will be reliant on the continuance of phone service through telephone pole wiring systems, which are also subject to fire damage.

WILDFIRE INFORMATION AND SITUATIONAL REPORTS

The National Interagency Fire Center (NIFC) publishes an Incident Management Situation Report (PDF, 747 Kb), updated each morning at 7:30 Eastern. They also frequently post the National Fire News – Current Wildfires, Monday through Friday: https://www.nifc.gov/fireInfo/nfn.htm

InciWeb is another site that updates active wildfire information. It includes the name of the fire, type, state, acres burned, and when it was updated. The site also includes maps of fire areas, pictures and video, announcements, road closures, and links to their social media pages: http://inciweb.nwcg.gov/

WEATHER, TOPOGRAPHY, AND FUEL FOR WILDLAND URBAN INTERFACE FIRES

Critical factors for wildland firefighting include the weather, topography, and fuel supply. The considerations of weather and fuel with wildland fires are a much greater concern than found at a normal structure fire. This is due to the many available fuels, which have different burning characteristics, such as trees, leaves, branches, and underbrush, to readily supply a fire. These plentiful fuels are affected by fuel moisture. The lack of moisture due to limited rainfall can add to the intensity of a fire and increase the difficulty in gaining fire control. Dry weather can reduce streams to trickles, eliminating the fire department's ability to utilize these water supplies by drafting from these streams. Heavy growth can restrict movement of firefighters when attempting to contain a wildland fire in close proximity to structures.

Wind can play havoc with a wildland fire. It can create severe conditions and cause a fire to spread rapidly by being pushed by the wind. In a strictly wildland environment, occasionally firefighting is minimized at night due to safety issues, such as the possibility of changing wind conditions, poor visibility in hazardous terrain, and the inability of firefighters to be able to identify fire behavior, making firefighting too dangerous to accomplish.

Topography or terrain plays a big factor in fire behavior, the direction in which a fire spreads and the speed in which a fire travels. Topography or terrain affects the ability of firefighters to access the fire-line, and also their ability to evacuate or escape in a timely manner if necessary due to changing conditions.

STRUCTURAL VERSUS WILDLAND FIREFIGHTING

At a WUI incident, after life safety is accomplished, the role of the IC is fire attack, and he or she must decide whether firefighters can contain and extinguish the fire. If there are not sufficient resources on scene or available locally, then a decision must be made as to how to best utilize the available personnel in protecting the endangered structures until additional personnel and equipment can arrive from other agencies or further distances.

The speed that these fires travel can involve numerous communities in a relatively short period of time. A WUI fire can occur in areas that may be protected by state or federal firefighters. These agencies can be of tremendous assistance to a local fire department. Their experience and training can help the IC in securing sufficient resources, predicting fire spread, developing strategy and tactics, deploying resources, structuring an incident management system, and many other factors. A structural fire officer must realize that unless he or she has extensive training and experience in wildland firefighting, there will be a genuine need for a wildland specialist to act in a role of a technical specialist. These incidents will demand expansion of the incident command system, including implementation of Unified Command, and coordination with firefighters from other jurisdictions and agencies.

OBJECTIVES, STRATEGY, AND TACTICS

To be effective, the IC must constantly monitor all reports and working conditions in areas that firefighters are operating. Status reports will help to determine the immediate fire areas involved and predict how and where the fire can reasonably be stopped. In general terms, the objective is to box in the fire or determine the acceptable perimeter using roads, streams, rivers, rock outcroppings, or man-made fire breaks using heavy equipment and ground firefighters to cut a fire-line. In many cases, contingency lines or secondary lines are determined and may also be prepared. Of course, to determine accurate boundaries, the IC must consider resource availability, estimated time needed to get resources in place, and create appropriate defendable fire-lines versus the direction and speed of fire spread.

FIGURE 7-33 Firefighters can be overwhelmed attempting to control fires that threaten structures adjacent to wildland areas. *Used with permission of Don Forsyth., Batt. Chief, Orange County Fire, CA.*

Systematic aerial surveillance with visual and thermal imaging, along with reports from divisions and groups, can show the progress that is being made and can be used to attack and contain the fire. Reports from the dispatcher or the emergency operations center (EOC) can be introduced into the overall equation. The dispatcher or EOC can keep the IC abreast of other incidents in the area that could affect their ability to provide sufficient resources in a timely manner. They can also provide updated weather reports. These reports could be of impending rain that could assist in fire control, or the prediction of high winds and a continued dry spell that could be detrimental to the situation.

If available, aircraft can be utilized to assist in arresting the spread of fire through the use of aerial drops of fire retardant, water, or a water-foam mixture. Retardant is typically applied to unburned fuel in front of the fire to decrease fire behavior and spread. When fire reaches the treated area, the fire behavior should be reduced to a fraction of what it was. However, ground personnel must follow up in these drop areas to fully suppress the remaining fire and smoldering material.

DETERMINING WHICH STRUCTURES CAN BE PROTECTED

A decision that must be made in WUI incidents is allowing some structures to burn and trying to defend others. This decision making is not a routine function for structural firefighters and can be very difficult. Yet, if it is not made, it can set up a situation where firefighters will be playing catch up and lose many more properties than if they had been realistic about what they could reasonably accomplish. A structure that is built into the mountainside that has a fire running up the mountain face from below, with flammable vegetation around and close to the structure, would be a situation where it would probably be best to evacuate the residents and write the dwelling off. These decisions will need to consider the risk of fighting the fire and the gain that could be achieved. A running mountainside fire would normally present too high a risk with little chance of protecting the property and seriously endangering firefighters. Another factor is the proximity of vegetation and natural fuels to the structures. If a clear area is not maintained around the structures, it will reduce the chances of being able to protect it.

The decision on which structures have the potential of being saved will be made at different levels. The decision makers will include the IC, Operations Section Chief, and the Division and Group Supervisors operating in the fire areas. The Division Supervisor will need to make decisions based upon how the current conditions can impact on the safety of personnel. They can, at times, find themselves in some very difficult and argumentative situations with a property owner whose property has been written off as being indefensible. The Division Supervisor may have received information from Operations that their position will become untenable and has ordered those resources out of the area. The property owner's contention may be that he has paid taxes and expects full protection of his property, if and when it is needed. The Division Supervisor can only present the owner with the facts and recommend that they evacuate immediately.

From a risk versus gain perspective, buildings can be placed into one of three categories:

- Structures that are not threatened
- Structures that are threatened but defensible and have the potential of being saved
- Structures that are threatened but not defensible and too dangerous to protect

The decision-making process on whether a property can reasonably be saved will rely upon many factors including:

- The type of building construction (frame construction can and will burn readily)
- The roofing system (could consist of wooden shingles that can collect sparks and readily ignite, or could be masonry tiles or other nonflammable material that can be very protective)
- Ignitable exterior walls of vinyl or wood siding with little clear space from the on-coming fire versus masonry walls that will help protect the structure
- Large wooden decks with combustible patio furniture
- Firewood stacked under or near the wooden decks
- Wooden out-buildings or detached frame garages
- The adequacy of the immediate water supply and its dependability (in many cases, the only available water supply is that which is brought by the fire department and a garden hose from the threatened structure(s) if the electricity can be maintained)
- The fuel supply of the wooded surroundings
- The presence of oil or propane tanks adjacent to the structures
- The presence of automobiles or trucks and the potential rupture of their fuel tanks
- The presence of hazardous materials, including pesticides and herbicides
- Extreme fire behavior
- Limited available resources

It has recently been determined that many more structures ignite due to ember intrusion rather than actual flame front contact.

In the wildland fire environment, there is a common rule of thumb during a wind-driven fire: If one quarter or more of a roof is on fire when you pass or arrive, do not engage this structure but move to the next exposure. This is due to set-up time and the probability of not being able to save this structure while losing others.

A staffing study of California wildfires shows the difference between effectiveness and overall efficiency of various crew sizes when stretching hose-lines. The report is available at: https://www.iaff.org/10news/pdfs/cdffirereport.pdf.

PROTECTING A BUILDING AT A WILDLAND URBAN INTERFACE FIRE

Once firefighters determine that a structure can be saved, there are a few things that the homeowner can do to assist them. They can place the homeowner's ladder to the roof on the opposite side from which the fire will threaten the structure and clear the roof of leaves, pine needles, and any other combustible material that has accumulated on the roofs or in gutters. This will remove a ready supply of fuel that can be easily ignited by flying brands. Wooden fences and wood piles can be removed from around the structures. Another consideration would be to move combustible outside furniture to inside of the garage or structure. Close all windows and doors, remove curtains from windows, and leave the garage door unlocked in case you need to use the structure as an emergency shelter.

If units have sufficient time to prepare a structure in advance of a fire threat, they will then have to decide whether to stay or leave as the fire approaches. They must rely on the size of their safety zone and gauge the fire's intensity. If they decide to stay, pre-positioning a charged exposure hose-line is recommended, and they can use the structure as a shield from the heat as they extinguish fire in the protected zone around the structure. They can go inside the structure until the main body of the fire passes and then exit to knockdown any fire that has ignited the exterior. If it is decided not to remain at the structure, and time permits, Class A foam or protectant gel may be applied before leaving. After the fire passes, the units can return and control and extinguish any fire that has attacked the building. Decisions on whether to stay or leave must be based upon firefighter safety.

NIOSH FIREFIGHTER FATALITY REPORT F2003-36

On October 29, 2003, a 38-year-old male career firefighter (the victim) was killed and a 48-year-old male career captain was severely injured when fire overran their position. The incident occurred during the protection of a residential structure during a wildland fire operation that eventually consumed more than 280,000 acres. The victim and his crew were part of a task force assigned to protect a number of residential structures located along a ridge on the flank of the fire. The victim's crew was in the process of preparing to defend the structure when the fire made a slope and wind-driven run through heavy brush directly toward their position. The crew retreated to the residential structure to seek refuge from the oncoming fire. Two of the four crew members were able to get into the structure while the captain was attempting to assist the victim as the fire reached their position. The victim died near the structure and the captain, who was seriously burned, had to be assisted into the structure by the other crew members.

NIOSH investigators concluded that, to minimize the risk of similar occurrences, fire departments and fire service agencies should:

- Ensure that the authority to conduct firing out or burning out operations is clearly defined in the standard operating procedure (SOP) or Incident Action Plan (IAP) and is closely coordinated with all supervisors, command staff, and adjacent ground forces
- Ensure that all resources, especially those operating at or near the head of the fire, are provided with current and anticipated weather information
- Stress the importance of utilizing LCES (lookouts, communications, escape routes, and safety zones) to help identify specific trigger points (e.g., extreme fire behavior, changes in weather, location of fire on the ground, etc.) that indicate the need for a crew to use their escape route(s), and/or seek refuge in a designated safety zone
- Ensure that, at a minimum, high-risk geographic areas are identified (e.g., topography, fuels, property, etc.) as part of the preplanning process and provide this information to assigned crews
- Ensure that incident command system (ICS) span-of-control recommendations are maintained
- Consider the implementation of a carbon monoxide–based monitoring program for wildland firefighters **See Appendix B: NIOSH Reports in www.pearsonhighered.com/bradyresources to read the complete report and recommendations.**

HOSE-LINE AND CLASS A FOAM FOR WILDLAND URBAN INTERFACE

Wildland firefighting utilizes 1½-inch hose-line. Adjustable nozzles should be on the lowest gallon setting since water must be used judiciously. Similar to structural firefighting, a second hose-line should be stretched around the structure when attempting to protect it. This provides two hose-lines to control the fire and acts as a safety factor should one hose-line fail. Realize that the use of the hose-lines is to protect the structure and not necessarily to fight the oncoming fire. Water needs to be used judiciously. The protection of the property in a WUI incident is a very different operation than structural firefighting where the aim is rapid and total fire extinguishment.

Prewetting the exterior of a structure and surrounding fuels has had limited success since wildland fires burn with such intensity that water dries up quickly. Class A foam with a compressed air/foam system or firefighting gels for wildland firefighting can be used to pretreat a structure. The foam will adhere to the building and can protect it from an oncoming fire. This can be advantageous if there are limited resources since the structure can be pretreated and the firefighters can move on to other assignments. A drawback to applying Class A foam is that it will break down due to wind and ambient temperature if a structure is pretreated too soon.

RESOURCE NEEDS FOR A WILDLAND URBAN INTERFACE FIRE

The IC must consider how many companies will be needed. Realistically, from a structural firefighting point of view, we usually have multiple companies operating on the interior of a working fire. Should a WUI fire threaten a subdivision of 20 homes, how many companies would be needed to protect them? Many considerations need to be factored into the equation. However, if a fire front directly threatens 10 of the homes, would a minimum of five companies be realistic? In reality, depending upon the water supply, distance from the wooded area, amount of underbrush present, and numerous other factors, it may be more reasonable to require 10 companies. Yet in rural areas, this amount of resources will take a period of time to secure unless the fire has been burning for some time and mutual aid, state, and/or federal resources are involved. In this case the resources may be obtainable, yet the IC will need to prioritize the utilization of the resources. Realize that the time it can take to obtain the requested resources and have those resources reach the scene can be quite lengthy.

To anticipate the need for additional resources, a rule of thumb that can be used is deploying one engine per structure if it is mostly surrounded by wildland fuels and one additional engine for every five structures. If the structures are close together (50 feet or less), one engine may be adequate to protect two structures. Another consideration is whether aircraft are available to assist in fire control. However, a word of caution here: the success of an operation should never be based solely on the use of aircraft because they are often diverted or shut down due to weather, fire behavior, or smoke conditions in the area.

 NIOSH FIREFIGHTER FATALITY REPORT F2011-17

On June 23, 2011, a 23-year-old male seasonal wildland firefighter (FF) on an interagency hot shot crew (IHC) deployed from his duty station in Utah to fight wildland fires in Georgia and Texas. After fighting fires in Georgia for four days, the crew was dispatched to Texas. After travelling for three days, then staging for three days, the crew began firefighting on July 4, 2011.

On the morning of July 7, 2011, the FF was assigned swamper duties (clearing limbs after tree-cutting) to construct a fireline, followed by cold trail operations (a component of mop-up) with a hand tool. After lunch, the FF refilled his water supply and continued securing the fireline and mopping up for about 1.5 hours. After being left alone for a short period of time, the FF was found unconscious at approximately 1550 hours. The weather was sunny and hot: a temperature of 105 degrees Fahrenheit (F), relative humidity of 24 percent with minimal wind (one to three miles per hour).

Initial assessment by the crew's emergency medical technician (EMT) suggested the FF suffered from heat-related illness (HRI). Air Attack was notified as the crew EMTs provided basic HRI care at this remote location (the FF's pack and shirt were removed, he was doused with water, and a tarp was held up for shade). Local emergency medical service (EMS) units (ambulance and air evacuation helicopter) were not notified of the incident for about 20 minutes due to uncertain drop point coordinates. This delay, however, did not delay advanced life support (ALS) treatment because it took 45 minutes to extract the FF to the drop point where the local EMS units were waiting.

Approximately 30 minutes after his collapse, the FF's condition deteriorated; respiratory arrest was followed by cardiac arrest, and cardiopulmonary resuscitation (CPR) was begun.

Approximately 15 minutes after his cardiac arrest, the FF arrived at the drop point and the local, ambulance, and air evacuation units initiated ALS, but their treatment protocols for exertional heatstroke did not include cold/ice water immersion therapy. When the FF arrived at the hospital ED, a core temperature of 108 degrees F was documented and ALS continued for an additional five minutes. At 1703 hours, the attending physician pronounced the FF dead and resuscitation efforts were stopped.

The autopsy report listed the cause of death as "hyperthermia." NIOSH investigators agree with the medical examiner's assessment. NIOSH investigators conclude that the FF's hyperthermia was precipitated by moderate to heavy physical exertion in severe weather conditions. These factors led to exertional heatstroke.

Fatal exertional heatstroke is extremely rare among wildland firefighters; this was the first reported case in the agency's 65-year history and only the second reported federal wildland firefighter to die from heatstroke according to wildland fire service records.

NIOSH investigators offer the following safety and health recommendations to reduce heat stress, heat strain, and prevent future cases of HRI and exertional heatstroke among wildland firefighters, including:

- Strengthen the agency's current heat stress program.
- Always work in pairs and/or be in direct communication with crew members.
- Promptly alert local EMS units of a medical emergency per Incident Command protocols.
- When exertional heatstroke is suspected, inform responding EMS units of the potential need for cold/ice water immersion therapy.
- Seek input from crew members and frontline supervisors about removing barriers, real or perceived, to reporting or seeking medical attention for heat strain or HRI.
- Consider cases of HRI, particularly severe cases such as heatstroke or rhabdomyolysis that result in death or hospitalization, as a sign that the current heat stress program is inadequate.
- Consider incorporating members of the department's Safety Office into the Operations Management Team. **See Appendix B: NIOSH Reports in www.pearsonhighered.com/bradyresources to read the complete report and recommendations.**

SAFETY CONSIDERATIONS AT WILDLAND URBAN INTERFACE FIRES

The assigning of an Incident Safety Officer should occur early at a WUI fire. A key to safety is that personal protective gear must be used at all times. Realize that there are distinct differences between structural and wildland firefighting gear. The wildland firefighter's gear is designed to be lightweight but with limited thermal protection. It is not suited for structural firefighting, and interior firefighting should be seriously scrutinized or not attempted with that gear. The protection provided by structural firefighting gear will give a higher thermal protection, but firefighters will need to be rotated more frequently to avoid heat exhaustion due to the increased protection.

Wildland firefighters carry a fire shelter that is made of an aluminized material over fiberglass that, when deployed, can help a wildland firefighter survive direct flame contact. However, deploying a fire shelter is a last resort effort that is only used when a firefighter finds he or she is faced with a life or death situation and all escape routes are cut off. Because fire shelters are life-saving tools, they require a minimum of annual training to maintain proficiency.

Firefighter safety should include:

- Designating a safety zone or refuge area available to the firefighters
- A safe location to park the apparatus and then backing the apparatus into that location to facilitate a rapid exit if necessary
- Not parking under power lines or trees nor on top of septic systems, which might collapse due to the weight of the apparatus
- Keeping the apparatus' water tank full if possible, and maintaining at least 100 gallons in reserve at all times
- Considering the amount of clearance from the wooded areas and undergrowth to the dwellings
- Recognizing that if the exterior of the building is already heavily involved that it is probably past the ability of the fire department to save it
- Whether fire has spread to the interior through attic vents, cracks, broken windows, etc., and has control of the structure
- The actions that you need to take if your escape route is starting to be comprised and an immediate exit is necessary

For wildland firefighters there are what are commonly known as the "10 Standard Fire Orders" and the "18 Situations That Shout Watchout." These can be found with explanations at http://www.fs.fed.us/fire/safety/10_18/10_18.html www.nifc.gov.

On June 30, 2013, 19 smoke jumpers from the Granite Mountain Interagency Hotshot Crew from the Prescott, Arizona, Fire Department died while fighting the Yarnell Hill wildland fire in Arizona. The Arizona State Forestry Division investigated the incident. Some of the conclusions found were:

- The fire destroyed more than 100 structures.
- Radio problems were encountered.
- The complexity of the fire increased in a very short time period.

LOOKOUTS, COMMUNICATIONS, ESCAPE ROUTES, SAFETY ZONES (LCES)

Wildland firefighters use the acronym LCES to remember key safety points. These key points are emphasized to improve and maintain firefighter safety. LCES are factors that can easily be applied to structural firefighting. LCES denotes:

■ _L_ookouts that observe fire conditions and report to the IC and/or personnel on the line the fire conditions that they see and any changes they observe in fire behavior. In structural firefighting, lookouts would be Division or Group Supervisors giving progress reports from the interior or rear of the fire building, or exposures out of the view of the IC.

■ _C_ommunications in any type of firefighting endeavor is always a critical factor on an incident scene. (Communications is discussed at length in Chapter 2.) The great distances spanned by a WUI fire demands that a communication link be established to allow all units access to available information. This will assist them in addressing safety factors, accountability, and updated safety messages.

■ _E_scape Routes must be pre-established to ensure a safe exit to the predetermined safety zone, should fire conditions accelerate to dangerous levels. It is also vital to determine the travel time required if the need arises to use the escape route under smoky and hazardous conditions, so that an appropriate trigger point can be identified. If the fire approaches from a different direction firefighters should know the escape routes that can be immediately implemented. Likewise, in structural firefighting, every firefighter entering a fire building should personally observe the structure and memorize secondary means of egress that the building presents. This could be roofs that are accessible from windows on the upper floors or ladders placed to windows. The company officer will also be doing a similar size-up and pointing out to company members the secondary means of egress should conditions warrant it in the wildland environment.

■ _S_afety Zones are locations where the personnel and/or apparatus are safe from direct flame front contact. It should be large enough in size and contain little to no flammable vegetation or other flammable materials. Firefighters need to maintain a safe means of reaching the safety zone should conditions rapidly change.

The Wildland Fire Lessons Learned Center website allows readers to learn from others. Though it is basically for wildland fires, there are many instances where the lessons learned apply to other all-hazard types of incidents. This site can be accessed at www.wildfirelessons.net.

PREPLANNING FOR A WILDLAND URBAN INTERFACE FIRE

The homeowner, the fire department, and dispatch centers need to ensure that each property has an address that is understandable and noted at the dispatch center and that any requests for resources by the homeowner use that address. In some rural and remote locations, this may be accomplished by using global positioning systems (GPS) to identify latitude and longitude. This seemingly minor step can pay dividends should an emergency occur and the local responders are not available to respond due to another incident. Likewise, the fire department will need to have mapping to ensure that alternate means of exiting, if they exist, are known. If no other means of exiting exist, that should be part of the strategic considerations on what is defensible.

Preplanning can help identify potential areas for a WUI fire. Once these areas have been identified, public education can be utilized in educating the residents of the hazards that exist. This public education can be used to warn residents of the steps that they can take to lessen the risk to their properties. They should be advised that even if they take all of the appropriate steps, there can be situations where their life could still be threatened, that they can still lose their home, and the fire department may be helpless to assist them. These developments could occur where the fire department can be faced with situations and locations that are too dangerous for firefighters to enter or operate in since it would seriously threaten their safety. Fire agencies in the western United States, with significant WUI areas, are currently implementing a comprehensive program called "Ready, Set, Go!" in which firefighters visit homes, provide a property evaluation, and educate the homeowners and residents in the steps they can take to reduce their risks, and provide adequate protection for their property and families.

LOGISTICAL CONSIDERATIONS AT A WILDLAND URBAN INTERFACE FIRE

Logistic assignments for structural firefighting can often consist of an air unit for refilling SCBA cylinders, water supply considerations, and refueling of apparatus on extended fires. At WUI fires, there will also be a need to consider medical attention and feeding of personnel, fuel, sleeping arrangements, as well as common communication needs and establishing facilities for an Incident Command Post, Base, and Staging Areas.

Logistics will need to provide specific tools that are utilized in wildland firefighting to supplement the tools carried by the structural companies.

SIZE-UP FACTORS FOR WILDLAND URBAN INTERFACE FIRES

Water
- Water supplies will need to be found and apparatus should, when possible, maintain a full water tank, but at least a minimum of 100 gallons.
- Drafting sites may exist along the banks of rivers, streams, or other waterways.
- Swimming pools can be used for water supplies.
- Garden hoses can be used to supplement water supplies or refill fire department apparatus tanks.
- Class A foam with a compressed air/foam system or firefighting gels for wildland firefighting can be used to pretreat a structure.

Area
- Consider the wildland and/or flammable growth impinging upon the structures and lack of defensible spaces.
- The size of the structure can be a small cabin to very large structures with numerous combustible decks, fences, and outbuildings.

Life Hazard
- Life hazard can be severe, and evacuations can range from difficult to impossible.
- The wildland fire will generate high heat conditions, significant blowing embers, severe smoke conditions, and can travel rapidly.
- TICs can assist in the primary and secondary searches.
- Firefighters' safety must be a prime concern due to the potential for rapidly changing conditions at these fires.

Location, Extent
- The location and intensity of the fire and the distance from the structure will assist in deciding whether a building is defensible.

- The proximity to the vegetation or tree lines, the amount of underbrush, the presence of wooden fences, wood piles, and outbuildings must be considered.
- Anticipate a fast-moving, extremely hot fire.
- TICs can assist in the overhaul stages of a WUI fire to check roofs, walls, ceilings, and floors for hidden fire.

Apparatus, Personnel
- Call for help early. A large number of resources will be needed to control these fires.
- Apparatus placement must be in a clear area and be backed in to allow a rapid exit.
- Avoid parking over septic systems to prevent their collapse due to the apparatus' weight.

Construction/Collapse
- Outbuildings may be flimsy and can contribute fuel to intensify a fire.
- Lightweight constructed buildings being attacked by fire can fail readily.

Exposures
- Large flying brands and embers are common. Fires must be anticipated downwind for great distances.
- Tremendous amounts of radiant heat must be anticipated.
- Structures can be wet down to cool them or Class A foam or gel applied to protect them.

Weather
- Wind will drive this fire and predictions of wind direction and speed can assist in determining the direction that a wildland fire will travel.
- Anticipate changes in wind direction and speed that could endanger firefighters and exposures.
- Low humidity will mean low moisture content of the fuel and a fast-moving fire.
- Heavy rain can assist in controlling these fires.
- A high percentage of wildland fires are ignited naturally as a consequence of lightning.

Auxiliary Appliances
- Some properties may have exterior fire protection systems; use them to your advantage.

Special Matters
- Outbuildings can contain flammable products that will supply fuel to the fire, or toxic/hazardous materials affecting personnel when involved in fire.
- Large outdoor fuel oil or propane tanks can be present and must be protected.
- There may be tanks of gasoline or storage of gasoline in smaller containers in outbuildings.
- Access may be via narrow or weight-restricted bridges or a single roadway. Apparatus for structural firefighting may find access difficult and off-road vehicles may be necessary.
- The location of the fire may cut off secondary roadways.

Height
- Buildings can range from one-story and higher. Sloped areas may contain additional floors/stories not visible from the front of the structure.
- Outbuildings consisting of garages are usually one-story.

Occupancy
- Single-family dwellings may be quite large in size.
- In the WUI there may be significant numbers of single- and double-wide mobile homes.
- Many buildings may be vacant, unoccupied, or abandoned.

Time

- Long response distances can impact upon the fire department's ability to protect properties in remote locations.

CONSIDERATIONS FOR WILDLAND URBAN INTERFACE FIRES

These fires can be broken down into two stages. The first stage would be attempting to prepare and protect a structure that is faced with an impending attack from a wildland fire. The second stage can occur to address any extension of fire to the structure if the units that prepared the structure in the first stage have left the scene and returned once the fire has passed. At that point they would extinguish all exterior fire, and if the structure is deemed savable, then interior firefighting can commence if needed. If fire damage is deemed too extensive, the unit's actions will be decided based on whether it would be best to deploy elsewhere or continue to final extinguishment.

Strategic Goals and Tactical Priorities to Prepare a Structure for an Impending Wildland Urban Interface Fire Attack

- Perform search and rescue for building occupants.
- Strongly recommend that any occupants evacuate to a safe area if time permits. If occupants intend to stay, find out where they will seek shelter to enable responders to check on their safety should they be able to return.
- Call for additional resources if needed and available.
- Close all doors and windows.
- Leave garage door unlocked should it be needed as an emergency shelter.
- Remove curtains and any window dressings and move combustibles away from windows.
- Use 1½-inch or 1¾-inch hose-lines, or Class A foam/gel if available to wet down/protect the exterior of the structure.
- Use the occupant's ladder to clear leaves or other combustibles from the roof and gutters.
- Remove wood piles and wooden fences near the structure that could contribute fuel to a fire. Place combustible lawn or deck furniture on the interior, or far away from the structure.

Incident Management System Considerations/Solutions for an Impending Wildland Urban Interface Fire Attack

- Incident Commander
- Safety Officer
- Rapid Intervention Crew(s) (if interior firefighting)
- Staging
- Search and Rescue/Evacuation Group
- Division A, B, C, D for an outside operation or a Division 1, 2, etc., for an interior attack
- Technical Specialist in wildland firefighting
- Specialized equipment, resources, and aircraft

Strategic Goals and Tactical Priorities for Units Returning to a Structure after an Attack by a Wildland Fire

- If residents remained at the scene, attempt to locate them. This may involve an interior attack to perform search and rescue.
- Decide if the building is structurally sound and whether any fire has entered the building and if it can be safely extinguished.
- If structurally sound and it is decided to enter and fight the fire, ensure that a sufficient number of firefighters are on scene with proper safety gear.
- If the building is beyond saving, a decision on whether to remain at the scene or deploy elsewhere must be made.
- Request additional personnel if needed and available.
- Monitor the building's exterior for hot spots or embers that could flare up.

Incident Management System Considerations/Solutions for Units Returning to a Structure after an Attack by a Wildland Fire

- Incident Commander
- Safety Officer
- Rapid Intervention Crew (if needed for an interior attack)
- Staging
- Logistics (for water supply, if needed)
- Division A, B, C, D for an outside operation or a Division 1, 2, etc., for an interior attack
- Search and Rescue Group
- Technical Specialist in wildland firefighting
- Specialized equipment, resources, and aircraft

Summary

Special situations and occupancies will challenge firefighters. Statistics have shown that increased firefighter injuries and deaths occur when fighting fires in renovated and vacant buildings. Cellar or basement fires are dangerous for firefighters. Row and town houses, garden apartments, and hotels and motels must be considered occupied at all times, and civilian life hazard will require searches that commit firefighters to interior operations. Wildland urban interface fires create new and different challenges for structural firefighters since they are dynamic and challenging types of incidents.

Review Questions

1. Discuss the importance of ventilation at a cellar fire.
2. Discuss the basic considerations for the first hose-line at a cellar fire.
3. If you have basements or cellars in your jurisdiction, describe the types of basement entrances that are common.
4. If you have a basement or subbasements in any high-rise building, describe the methods of entry and whether any fire protective systems protect these locations.
5. Discuss the location of garden apartments in your response district and in adjacent mutual-aid districts.
6. Describe the benefits and problems associated with eaves on garden apartments.
7. What cues can firefighters utilize to predict fire travel in garden apartments?
8. List the reasons why garden apartment fires can burn out of control.
9. What factors should be considered to decide whether an offensive attack is the correct operational mode at a garden apartment?
10. What access problems can be found at garden apartments in general? What specific problems can occur at garden apartments in your response district?
11. What problems are associated with the newer type of row houses or town houses that have garages built under the houses?
12. List some of the differences between row houses and town houses.
13. Describe truck company operations at row house or town house fires.
14. Discuss ladder placement at row house fires. Can some of these methods be used at garden apartment fires? Explain.
15. What problems in relation to life safety will firefighters confront at a hotel or motel fire?
16. What exterior access problems could firefighters confront at a hotel or motel fire?
17. What actions are required of firefighters to restore a hotel or motel to the condition it was in prior to a fire?
18. What is the benefit of marking vacant buildings to categorize their structural stability? What are the drawbacks?
19. Discuss why vacant buildings should be searched for occupants.
20. Discuss the pros and cons of an offensive attack in a vacant building.
21. How do large-scale building renovations affect firefighting?
22. What problems are associated with lightweight roofs built over a timber truss roof? Are there any benefits? If so, to whom?
23. What are the signs of a partition fire?
24. At a wildland urban interface fire, what are the three categories that structures can be placed into when considering risk versus gain?
25. What are the factors that must be considered when deciding whether a property can be saved at a wildland urban interface fire?
26. List the firefighter safety actions at a wildland urban interface fire.
27. Select one of the occupancies presented in this chapter, and in reviewing the size-up factors, add five additional potential problems that could be present.

Suggested Reading, References, or Standards for Additional Information

Rahn, M. 2010. "Initial Attack Effectiveness: Final Report." *Wildfire Research Report* 2: https://www.iaff.org/10news/pdfs/cdffirereport.pdf.

US Fire Administration (USFA)

Jennings, C. "Nine Elderly Fire Victims in Residential Hotel." Technical Report 041.

"Civilian Fire Fatalities in Residential Buildings (2011-2013)." *Topical Fire Report Series* 16, no. 2 (July 2015): http://www.usfa.fema.gov/downloads/pdf/statistics/v16i2.pdf.

"One- and Two-Building Residential Basement Fires (2010-2012)." *Topical Fire Report Series* 15, no. 10 (March 2015): http://www.usfa.fema.gov/downloads/pdf/statistics/v15i10.pdf.

"Vacant Residential Building Fires (2010-2012)." *Topical Fire Report Series* 15, no. 11 (March 2015): http://www.usfa.fema.gov/downloads/pdf/statistics/v15i11.pdf.

Related Courses Presented by the National Fire Academy, Emmitsburg, Maryland

Command and Control Decision Making at Multiple Alarm Incidents

Strategy and Tactics for Initial Company Operations

Cooperative Leadership in Wildland Urban Interface Operations

Wildland Urban Interface Firefighting for the Structural Company Officer

Command and Control of Wildland Urban Interface Fire Operations for the Structural Chief Officer

Wildland Urban Interface: Fire-Adapted Communities—Introduction and Leadership

Wildland Urban Interface: Fire-Adapted Communities—Strategies for Developing a Fire-Adapted Community

Wildland Urban Interface: Fire-Adapted Communities—Developing a Community Wildfire Protection Plan

8

Health Care and High-Risk Populations

> ❝I put higher expectations on myself than anyone else ever could. I expect a perfect game. I keep working on everything. You want that game where you have no turnovers and where your man doesn't score. It may never happen, but that's what I want every night. For great players the game is easy. I want to get to that point. ❞
>
> —Grant Hill, NBA player

Church fires are difficult to control and often cause the destruction of the building. *Used with permission of Joseph Hoffman.*

KEY TERMS

OBJECTIVES

Upon completion of this chapter, the reader should be able to:

- Understand the special occupancies discussed and the types of fires that may confront firefighters.
- Identify pertinent characteristics of special occupancies.
- Recognize the 13 points of size-up that pertain to these special occupancies.
- Recognize the strategic considerations for these special occupancies.
- Understand the incident management considerations of special occupancies.

Chapter 8 discusses health care and high-risk occupancies. Hospitals, nursing homes, assisted living facilities, penal institutions, and schools have occupants who typically cannot assist themselves should a fire occur in these facilities. The elderly and the very young will require assistance from firefighters should a fire occur. Houses of worship and public assembly buildings, when occupied, can have a large number of occupants who may not be familiar with the building.

Mass casualty incidents will tax every fire and emergency medical system in their quest to save lives and minimize injuries.

School incidents include firefighters' responsibilities at violent situations created by individuals that the police refer to as *active shooters*, meaning those who have shot, killed, and injured students, both at schools and on college campuses. Active shooter incidents now have been extended to other venues, including the workplace, movie theatres, and other locales.

At the end of each section are size-up factors as well as strategic goals and tactical priorities that need to be taken into account when confronted with fires or incidents in these occupancies. These lists are not meant to be all-inclusive. Other factors can and will exist that must also be considered.

The strategic goals and tactical priorities are for both offensive and defensive attacks with suggested considerations for implementing an incident management system to meet the potential problems. The strategic goals and tactical priorities that are noted are in response to some of the potential problems that may occur. The suggested incident management positions do not mean that they must be used or that using them is the only correct method. These are initial positions that should be considered. Depending on the problems presented by each individual situation, the Incident Commander can implement these or other incident management positions to deal with the problems.

The mass casualty incidents contain suggested size-up factors, strategic considerations, and incident management system/solutions.

Hospitals

Hospital fires will present numerous challenges to ensure the safety of the occupants. Patients encompass all age groups and are either sick or in need of treatment, and they rely on the doctors, nurses, and hospital staff for their well-being. The fact that these structures are occupied around the clock and contain a security staff is a benefit that should allow for early discovery of a fire.

OCCUPANCY

hospitals ■ Range from full-service facilities handling hundreds of patients daily, to those with a few beds and limited care capabilities.

Hospitals range from full-service facilities handling hundreds of patients daily, to those with a few beds and limited care capabilities. Specialty hospitals include pediatric facilities and those specializing in mental health, cardiac, and orthopedics.

Hospitals can be comparable to a small city. The structure may be a high-rise building with a below- or above-ground multilevel parking garage. The hospital may be a single structure or a group of buildings housing patients and numerous support activities.

Parking garages and interconnected buildings can limit access, which may impede apparatus placement. In some situations the hospital may be part of a sprawling complex of buildings. In addition to patient rooms, emergency rooms, and operating rooms, there can be auditoriums, classrooms, laboratories, carpentry shops, and large heating plants. Loading docks may be receiving supplies on a daily basis, while laundry facilities can involve a large operation, and kitchens can be quite extensive.

PREPLANS

Hospitals are required to have emergency operating plans (EOP). These plans describe actions for typical problems that a hospital could encounter. The hospital's goal is continuity of operation in a safe manner. Likewise, firefighters attempt to handle an incident with minimal disruption. By working together, the fire department and hospital staff can be successful in mitigating emergencies.

The hospital's EOP should be reviewed by the fire department when it prepares the preplans. In addition to the basic information, the preplan should include the various hospital codes that are used. These codes identify anticipated problems and actions required to mitigate them.

Once preplans are developed, drills at the hospital will determine how effective the written plans are to handle an emergency. Drills allow responding units to gain familiarity with the building, the plan, and the hospital staff.

Fire drills should be scheduled monthly. They are intended to educate the hospital staff on proper procedures. They should be monitored at least semiannually by the local fire department.

HOSPITAL CODES

The hospital's EOP should contain an alerting system for emergencies. They are sometimes called *codes* or *operations*. These alerts are transmitted to the staff as announcements over the public address systems (PASs). They are used to notify them of current or impending problems while not exciting the patients and visitors within the hospital.

An example is the common use of the term *Code Blue* to indicate a medical emergency. This may also be referred to as Operation Blue or Blue Alert. Each code has a basic set of instructions that the hospital employees will attempt to accomplish. The following is a sampling of codes and abbreviated actions or information.

- *Code Red, Fire:* The critical factor is the safety of the patients. If possible, the patient(s) should be removed from the fire area. Sound the alarm if it has not already been activated and close windows and doors to the fire area. If you are sufficiently trained, extinguish the fire with a fire extinguisher. Stuff wet towels, sheets, or blankets under doors in nearby rooms to prevent smoke from entering them. Calm the patients and await the arrival of the fire department.

- *Code Blue, Adult medical emergency:* Personnel assigned to the medical teams needed to treat the patient will report to the announced area.

- *Code White, Pediatric medical emergency:* Personnel assigned to the medical teams needed to treat the patient will report to the announced area.

- *Code Amber, Infant or child abduction:* This may be preceded by an alarm activation signaling the abduction. Nurseries in many hospitals place an electronic bracelet around an infant's ankle. This will sound an alarm if a child is taken from the nursery and through the alarmed doorways. If an infant is missing, an immediate head count must be made and security alerted. Security must then be assigned to the hospital's outer perimeter and police notified. Recognize that this is a crime scene. Fire alarm systems may be activated by the abductors to add confusion to the kidnapping in their attempt to exit the hospital without detection. The activation of a fire alarm will unlock some doors that are normally kept locked.

- *Code Yellow, Bomb threat:* A bomb threat should be handled as a police incident until the bomb has detonated, in which case the fire department will attempt to mitigate the problems found. Hospital personnel can assist in searching for the suspected bomb. No radio transmissions, cell phone usage, or beeper or pager operation should take place in the danger area since this could possibly trigger an explosive device.

- *Code Gray, Security emergency/patient elopement:* The disappearance of a patient with dementia or Alzheimer's will alert hospital security to check the hospital proper and outer areas. Firefighters may be called to assist in the search.

- *Code Silver, Hostage situation:* A hostage threat is a police incident that could require support from the fire department. Under no circumstances should a firefighter be placed in areas of danger. The placement of ladders where there is no threat of harm to the firefighters can be performed, but the climbing of the ladders must be performed by police personnel.

- *Code Orange, Hazardous materials situation where decontamination is needed:* A hazardous materials incident occurring at the hospital requires notifying the local fire department and the hazardous materials team. The area involved must be cordoned off and the patients and hospital staff protected. Individuals requiring decontamination should not be allowed to enter the hospital until decontamination has been performed.

- *Code Triage, External disaster situation:* This is for a situation outside of the hospital that could require a large involvement by the hospital, such as a mass casualty incident.

- *Code Clear or Condition Green:* The situation has been resolved.

These codes can vary from region to region, or within individual cities. Part of each fire department's preplan should include a list of the codes/operations/alerts for each site unless it is known that all hospitals in your response district conform to the same codes. In this case, one set of codes will suffice. (In Philadelphia some hospitals in the same network differentiate between the terminology and the color coding system.) Some progressive states have specified systems that the hospitals must institute.

In certain areas of hospitals, nursing homes, and other treatment facilities, door locks may be installed that prevent egress under normal circumstances. Electric locking systems, including electromechanical locking systems and electromagnetic locking systems, by code are permitted to be locked in the means of egress in certain occupancies where the clinical needs of persons receiving care require their containment. Maternity wards may have infant abduction systems that use a sensor to activate a delayed egress lock when a baby is taken near the exit. Though many locking systems need to be released during certain events, this may not apply to maternity or obstetric care areas. Check with local officials to clarify requirements in your response area.

HOSPITAL STAFF

The actions of the hospital staff are one of the most important factors concerning the life safety of the patients. Immediate notification of the hospital security for an emergency can allow early notification of the staff via the PAS and outside agencies (fire and police) that need to respond. The hospital staff should be trained and their incident management system should interact with fire and police.

In the event of a fire in the incipient stages, the occupants of the fire room should be removed and the door to the room closed. The nurses and staff members must ensure that doors to other patient rooms are closed. The patients in nearby rooms should be safe if they remain in their rooms. The fire department on their arrival will give further direction to the hospital staff on how they can best assist.

LABORATORIES

Laboratories may be located in various areas of the hospital. Since the laboratories may be near patient rooms, a fire or chemical release can threaten them.

Firefighters entering laboratories can be exposed to biohazards and radioisotopes. Doors to these rooms should be marked to warn those entering the room. A common marking is "Laboratory—Potentially Hazardous Substances." Other signs may warn of a hazard peculiar to the materials used within the lab. Firefighters should read any postings on doors before entering to ensure that proper personal protection is in place.

OXYGEN AND MEDICAL GASES

Oxygen, nitrogen, and other medical gases may originate in exterior bulk storage tanks and be piped throughout the hospital. The piping should be labeled and color-coded. Remote emergency shutoff valves should be marked to indicate the areas or rooms that they supply. Staff training should emphasize the importance of shutting down these valves should a fire occur within a patient's room. Realize that portable oxygen cylinders may be in use throughout the hospital, which can compound a situation if exposed to fire conditions.

MAGNETIC RESONANCE IMAGING (MRI)

The MRI is a diagnostic tool used for testing. It allows trained personnel to scan parts of the body to obtain precise pictures of soft body tissue. The technology combines the use of radio frequency transmissions with the unit's ability to generate and sustain high and uniform magnetic fields. In order to accomplish this superconductivity, a superconductive magnet is used, which operates at temperatures usually below −255 degrees Celsius. This is accomplished by immersing the system in liquid helium. These extremely low temperatures eliminate electrical resistance so that their magnetic fields persist indefinitely.

The greatest danger for patients, hospital staff, and firefighters is the tremendous power of the magnet. If ferromagnetic objects are allowed to enter the room containing the MRI machine, serious problems can occur. Firefighters can be endangered with their personal protective ensemble. Self-contained breathing apparatus (SCBA) and most firefighting hand tools will contain ferrous metals that will be forcibly pulled from their hands, or in the case of an SCBA, the firefighter can be pulled into the MRI machine.

Doors to the rooms containing MRI machines will be well marked and have warning signs posted. Firefighters must ensure that before entering any room they read and comply with the warnings.

To reduce the strong magnetism of these scanners, they are equipped with quench buttons that expel the liquid helium that is used to cool the magnets. To shut down the magnet, the helium must be vented to the exterior—a process that takes several minutes and contains certain hazards. The helium when vented to the exterior erupts in a frigid blast as it is rapidly heated and expands 760 times. Anyone exposed to the sudden blast on the building's exterior would be in danger.

Operational considerations when working or performing investigations at hospital fires that contain MRI equipment:

- Do not enter the scan room with any ferromagnetic items: hand tools, walkie-talkies, or jewelry.
- Ensure that firefighters do not wander into the scan room.
- Utilize the technicians who operate the MRI for information on its operation.
- Unless there is an immediate life safety factor within the scan room, quenching should not be necessary.
- Radio communications are not possible in the scan room due to the copper sheathing used to prevent interference from outside radio frequencies.

HOSPITAL FIRES

Investigations of past deadly hospital fires found common problems:

- Fire that spread through open doors leading to the corridors exposed occupants of both nearby and distant rooms to the products of combustion.
- Combustible materials in corridors caused rapid spread of fire.
- Delayed notification of the fire department compounded these fires once they started.

To correct these factors, the following code changes were recognized:

- The need for sprinklers in all hospitals
- The importance of automatic smoke-proof doors to segregate floor areas and contain smoke to a limited area
- The need for early notification of the fire department
- The value of self-closing doors

FIGURE 8-1A A tank truck carrying 6,000 gallons of milk lost control and drove through an exterior wall of a medical center into a room housing an MRI unit. *Used with permission of Randy Padfield, Chief, Cumberland County, PA, Technical Rescue Operations Team.*

FIGURE 8-1B When a firefighter entered the room containing an MRI unit, his metal halligan tool was ripped from his hand and attached to the powerful magnet of the MRI. *Used with permission of Randy Padfield, Chief, Cumberland County, PA, Technical Rescue Operations Team.*

These changes were incorporated into the National Fire Protection Association's Life Safety Code 101. To ensure that hospitals adopted these fire safety requirements, it was mandated that to receive reimbursement under Medicare and Medicaid these facilities had to comply with that code. The use and enforcement of these codes has drastically reduced fire deaths in hospitals.

PROTECTING THE PATIENT

Even with the changes made to the Life Safety Code, hospital fires still occur. A well-designed sprinkler system should quickly control and extinguish an incipient fire. Exceptions would be a sprinkler system damaged by an explosion, a fire occurring in a concealed area not protected by a sprinkler, or a fire that is not producing sufficient heat to activate the sprinkler system.

Smoke-proof doors should automatically close once an alarm system is activated. This will contain the horizontal spread of smoke and allow for the lateral evacuation of patients to areas of safe haven without having to move them to other floors.

Even with these safeguards, there are certain locations within hospitals that will not be able to evacuate their patients quickly. These include operating rooms, recovery rooms, intensive care, and critical care units.

EVACUATION

Evacuation/relocation of a hospital is a major endeavor. Since patients are often unable to help themselves, this places a heavy demand upon the responders and hospital personnel. The ratio of rescuers to patients can be as great as 2 to 1.

Difficult decisions will need to be made on what resources should be assigned to fight the fire and which should assist in the evacuation and rescue efforts. Keep in mind that fire extinguishment will solve most problems. The rapid control and extinguishment will reduce the heat and smoke. It will stop any further degradation of the structure and enable those firefighters who have been involved in the fire attack to be released to assist in the evacuation and the medical needs of the patients.

Likewise, the Incident Commander (IC) should consider assigning members to facilitate adequate ventilation to relieve pent-up heat and smoke. Ventilation considerations include utilizing the heating, ventilation, and air-conditioning (HVAC) systems to exhaust smoke to the exterior and prevent it from traveling to unaffected areas.

Evacuation of intensive care units, critical care units, nurseries, etc., can involve a greater ratio of rescuers to patients than other areas. Conversely, pediatric patients may be grouped onto one bed or gurney due to their small size, and one nurse, firefighter, or attendant can move a number of patients. Naturally this must consider the medical condition of the patients and what life-sustaining methods each patient requires. These lifesaving tools can include IVs with medications added, oxygen, and heart-monitoring machines.

Hospital beds should be on wheels and doorways and corridors should be of sufficient size to allow ready access to move patient beds horizontally to safe havens on the nonfire side of the smoke-proof doors. Controlling the smoke-proof doors is important. Heavy smoke conditions in corridors near the fire area may endanger patients taken into these smoke-filled hallways. It may be best to keep these patients in their rooms to determine if they do need to be moved immediately, or whether to wait until the smoke conditions in the hallway abate and it is safe to move them.

RESTRAINED PATIENTS

Mental health facilities may have patients locked in wards or individual rooms to protect the patient and other patients and hospital staff. Similarly, prisoner wards will have high levels of security. Assistance from security personnel in evacuating patients from these locations will be necessary.

Restraints may be found on some patients either to prevent harm to themselves or to avoid the dislodging of tubes or other medical equipment. Under fire conditions the patient will need to be released from these restraints by cutting them with a knife or scissors.

CALLING FOR HELP

When it is determined that a fire exists in a hospital, there should be an early request for sufficient resources. The first-due officer should not hesitate in calling for these resources since minutes can matter. A working fire in a hospital will require an adequate number of firefighters to accomplish the many tasks that must be performed simultaneously. Some tasks, such as evacuation and accountability of occupants, can be accomplished by the hospital staff. Firefighters will be needed to attack the fire, for ventilation, evacuation, rescue, and protecting the patients in place.

COMMAND AND CONTROL

Command and control at a hospital fire will demand a balance between fire control and the patients' medical needs. The relocation of patients will necessitate moving their records or charts to ensure continuity of treatment. The interaction between the fire department representative or emergency medical services (EMS) representative and the hospital staff can be facilitated by having a member of the hospital's staff at the Command Post (CP). Hospital concerns can then be shared with Command.

There may be a need to transfer patients to other hospitals to ensure proper care. The hospital staff must decide how many patients will need to be transferred and what level of care will be required. It must be determined exactly how many and what type of patients the receiving hospitals can effectively handle. Any type of evacuation or relocation of patients would require interagency support. This can involve EMS, health department personnel, private ambulances, and private bus companies.

US STRUCTURE FIRES IN HEALTH CARE PROPERTIES: FACT SHEET

According to the National Fire Protection Association, US fire departments responded to an estimated average of 6,240 structure fires in health care properties per year in 2006-2010. These fires caused an annual average of:

- 6 civilian deaths
- 171 civilian injuries
- $52 million in property damage

Additionally, it was found that:

- Most fires in these properties are small. Fire spread beyond the room of origin in only 4 percent of health care fires.
- Sprinklers were present in 55 percent of reported health care occupancy fires. The direct property damage per fire was 61 percent lower in properties with wet pipe sprinklers than in properties with no automatic extinguishing equipment.
- Leading causes, fire spread, and fire circumstances varied by specific health care occupancy.

Source: NFPA, "Fire in Health Care Facilities"

FIREFIGHTING

The first-arriving firefighters must determine the exact location of the fire, those immediately threatened, and the best method of attacking the fire. By checking the alarm panel, they can identify the areas from which the alarm is being transmitted. Hose-line placement should consider the closest fire stairs or standpipe location from which to attach the hose-line. Selection of the nearest stairway or location may be found by viewing a floor plot plan and identifying the fire's location by room number. In multistory structures, it is not uncommon for hospital floors to have diverse layouts, and the utilization of a floor plan will facilitate the selection of the correct location from which to attack the fire. Using the closest stairway or location to the fire area will minimize the number of smoke-proof doors that will be violated, maximizing the protection of patients in adjoining areas. (If possible, firefighters should avoid stretching hose-line through these doors to prevent spreading smoke to other areas.) Should the building contain individual standpipe risers, knowing which stairway is being used will allow the pressurization of the correct standpipe. Using the nearest fire tower to attack the fire will help contain the smoke and heat to as small an area as possible, drastically reducing the number of occupants exposed and needing to be moved from the fire area.

The supervisor of the division or group assigned to the fire area must check with the hospital staff about the number of individuals still in the fire area or unaccounted for. This will enable that officer to direct firefighting and search and rescue operations. Another determination that must be made is whether patients currently being protected in place are safe, or whether they need to be moved. Movement of any patient has to consider the actual threat posed by the fire and smoke and the potential medical problems should he or she be moved.

OPERATING ROOMS

Fires that occur near operating rooms can further complicate patient safety, evacuation, and ventilation. It may not be medically possible to abort an ongoing operation. This will place the operating team in jeopardy as the firefighters attempt to control the fire while protecting them.

A fire in an operating room can be supplied with fuel from within the operating room and accelerated by the high oxygen content in the room. Typically a fire can be slowed in other situations by the generation of smoke, reducing the oxygen content in the room. In operating rooms, the use of oxygen in conjunction with the ventilation system will allow a free-burning fire.

There may be additional problems if cylinders of oxygen or surgical gases are within the fire room. A fire impinging on these cylinders can cause their relief valves to operate, radically enriching the atmosphere and accelerating the fire.

HIGH-ENERGY ELECTROSURGICAL UNITS

High-energy electrosurgical units (ESU) and lasers are common tools used in operating rooms for both cauterizing and cutting. The ESU generates radio frequency energy that arcs to the tissue with enough energy to rupture the cell (cutting) or just enough energy to seal the cell (coagulation). It does the work of a knife while minimizing bleeding. The laser focuses light energy to heat up tissue for cutting or sealing. Lasers are typically more expensive and tend to be used for specialty surgical applications.

Most surgical fires involve electrosurgery and initiate when the ESU or laser is accidentally activated by a hand lever or foot switch in oxygen-enriched environments. Accidental activation can cause burns to the patient or ignite combustible material on or near the patient. Severe burns and death are associated with these types of fires. Surgical fires, though rare, are devastating when they occur. It is estimated that 50 to 100 of these fires occur annually in the United States, resulting in up to 20 serious injuries and one or two patient deaths. The areas of the body most often affected are the face, neck, chest, and abdomen.

The greatest concern with an operating room fire is that the patient is defenseless. If the patient is supplied with oxygen, it will be through plastic tubing, which can be quickly consumed in a flash fire, causing a rapid acceleration of the fire. Oxygen can be trapped and collect under surgical drapes in the head and neck area. A fire involving these materials can readily ignite due to the elevated oxygen level. The oxygen supply and exhalation of the patient creates an oxygen-enriched atmosphere, which becomes a ready source of rapid fire development and a deadly threat to the patient.

Nitrous oxide may be used as an anesthetic, which can increase the fire's intensity. Surgical drapes used in operating rooms, as well as surgical gowns used by the surgical staff and patients, are extremely flammable. They are designed to shed liquids, thereby inhibiting extinguishment by water should a fire occur. Other sources of fuel are alcohol, foam pads, cotton sheets, plastic drapes, trash bags, prep sponges, and hampers with sheets or blankets.

These fires, though a severe threat to the patient, may not generate sufficient heat to activate the sprinkler systems within the operating rooms. It is recommended that operating rooms be equipped with carbon dioxide fire extinguishers to control fires and minimize contamination. The problem is that if the sprinkler does not activate due to the low heat production, the time required to activate the extinguisher can be sufficient to produce severe injury to the patient.

VENTILATION

The only windows that may be present in an operating room will be in the door to the room or windows that look onto the common hallway. These windows will be of minimal value for ventilation purposes. Ventilation through the HVAC system will often be the only feasible method of smoke removal from an operating room. Should the HVAC system be shut down, ventilation can become a major problem.

SIZE-UP FACTORS FOR HOSPITALS

Water
- There must be an adequate water supply for both an offensive and, if necessary, a defensive attack.
- Hospitals located in rural areas may have a limited water supply, necessitating a water tender operation.

Area
- These buildings will vary in size.
- Large common areas will permit fire and smoke to quickly spread.
- Fire entering a hallway will threaten all occupants located adjacent to that hall area.

Life Hazard
- Severe; the occupants will often require assistance to evacuate.
- Most patients and visitors will not know what to do if an alarm is sounded.
- There may be visitors within the building at the time of the alarm who are unfamiliar with the exit locations.
- The visitors and patients will attempt to exit the building in their usual way regardless of smoke and fire conditions. This can typically be via an elevator.
- Though fire drills are routinely held, the staff will need to direct visitors to the proper exits.
- Exit doors to fire stairs may normally be locked, and unlock upon activation of the fire alarm.
- Fires may be supplied by oxygen or other medical gases that can accelerate the fire.
- Emergency lighting may reduce visibility.
- Patients may need continuous medication and oxygen administered.
- Patients' medical charts should be kept with them when evacuated.
- If an evacuation is ordered, patients may need to be transported to other hospitals.
- Firefighters may need to release patients who are restrained to beds or chairs.
- Thermal imaging cameras (TICs) can assist in the primary and secondary searches.

Location, Extent
- Locating the fire must include using the building's alarm system or indicator panels to pinpoint its location.
- HVAC systems can spread smoke throughout a building. If it cannot be determined if the system is assisting the fire department's operations, it should be shut down.
- As firefighters enter the fire floor, they should ensure that smoke-proof doors are closed.
- Thermal imaging can assist in checking for hidden fire in walls, floors, and ceilings.

Apparatus, Personnel
- Call for help early. A large number of resources will be needed for fire control; to achieve adequate ventilation; and accomplish search, rescue, and evacuation.
- A decision on how to fight the fire and assist in the evacuation will need to be based on the available resources.
- Determine the best method of utilizing the staff to assist in evacuation and obtaining medical charts for relocated patients.

Construction/Collapse
- All types of building construction can be found, including fire-resistive, noncombustible, ordinary, frame, and mixed construction.
- Older hospitals may contain renovations that can affect fire spread.

- There may only be one stairway for firefighting and evacuation, or there may be multiple stairs whereby the fire department will need to determine which to use for firefighting and which for evacuation.

Exposures
- Internal exposures can be severe.
- External exposures could involve adjacent buildings, bulk storage tanks of oxygen, etc.

Weather
- Cold weather will affect occupants evacuated from the building.
- Extremely cold weather may become life-threatening and an evacuation site must be found immediately. Consider buses as a temporary measure to house the patients if ambulatory.

Auxiliary Appliances
- Sprinklers are required in these occupancies, though some select areas may not be protected.
- Standpipes can be anticipated.
- Kitchens may contain special extinguishing systems.
- Portable carbon dioxide extinguishers should be used in operating rooms to control and extinguish fires to minimize contamination.

Special Matters
- Employees can assist the fire department in evacuation.
- Hospital employees typically receive incident command system (ICS) training, which allows them to interact with responders.
- Combustible materials and oxygen/medical gases can support an operating room fire and accelerate combustion.
- Heliports may be on the roof or grounds of the hospital. Procedures should be in place for takeoffs and landings.
- PASs should be made available to the fire department if requested.

Height
- Hospitals can be from one story to a high-rise.

Occupancy
- Congestion in hallways and on stairs can occur when occupants are utilizing canes, crutches, walkers, wheelchairs, etc.
- The slow pace of those exiting the building can compound the evacuation problem.

Time
- The critical time period is generally at night. There is reduced staffing and the occupants are asleep, leading to alarm systems that may not be heard and disorientation on the part of the occupants.

CONSIDERATIONS FOR FIRES IN HOSPITALS

Strategic Goals and Tactical Priorities for an Offensive Attack
- Evacuate and perform search and rescue for occupants.
- Consider lateral evacuation.
- Consider protecting occupants in place.
- Consult with the hospital staff and interact with them throughout the incident.
- Call for a sufficient amount of resources (fire and EMS).
- Ensure that hose-lines are not stretched through doorways being used as exits by occupants.
- Locate and confine the fire to the area of origin if possible.
- Hose-lines of 1½-inches or 1¾-inches may be ineffective; 2½-inch hose-line may be needed for large-area fires.
- Support sprinkler systems with a secondary water supply.
- Ventilation in conjunction with interior hose-line crews can act as a lifesaving tool.

- Horizontal ventilation should be sufficient for a room and contents fire. For fires involving large areas, vertical ventilation may be necessary. All ventilation must be coordinated by the IC or the interior crews.
- Roof vents may be present over common areas; ensure that they are open.
- Ensure that doors leading to the roof from the stairways are open.
- If multiple stairways are present, designate one for firefighting and the others for evacuation.
- Try to utilize the HVAC systems to assist in removing smoke from the building.
- Before using positive-pressure ventilation (PPV), determine the exact location of the fire, the location for smoke to exit, and whether any occupants would be endangered.

Incident Management System Considerations/Solutions for an Offensive Attack
- Incident Commander
- Safety Officer
- Public Information Officer
- Rapid Intervention Crew(s)
- Operations (if needed)
- Staging
- Ventilation Group
- Search and Rescue/Evacuation Group
- Division 1, 2, 3, etc. (for multistoried building can be implemented for any floor where deemed necessary)
- Logistics (if needed to assist in water supply and obtaining protection or sheltering for those evacuated)
- Medical Group or Branch (to set up triage, treatment, transportation, morgue, etc.)

Strategic Goals and Tactical Priorities for a Defensive Attack
- If changing from an offensive to a defensive attack, ensure that a personnel accountability report (PAR) is taken for accountability purposes.
- Set up and maintain a collapse/safety zone.
- Utilize master streams.
- Protect exposures if necessary.

Incident Management System Considerations/Solutions for a Defensive Attack
- Incident Commander
- Safety Officer
- Public Information Officer
- Operations (if needed)
- Staging
- Logistics (for water supply and/or sheltering of evacuees if needed)
- Divisions on the exterior on Sides A, B, C, D
- Exposure Divisions (if needed)
- Medical Group or Branch

Nursing Homes and Assisted Living Facilities

Most communities have residences that care for the elderly. They may be **nursing homes** or **assisted living facilities**. These types of occupancies will present problems should a fire occur. The assisted living residence is designed to allow the occupants the privacy of their own living quarters. There is a staff for housekeeping and the dispensing of medication at the proper time and dosage. The number of staff assigned to assist the occupants is usually minimal.

The nursing home requires a much greater hands-on approach for the occupants, who are often bedridden or lack mobility. Individual feeding may be necessary and dementia or advanced Alzheimer's disease is possible. They may be dependent on oxygen or intravenous medication.

nursing home ▪ A facility in which the occupants are often bedridden, lack mobility, and may suffer from dementia or Alzheimer's disease.

assisted living facilities ▪ Designed to allow occupants the privacy of their own living quarters, and normally have a minimal staff.

Fires in nursing homes and assisted living facilities will tax every fire department. Most of these fires are small and are controlled very quickly, without major damage or injuries. The ability to have a sufficient number of firefighters at the incident scene can mean the difference between life and death. The labor intensiveness of these assignments will necessitate the rotation of personnel due to the firefighting demands and the exertion necessary to move a large number of occupants.

For firefighters to be successful at a fire in a nursing home or assisted living facility requires:

- Preplanning
- Fire prevention education
- Staff training
- Properly operating building protective systems
- Early fire department notification
- Implementation of proper strategic and tactical considerations

An initial consideration at a nursing home fire is that all occupants needing evacuation will have to be physically moved by the nursing home staff or the firefighters. Evacuation of residents in assisted living facilities under fire conditions may also require the hands-on removal by staff and firefighters. Since these facilities may have a reduced number of staff in comparison to nursing homes, a greater response of firefighters for evacuation will be required.

> A nursing home fire in Hartford, Connecticut, took the lives of 16 residents in February 2003. In September 2003, in Nashville, Tennessee, 15 residents died in a nursing home fire. Neither facility had sprinklers.

REGULATION

Federal and state agencies regulate and license these facilities. Insurance providers routinely conduct inspections to ensure that they are conforming to applicable standards. Yet, illegal nursing homes and assisted living facilities can be found that masquerade as rooming or boarding homes.

Firefighters will be facing a dangerous and potentially deadly situation if confronted with a fire in an illegal facility.

PREPLANNING

Critical considerations for fighting a fire in a nursing home or assisted living facility should be made in the preplanning stages. This allows careful analysis of all facets of the structure, systems, and most important, the life safety of the occupants. Interaction between the fire department and the facility staff can mean the difference between success and failure.

The preplan should be comprehensive enough to contain the data that will be needed. Basic information would include:

- Construction type
- Whether the building is fully or partially sprinklered; if partially, what areas?
- Whether there are any special extinguishing systems in kitchens or other areas
- The presence, location, and type of standpipe system
- Number of staff on duty during the day and night
- Needed fire flow
- Location of nearby hydrants or other available water supplies
- Location of utility and oxygen shutoffs
- Special attention required for the occupants due to mental or physical incapacities
- Type of HVAC systems; can they be utilized for smoke removal?
- Telephone numbers of emergency contact personnel
- Potential for contagious disease contact
- Necessity of any special protection against bodily fluids

CONSTRUCTION

The types of construction found in nursing homes and assisted living facilities are as diversified as are communities. These buildings can be built of fire-resistive material and be fully sprinklered, or of ordinary, frame, or mixed construction with minimal protective systems. It is common practice in many areas to convert large single-family dwellings for this use.

Construction type refers to the basic structural elements and not the contents and furnishings. Since the construction types vary, so do the interior layouts. The layout of newer buildings will permit ease of movement of the occupants during everyday activities and contain wide doorways to allow firefighters to evacuate the residents under fire conditions. The layout will also undoubtedly contain proper compartmentation.

Older, converted buildings may be designed in a dormitory style with rooms or apartments containing multiple occupants. This layout allows the carelessness of one occupant to threaten the lives of his or her roommates.

Converted structures can contain vertical voids that carry the utilities upward through the structure. These voids are seldom fire-stopped, and should a fire extend into these spaces, it will threaten all areas above.

PROTECTIVE SYSTEMS

Nursing homes and assisted living facilities may contain sprinklers, standpipes, alarm systems, and self-closing doors on rooms. There also may be doors within corridors that will close automatically when an alarm is activated. This will prevent the movement of smoke throughout a floor area by compartmentizing the floor.

Firefighters should ensure that the sprinklers are operating and that they have a sufficient water supply. This may require the responding firefighters to pressurize the system by utilizing their apparatus and securing a separate water supply. It is important to identify if the entire building is sprinklered or only partial protection is afforded. In some instances, only public areas and basements may receive this protection.

Coded alarm systems can pinpoint the problem area. The bells or klaxons are sounded in a numerical pattern; when counted, these can identify on a chart the exact location of the alarm. The next step for responders after identifying the location is to refer to a floor plan of the facility to determine the best method of accessing the area. It is imperative that firefighters check alarm system panels in lobby areas to pinpoint the exact location of the alarm to assist them in determining the best stairway or access point to attack the fire while protecting the occupants in other areas.

LIFE SAFETY

As with most fires, one of the best methods of saving lives is by an aggressive attack on the fire to control and extinguish it in the incipient stage. To successfully accomplish this requires:

- A thorough knowledge of the building's layout, which can be achieved by prior inspections and the development of a comprehensive preplan
- Knowing who is in the building at the time of the fire, their physical condition, and their approximate location
- A response of a sufficient number of firefighters to permit fighting the fire and assisting staff in removing occupants who are in harm's way
- An aggressive attack on the fire initiated from the proper location/stairway for containment and extinguishment
- Coordination between the interior crews and the exterior ventilation crews
- An adequate water supply
- An understanding of the building's emergency evacuation plan

There is a tremendous potential for loss of life in nursing homes and assisted living facilities. The size of the building, the number of occupants, their age, lack of agility, and diminished mental capacity can all impact on being able to protect them under emergency scene conditions. The ambulatory often need the assistance of walkers or canes. Leg prostheses normally worn during the day may be removed at night and prevent those

Area
- These buildings will vary in size.
- Large common areas will permit a fire to spread quickly.
- Fire entering a hallway will threaten all occupants located in that floor area.

Life Hazard
- Severe; the occupants will often require assistance to evacuate.
- There will most likely be disorientation of the occupants.
- There may be visitors within the building at the time of the alarm who are unfamiliar with the exits.
- Exit doors may contain a locking mechanism to prevent occupants from wandering.
- Fast-moving fires will necessitate a prompt response and attack on the fire.
- Emergency lighting in the building may reduce visibility.
- Many occupants may be unaware of the sounding of the alarm system or not know what to do.
- The occupants will attempt to exit the building in their usual way regardless of smoke and fire conditions. This can typically be via an elevator.
- Patients who have been evacuated may require oxygen or medication.
- Immediate transportation to hospitals or other facilities may be required for some nursing home patients.
- Though fire drills are routinely held, the occupants may be oblivious to the alarm or forget what they should do.
- Patients may be restrained to beds or chairs and be unable to release themselves.
- TICs can assist in the primary and secondary searches.

Location, Extent
- Locating the fire must include using the building's alarm system or indicators to pinpoint the fire's location.
- HVAC systems can spread smoke throughout a building. If it cannot be determined if the system is assisting the fire department's operations, it should be shut down.
- HVAC systems may be capable of being reversed and assist in removal of smoke from the building.
- Thermal imaging can assist in checking for hidden fire in walls, floors, and ceilings.

Apparatus, Personnel
- Call for help early. A large number of resources will be needed to control these fires; achieve adequate ventilation; and accomplish search, rescue, and evacuation.
- A decision on how to fight the fire and assist in the evacuation will have to be made based upon the available resources.
- Determine the best method of utilizing the staff to assist in evacuation.

Construction/Collapse
- All types of building construction can be found including fire-resistive, ordinary, frame, and mixed construction.
- Renovations may be commonplace and can affect fire spread.
- There may only be one stairway for firefighting and evacuation, or there may be multiple stairs whereby the fire department will need to determine which to use for firefighting and which for evacuation.

Exposures
- Internal exposures will be severe.
- No specific external exposure problems.

Weather
- Cold weather will affect occupants evacuated from the building.
- Extremely cold weather may become life-threatening and an evacuation site to temporarily house the occupants must be found immediately. Consider a nearby school or government building.

Auxiliary Appliances
- Sprinklers will be present in some of these occupancies, though only selected areas may be protected.

- Standpipes can be expected in high-rise buildings.
- Kitchens may have special extinguishing systems over deep fryers.

Special Matters

- Employees trained in evacuation procedures can assist the fire department, but can be immediately overwhelmed.
- Combustible furnishings can contribute to a fast-burning fire and can generate large volumes of deadly smoke.
- Oxygen can accelerate combustion and is commonplace in many of these facilities.

Height

- Nursing homes and assisted living facilities can be from one story to a high-rise.

Occupancy

- Congestion in hallways and on stairs can occur when occupants are utilizing canes, walkers, or wheelchairs.
- Exit doors can become congested with people trying to leave the building, restricting egress. Their slow pace will compound the evacuation problem.
- Restoration of elevator service will assist in evacuation and the firefighting efforts.

Time

- The critical time period is generally at night. There is reduced staffing and the occupants are asleep, meaning alarm systems may not be heard and occupants may be disoriented.

CONSIDERATIONS FOR FIRES IN NURSING HOMES AND ASSISTED LIVING FACILITIES

Strategic Goals and Tactical Priorities for an Offensive Attack

- Evacuate and perform search and rescue for occupants.
- Consider protecting occupants in place.
- Consider lateral evacuation if practical.
- Ensure that all exits are unlocked and unobstructed.
- Recognize the potential for a mass casualty incident.
- Call for a sufficient amount of resources (fire and EMS).
- Ensure that hose-lines are not stretched through doorways being used as exits by occupants.
- Locate and confine the fire to the area of origin if possible.
- Hose-lines of 1½-inches or 1¾-inches may be ineffective; 2½-inch hose-line may be needed for large meeting rooms or auditoriums.
- Ventilation can act as a lifesaving tool.
- Horizontal ventilation should be sufficient for a room and contents fire. For fires involving large areas, vertical ventilation may be necessary.
- Roof vents may be present over common areas; see if they are open.
- Ensure that doors leading to the roof from stairways are open.
- If multiple stairways are present, designate one for firefighting and the others for evacuation.
- Try to utilize the HVAC systems to assist in removing smoke from the building.
- Before using PPV, determine the exact location of the fire, the location for smoke to exit, and whether any occupants would be endangered.

Incident Management System Considerations/Solutions for an Offensive Attack

- Incident Commander
- Safety Officer
- Public Information Officer
- Rapid Intervention Crew(s)
- Operations (if needed)
- Staging
- Ventilation Group
- Search and Rescue/Evacuation Group

- Division 1, 2, 3, etc. (for multistoried building can be implemented for any floor where deemed necessary)
- Logistics (if needed to assist in water supply and obtaining protection or sheltering for those evacuated)
- Medical Group or Branch (to set up triage, treatment, transportation, morgue, etc.)

Strategic Goals and Tactical Priorities for a Defensive Attack
- If changing from an offensive to a defensive attack, ensure that a PAR is taken for accountability purposes.
- Set up and maintain a collapse/safety zone.
- Utilize master streams.
- Protect exposures if necessary.

Incident Management System Considerations/Solutions for a Defensive Attack
- Incident Commander
- Safety Officer
- Public Information Officer
- Operations (if needed)
- Staging
- Logistics (for water supply and/or sheltering of evacuees if needed)
- Divisions on the exterior on Sides A, B, C, D
- Exposure Divisions (if needed)
- Medical Group or Branch

Mass Casualty Incidents (MCIs)

mass casualty incident (MCI) ■ Occurs when the number of patients exceeds the level of stabilization and care that can normally be provided in a fast and timely manner.

A **mass casualty incident (MCI)** can be the result of a manmade or natural disaster. MCIs are often associated with earthquakes, tornadoes, hurricanes, floods, fires, and other catastrophic events. The definition of an MCI can vary depending on the level of EMS that serves a community. An incident in a major city that involves several injuries would typically be considered a multiple casualty incident. Likewise, the same number of injured in an area with limited EMS could be defined as a MCI.

An MCI is one in which:

- The number of patients exceeds the level of stabilization and care that can normally be provided in a fast and timely manner.
- Responding units will not be sufficient in nature to handle the many injured.
- Nearby hospitals will be overloaded.
- The number of patients involved will be sufficient to require significant implementation of EMS personnel in the affected community.

Communities can classify MCIs at different levels.

- A Level 1 MCI is an incident that can be handled completely by the local agency with minimal or no outside response.
- A Level 2 MCI requires significant assistance from outside agencies. There would probably be a need to transport patients to hospitals outside of the area. In remote areas, medical evacuation (MEDEVAC) helicopters might be required to transport patients.

INCIDENT REQUIREMENTS

Whether a jurisdiction can successfully handle an MCI will depend on:

- Having and implementing a standard operational guideline (SOG) detailing the responsibilities and procedures required of personnel
- A good working relationship between area hospitals and responding agencies
- Having an alerting system in place that will notify the area hospitals of the potential of an MCI in progress and ongoing communications with those hospitals to continually update their ability to accept patients
- Implementing a command structure with modular expansion capabilities to address the unique problems that accompany an MCI

- A well-trained fire department that can initially treat patients at a medical emergency while interacting with EMS personnel
- Well-trained and disciplined EMS personnel who will interact with the local fire department
- An EMS command structure that is either a part of the fire department or has a close working relationship with it
- Simulated exercises at which responders can receive realistic training to prepare them for an MCI
- A mass casualty vehicle outfitted with medical equipment intended for use at an MCI

 ON SCENE

On May 12, 2015, an Amtrak Northeast Regional passenger train, traveling from Washington, DC, and bound for New York City, crashed in the Port Richmond neighborhood of Philadelphia. The train was carrying 238 passengers and a crew of five. Eight passengers were killed and more than 200 were injured. It was reported that the train was going in excess of 100 mph in a 50-mph zone of curved tracks.

The Philadelphia Fire Department initiated its mass casualty procedure, which proved to be highly effective in handling the overwhelming number of injured and dead. They were greatly assisted by interagency response and the availability of numerous hospitals both in the city and regionally, including more than 10 trauma centers.

Oddly, a previous accident had occurred at the same location in 1943, killing 79 and injuring 117. The 2015 crash was the deadliest on the Northeast Corridor since 1987, when 16 people died in a crash near Baltimore. (See Figure 8-2.)

A

B

FIGURE 8-2 A-B An Amtrak train reported to be traveling over 100 mph on a curve in a 50-mph zone crashed, killing eight passengers and injuring more than 200 others. *Used with permission of Greg Masi. These photos first appeared in Fire Engineering.*

STANDARD OPERATIONAL GUIDELINE

The local fire department, EMS agency, police, and area hospitals should develop a SOG that can be used as a blueprint for handling an MCI. The document should spell out the basic areas of responsibilities for each agency involved. It should stipulate how these agencies need to interact in the development of a command system to address the many and varied problems associated with an MCI.

An important part of any written document must include a system that will be in place to alert the area hospitals in the event of an MCI. It can be based on levels that can be alphabetical or numerically coded and are used to alert hospitals at different junctures:

Level A or Level 1: Alerts hospitals that an incident has occurred and there is the possibility of mass casualties. The affected hospital(s) can then activate the necessary personnel and equipment to handle the anticipated patients.

Level B or Level 2: Alerts hospitals that a confirmed MCI has occurred and the affected hospital(s) will be receiving patients.

Level C or Level 3: Alerts hospitals that an all-clear has been declared, meaning that the incident has been resolved or that all patients have been transported from the incident scene.

TRAUMA CENTERS

Trauma centers were created after it was recognized that patients suffering severe trauma needed immediate and often complex surgery in order to survive. That type of surgery was not previously available in most emergency rooms in the United States. A trauma center is a hospital that provides the best possible emergency care for traumatic injuries.

A trauma center must be able to meet a number of complex capabilities. These include a strong emergency medicine department with a well-trained nursing staff supervised by a highly qualified specialist, a team of trauma surgeons, the support of advanced diagnostic equipment that is immediately available, surgical specialists on duty, and a well-stocked emergency room. The trauma centers must also have a helipad for receiving patients via MEDEVAC helicopters. Additionally these centers must have well-developed contingency plans for disasters.

There are four levels of trauma centers in the United States:

- Level 1—Has a full range of specialists and equipment available at all times and must admit a minimum volume of severely injured patients annually.
- Level 2—Works in collaboration with a Level 1 center. It provides comprehensive trauma care and supplements the clinical expertise of a Level 1 institution. It too provides service at all times for all essential specialties, personnel, and equipment.
- Level 3—Does not have the same availability of specialists as Level 1 and 2 centers, but does have resources for the emergency resuscitation, surgery, and intensive care of most trauma patients. A Level 3 center has transfer agreements with Level 1 and/ or Level 2 trauma centers for the care of the most severe injuries.
- Level 4—Provides for the stabilization and treatment of severely injured patients in remote areas where no alternative care is available.

HAZARDOUS MATERIALS

MCIs that involve hazardous materials elevate the threat to the injured and the responders alike. There is an immediate need to identify the type of hazardous materials and take steps to mitigate their impact on the injured. This can include removing the injured from the immediate area and decontaminating when needed.

Responders must limit the handling of anyone who is contaminated with hazardous materials and ensure that their contamination is not allowed to spread to other victims who may not have been exposed, nor to the responders or their equipment. If a contaminated person is transported to a hospital, the problem will be compounded by spreading the hazardous materials to the ambulance that did the transportation and to the hospital that received the patient. Members of an ambulance crew who must transport a contaminated person should attempt to have that individual decontaminated prior to transport to a hospital. If a local hospital has the facilities for decontaminating a patient, the ambulance could be redirected to that hospital; otherwise the hazardous materials unit will need to perform the decontamination prior to the patient entering the hospital. (Decontamination is discussed further in Chapter 10 under Hazardous Materials and the Initial Responder, and Terrorism Incidents.)

When hazardous materials are involved in an incident, the IC or Operations Section Chief should consider establishing a Hazardous Materials Group or Branch to address the resultant problems that will be necessary to achieve mitigation, decontamination, and recovery.

TERRORISM-RELATED MCI

The potential of an MCI associated with terrorism is a strong possibility, as witnessed at the bombing of the Murrah Federal Building in Oklahoma City and the deadly occurrences at the World Trade Center in New York City.

The use of chemical or biological weapons in a terrorist attack can also result in an MCI. The use of these weapons can make it very difficult for the first-arriving fire department or EMS personnel to identify that a terrorist event has occurred. This lack of immediate identification can result in responders being exposed to, and affected by, these weapons.

Effective actions when dealing with chemical or biological weapons are early identification by the responders; proper use of protective equipment; setting up hot, warm, and cold zones; and timely decontamination of patients.

MASS CASUALTY VEHICLES

Mass casualty vehicles can be designed to carry equipment for handling conventional MCIs, such as cervical collars, spine boards, splints, gauze, tape, etc. They can also be utilized for responding to a terrorist attack causing an MCI. Specialized equipment can include nerve agent antidote kits that can be utilized for civilians and responders. These units can be motorized or the equipment can be stored in a trailer that is towed to the scene.

COMMUNICATIONS

To exercise command and control of an MCI, it will be necessary to have clear and accurate communications. This can be a very simple system that is utilized daily or one that needs to be implemented at the incident scene. A fire department that provides EMS should have few problems and will need only to assign specific channels for the various duties. Consideration should be given to:

- A command channel
- An operations channel
- A channel for transportation
- A logistics channel

When the fire department and EMS are separate entities, greater coordination will be required. Prior designation of specific radio frequencies for use between fire and EMS will allow a smooth overall operation. The use of multiple channels may require using state or countywide systems. If multiple channels are used, a communications plan should be created. This plan should specifically designate which frequencies are used by which functions.

Fire and EMS departments utilizing only one radio frequency will require strict radio discipline to avoid overloading the one channel and preventing necessary messages from being transmitted and/or received.

CRITICAL INCIDENT STRESS DEBRIEFING

The traumatic injuries witnessed at an MCI by the emergency responders can have a debilitating effect on them. All responders exposed to these scenes need to attend a critical incident stress debriefing. (See Chapter 11, Critical Incident Stress.)

SIMULATED EXERCISES

Live simulations are a tremendous tool to test the MCI components contained in a written procedure. A complete test of the system is usually not feasible with one exercise, but through a series of drills, the proficiency of all components can come together and be properly implemented when a real MCI occurs.

Some components that could be tested in an exercise are:

- The command and control procedures of all agencies that would normally respond to an MCI
- The implementation of specific procedures as it pertains to the various types of MCIs, such as building collapse and natural and manmade disasters
- Regional medical plans and individual hospital plans
- Interaction between the MEDEVAC helicopters and ground units, including communications and landing procedures
- Interaction with military facilities willing to participate in the exercises
- Testing to see if hospital alerting systems work properly or if revisions are needed
- Testing the triage skills, rapid treatment, and transportation of patients by the EMS system
- Testing the ability to select and implement a casualty collection area and casualty transportation area to meet the needs of the MCI
- Assembling the appropriate vehicles for patient transport and seeing how they are integrated into the staging area
- Determining whether all patients can be transported from the simulated disaster site in a reasonable (predetermined) time frame
- Testing the EMS system for proper distribution of patients throughout the available hospital system; if trauma centers are available, checking that they receive the correct type of patients

For an exercise to be beneficial, controllers should coordinate the overall exercise. These controllers should also be evaluators who submit written comments on the exercise for use at the critique. Their critical analysis of the exercise will permit improvements in procedures and guidelines to better the system. A weak controller who fails to recognize or ignores problems that develop does an injustice to all the participants. Constructive criticism is the basis for improvement of fire and EMS operations.

COMMAND

Organization and control are necessary ingredients for success at an MCI. The implementation of an incident management system starts with the initial arriving responder. This can be a firefighter, emergency medical technician, paramedic, or police officer. This individual should give an initial report of conditions found and establish Command. If multiple organizations will be required to participate in the MCI, it may be necessary to establish a Unified Command structure with officers from the affected departments sharing the command role.

OPERATIONS

The complexity of MCIs will often necessitate the designation of an Operations Section Chief. This position is assigned when the IC cannot dedicate sufficient time to the operations function. Operations should be delegated very early in the incident to allow the development of an organization to address the problems found. Operations should establish:

- A Medical Unit that can keep Operations abreast of the problems found and the resources that will be required
- One or more staging areas for fire and EMS units
- A location designated to be used as a casualty collection area (CCA); this location will be utilized for triaging and treatment of patients
- A location to be used as a casualty transportation area (CTA), where patients are taken once triage and treatment are performed and they are awaiting transportation to the hospital

The Operations Section Chief will keep the IC informed of conditions through timely progress reports.

STAGING

A Staging Area Manager should be assigned. Depending on the complexity of the incident and the available resources, the Staging Area Manager could be an apparatus driver, company officer, or chief officer. Staging should consider one area for fire department apparatus and another for patient transport vehicles. The patient transport vehicle area may be further delineated between ambulances, paramedic units, and buses. The staging area can be secured through the assistance of the police. Vehicle drivers should leave their units unlocked and their keys with the vehicles. This will facilitate movement of vehicles and prevent delays in transporting patients.

Once the ambulances and medic units are at Staging, the Staging Area Manager should ensure that a sufficient number of drivers are available to drive the units. Paramedics should not be used as drivers. Typically they will be in short supply and should be assigned to the patient compartment. Firefighters, police officers, and other responders should be utilized as drivers.

LOGISTICS

The Logistics Section Chief can work with the Medical Unit Supervisor and police to secure areas for setting up the CCA and CTA. The location of Staging should be considered for ease of movement of ambulances.

If helicopters will be used, a landing zone should be secured. The considerations for the landing zone will be similar to the staging areas with the exception that the noise and backwash of the helicopters, as well as any overhead obstructions, must be considered. Some fire departments require that an engine company with foam operations capabilities be on location for all helicopter landings and takeoffs. It is also beneficial if communications can be established between the ground and air units.

Logistics will work closely with the Medical Group Supervisor and can keep abreast of the need to secure additional medical supplies or special equipment to assist rescuers in extricating patients.

MEDICAL GROUP OR MEDICAL BRANCH

The Medical Group should be established and the assignment of the Medical Group Supervisor should be an individual with medical training. Ideally it will be a paramedic if one is available. Many fire or EMS departments utilize the position of Emergency Medical Control Officer (EMCO). This individual can be an EMS supervisor who oversees the daily operations of paramedic units. In an MCI that individual will assume a major role in controlling the incident. He or she will be the medical expert for the Operations Section Chief.

The EMCO can become the Medical Group Supervisor or Medical Branch Director and run the medical portion of the incident. Whether to designate the function as a group or a branch should be determined by the need for expansion of its activities. If numerous activities will be needed, then designating the function as a Medical Branch will allow for expansion. If the medical duties will encompass only triage, treatment, and transportation within a limited area, then the designation of Medical Group should be sufficient.

Timely reports given to Operations or Command will allow for a successful conclusion to the incident. The Medical Group Supervisor will establish the portions of the incident management system that are needed to address the problems found. The Medical Group Supervisor should:

- Continually assess the medical needs of the incident and request additional resources if needed
- Ensure that the CCA and CTA are of sufficient size and at the proper locations to meet the incident needs
- Designate the positions of the Triage, Treatment, and Transportation Unit Leaders and assign those responsibilities to qualified individuals
- Request a hospital emergency response team through Command, if needed and available
- Keep abreast of the number and type of injuries and the vehicles necessary to transport them and request additional resources as needed

TRIAGE

The first steps taken at an incident scene can mean the difference between success and failure. At an MCI, proper triage can literally be the difference between life and death. Triage is a procedure that paramedics, emergency medical technicians, and firefighters are not normally accustomed to. Our first instinct with injured people is to give them immediate treatment and stay with them. We try to comfort and reassure them that we are doing everything possible for them.

An injured person contaminated with hazardous materials will need to be decontaminated. This would be critical at a terrorism event where nerve gas, cyanide, or radiological exposures could occur. This decontamination would protect the injured and those treating the patient.

In contrast, to properly triage the injured at an MCI involves a totally different procedure. The first step that is needed is to perform a primary scan of all victims. This should be done by the first-arriving emergency personnel. This step will locate those whose lives are threatened and for whom rapid treatment can save them. It could involve a simple exercise of opening a closed airway, stopping a hemorrhage, or moving someone from a dangerous area to a secure place. The person giving this primary scan cannot take the time to deal with victims with multiple injuries that would entail a considerable amount of time.

To indicate that the primary scan has been performed, the emergency personnel should place a triage tag on each person scanned. At this juncture the emergency personnel will not take the time to classify the injured in classes red, green, or white, but rather place an entire triage tag on the individual that can be separated at a later time when the classification of the injuries is made.

The second triage step is used to treat those injuries where the person has a chance of survival. Again, this should not be an in-depth examination, but in preparation for moving the injured to the CCA. In the initial stages of an MCI, there will be a shortage of trained personnel to handle the massive amount of medical problems, so time management is critical. For this reason, victims with no chance of survival should be treated only after all others.

The third triage step should include a thorough examination followed by tagging of the individual. This can be accomplished at the CCA. The tagging should list the injured according to color classifications:

- *Red tag*—1st Priority—Immediate transport
  - Unconscious
  - Open chest or abdomen wounds
  - Severe medical conditions
  - Severe or uncontrolled bleeding
  - Severe shock
  - Several major fractures
  - Burns—respiratory damage—unconscious
- *Yellow tag*—2nd Priority—Transport as soon as possible
  - Head injury—conscious
  - Loss of distal pulses—extremity
  - Severe burns—unconscious
  - Spinal injuries
  - Moderate blood loss
- *Green tag*—3rd Priority—Minor/Delayed transport
  - Minor fractures
  - Minor injuries that are controlled
- *White tag*—4th Priority—No injury, medical transport not required
  - Assessed—no medical problem
  - Patient refused assessment/treatment
- *Black tag*—Deceased—No transport
  - Obviously mortal wounds
  - Obviously dead

TRIAGE FORM

The Triage Unit Leader will assign incoming EMS personnel to assist in triaging of patients. He or she will assure that each casualty is triaged and appropriately marked for treatment.

There are different types of triage forms, but they contain the same basic information. Each form is serially coded for tracking the casualties, the transporting unit, and the hospital destination. There are appropriate spaces for specific data, including name, address, age, sex, vital signs, and treatment provided. Because trained EMS personnel will initially be limited, the person in charge of the CCA can assign other emergency personnel to assist in filling out the necessary data on the form. This will allow a faster evaluation and treatment of the injured, and the data will allow better tracking when a person is transported to a hospital.

After the form is filled out, the doctor, paramedic, or EMT who has done the actual triage should check the completed form to ensure that the data are accurate. On the reverse side of the form, indicate the chief complaint. Depending upon the form used, there may be the need to provide additional information on the injuries by noting them on an anatomical diagram.

TREATMENT

The Treatment Unit Leader will ensure that patients who have been triaged receive the treatment needed to stabilize them. Patients should be categorized as those needing immediate treatment and those who can have their treatment delayed. The Treatment Unit Leader should request sufficient personnel to establish medical teams to perform the necessary functions. Each triage form should be checked to ensure that all pertinent data are entered on the form. Once patients have received treatment, they should be prioritized for transport and the Transportation Unit Leader notified. The movement of patients should be performed by firefighters or other personnel; medical personnel can be utilized in other capacities.

TRANSPORTATION

The greatest asset of a Transportation Unit Leader is coordination. Area hospitals must be contacted to determine their ability to accept patients and level of trauma they can handle. This can be accomplished by the dispatcher, who can relay the information to the incident scene, or the information can be collected at the scene by someone designated by the Transportation Unit Leader. In either case, these data will be utilized to develop a transportation plan. The data can be categorized by distances to the various hospitals, and whether they are specialized facilities—e.g., trauma or burn centers.

The Transportation Unit Leader ensures that a record is maintained of all patient transports and their allocation to hospitals. A current list of hospital capabilities must be continually updated since admissions from other emergencies can change the hospital availability of beds.

Ambulances and medic units must be directed to a specific hospital for the transportation plan to be effective. This plan should be utilized to prevent the overloading of any hospital through the proper distribution of patients. A consideration to reserve a nearby hospital for red tag patients who are currently being treated or still in the process of being extricated can be implemented to lessen their transportation time.

Ambulances may be able to provide transportation for more than one patient. The configuration of certain ambulances allows for the handling of more than one red or yellow tag. Consideration should be given to utilizing each ambulance for a combination of a red or a yellow tag, and one or more green tags.

NEED FOR COORDINATION

Interaction among the Triage, Treatment, and Transportation Unit Leaders is needed to determine:

- The overall number of patients who will require transportation
- The classification of those patients
- Injuries that could require a specific type of hospital treatment

Treatment should notify Transportation when patients have been triaged, treated, and are ready for transport. Transportation will need to coordinate with Staging to keep abreast of the number and types of transport vehicles available. A checkpoint should be established through which all departing vehicles exit the scene. At the checkpoint the Transportation Unit Leader will ensure that data on the patient(s) being transported are recorded. This may be accomplished by removing part of the triage tag with the data already inscribed.

Ambulances should be directed to designated hospitals and, if necessary, directions given to the drivers. If possible, notify Dispatch of the transport so that Dispatch can relay the data to the receiving hospital. The driver of the ambulance should also be directed to return to the staging area upon leaving the hospital if necessary.

MASS CASUALTY INCIDENT CONSIDERATIONS

This information on MCIs is not meant to be an all-inclusive procedure for handling an MCI, but a guideline for initial responders. It is intended to allow responders to develop a command system to handle the initial problems and to perform their duties to their level of training. It will enable them to recognize the actions of those with higher levels of training and understand why certain steps are taken. It makes the assumption that there are multiple hospitals that are a timely distance from the incident scene. In many areas, only one or two hospitals or medical centers are available. This will severely restrict the patient treatment available. Likewise, the likelihood of a surgical team being available to respond to the scene may occur only via helicopter from distant areas. Some military installations may be able to assist by providing assistance in dispatching helicopters and other medical personnel.

SIZE-UP FACTORS FOR MASS CASUALTY INCIDENTS

Water
- Typically water will be needed if decontamination is necessary and for washing down an incident after the transport of all victims to remove blood and other bodily fluids at the scene.

Area
- An MCI can occur anywhere.
- Large common areas and public assemblies may be a target for a terrorist attack causing an MCI.
- An MCI occurring in conjunction with a natural or manmade disaster could encompass a widespread area.

Life Hazard
- Severe. An MCI will place numerous patients in danger, and immediate actions will be needed to minimize further injury or death.
- Injured persons may leave the incident scene and go directly to the hospital, making it difficult to accomplish accountability and changing the hospital availability for other patients.
- Transportation to hospitals will be delayed until triage and treatment are performed.
- A sufficient number of vehicles will be required to accommodate all of the injured.

Location, Extent
- An MCI can occur anywhere. It could happen due to a fire, building collapse, vehicle accident, or a natural or manmade disaster.

Apparatus, Personnel
- Call for help early. A large number of both fire and EMS resources will be needed.
- Determine the best method of utilizing personnel to assist in triage, treatment, transportation, and extrication.
- Call for MEDEVACs if needed.

Construction/Collapse
- This can be a concern if a building collapse or severe damage to a structure has occurred.

Exposures
- There are none predictable.

Weather
- High heat could dehydrate victims and tire responders.
- Cold weather will affect the injured if they are outside without protection.
- Extremely cold weather may become life-threatening, and a casualty collection area may need to be considered in a nearby school, government building, or public assembly.

Auxiliary Appliances
- There are none predictable.

Special Matters
- Emergency lighting should be considered for nighttime operations.
- Private tow trucks could assist at an MCI involving vehicles.
- The police can screen volunteers at the incident scene and then make recommendations to fire and EMS personnel as to their ability to assist. Volunteers can be utilized as litter bearers.
- There should be a listing of any hospital or trauma center that has the ability to decontaminate a patient.
- The Transportation Unit Leader should balance the distribution of patients among the various hospitals and trauma centers.

Height
- Height could be a factor if the MCI occurs in a building above grade since it would necessitate the movement of patients from upper floors. It could cause congestion in hallways and on stairways as equipment is moved into the building and as patients are removed.

Occupancy
- Occupancy could be an impact if an MCI occurs within a building. The responders would need to ascertain if any special groups could be involved, including the elderly, children, physically or mentally challenged people, or special needs groups.

Time
- Time could be a concern in areas that depend on volunteer or on-call responders. There may be certain time periods where there is difficulty in obtaining a sufficient number of personnel.
- EMS organizations would be impacted during peak response periods when there would normally be few ambulances and medic units available.

CONSIDERATIONS FOR MASS CASUALTY INCIDENTS

Strategic Goals and Tactical Priorities
- Recognize that an MCI may need to be initiated while firefighting or other emergency operations are still taking place.
- Call for a sufficient amount of resources (fire, EMS, and transport vehicles).
- If operating under hazardous conditions (fireground, hurricane, tornado, etc.), ensure the safety of responders.
- Ensure that established collapse and/or safety zones are not violated by responders.
- Consider taking a PAR of units that are operating in hazardous areas at timed intervals.
- Request the response of a mass casualty vehicle that contains medical supplies for use at an MCI.
- Consider requesting MEDEVACs for transporting patients.
- Consider requesting assistance from military installations, especially when operating in remote areas.
- Set up a CCA.
- Set up a CCT.
- Set up an area to be used for staging. Consider setting up a separate area for fire department apparatus and one for ambulances and medic units.

FIGURE 8-3 This church was erected in 1884. Like many Gothic churches built in the nineteenth century, it is still in use today. *Used with permission of William V. Emery.*

- Document all activities for current and future use. This is especially critical if there is a chance that a disaster may be declared, and the recovery of funds may be a possibility.

Incident Management System Considerations/Solutions
- Incident Commander
- Unified Command (if needed)
- Safety Officer
- Operations (if needed)
- Rapid Intervention Crew (if operating in hazardous areas)
- Staging
- Logistics (if needed)
- Extrication Group (if needed)
- Medical Group or Branch [to set up triage, treatment, transportation, fatality management (morgue), etc.]
- Division 1, 2, 3, etc. (for multistoried building can be implemented for any floor where deemed necessary)

Houses of Worship

Gothic-style churches, many of which were built in the nineteenth and early twentieth centuries, are still in use today. (See Figure 8-3.) They are one-story buildings equal to four or more stories in height. The beautiful edifices contain masonry walls and roofs constructed of large wooden beams formed into timber trusses. Typically, this style of construction creates a hanging ceiling 50 feet or higher above the nave or pew area. These artistic hanging ceilings are formed with plaster spread over wood or metal lath. Though the religious denominations may change, these older churches still have active memberships. It is rare to see Gothic-style churches demolished. Even after disastrous fires, they are usually rebuilt.

> The nave is the seating or pew area in a church.

BUILDING INSPECTION

houses of worship ■ Church. Gothic-style churches are one-story buildings that can be equal to four or more stories in height.

How often are **houses of worship** inspected by fire companies? When they are inspected, are comprehensive inspections done? Do the firefighters climb into the hanging ceiling to see what hazards exist there? Is the alarm system manually activated to ensure that it can be heard throughout the building? Do the repairs or renovations made meet the current codes?

Though today's codes incorporate life-safety features, many houses of worship were built before the codes were written and do not conform to them. Many churches have undergone changes over the years. Repairs and renovations have been performed by the church's faithful. Though well meaning, these workers are often less qualified than professionals. The alterations may be substandard and can create future problems.

Routine inspections can assist in recognizing hazardous situations or dangerous housekeeping practices. One example is the use of votive candles. They produce an open flame that requires attention to surroundings. Banners or draperies, though seemingly a safe distance away, can be windblown when a door is opened and come in contact with the candles. Many houses of worship have changed to an electric bulb–type candle to eliminate this hazard.

During a building inspection, firefighters should gather as much information as possible about the building and its surroundings. They should physically check the entire structure. This includes entering the attic to see what potential problems may be hidden there. A diagram should be drawn that includes the location of the heater room, access points for the hanging ceiling, and the location of utility shutoffs. It is not uncommon to find heating plants and utility shutoffs located in interconnected buildings.

The fire officer, along with a representative of the church, should decide what valuables must be protected or removed in the event of a fire. This includes placing a value on the stained glass windows. Stained glass is expensive to replace; in the preplanning stage, thought must be given whether to break or try to protect these windows. This tactic should be discussed with the minister, rabbi, or priest. Imported stained glass windows may have been painted by a famous painter and could be worth a fortune. Locally produced windows, though expensive, may be easily replaced. This prior determination will assist the IC in determining strategy and tactics.

 ON SCENE

Renovations can include major changes to the exterior to transform its appearance and make the church appear to be of newer construction. In one instance, as I arrived on scene at a fire in a church that was heavily charged with smoke, I found that a new façade of brick had been installed over the old front stone wall. This included a new cornerstone that was marked 1957, indicating the year of the renovation. Without doing a 360-degree walk-around, one might assume that the year of construction was indeed 1957. However, turning the corner from the A to the D side revealed another cornerstone, dated 1854. The fast-moving fire within the church had not been detected in the early stages, and before hose-lines could have any effect on it, the structure was quickly consumed.

CONSTRUCTION FEATURES

The footprint of these buildings can differ. Some are built with a cruciform design or in the shape of a crucifix, others in a rectangular design. If a basement exists, it may be under only a portion of the building. It can consist of a large open area, or it can be divided into many small rooms. Each type presents problems for firefighters. The large open area can contain large quantities of combustible material that, once ignited, will be difficult to control. The presence of many small rooms can create a mazelike situation, causing a delay in locating the fire and proving deadly to firefighters operating in reduced visibility.

An inherent construction feature of a Gothic-style church is the existence of large void spaces within the walls. These spaces can be 12 inches deep by 16 inches wide. It is common for these void spaces to extend from the basement to the hanging ceiling. These concealed spaces make it very difficult to contain and control church fires because fire entering these voids can rapidly extend into the hanging ceiling.

Older wood-frame churches often have balloon-frame construction containing similar void spaces in the walls.

Building features may include entry to cellars or basement areas through crawl spaces that have to be accessed via a trap door. A firefighter who was trapped in a crawl space area of a church, due to heavy fire conditions on the first floor, had to be removed through the same trap door he had entered in the sanctuary. Unfortunately, due to incident scene conditions, the rescue was unsuccessful. Encountering trap doors should raise a red flag of concern, since using this method of entering an area will often mean it is the only avenue to exit.

INTERCONNECTED BUILDINGS

The church may be only one building in a complex surrounded by other buildings. There can be interconnections between the house of worship and the rectory, parsonage, convent, or school. Buildings may be linked together by doorways, tunnels, or aboveground walkways.

Where doorways exist between these buildings, the possibility of fire starting in one building and communicating to the other must be considered. Doors should be fire-rated and self-closing. There have been instances in which a door between buildings has been propped open and fire has extended to the other building. (See Figure 8-4.)

 **ON SCENE**

It is critical to check any doorways between the house of worship and interconnected buildings. A stubborn fire in a rectory that was interconnected to the church via a walkway was brought under control, but the fire had spread through an open doorway into the church and destroyed it.

ACCESS

Access to the building may be a problem. The church may be set back from the street. Obstructions such as trees or fences may restrict apparatus access. Parked cars in parking lots can hinder apparatus placement. Fire apparatus will have difficulty entering a parking lot if cars attempt to exit.

DELAYED ALARM

The leading contributing factors of major fires in churches are delay in detection and delay in notification of the fire department. A delayed alarm can be caused by lack of discovery or by attempts to extinguish the fire by well-meaning individuals. In either case, the delay can give the fire a sufficient head start that firefighters may not be able to overcome.

BUILDING USAGE

The use of houses of worship has changed. In the past, civilian life safety concerns were restricted to the hours of worship. Today, churches and synagogues have expanded programs and are utilized as day care centers for young children, meeting places for senior citizens, thrift stores where clothing is collected and distributed, and food kitchens to feed the homeless.

Some congregations have reached out to the homeless and provided them with shelter. Meeting rooms and banquet halls have been converted to dormitories.

There have been instances when those being helped have stolen valuable religious items. To counteract these thefts, church officials have been forced to restrict entry into certain areas of the buildings. At times, this has been accomplished by locking doors. This changes the normal means of egress and compounds life safety problems.

Social events held in church halls can exceed safe occupancy loads for the building. Aisles and exit ways may be blocked with tables and chairs hastily erected to accommodate larger-than-expected crowds.

Problems confronting firefighters will involve evacuation of elderly attendants and children. A major issue is that fire drills are rarely held in houses of worship.

FIGURE 8-5A This church was converted to a residential apartment building by leaving the exterior walls intact and creating nine interior levels. *Used with permission of William V. Emery.*

CONVERTED BUILDINGS

Some church buildings have been converted to residential and commercial use. (See Figures 8-5A, 8-5B, and 8-5C.) One particular building has maintained its outward appearance, yet it has been converted to apartments. The roof was changed, and a building was built within the old church structure. Nine levels of apartments with 10 apartments per floor have been installed. Likewise, buildings constructed for other uses are now houses of worship.

FIRE PREVENTION

Houses of worship built today should contain sprinklers, as well as monitored smoke and heat detection systems. A monitored alarm system permits immediate notification of the fire department, reducing delayed alarms.

Though modifications have been made to many older church structures, financial considerations have limited the installation of sprinkler systems. My experience has shown, with the exception of new construction, the only churches that have sprinklers are those that have been destroyed by fire and rebuilt.

New church construction techniques have placed an emphasis on compartmentation by fire-stopping void spaces. This construction feature gives responding firefighters a chance to control the fire.

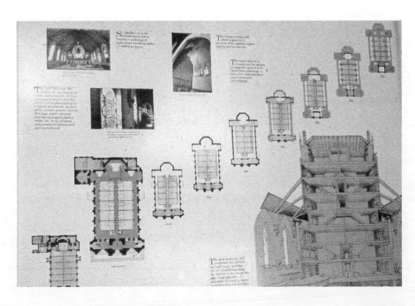

FIGURE 8-5B A floor layout shows that 10 apartments per floor have been created. The side view shows how this was accomplished. *Used with permission of William V. Emery.*

FIGURE 8-5C The stairs are seen adjacent to the large masonry walls. This design created a space between the original walls and the new construction.
Used with permission of William V. Emery.

Another beneficial construction feature uses exposed wooden ceiling beams, eliminating the hanging ceiling and attic space and removing a critical firefighting problem.

FIRE CAUSES

The most common cause of church fires is arson. Many houses of worship are unlocked to allow the faithful to enter and pray at any hour. This ready access has led to the theft of valuable artifacts and the pilfering of poor boxes. Fires have been started to cover up these thefts.

Electrical problems are the second leading cause of church fires. Old wiring with broken insulation and overloaded circuits is a contributing factor.

Another major cause of church fires is associated with the heating plant in the building. Many fires are caused by a pyrolytic effect of wood in contact with the heater or flue. Over an extended period, these wooden beams are constantly heated. This heating causes the wood to turn to charcoal. Charcoal has a much lower ignition temperature than a wooden beam.

In climates with severe winter weather, heating systems may be used minimally during the week and operated constantly on the weekends. During a prolonged cold spell when the heater operates for extended periods, the charcoal beam can ignite. Because this often occurs in concealed spaces, the fire can burn undetected. If there is a sufficient amount of air to sustain the fire, it can have a tremendous head start before the notification and arrival of the fire department.

Though housekeeping is usually good, another fire cause is improper storage. Hazards can include paints, thinners, power lawn mowers, snow blowers, and used clothing collected for distribution to needy families.

Pyrolysis is a chemical change in wood caused by the heating of wood over a prolonged period of time. This process changes the wood to charcoal, which has a lower ignition temperature than wood.

NIOSH FIREFIGHTER FATALITY REPORT F2011-14

On June 15, 2011, a 40-year-old male career firefighter (the victim) lost his life at a church fire after the roof collapsed, trapping him in the fire. At 1553 hours, the victim's department was dispatched to a report of a church fire at an unconfirmed address. Units arriving on scene observed visible flames and heavy smoke coming from the roof of the church. A second alarm was immediately requested due to the lack of hydrants in this area. Initially, the IC sent in a truck crew consisting of an officer and four firefighters, followed by two firefighters (including the victim) from the arriving engine company, for search and suppression activities. The interior crew was initially met with visible conditions, light smoke, and no visible fire within the church. Conditions quickly changed after walls and areas of the ceiling were opened, exposing a fire-engulfed attic space. A decision was then made to evacuate the building, due to the amount of fire burning above the firefighters. At this same moment (approximately 1610 hours), the roof began to collapse into the church where the firefighters were working, trapping the victim and injuring others as they exited out of windows or ran from the collapse. Due to the magnitude of the fire, the fire department was unable to return to the collapsed area to rescue the victim. The victim's body was later recovered after the fire was extinguished.

Contributing factors included inadequate initial size-up that did not fully consider the impact of limited water supply, available staffing, occupancy type, and lightweight roof truss system; risk management principles not effectively used; high risk-low frequency incident; rapid fire progression; offensive versus defensive strategy; failure to fully develop and implement an occupational safety and health program per NFPA 1500; fire burned undetected within the roof void space for unknown period of time; and roof collapse. **See Appendix B: NIOSH Reports in www.pearsonhighered.com/bradyresources to read the complete report and recommendations.**

PROBLEMS

More than three-fourths of fires in houses of worship occur when the buildings are unoccupied. About half of all church fires occur between 2300 hours and 0700 hours. A fire occurring during religious services can be compounded by overcrowding at certain religious celebrations because the occupancy load may greatly exceed a safe level.

Some denominations have separate services for adults and children. Should a fire occur, the adults might complicate the search and rescue problem by attempting to locate children who have already been evacuated. A difficult part of search and rescue is knowing who is still in the building. As occupants evacuate the church, they may leave the area, and unknowingly they may be reported as still missing.

The firefighters' initial actions must be to calm and evacuate civilians. Hose-lines must not be stretched through doorways that are being used by fleeing occupants.

FIGURE 8-6 When arriving at a church fire, we must ascertain what is burning. This fire has control of the hanging ceiling and roof and is past the point of an offensive attack. *Used with permission of Joseph Hoffman.*

Houses of worship often contain an abundance of combustible building components to create a hot, fast-burning fire that will be difficult to control. Decorations for religious celebrations during the year can add to the problem. An example is the placement of live evergreen trees in the altar area at Christmastime.

ON SCENE

At one fire started by burning candles, live evergreen trees supplied the fuel for a fast-spreading fire in the altar area. The maintenance man attacked the fire with fire extinguishers. As firefighters arrived on the scene, he reported that he had extinguished the fire. The large number of spent fire extinguishers attested to the fact that a long delay had occurred before the fire department was notified. Unbeknownst to the maintenance man, the fire had extended into the hanging ceiling. Six alarms were required to control the fire. The building was heavily damaged.

FIREFIGHTING

Firefighters must first ascertain what is burning. (See Figure 8-6.) Can it be controlled with handheld hose-lines or is it already past the point of control of an offensive attack? As firefighters, we base our firefighting strategies and tactics on time factors. On arrival at a building fire where only smoke is showing and no fire is visible, we attempt an interior attack on the fire. A fire that is burning in concealed spaces is difficult to locate, and it is especially difficult to determine how extensive the fire is.

Locating and containing the fire in the early stages will be necessary for an interior attack to be successful. Thermal imaging can prove to be a valuable tool in this endeavor.

ON SCENE

If the building is unoccupied at the time of the fire, the fire department may experience problems in gaining entry. For example, three-inch-thick oak front doors may be locked. On one occasion, a minister pleaded with firefighters not to damage a church's ornate front doors. He quickly left to get the keys, which he was unable to locate. Firefighters were able to gain entry through a side door, minimizing damage but delaying their attack on the fire.

The first hose-line should be stretched to the seat of the fire. The second hose-line should be stretched to directly above the fire area. Floors and walls should be checked with a TIC and, if necessary, opened to check for extension from below. The next hose-line can be used as a backup for the first or second hose-line or can be taken to another location, depending on the problems encountered. A fire involving a large amount of combustibles will require the use of 2½-inch hose-line to ensure a sufficient volume of water for a quick knockdown.

During an interior attack, the exterior of the building must be monitored for changing conditions. The smoke emitting from the building can give telltale signs. Is it being pushed as if under pressure? Is dark smoke turning to light smoke, indicating that hose-lines are reaching the seat of the fire? Are conditions improving or worsening? Are the reports received from the interior coinciding with what is seen on the exterior? If there is a difference in interior and exterior reports, it could indicate that the fire has control of void spaces and there is the need to abandon an interior attack and go defensive.

A fast-moving interior fire will be difficult to control. Handheld hose-lines will be ineffective on fires past the incipient stages. Rapid fire development and deteriorating conditions could dictate the need for an immediate withdrawal of firefighters and a change to an exterior attack.

ON SCENE

The fire attack may generate enough steam to extinguish fire that has extended into the hanging ceiling. In one instance, a fire that started in the sacristy alongside the altar had smoldered for approximately 12 hours. Hose-lines attacked the fire, and it was quickly knocked down. During overhaul, a large amount of char was found on the rafters above the ceiling. The ensuing investigation found that we were able to extinguish this fire because the attic area had few vents. This caused the fire to smolder for a long period of time. Our attack on the fire came before the attic fire had burned through the roof, which would have introduced sufficient air to accelerate the fire and destroy the church.

COMMUNICATIONS

Once inside the building, there is a tendency to refer to the altar area as the front of the building. When transmitting radio reports, it may be necessary to clarify the interior locations by referring to sides A, B, C, and D or by using exact locations, such as the "altar area" or "directly inside the front door on side A." Otherwise, the message "Chief, we are in the front of the church and have knocked down all visible fire" could be interpreted to mean different locations.

CEILINGS

The ceiling in a Gothic-style church is referred to as a *hanging* ceiling. (See Figures 8-7 through 8-10.) It is suspended from the timber truss by steel supports, creating an attic area. The hanging ceiling depends on the truss for its support. Once fire enters the space above the hanging ceiling, it will spread quickly on the building components and any storage material located within the attic. There is a possibility of ceiling collapse. The large, ornate plaster decorations found in church ceilings become deadly projectiles should they strike a firefighter if parts of the ceiling start to fall.

FIGURE 8-7 This ceiling in a Gothic-style church is referred to as a hanging ceiling. *Used with permission of William V. Emery.*

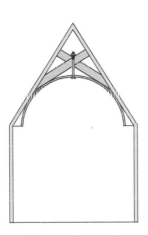

FIGURE 8-9 This diagram shows how the hanging ceiling is supported. *Used with permission of Pearson Education.*

FIGURE 8-10 This photo shows the timber truss supporting a hanging ceiling by means of steel plates and rods in the church attic. *Used with permission of William V. Emery.*

The problems associated with hanging ceilings include:

a. There may be no access into the ceiling, or it is provided through trapdoors that must be reached from portable ladders or narrow staircases.
b. Exposed wiring within the attic can cause an electrical hazard to firefighters.
c. Firefighters operating with SCBA and hose-lines have limited mobility.
d. Walkways are often planks laid between the ceiling rafters. These planks are as old as the building and can be dry-rotted. The weight of a firefighter and firefighting equipment may cause the planks to fail.
e. Lighting is usually nonexistent.
f. The amount of wood present in the hanging ceiling contributes to a hot, fast-moving fire that is difficult to control.
g. The reach of handheld hose streams is restricted, and these hoses are usually ineffective.
h. Storage of combustible materials in the attic can provide additional fuel for the fire.

In reality, entering a hanging ceiling to extinguish a fire past the incipient stage is too great a risk for firefighters. *After fire has control of the hanging ceiling, it is past the control efforts of an interior attack.* Firefighters must be withdrawn from the building, and the fire must be fought from the exterior.

NIOSH FIREFIGHTER FATALITY REPORT F2004-17

On March 13, 2004, a 55-year-old male career battalion chief (Victim #1) and a 51-year-old male career master firefighter (Victim #2) were fatally injured during a structural collapse at a church fire. Victim #1 was acting as the Incident Safety Officer and Victim #2 was performing overhaul, extinguishing remaining hot spots inside the church vestibule when the bell tower collapsed on them and numerous other firefighters. Twenty-three firefighters injured during the collapse were transported to area hospitals. A backdraft occurred earlier in the incident that injured an additional six firefighters. The collapse victims were extricated from the church vestibule several hours after the collapse. The victims were pronounced dead at the scene.

NIOSH investigators concluded that, to minimize the risk of similar occurrences, fire departments should:

- Ensure that an assessment of the stability and safety of the structure is conducted before entering fire- and water-damaged structures for overhaul operations
- Establish and monitor a collapse zone to ensure that no activities take place within this area during overhaul operations
- Ensure that the IC establishes the CP outside of the collapse zone
- Train firefighters to recognize conditions that forewarn of a backdraft
- Ensure consistent use of personal alert safety system (PASS) devices during overhaul operations
- Ensure that preincident planning is performed on structures containing unique features such as bell towers
- Ensure that ICs conduct a risk-versus-gain analysis prior to committing firefighters to an interior operation, and continue to assess risk-versus-gain throughout the operation including overhaul
- Develop SOGs to assign additional Safety Officers during complex incidents
- Provide interior attack crews with TICs

See Appendix B: NIOSH Reports in www.pearsonhighered.com/bradyresources to read the complete report and recommendations.

VENTILATION

Steep, sloped roofs are too dangerous to operate on. If roof venting is required, it should be performed by firefighters operating from a platform or a main ladder.

During an offensive attack, ventilation is a must. (See Figure 8-11.) All ventilation efforts must be directed by the units operating on the interior of the building, communicating with the outside ventilation group. There may be hesitancy on the part of firefighters to break stained glass windows. Breaking these windows may or may not be of assistance to the firefighting effort. To determine the benefit of ventilation through window openings, firefighters can break one or two stained glass windows. If this ventilation assists in relieving the smoke conditions and allows the interior crews to attack the fire, then additional windows can be broken. Likewise, if this type of ventilation proves unsuccessful, then no more windows should be removed, and other methods of ventilation should be sought.

Firefighters may find difficulty in breaking these windows. They may be protected by a Lexan covering or a wire screen installed to protect the stained glass from breakage and perching birds. This screen can be of heavy-gauge steel. Both the wire screen and Lexan will require the use of a power saw to remove them. This can be a time-consuming operation.

PPV may be beneficial if the fire is not burning in concealed spaces. Should the fire reach these areas, positive pressure can rapidly accelerate the fire in these voids.

FIGURE 8-11 During an offensive attack, ventilation is a must. At a church fire there may be some hesitancy on the part of firefighters to break stained glass windows. *Used with permission of William V. Emery.*

VALUABLE ARTIFACTS

A concern of the congregation is the valuable artifacts contained within the house of worship. An attempt should be made to remove chalices, Torahs, relics, and other irreplaceable items. Firefighters should be aware that well-meaning persons might attempt to enter the burning structure with no concern for their own safety to retrieve these items.

DEFENSIVE ATTACK

A defensive attack must consider exposed buildings. Collapse zones will have to be set up and maintained. Large-diameter hose-lines can assist in securing an adequate water supply. Interconnected buildings will require monitoring for fire extension. After fire has broken through the roof, large embers can be taken downwind, starting other fires.

One tactic that has been successful in a defensive operation is to break out the front window above the hanging ceiling. Playing a master stream into that opening from a ladder pipe or platform can be effective in controlling and extinguishing fire in the hanging ceiling.

FIGURE 8-12 A hanging ceiling is still intact after the roof was attacked by fire, broke its bond with the timber truss, and slid off in large sections to the exterior. *Used with permission of William V. Emery.*

Firefighters must be concerned about collapse. A church steeple will not only attract lightning, but it can fail. In Houston, a steeple collapsed, striking a pumper that had stopped in front of the fire building, injuring the officer.

The wooden supports in the bell tower can be weakened by the weight of the bells and the fire and can collapse during the fire or in the overhauling stages.

ROOF FAILURE

Under fire conditions, a roof set on a timber truss will react in one of three ways. (See Figure 8-12.)

- It can burn in place and not collapse.
- It can collapse into the church proper, spreading the fire into the nave or seating area.
- The weight of the roof can break the bonds with the timber truss, and the planks that the slate or shingles are attached to can slide off the truss in one piece or in large sections, threatening firefighters operating on the building's exterior.

LESSONS LEARNED AND REINFORCED

In reviewing lessons learned and lessons reinforced from many fires in houses of worship, some points were consistently cited. The following are some of those critical considerations:

Get a Full 360-Degree View of the Church and Exposures This will allow the IC to understand the problems of building layout, exposures, and fire and smoke conditions.

Call for Help Early Having enough personnel to address areas of concern can help stabilize the incident and address minor problems before they become critical situations.

Stop the Upward Spread of Fire in Wall Voids Open suspected areas. Only by containing the fire and preventing extension into the hanging ceiling will an interior attack be successful. This operation will be enhanced with thermal imaging.

Overcome the Fear of Breaking Stained Glass Windows Ventilation is a must for an interior attack. If the building remains unvented, an interior attack will fail.

Feedback Is Essential Critical information must be promptly given to the IC. Knowledge of what is happening at the incident scene is realized only through effective communication.

Set Realistic Goals Fires in houses of worship are dangerous. Fire conditions can change rapidly. As they change, we must adjust our firefighting methods. We must decide what can be accomplished. This will depend on the size of the building, amount of fire, fire location, available personnel, and water supply.

If an Offensive Attack Does Not Quickly Control the Fire, Consider Switching to a Defensive Attack Fire in walls will quickly spread into the hanging ceiling, and the building will be lost. Observing smoke and fire conditions on the exterior can be utilized as an indicator of whether interior operations are being successful. Worsening smoke conditions and visible fire from roof ventilators are negative factors indicating that the fire has taken control of the building and that a defensive attack is warranted.

Monitor the Amount of Time Involved until the Fire Is under Control In reviewing numerous fires in houses of worship, there seem to be some constants. It was noticed in a large number of incidents that it took approximately one hour to bring the fire under control. This fact is important because the majority of serious firefighter injuries and deaths occur during this period.

CHURCHES OF FRAME CONSTRUCTION

In addition to the Gothic-style churches, many churches are of frame construction. These structures contain their own inherent problems. Very similar to the Gothic-style, masonry-constructed churches, many of these buildings date to the nineteenth century. The nature of the problems facing firefighters when fighting fires in these buildings is often predicated on when the church was built. These older structures can be massive in size and have both large wall voids and attic spaces. Balloon-frame construction that utilizes a continuous wall stud from the lowest level through the roof area is prevalent. It interconnects the void spaces created by the floor joist at each level of the building with the wall void spaces. Fire-stopping is rare. In actuality, the entire void space within the building is one large interconnected area. A fire starting on a lower level and entering wall or floor voids can quickly extend to the attic area.

 ON SCENE

In Commerce, Georgia, a wood frame church built in 1855 was ravaged by a fire started by an arsonist. As firefighters fought to save the building, the roof collapsed, trapping four firefighters, killing a fire captain, and injuring the other three.

WOOD FRAME CHURCHES WITH LIGHTWEIGHT TRUSSES

Lightweight constructed houses of worship have all of the problems associated with other buildings using similar materials. Fast-spreading fires and very short failure times can make them death traps for firefighters.

 ON SCENE

In Memphis, Tennessee, a fire in a church that was less than 20 years old and constructed with lightweight trusses took the lives of two Memphis firefighters when the roof collapsed into the interior, burying two firefighters and critically burning them. This collapse occurred within seven minutes of arrival of the first due company. Basic tactics that would be employed by any fire department were initiated. The fire was knocked down in the sanctuary, and then as all firefighters would do, they checked the ceiling area for spread of fire. As the ceiling was being opened, it collapsed onto them.

In Lake Worth, Texas, a fire that started in an outside shed extended to involve a church of lightweight construction. The church became heavily involved, and a firefighter fell through the fire-weakened roof but was able to escape. At the same time, five firefighters were trapped in the interior. Two of the firefighters were able to escape, but three firefighters were trapped and succumbed within the building.

SIZE-UP FACTORS FOR HOUSES OF WORSHIP

Water
- Large amounts of water and large-diameter hose-line will be needed if a defensive attack is utilized.

Area
- Access to the building may be hindered by parked automobiles, fences, trees, and buildings that are set back from the street.
- Large wall voids and a tremendous void space created by the hanging ceiling can conceal a fast-moving and dangerous fire.
- Holiday decorations (e.g., live evergreens) can fuel a fire.

Life Hazard
- There will be severe life safety problems if the building is occupied at the time of a fire.
- TICs can assist in the primary and secondary searches.
- Life safety problems can occur when religious services are going on, during day care activities, senior citizen functions, social events, food kitchens servicing the homeless and hungry, and at houses of worship that extend their facilities to the homeless as a shelter for temporary sleeping quarters.

Location, Extent
- The most common cause of church fires is arson. Flammable liquids may be used.
- Minor fires in the incipient stages can be controlled and extinguished with an offensive attack.
- A common cause of church fires is the building's heating plant. Fire can extend to wall voids and upward to the hanging ceiling.
- To control a fire in the incipient stages, walls must be opened above the fire area to extinguish any fire extending upward through wall voids.
- TICs can assist in checking walls, ceilings, and floors for hidden fire.
- Fires past the initial stages, and especially a fire that has control of the hanging ceiling, will require a defensive attack utilizing master streams.

Apparatus, Personnel
- An interior attack will require long hose-line stretches, and a heavier than normal response will be required.
- If life safety of occupants is a concern, a sufficient number of personnel will be needed to assist in the evacuation and search and rescue.
- If a defensive attack is employed, numerous master streams will need to be implemented, and a sufficient amount of personnel, special equipment, and apparatus will be needed.

Construction/Collapse
- Repairs and renovations performed by members of the church can be substandard and create future problems, especially when under attack by fire.
- Stained glass windows may be protected with a Lexan covering or a wire screen that will need to be removed with a power saw.
- Collapse of building walls and steeple must be anticipated and a collapse zone established.
- Converted church buildings may house apartments or other businesses.
- Converted buildings may be utilized as houses of worship.

Exposures
- Interconnected buildings (church, rectory, parsonage, convent, school, etc.) can permit a fire in one structure to spread to other buildings if fire doors are not closed.

Weather
- Wind can impact a defensive attack and can carry flying brands downwind, starting other fires.
- Many fires start in the heating plant; prolonged cold weather may overwork the building's heater.

Auxiliary Appliances
- Sprinklers can be found in newer churches and those rebuilt after a fire.
- Alarm systems may not sound in the assembly area for an alarm elsewhere in the building.

Special Matters
- Many houses of worship were built before life safety codes were written.
- Votive candles burn unattended, and the open flame can ignite nearby combustible material.
- Stained glass windows may be priceless. Their value should be indicated on the pre-plan and the firefighters should be guided accordingly.
- Delayed alarms are common, either due to a delay in detection or to church members attempting to fight the fire themselves prior to notifying the fire department.

Height
- Houses of worship are typically one-story structures equal to four or more stories in height.
- High plaster ceilings on the interior can collapse from heights of over 50 feet, causing severe injury or death to firefighters operating below.

Occupancy
- Valuable artifacts should be removed, if possible.

Time
- Most fires occur between 2300 and 0700, hours when the church is unoccupied.
- Depending on the time of year, there can be flammable interior decorations for holy day celebrations.

CONSIDERATIONS FOR FIRES IN HOUSES OF WORSHIP

Strategic Goals and Tactical Priorities for an Offensive Attack
- Ensure the life safety of occupants.
- Call for a sufficient amount of resources and be prepared to go defensive if the fire is past the control of an offensive attack.
- Attempt to confine and control the fire with handheld hose-lines.
- Immediately check for upward extension of fire in wall voids above the fire area.
- Effect ventilation by positive pressure or by opening doors and windows. If this is insufficient, and the interior crews order it, then break windows, including stained glass windows.

Incident Management System Considerations/Solutions for an Offensive Attack
- Incident Commander
- Rapid Intervention Crew(s)
- Division 1, Basement Division
- Ventilation Group
- Search and Rescue Group
- Staging
- Medical Group

Strategic Goals and Tactical Priorities for a Defensive Attack
- If changing from an offensive to a defensive attack, ensure that a PAR is taken.
- Set up and maintain a collapse zone.
- Set up master streams to protect the most endangered exposure(s).
- Be prepared for flying embers downwind.
- Be prepared for roof failure.
- Protect interconnected and other exposed structures.

Incident Management System Considerations/Solutions for a Defensive Attack
- Incident Commander
- Safety Officer
- Operations (if needed)

- Staging
- Logistics (for water supply, if needed)
- Divisions on the exterior on Sides A, B, C, D
- Exposure Divisions (for immediate exposures and downwind, if needed)

Penal Institutions

Penal institutions can be prisons or detention centers. They can be located in urban areas attached to or situated near courthouses or in rural areas as state and federal prisons once were. In many instances, in these once rural locations the surrounding areas have become built up and these prisons now find themselves situated in highly populated areas. These facilities may include a section for male and female inmates. They can be of different configurations. A common shape is the wheel. Each wing starts at a center hub and extends outward, appearing to form the spokes on a wheel.

penal institutions ■ Prisons and detention centers.

Local fire departments can easily become overwhelmed should a major emergency occur at a penal institution. The warden, though wishing to be cooperative, must constantly be concerned about the potential of civil disobedience of the inmates. These actions could cause riots while inflicting injury upon fire department members, correctional officers, prison workers, and the prisoners themselves. Many arson fires are started when there are periods of unrest in prisons.

Firefighters entering these facilities to conduct an investigation, treat a medical emergency, or fight a fire can be truly intimidated. Verbal threats are often directed at the emergency responders. The name tags that firefighters wear on their bunker gear make them direct targets for verbal abuse by the inmates.

TYPES OF INSTITUTIONS

The threat to the safety of firefighters can increase depending on the type of facility. These institutions can range from minimum-security buildings of frame or ordinary construction with relatively no confining walls around the facilities, to high-security institutions consisting of large, massive stone and concrete structures that are meant to be impenetrable. The concrete in these structures is typically built with larger diameter reinforcement bars. The windows often consist of Lexan with steel bars.

The types of penal institutions can be categorized as:

First Generation Confinement This involves indirect, intermittent supervision. These penal institutions contain bars and corridors. Correctional officers view prisoners only periodically. Inmates are isolated within a cellblock. There is widespread use of cameras and audio receivers. This is often called maximum security.

Second Generation Confinement This involves indirect supervision. There is little contact between the correctional officers and the prisoners. These penal institutions contain Plexiglas by which the correctional officers are separated from the inmates. This is often called medium security.

Third Generation Confinement This involves direct supervision. There are correctional officers within the units. Interpersonal skills are required of the correctional officers in interaction with inmates. This is often called minimum security.

Each type of facility has associated problems. The higher the security risks of the population, the more dangerous will be the assignment for emergency personnel to enter and perform the function that they were called to do. Though there are different levels of security risks, firefighters should realize that inmates can be unpredictable and potentially hostile, regardless of the security classification, and contact with prisoners should be avoided at all times.

EMERGENCY OPERATING PLAN

Most prisons have an EOP. The emergency responders must know these plans. The plans should detail the location of the CP where outside agencies will report. It should explain in detail what is expected of the emergency responders and the interaction and protection that will be provided by the prison personnel. Included in these plans will be benchmarks, cues, or triggers that the officials at the penal institutions will use as guidelines for moving prisoners within the facility or to evacuate the facility.

EOPs must take into consideration the communications that will be needed at an incident. There are numerous methods of establishing effective communications. One method is the implementation of an existing radio channel(s) that could be used by the various agencies. If the radios are not compatible, close contact between correctional personnel and fire personnel must be maintained at all times so that the utilized radio frequencies will be constantly monitored. In the locations where firefighting will be ongoing, this should always occur due to the close interaction that will be required. However, there will be a need to have radios that monitor the required channels at Staging, Base, and for use of the Rapid Intervention Crews (RICs).

Radio transmissions within these structures can experience problems due to the building's steel and concrete construction.

FIRE DEPARTMENT PREPLANS

The need to preplan these facilities allows many decisions to be made in a nonemergency type of atmosphere, rather than under the tremendous pressure that can exist at the emergency scene. It will allow firefighters to develop plans A, B, C, D, etc.

The preplan must start with a total understanding on the part of the prison officials and fire department members of the potential problems and possible solutions to those problems. There is a need on the part of firefighters to understand such basic terminology as *total lockdown*. Additionally, prison officials need to understand the basic concepts of firefighting. Fire officials want the safety of their personnel entering these institutions to be ensured, which can only occur by prison correctional officers accompanying them and creating clear paths through which fire department members can enter and operate.

In one incident fire department responders demanded a total lockdown. Prison officials balked at this request. Total lockdown to a prison official is an involved process in which each cell must be physically checked and every inmate totally accounted for before it can be accomplished. This involves inmates who may be remote from the problem area. It would have been better if the firefighters had requested that no inmates would have been able to make contact with the firefighters as they proceeded to the fire area.

The inclusion of basic preplan information is always necessary and could include:

- The location of sprinkler or standpipe intakes
- The location of standpipe outlets
- Hydrant locations
- The number of inmates typically housed in each area
- The security measures in place and how firefighters can find secondary means of egress if they exist
- How to gain access to the exterior of the building to effect ventilation
- Building plot plans
- Types of storage in outbuildings
- A copy of the EOP for the facility
- Fire and emergency evacuation plans
- HVAC systems
- Location of any hazardous storage and SDSs sheets
- Location of utility shutoffs and any backup systems that are available

 **ON SCENE**

The preplan affords fire officials the opportunity to discuss how they will attempt to fight fires occurring in different parts of the facility. As a lieutenant, I was doing a preplan at a local detention center. Some rather long wings contained prisoner cells and had experienced a rash of mattress fires. The prison officials wanted the responding firefighters to break out the window of the cell from the exterior of where the fire was and place a hose-line into the window to extinguish the fire. This tactic was never going to happen; however, it gave me an opportunity to discuss the fire department's operation with the prison officials. These discussions led to future cooperative operations where interaction between both agencies achieved mutual respect and support.

The preplan should have contingencies for a variety of situations that could occur at a penal institution. These include:

- Riots
- Fires
- Medical emergencies, including MCIs
- Power outages
- Lockdowns
- Movement of prisoners during emergency situations
- Utilization of outside police personnel
- Contracting of buses for the emergency movement of prisoners

DELAYED ALARMS

Though constantly reinforced by fire officials that the fire department should be immediately called to all alarms, this often does not happen. There are numerous false alarms or minor trash can fires that are quickly extinguished by the correctional officers. In newer facilities that contain sprinklers, these fires may be extinguished or held in check until the arrival of the fire department. In other facilities this can present a problem if the fire cannot be extinguished in the initial stage.

Though there is a large life consideration within these structures, there is typically a light fire load. The exception would be in:

- Kitchens
- Cooking facilities
- Laundry (cleaning solvents, drying equipment)
- Shops containing machinery and other equipment
- Warehouses

FIRES

Many fires in penal institutions start with the careless handling of smoking products and fires intentionally set in bedding. Typically, the bedding contains flame-resistant mattresses, but the blankets and sheets will sustain a fire. It should be noted that flame-resistant mattresses will burn if exposed to enough flame. These fires impact on inmates in the immediate area due to their inability to remove themselves because of the security measures in place.

There are different locales that have banned smoking in penal institutions. The intent of the ban on smoking is to lessen the amount of money that will be spent on medical care due to the direct effects of smoking and secondhand smoke. From a firefighting point, it should also reduce the number of accidental fires caused by smoking.

RIOTS

Fights between prisoners occur in many penal institutions. If these altercations are not quickly quelled, they can escalate, creating serious problems. Occasionally these situations will develop into riot conditions.

Firefighters should take no part in controlling riots in any penal institution. Firefighters should not be in areas where the potential for bodily harm from prisoners is possible. Should conditions deteriorate once firefighters have entered an area, they should be withdrawn until their safety can be guaranteed.

Rioting is often accompanied by arson. Inmates may use fire as a method of attracting attention for demands that they feel have gone unheeded. They may also be used as a diversion to cover an escape scheme.

Riots in penal institutions could involve a MCI. This type of MCI varies from others due to the potential for violence of the inmates/patients and their attempt to escape.

Confrontations between correctional officers and rioting prisoners can involve hand-to-hand combat with serious injuries occurring. The need to treat these injuries by emergency responders should be accomplished after the combatants are removed to a secure area. The attempt to use hose-lines to control prisoners, if undertaken, must be performed by law enforcement personnel and not firefighters.

Firefighters must realize that wardens and prison officials frown on having to utilize outside law enforcement personnel to maintain order in their facilities. If at all possible, it is necessary for the correctional officers to regain control of disturbances and to maintain the order within the walls of these institutions. This is essential for the correctional officers to save face and be able to regain control of the inmates on a daily basis.

 **ON SCENE**

Rioting inmates set fires, destroying one living unit and extensively damaging four others, at a medium-security correctional facility in Colorado. The riot involved several hundred prisoners. At the time of the riot, there were more than 1,100 prisoners in the facility. No correctional officers were injured, but 13 inmates were transported to area hospitals for injuries sustained. The incident took five hours to quell.

A fire in a Honduran prison killed 103 prisoners and injured 25 others. Some prisoners burned to death, while others died from smoke inhalation. It was reported that the cause of the fire was due to an air conditioner short circuit. The prison had a capacity of 800 prisoners, but held 1,960 prisoners at the time of the fire.

EVACUATION

Some penal institutions have sent correctional officers to receive firefighting training with municipal fire departments. This has been accomplished by their participating in firefighting school with fire department recruits. There are also correctional officers who are volunteer firefighters. During a fire incident in their institution, these trained professionals can often intervene in the situation at an early point in the emergency and mitigate the problems.

For outside units to arrive at the facility, enter through the stringent security measures, and take their equipment to what can be a remote location within the facility will consume precious minutes. In many cases, the prompt response of the correctional officers will control and often extinguish the fire prior to the firefighters reaching the fire area. This can be critical since you are dealing with a population that cannot be easily evacuated. In most cases, the best type of evacuation that can be achieved is to have the inmates exit to an enclosed yard area or to another area within the facility. A problem that can exist is that structures that are built as detention centers for prisoners awaiting trial may have no outside areas to which they can be moved.

The total evacuation of a penal institution is a tremendous undertaking and will typically not occur until after a situation has been stabilized. The EOP should spell out the exact method as to how it will be carried out. Normally the warden or his/her designee will approve the need for evacuation. It must be carried out with maximum security at all times. It will require a large number of correctional officers, police, buses, and a secure facility that they can be transported to. No matter how small these institutions are, this would be a massive undertaking.

UNIFIED COMMAND

The need for implementing Unified Command under the ICS may be necessary at a working fire in a penal institution. Lack of knowledge about ICS can limit the willingness of some jurisdictions or agencies to participate in a Unified Command incident organization. It is impossible to implement Unified Command unless agencies have received training in ICS and Unified Command and agree to participate in the process. (Chapter 2 discusses Unified Command.)

Once it has been decided to implement Unified Command, it is important to select a spokesperson to provide a designated channel of communications for Command and General Staff members into Unified Command. Realize that contrary to the full authority that the Fire IC has over most fire situations, the authority at a fire in a penal institution will normally be controlled by the warden or his or her designee. Strict adherence to the safety parameters that are set up by the warden must be enforced to ensure the safety of all responders. Failure to heed their direction could jeopardize firefighters and correctional officers.

The goals of life safety, incident stabilization, and property conservation will be foremost as always, but recognize that life safety must consider the potential of injury to firefighters and other responders that could be inflicted upon them by the inmates. In addition to the warden or his or her designee, the fire, state, and local police could also be a part of the Unified Command structure. Additionally, under Unified Command, the Operations Section Chief will typically be someone from the prison staff. The Operations Section could also include a Deputy Operations Chief that could be staffed by a fire officer.

The use of an ICS at a prison fire can often resemble operations at a fire in a high-rise building. A Base should be established on the exterior for the marshaling of equipment and personnel. The CP can be positioned either on the exterior of the facility, or in a safe location within the prison proper. In either case, it should be spelled out in the EOP.

The designation of an Operations Section Chief who will operate on the interior will allow better control of the situation, as well as a better understanding of the problems presented by the incident. Once Operations is established, Staging and the RICs would report to Operations. Ideally, a correctional officer assigned to the RIC should have a firefighter background with training in the use of SCBA.

HANDLING OF EMS CALLS

Similar to any medical call, emergency personnel must ensure that they protect themselves from contagious disease exposures. Many penal institutions have found high numbers of inmates who have tested positive for HIV and hepatitis. The inmates can become hostile toward the firefighters and may resist treatment. Security must include the securing of the normal tools utilized by paramedics. Though they may routinely have scissors or pocketknives in clear view on their belt, when entering a penal institution they should place these potential weapons out of sight.

Responders must not assume the role of the correctional officer. When tending to any inmate, there must always be a correctional officer at their location to ensure their safety from the patient and other potential threats. Should an inmate need to be transported to a medical facility, the correctional officer must accompany the inmate within the medic unit or ambulance. Some fire departments and ambulance agencies require a minimum of two correctional officers or police officers to perform this function. In addition to the presence of the correctional officers, the prisoner should be handcuffed on both the hands and feet unless medically detrimental to the patient's health. Likewise, scrutinize the interior of the medic unit to ensure that the patient cannot use any instruments or objects as a potential weapon.

CRITICAL INCIDENT STRESS DEBRIEFING

Riots and serious fires at prisons started by prisoners can lead to firefighters witnessing serious confrontations between the correctional officers and the inmates. These encounters can be quite disturbing to the firefighters. It is recommended that responders who have been subjected to witnessing the brutality that often occurs in quelling prison riots be debriefed for critical incident stress. As with any incident, some responders may be more affected than others. Those who request additional medical treatment should be provided the needed care so that they are able to put these events behind them. (Chapter 11 discusses Critical Incident Stress.)

SIZE-UP FACTORS FOR PENAL INSTITUTIONS

Water
- There must be an adequate water supply for both an offensive and, if necessary, a defensive attack.
- Penal institutions may be located in rural areas and have a limited water supply.
- Preplans should indicate secondary means of water supply available.

Area
- Penal institutions will vary drastically in size, from large inner-city facilities to small frame buildings in rural areas.
- Mazelike configurations can be confusing under heavy smoke conditions.

- Large open areas will permit the spread of smoke to uninvolved areas, complicating life safety problems.
- Warehouses may contain large amounts of combustible/flammable material, creating a large fire load.
- Barns or outbuildings may be of frame construction and contain hay or feed for livestock that could contribute to a fast-moving fire.

Life Hazard
- Severe. Confined occupants will be at the mercy of the correctional officers to make the right choices on movement of inmates from seriously exposed locations.
- TICs can assist in the primary and secondary searches.
- Locked doors will require that correctional officers accompany firefighters with the proper keys.
- Smoke conditions must be monitored and if possible ventilated as early as possible.
- Life safety of firefighters includes potential harm that could be inflicted on them from inmates.
- A decision on whether the safety of the firefighters can be assured must be made before fighting a fire occurs. This should include a partial lockdown that will permit the firefighters, accompanied and protected by the correctional officers, access to the fire area without encountering any inmates.

Location, Extent
- The location of the fire will dictate operations. A fire in an outbuilding containing no inmates will be fought utilizing normal firefighting considerations. A fire in areas containing prisoners will be fought differently.
- TICs can assist in checking walls, ceilings, and floors for hidden fire.
- HVAC systems can spread smoke throughout a building. If it cannot be determined if the system is assisting the fire department's operations and the life safety concerns of the inmates, then it should be shut down.

Apparatus, Personnel
- Call for help early. Depending upon the situation, a large number of resources may be needed for fire control at these facilities.

Construction/Collapse
- The type of building involved will vary, from high-rise buildings of fire-resistive construction to one-story frame structures. There could be collapse potential in older frame or ordinary constructed buildings. The use of heavy timber or noncombustible outbuildings used for storage or shops could also present a collapse potential.
- Buildings that are windowless or have limited window openings will impact ventilation operations.
- Due to the nature of penal institutions, exits are limited and forcible entry may not be feasible to accomplish a means of egress.
- Lexan windows are typically small and will need to be cut with power saws to provide window ventilation.
- Roof ventilation will normally be nonexistent. The trimming of roof ventilators may assist in the ventilation efforts.

Exposures
- Interconnection of penal institutions with courthouses can threaten the other structure should a fire occur.
- Outbuildings can pose a tremendous exposure problem to other buildings in the prison complex, especially buildings housing prisoners.

Weather
- Extremely cold weather could impact on prisoners should they be evacuated to outside areas. These considerations should be addressed in the EOP.

Auxiliary Appliances
- Tamper-resistant sprinklers will be present in some occupancies.
- Standpipes can be expected in high-rise buildings and low-rise sprawling facilities.
- Kitchens should have special extinguishing systems over deep fryers.

Special Matters

- Correctional officers must be familiar with the EOP especially when dealing with evacuation procedures.
- Prisoner control must remain a constant consideration.
- Contracts with bus companies must be current in case evacuation is required.
- Courthouses and sheriff's offices may have buses that offer a more secure method of transportation of prisoners than a typical bus.

Height

- Penal institutions can be of any height.

Occupancy

- Overcrowding is a problem in many prisons. It is not uncommon to find a prisoner population double or triple what the institution was built to hold.
- Exit locations are restricted and correctional officers with keys will be required.

Time

- Nighttime operations may find fewer correctional officers than daytime operations.
- Lag time or reflex time (the time it takes from when an order is given until it is accomplished) can be considerably longer due to distances that will need to be traversed and security measures that will be in place.
- The location of the fire will dictate the amount of time needed for firefighters to reach the fire area.

CONSIDERATIONS FOR FIRES IN PENAL INSTITUTIONS

Strategic Goals and Tactical Priorities for an Offensive Attack

- Require a partial lockdown if the fire is located in prisoner areas.
- Require that a correctional officer accompany firefighters at all times for their protection.
- Require as soon as possible a list of prisoners that are missing and possibly overcome by the fire.
- Ensure an adequate water supply.
- Recognize the potential for an MCI.
- Recognize the possibility that the fire is a ploy to lure correctional officers and firefighters into a position where they can be taken hostage.
- Keep the levels of communication open between the warden and the IC especially in terms of the situation status and evacuation and/or movement of prisoners within the facility.
- Call for a sufficient amount of fire and EMS resources.
- Utilize standpipes to reduce hose-lines stretches.
- Locate and confine the fire to the area of origin if possible. If located in a wing containing prisoners, attempt to prevent the smoke from contaminating other occupied wings. This can limit the exposure of other prisoners and reduce the amount of evacuation or relocation that is necessary.
- The location of the fire will dictate the size of hose-lines. In living quarters or cells, 1½-inch or 1¾-inch hose-lines will be effective. Fires in shops or the laundry may require the use of 2½-inch hose-lines. Fires of a defensive nature will require master streams.
- PPV may be ineffective due to the inability to create adequate openings for the smoke to escape. Ventilation in general will be difficult. Utilize the HVAC systems when possible.
- Trim roof ventilators and cut Lexan windows from the exterior to afford some means of ventilation. Close coordination of ventilation operations between units operating on the exterior and the interior is needed.
- Ensure that a PAR is taken at timed intervals for accountability purposes.

Incident Management System Considerations/Solutions for an Offensive Attack

- Incident Commander(s) (Unified Command should be considered)
- Liaison Officer (if needed)

- Safety Officer
- Rapid Intervention Crew(s)
- Operations (if needed)
- Staging
- Ventilation Group
- Search and Rescue/Evacuation Group
- Division 1, 2, 3, etc. (for multistoried building can be implemented for any floor where deemed necessary)
- Planning (if needed)
- Logistics (for water supply and establishing Base)
- Base (to marshal equipment and personnel)
- Medical Group or Branch (to set up triage, treatment, transportation, morgue, etc.)

Strategic Goals and Tactical Priorities for a Defensive Attack

- If changing from an offensive to a defensive attack ensure that a PAR is taken for accountability purposes.
- Set up and maintain a collapse zone if necessary.
- Utilize master streams.
- Protect exposures if necessary.
- Ensure the safety of firefighters at exterior operations that require the firefighters to enter the facility's grounds.

Incident Management System Considerations/Solutions for a Defensive Attack

- Incident Commander(s) (Unified Command should be considered)
- Liaison Officer (if necessary)
- Safety Officer
- Public Information Officer
- Operations (if needed)
- Staging
- Divisions on the exterior on Sides A, B, C, D
- Exposure Divisions (if needed)
- Planning (if needed)
- Logistics (for water supply and establishing Base)
- Base (to marshal equipment and personnel)
- Medical Group or Branch

 NIOSH FIREFIGHTER FATALITY REPORT F2007-32

On August 29, 2007, a 55-year-old male career firefighter (Victim #1) and a 52-year-old male career firefighter (Victim #2) died while conducting an interior attack to locate, confine, and extinguish a fire located in the cockloft of a restaurant. Upon arrival, fire was showing through the roof with negligible smoke and heat conditions in the main dining area. Victim #1 was on the nozzle flowing water on the fire in the ceiling area above the exhaust hood and duct work for the stove/broiler in the kitchen. His officer and the officer from the first arriving ladder company provided back-up on the 1¾-inch handline. Victim #2 was in the main dining area searching for fire extension above the suspended ceiling. Approximately five minutes after the first crew arrived on the scene, a rapid fire event occurred. Victim #1 was separated from his crew and was later found on the handline under debris with trauma to his head. Victim #2 had a lapel microphone with an emergency distress button that sounded a minute after the rapid fire event, likely from fire impingement. He was found in the area of the dining room where he was operating just before the rapid fire event occurred.

Key contributing factors identified in this investigation include an insufficient occupational safety and health program, ineffective incident management system at the incident, insufficient incident management training and requirements, insufficient tactics and training, ineffective communications, delay in establishing a RIC, inadequate building code enforcement and development, and inadequate turnout clothing and personal protective equipment. See Appendix B: NIOSH Reports in www.pearsonhighered.com/bradyresources to read the complete report and recommendations.

Public Assembly Buildings

There have been many disastrous fires in **public assembly buildings** over the years, including:

public assembly buildings ■ Include auditoriums, dance halls, bowling alleys, stadiums, movie theaters, restaurants, and museums.

- Cocoanut Grove nightclub in Boston, 1942 (492 dead)
- Beverly Hills Supper Club in Southgate, Kentucky, 1977 (164 dead)
- Happyland Social Club in Bronx, New York, 1990 (87 dead)
- Station Nightclub in West Warwick, Rhode Island, 2003 (100 dead)

Public assemblies consist of a variety of structures and occupancies:

- Amusement park buildings and grandstand tents
- Auditoriums, dance halls, poolrooms, and lecture halls
- Bowling alleys, gymnasiums, stadiums, and skating rinks
- Houses of worship
- Restaurants
- Nightclubs, movie theaters, legitimate theaters, and museums

The potential for loss of life in a public assembly fire is directly related to the number of people threatened. Success will depend upon a number of factors or special problems associated with public assembly buildings.

OCCUPANCY

Some public assemblies have the potential for overcrowding. This can be a common occurrence in popular nightclubs during special events. When overcrowding exists, exit doors can become congested with people, restricting egress. The number and placement of exit doors is based upon the number of people that can reasonably be expected to occupy a building. The inability to leave a building when confronted with smoky conditions can lead to panic.

 ON SCENE

A fire at the Cocoanut Grove Nightclub in Boston occurred on November 28, 1942. It is considered one of the worst disasters in American history. The fast-moving fire broke out and roared through the nightclub being fed by flammable decorations. There were approximately 1,000 patrons in the club at the time of the fire, which had a legal occupant capacity of 460. Panicked occupants were trapped by locked exits and jammed revolving doors at the front entrance. The loss of life reached 492 people plus 166 injured, many seriously. Major changes to the Life Safety Code occurred due to this fire including requirements for emergency lighting, exit lights, occupant capacity, and exit doors opening outward.

Though buildings have maximum occupancy loads, they are difficult to enforce. When compliance is a constant problem, some fire officials have been forced to tell the owner that the establishment appears to be overcrowded and that everyone must leave the building. As patrons reenter the structure, the fire official will count heads to ensure compliance. Though a stringent measure, this interruption of business often leads to future cooperation and compliance.

COMBUSTIBLE FURNISHINGS

Though an official can diligently inspect a building during construction for code compliance, furnishings can later be used that can contribute to a fast-burning fire. Management is concerned with the aesthetic effect created by furnishings. They rarely pay attention to the fire resistance of a material selected for their occupancies. Seasonal motifs can pose a deadly threat. Live evergreen trees, though beautiful, dry out quickly. When ignited, they release a tremendous amount of heat. Loosely hanging banners or crepe paper can ignite easily. Combustible interior furnishings can permit a fire to spread rapidly, often creating large amounts of deadly smoke. Decorative combustible furnishings will let fire spread at such a rate that the occupants may have little time to escape.

 **ON SCENE**

In February 2003, a deadly fire occurred in West Warwick, Rhode Island, at the Station Nightclub. The blaze killed 100 people and injured 200 others. The fire was started by a band's pyrotechnic display that ignited foam decoration and wood paneling. The nightclub patrons first recognized danger 24 seconds after ignition. The bulk of the crowd began to evacuate around the time the band stopped playing (about 30 seconds after ignition). The heat and smoke detection system was activated 41 seconds after the fire started. About 90 seconds into the fire, about two-thirds of the occupants attempted to leave via the main entrance; a crowd crush occurred at that location that almost entirely disrupted the flow of the evacuation through the front exit. There was fire observed breaking through the roof in less than five minutes after the fire started.

Measurements in a fire test conducted by the National Institute of Standards and Technology (NIST) in a mockup of the Station Nightclub's platform and dance floor produced—within 90 seconds—temperatures, heat fluxes, and combustion gases well in excess of accepted survivability limits. The tremendous speed of this fire, which was captured on film, shows how little time firefighters had to even attempt to make a difference.

When arriving on the scene of a fully involved occupied structure, there is little firefighters can accomplish in attempting to rescue trapped occupants. The fire safety of the occupants must occur in the preplanning stages with a properly designed and installed sprinkler system. Other factors impacting this fire were the use of pyrotechnics and the decorative foam and wood paneling, which easily ignited, causing the rapid fire spread.

 **ON SCENE**

On August 13, 2011, at the Indiana State Fair in Indianapolis, an outdoor concert was set up on a stage constructed for the event. Winds from an approaching thunderstorm hit the stage's temporary roof structure, causing it to collapse. The roof landed on a crowd of spectators, killing seven and injuring 58 others.

CONSTRUCTION AND DESIGN

Construction and design of a building must consider the life safety of occupants. Many nightclubs and restaurants are on the top floors of high-rise buildings. This distant location can delay firefighters from arriving and attacking a fire in a timely manner. If delayed, they cannot assist in initial rescue or evacuation efforts.

A public assembly may consist of large unbroken areas that allow a fire to spread quickly. A mazelike configuration can confuse people disoriented by smoke and trying to locate the door through which they entered the building.

Firefighters may find that building access is limited due to renovations that could include false fronts and windowless buildings. Some buildings have been converted from other occupancies. An example is a theatrical group that converts a barn to a theater. Firefighters unaware of such a change in building occupancy could be confronted with serious life hazard in a situation where they would expect a routine barn fire.

A person unfamiliar with a building most likely will not react according to a prescribed evacuation plan unless directed to do so by a person in authority. Under emergency conditions, this direction may not be given or the person may not understand the message. This leads to individuals reacting the way they feel will best safeguard them, causing evacuation problems for firefighters.

A building's HVAC systems can spread fire and smoke. Knowledge and control of these systems by the fire department may assist firefighting operations if these systems can be utilized to ventilate deadly smoke from the building. If this is not possible, they should be shut down immediately to prevent spread of fire and smoke.

A prime ingredient in life safety in public assemblies is the actions of employees should a fire start. The training that they receive enhances their reactions. Attempting to extinguish a minor trash can fire with a portable fire extinguisher can prevent a major problem. Promptly reporting a fire and alerting the occupants is critical. Employees have the ability to cause panic or reduce it.

 **ON SCENE**

EXITS AND EVACUATION

Conference rooms in places of assembly may have only single entrance and exit points, hindering rapid and orderly exit. The older style of panic hardware was a bar that was pushed toward the exit door, allowing the door to open. The newer style is secured flat against the door. This prevents a person's arm from sliding into the open bar, preventing the door from opening. An adequate number of properly marked and unobstructed exits is a must. Tables, curtains, large planters, and other decorative schemes may block exits. No perception of danger is realized by management or by most occupants by these actions. The problem is the general apathy toward fires. Most people feel that fires happen to other people, or they say, "I've been in business for 25 years and have never had a fire." This indifference is difficult to overcome.

A problem is found at auditoriums and stadiums when large sporting events or concerts attract teenagers. Exit doors may be chained or padlocked to prevent a ticket-holder from opening an unguarded door and allowing unpaid persons to enter.

Locked exits not only cut down the number of exits available but also can cause deadly delays. Anyone finding a locked exit door needs to locate another exit. A firefighter may be endangered by relying on an exit as a secondary means of egress, only to find the door locked.

 ON SCENE

EVACUATION

Some public assemblies are easier to evacuate than others.

- A bingo hall can be difficult. Patrons do not want to be disturbed unless confronted with conditions that they consider life-threatening.
- People told to leave a movie theater want a refund before exiting, adding confusion and delay.
- The ease of evacuating a restaurant depends upon whether the patrons have already eaten their meals. If so, they leave happily; if not, they may resist.
- Alarm systems in casinos may not ring in the gaming areas themselves. The owners realize that if patrons are made to leave, many fail to return. The winners rethink their lucky streak, and the losers think their luck will change at another casino.

 **ON SCENE**

While patronizing a casino, I witnessed a trash can fire that brought the local fire department, but no alarm ever sounded. The fire was in an area of slot machines and was rolling up the wall. The patrons playing nearby slot machines shielded themselves from the heat of the fire and continued to gamble, acting as if nothing was amiss. Though it was a minor fire, an audible alarm should have sounded and evacuation should have been initiated.

Response times can be critical for fires in public assembly buildings in remote rural areas. Their distant location will not only increase response time but also could hinder apparatus arrival and placement if large numbers of occupants try to drive their vehicles from the building when only one narrow road exists for entry and exit.

 NIOSH FIREFIGHTER FATALITY REPORT F2013-04

On February 15, 2013, a 36-year-old male career fire lieutenant (Victim #1) and a 54-year-old male career fire lieutenant (Victim #2) were killed and two career firefighters were injured (FF1 and FF2) at an assembly hall fire. At approximately 2320 hours, the fire department responded to a reported assembly hall fire with flames visible. Upon arrival, fire was observed by the fire department burning at the roof level in the area of the A/B corner. After arriving units conducted a 360-degree walk-around, offensive interior operations were employed to stop the fire's progression. The first interior attack crew (Victim #1 and his probationary firefighter) advanced a hose-line toward what they believed was the seat of the fire. After discovering the fire in the A/B corner and flowing water on it, Victim #1 and his probationary firefighter both became low on air. Victim #1 told his probationary firefighter they needed to exit. Both began following the hose-line out, with the probationary firefighter in front. However, for an unknown reason, Victim #1 was unable to follow the hose-line and he became separated from his probationary firefighter. Victim #1 radioed for help. The probationary firefighter called out to Victim #1 but did not receive an answer nor did he hear any alarms. He then continued following the hose-line to the outside. At this time, an engine company (Victim #2, FF1, and FF2), which was designated as the rapid intervention team (RIT), was immediately deployed into the structure to locate Victim #1. The RIT followed the hose-line in and was able to locate Victim #1, who was responsive. While dragging Victim #1 toward the exit, the RIT was caught in a flashover. Following the flashover, all of them were quickly removed and transported to local hospitals. Unfortunately, Victim #1 did not survive his injuries. FF1, FF2, and Victim #2 were transferred to a regional burn center for extensive treatment where Victim #2 later succumbed to his injuries.

Contributing factors include a nonsprinklered commercial building; risk management principles not effectively used; high-risk, low-frequency incident; fireground strategy, tactics, and ventilation; rapid fire progression; fire burn and spread undetected above the ceiling; crew integrity; SCBA air management; fireground communications; and flashover. **See Appendix B: NIOSH Reports in www. pearsonhighered.com/bradyresources to read the complete report and recommendations.**

STRATEGIC CONSIDERATIONS

Overcrowding will compound evacuation problems. There may be a large number of occupants unfamiliar with the building hurrying to escape, but in the wrong direction. Lighting may be dimmed for an aesthetic effect. Smoke may further cut visibility. It is not always possible to determine if everyone has been evacuated or if some individuals are still trapped within a building. A parking lot full of cars could indicate a large number of occupants, but this is not always the case, since the owners of the vehicles may be patronizing other nearby businesses, or the patrons may have exited the building and be unable to move their vehicles due to the fire department blocking the parking lot exits.

Familiarization with suppression systems lets firefighters know what building protective systems can assist them.

A decision must be made on whether firefighters should be committed to fight the fire or assist in evacuation. A direct attack, which controls a fire, is often the best method of saving lives. Coordination is needed to assure that firefighting efforts do not endanger occupants trapped or evacuating the building. Immediate ventilation is one of the best strategies for saving lives. The removal of smoke and gases, though initially accelerating the fire, permits an effective evacuation and fire attack.

SIZE-UP FACTORS FOR PUBLIC ASSEMBLY BUILDINGS

Water
- There must be an adequate water supply for both an offensive and, if necessary, a defensive attack.
- Public assembly buildings located in rural areas may have a limited water supply available.

Area
- Public assemblies will vary drastically in size.
- Mazelike configurations can confuse patrons. They will most often attempt to exit the same way they entered.
- Large open areas will permit a fire to spread quickly.

Life Hazard
- Life hazard is severe if the building is occupied. It is difficult to know who the occupants are and how many may still be in the building.
- Many people may be unfamiliar with the building.
- The parking lot may be full of cars, yet not many patrons may be visible on the exterior.
- TICs can assist in the primary and secondary searches.
- Occupants of nightclubs may have consumed alcoholic beverages that can cloud their judgment.
- Exit doors may not be adequately marked or may be locked or chained shut.
- Fast-moving fires will necessitate a prompt response and attack on the fire.
- Lighting may be subdued for aesthetic purposes, reducing visibility. Responders should ensure that during emergency situations lighting is raised to the maximum level to assist in evacuation efforts.

Location, Extent
- Locating the fire can be difficult if a building is full of people trying to evacuate at the same time. Light smoke conditions can worsen in a short period of time, and panic is possible.
- TICs can assist in checking walls, ceilings, and floors for hidden fire.
- HVAC systems can spread smoke throughout a building. If it cannot be determined if the system is assisting the fire department's operations, then it should be shut down.

Apparatus, Personnel
- Call for help early. A large number of resources will be needed in an attempt to control these fires, achieve adequate ventilation, and accomplish search, rescue, and evacuation.
- A decision on whether to fight the fire or assist in the evacuation may have to be made if resources are limited.

Construction/Collapse
- Many of these buildings started out small, but were constantly enlarged to meet the growing demands of the business.
- Windowless buildings will impact on ventilation operations.
- Renovations are commonplace in some public assembly buildings, and they can affect fire spread.
- There may only be one exit from some areas.

Exposures
- There are no specific external exposure problems.

Weather
- Extremely cold weather may delay some evacuation from these structures until the patrons gather their overcoats.

Auxiliary Appliances
- Sprinklers will be present in some of these occupancies. Standpipes can be expected in high-rise buildings.
- Kitchens may have special extinguishing systems over deep fryers.

Special Matters

- Employees trained in evacuation procedures can assist the fire department.
- Combustible furnishings can contribute to a fast-burning fire, creating deadly smoke.
- Tents are considered public assemblies.

Height

- Public assemblies can be found in any size of building.

Occupancy

- Overcrowding is a problem in some occupancies.
- Exit doors can become congested with people trying to leave the building, restricting egress and leading to panic.
- Maximum occupancy loads are rarely enforced.

Time

- The critical time period is when the building is occupied, especially if overcrowded.

CONSIDERATIONS FOR FIRES IN PUBLIC ASSEMBLY BUILDINGS

Strategic Goals and Tactical Priorities for an Offensive Attack

- Evacuate and perform search and rescue for occupants.
- Ensure that all exits are open.
- Recognize the possibility of an MCI.
- Call for a sufficient amount of resources (fire and EMS).
- Hose-lines cannot be stretched through doorways being used as exits by occupants.
- If subdued lighting is present, ensure that it is raised to facilitate evacuation.
- Locate and confine the fire to the area of origin, if possible, while attempting to protect avenues of egress for the occupants.
- Depending on the size of the public assembly building, 1½ -inch or 1¾ -inch hose-lines may be ineffective; 2½ -inch hose-lines may be needed.
- Ventilation can act as a lifesaving tool. With limited window openings in many of these buildings, the roof should be an early consideration. Roof vents may be present; see if they are open. Try to utilize the HVAC units to assist in removing the smoke.
- To use PPV, you need to know the exact location of the fire, the location for smoke to exit, and whether any occupants would be endangered. This information will be difficult to obtain in the early stages of a fire, making PPV a poor initial choice.

Incident Management System Considerations/Solutions for an Offensive Attack

- Incident Commander
- Safety Officer
- Rapid Intervention Crew(s)
- Operations (if needed)
- Staging
- Ventilation Group
- Search and Rescue/Evacuation Group
- Division 1, 2, 3, etc. (for multistoried building, a division can be implemented for any floor where deemed necessary)
- Medical Group or Branch (to set up triage, treatment, transportation, morgue, etc.)

Strategic Goals and Tactical Priorities for a Defensive Attack

- If changing from an offensive to a defensive attack, ensure that a PAR is taken.
- Set up and maintain a collapse zone.
- Utilize master streams.
- Protect exposures, if necessary.

Incident Management System Considerations/Solutions for a Defensive Attack

- Incident Commander
- Safety Officer
- Public Information Officer

- Operations (if needed)
- Staging
- Logistics (for water supply, if needed)
- Divisions on the exterior on Sides A, B, C, D
- Exposure Divisions (if needed)
- Medical Group or Branch

 ON SCENE

The school fire that gained much notoriety due to the seriousness of the fire and the number of fire deaths was the Our Lady of the Angels School fire in Chicago on December 1, 1958. This grade school was occupied by 1,635 students, including 329 who had their primary escape route cut off. Only through the valiant efforts of the firefighters were more than 160 children rescued. Yet 92 children and three nuns died.

In the aftermath of the Our Lady of the Angels fire, many drastic recommendations and code changes were made for fire safety in schools. These included enclosing open stairways, providing self-closing fire doors at all corridor openings, more stringent exit requirements, stricter interior finishes, and the installation of automatic sprinklers. These recommendations have been enacted to varying degrees by schools throughout the United States.

School Fires

The US Fire Administration (USFA) gathers and analyzes information on the nation's fire problem through the National Fire Incident Reporting System (NFIRS). The causes of school fires can range from accidental to incendiary. Statistics on school building fires and nonconfined school building fires from 2009 to 2011 are quite significant.

school fires ■ A fire in an institution of learning.

SCHOOL BUILDING FIRES: FINDINGS (2009–2011)

- An estimated 4,000 school building fires were reported by US fire departments each year and caused an estimated 75 injuries and $66.1 million in property loss.
- Fatalities resulting from school building fires were rare.
- There was a general increase in school building fires toward the beginning and end of the academic year.
- The three leading causes of school building fires were cooking (42 percent), intentional action (24 percent), and heating (10 percent). Intentional action was the leading cause of nonconfined school building fires (41 percent).
- The leading area of fire origin in nonconfined school building fires was the bathroom at 25 percent.
- In 75 percent of school building fires, the fire spread was limited to the object of origin.
- Smoke alarms were reported as present in 66 percent of nonconfined school building fires.

Source: USFA Topical Fire Report Series, "School Building Fires (2009–2011)"

If properly functioning, smoke and heat detectors will allow early detection. Naturally, properly functioning sprinklers will control and quickly extinguish most fires in the incipient stage.

Schools that have laboratories can contain myriad dangers and can set the stage for a hazardous materials incident. It is not unusual to find storage cabinets with dangerous chemicals stored next to one another, or to find containers of hazardous chemicals that have crystallized.

EVACUATION PLAN

Life safety in schools demands designing and practicing a comprehensive plan for the safe evacuation of all occupants. This plan must include primary and secondary exit routes for each classroom. The secondary exit route would be used if the initial exit point were inaccessible due to smoke or fire conditions. The exit points should be designated to provide for an even distribution of the school's occupants. A drawing of the evacuation plan should be posted near the doorway in each classroom.

School fire drills should be conducted monthly. The first drill should occur as early as possible at the beginning of the school year. School fire drills should be initiated by sounding the school's fire alarms to familiarize students and teachers with the alarm system.

At least once each semester, the drills should be monitored by the fire department. The presence of firefighters will reinforce the importance of the drills. Through their observation, the firefighters can see if any deficiencies exist. Problems observed by the fire department should be brought to the attention of the school administrators and potential solutions discussed.

The emergency evacuation plan should designate specific locations for assembling classes on the exterior that are away from the school building. These exterior assembly points should not be located where they could interfere with firefighting operations.

Roll call books must be taken from every classroom by the teachers to ensure that an accurate head count is taken on arrival at the exterior assembly point. The results of the roll call should be reported to the school principal or administrator. If anyone is missing, the fire department must be notified immediately. This notification should include the last known location of the missing person in the building and the approximate age, if a student.

School personnel must not reenter the building to search for missing students. After the roll call is taken, the students should be kept at a safe location on the exterior or at an assembly point in another building. Any search will be accomplished by the firefighters.

Should a fire force the dismissal of the students for the day, the following points should be considered:

- The age of the children. If younger students are being dismissed earlier than normal, parental notification will be required.
- High school students can usually be dismissed without parental notification.
- In all cases a thorough roll call must be taken and no students may be allowed to return to the school building while an active fire operation is underway.

There should be a concern that radio and television reports can set off panic on the part of parents who will drive to the scene for their children. If the children are not still on location, it can cause numerous problems for the school administration and the fire department.

EVACUATION PROBLEMS

Many codes require that kindergarten and first-grade students be housed on the first floor. This will prevent them from having to traverse stairs where they would be prone to being overrun by older students evacuating the building. Evacuation of physically disabled students, mentally challenged students, teachers, and teacher aides will need to be addressed.

A problem may exist where smoke detectors have been tampered with so they won't be activated by smoke from cigarettes. This condition can be found near bathrooms where illegal smoking occurs. Malfunctioning smoke detectors will cause a delay in the sounding of an alarm should a fire take place.

BUILDING DESIGN

Buildings designed with an open-classroom concept utilize movable partitions to separate classrooms. These partitions are similar to those found in business occupancies where partitions separate offices in large rooms. This arrangement provides no fire-stops and could threaten a much larger number of students in the one room. The benefits of a large open classroom are that it may provide for early detection, immediate evacuation, and extinguishment of a minor fire.

A fully circular school building can be confusing under heavy smoke conditions. Students attempting to evacuate the building may become disoriented and have difficulty in finding the exit doors. Firefighters may have difficulty in determining their location within the building. Getting your bearings often requires orientation by looking through windows to the outside.

Windows in older classrooms may be four feet or higher above the floor level, making it difficult to reach for younger children. Window design may restrict students from climbing through them to reach the exterior.

EXITS AND FIRE DOORS

Firefighters conducting school inspections may find that exit doors are chained or locked. This is often done for the safety of the students to prevent outsiders from entering the building and causing disturbances. It may also be used as a tool to keep students from leaving the building at unauthorized times without being observed.

Regardless of the good intentions of school security and administrators, all emergency exit doors must be fully functional whenever the school is occupied. This may demand a larger security force or the installation of cameras to monitor these locations.

Another problem is fire doors that are propped open to allow for ease of travel for the students. This situation will negate the containment of smoke and fire, allowing a minor fire to affect a larger area. Fire doors should contain mechanisms to close automatically when an alarm is activated.

School social events held in gymnasiums and auditoriums can involve a large number of students who will need to be immediately evacuated in the event of a fire.

FIREFIGHTING

The actions and decisions that need to be made by the first-arriving firefighters at a school fire will be difficult. There is a natural instinct for all units to become involved in the rescue effort and ignore the fire. (See Figure 8-13.)

Realize that assigning all resources to rescue efforts and allowing a fire to continue to burn freely will cause rapidly deteriorating conditions. If all rescues cannot be accomplished before the conditions become intolerable, more lives may be lost than could have been saved with an attack on the fire. Those initial units must base their decisions upon the following considerations.

- The fire department resources available on the scene
- The amount of time before additional units will arrive
- The number and location of the students still in the school
- The time it will take for a total evacuation
- The actions that will best protect the lives of the students and teachers
- The smoke and fire conditions
- Whether the fire doors have been shut
- Whether there are sprinklers operating that are containing the fire

The best method of protecting trapped occupants is often a quick extinguishment of the fire. Locating the fire and deciding on the correct entrance from which to attack the fire without interfering with the evacuation must be made. The hose-line should be placed between the fire and those attempting to escape in order to provide time for their removal.

A primary search must be initiated as soon as possible. Determining the approximate age of a missing student may assist in locating them. An older student may have just left the school rather than participate in the evacuation plan. A younger student may be afraid and hiding. Children tend to hide from the fire or become afraid due to the noise and commotion caused by the situation. Favorite hiding places include closets and bathrooms.

Ventilation must be an initial consideration. Performed properly, it can channel smoke and fire away from the occupants, protecting them until rescues can be made.

FIGURE 8-13 School fires will challenge every fire department. *Used with permission of Greg Masi.*

LADDERING

Portable ladders, main ladders, and tower ladders can be used to gain access to the upper floors of a multistory school. To achieve maximum benefit of an aerial apparatus, it must be able to approach the building and be spotted in the most advantageous location to reach the upper floors. In newer schools with paved surfaces and ample driveways, this may be achieved easily. The school, however, must enforce no-parking restrictions and not permit illegally parked vehicles that could interfere with apparatus placement.

Some schools are fenced in and surrounded by a schoolyard. This can restrict access of apparatus if only pedestrian gates are provided. There must be methods of permitting apparatus access to reach the building.

Multistory schools built on sloping terrain and containing unpaved areas can restrict access or make it difficult or impossible for apparatus placement. Soft soil could cause the apparatus to sink to their chassis before reaching their intended location.

 **ON SCENE**

Any Philadelphia school that is set back from the street with a schoolyard that does not contain an entrance gate wide enough to accommodate fire apparatus must maintain a removable section of fencing. A center post painted red identifies this removable section of fence. This allows the fire department to quickly remove the necessary section of fence and gain access in a timely manner.

PREPLANNING

A plot plan of the building will allow the easy identification of common assembly areas, such as gymnasiums, cafeterias, libraries, and auditoriums. Firefighters should anticipate that a large number of evacuations will be required from these locations.

The kitchen or cafeteria areas can have the same problems associated with any restaurant, and they should be protected the same way as the codes stipulate for restaurants.

SIZE-UP FACTORS FOR SCHOOL FIRES

Water
- There must be an adequate water supply for an offensive and, if necessary, a defensive attack.
- School buildings in rural areas may have a limited water supply available.

Area
- Schools can vary drastically in size. The common denominator is the extreme life hazard.
- Fire doors should be installed and functioning properly.
- Combustible furnishings in storage and stage areas can contribute to a fast-burning fire, creating deadly smoke.

Life Hazard
- This is severe if the school is occupied.
- Fast-moving fires will necessitate a prompt response and attack on the fire.
- Children who are not in their classroom will attempt to exit the building the same way they entered.
- Exit doors may be locked or chained shut.
- Parents will immediately respond to the school on the report of a fire, looking for their children and possibly interfering with fire department operations. Someone must be prepared to handle this influx of parents, answer their questions, and reunite them with their children.
- TICs can assist in the primary and secondary searches.
- A fire may initiate an MCI and place numerous students in danger.

Location, Extent

- Locating the fire can be difficult if a school is fully occupied and school officials are trying to evacuate at the same time. Light smoke conditions can worsen quickly and panic is possible.
- HVAC systems can spread smoke throughout a building. If they are not assisting the fire department's operations, they should be shut down.
- Thermal imaging can assist in checking for hidden fire in walls, floors, and ceilings.

Apparatus, Personnel

- Call for help early. A large number of resources will be needed to control these fires; achieve adequate ventilation; and accomplish search, rescue, and evacuation.
- A decision on how to fight the fire and assist in the evacuation will have to be made based upon the available resources, amount and location of the occupants, and the fire and smoke conditions found.
- Determine the best method of utilizing the school's staff to assist in the accountability of the students, teachers, and other occupants of the school.
- Consider requesting MEDEVACs for patient transport in the event of an MCI.

Construction/Collapse

- Many schools are overcrowded and use trailers as classrooms on the exterior. These trailers are permanent in nature and immobile, and they may restrict apparatus access and ladder placement.
- Sloping terrain or unpaved surfaces may restrict apparatus placement for access to upper floors.
- There may be only one exit from some areas.
- All types of building construction can be found, including fire-resistive, ordinary, frame, and mixed construction.
- Renovations may affect fire spread.
- In multistoried schools, there may only be one stairway for firefighting and evacuation, or there may be multiple stairs. The fire department will need to determine which to use for firefighting and which for evacuation.

Exposures

- Internal exposures will be severe.
- A religious school may have interconnected buildings with a church, rectory/parsonage, or convent.

Weather

- Cold weather will affect occupants evacuated from the school; a nearby indoor area will need to be secured to protect the students and conduct a comprehensive roll call.
- Triage and treatment areas, if established, will be affected by cold and inclement weather.

Auxiliary Appliances

- Sprinklers and standpipes will be present in some schools. These systems should be pressurized by the fire department to ensure a continuous and adequate supply of water.
- Kitchens or cafeterias may have special extinguishing systems over deep fryers.
- Intentionally set fires can activate sprinkler systems for fire control.

Special Matters

- Closing classroom doors after evacuating will assist in preventing the spread of fire into those areas.
- If multiple victims exist, develop a list of hospital availability for the transport of patients.

Height

- One-story buildings will allow evacuation through hallways and via windows if necessary.
- Multistory buildings may require rescue/evacuation via fire department ladders.

Occupancy
- Under fire and smoke conditions, one excited student yelling and screaming can become contagious, and panic is possible.
- Exit doors can become congested with children trying to leave the building, restricting egress, and possibly panicking others.
- The school building may be occupied at various times, when used for sporting events, parent-teacher meetings, community education, and social events.

Time
- The critical time period is when the building is occupied. Heavy smoke and fire conditions will require immediate evacuation assistance from the fire department.

CONSIDERATIONS FOR FIRES IN SCHOOLS

Strategic Goals and Tactical Priorities for an Offensive Attack
- Evacuate and initiate search and rescue for occupants.
- Ensure that all exits are unlocked and unobstructed.
- Recognize the potential for an MCI.
- Call for a sufficient amount of resources (fire and EMS).
- Ensure that hose-lines are not stretched through the doorways being used as exits by students.
- Ventilation can act as a lifesaving tool.
- Large, sprawling schools should contain fire doors that must be checked to ensure that they are closed.
- Locate and confine the fire to the area of origin if possible while attempting to protect avenues of egress for the students.
- Depending on the location and size of the fire, 1½-inch or 1¾-inch hose-lines may be ineffective; 2½-inch hose-line may be needed for large meeting rooms or auditoriums.
- Horizontal ventilation should be sufficient for a room and contents fire. For fires involving large areas, vertical ventilation may be necessary.
- Try to utilize the HVAC system to assist in removing smoke from the building.
- Roof vents may be present over common areas; ensure that they are open.
- To use PPV, you need to know the exact location of the fire, the location for smoke to exit, and whether any students/teachers would be endangered. This information will be difficult to obtain in the early stages of a fire. It will also be difficult to control building openings since the exit doors should all be opened to permit a safe and timely evacuation. These factors impact PPV, making it not a good initial choice.

Incident Management System Considerations/Solutions for an Offensive Attack
- Incident Commander
- Safety Officer
- Liaison Officer (to interact with school and police personnel on accountability, and assist in communicating information to concerned parents)
- Rapid Intervention Crew(s)
- Operations (if needed)
- Staging
- Ventilation Group
- Search and Rescue/Evacuation Group
- Division 1, 2, 3, etc. (for multistoried building can be implemented for any floor where deemed necessary)
- Exposure Divisions A, B, C, D (if needed)
- Logistics (if needed to assist in water supply and/or sheltering of students)
- Medical Group or Branch (to set up triage, treatment, transportation, morgue, etc.)

Strategic Goals and Tactical Priorities for a Defensive Attack
- If changing from an offensive to a defensive attack, ensure that a PAR is taken for accountability purposes.
- Set up and maintain a collapse/safety zone.
- Utilize master streams.
- Protect exposures if necessary.

Incident Management System Considerations/Solutions for a Defensive Attack

- Incident Commander
- Safety Officer
- Liaison Officer (to interact with school and police personnel on accountability, and assist in communicating information to concerned parents)
- Public Information Officer
- Operations (if needed)
- Staging
- Logistics (for water supply and/or sheltering of students if needed)
- Divisions on the exterior on Sides A, B, C, D
- Exposure Divisions (if needed)
- Medical Group or Branch

 ON SCENE

A student starting a fire usually does so without concern for the other students and teachers. Realize that a rather small fire can cause a tremendous amount of smoke. A problem found in one high school occurred when just one hard plastic toilet seat was wrapped with toilet paper and ignited. The doors to the bathrooms were propped open. The resultant smoke was enough to bank down an entire floor of a school that contained wide hallways, a few hundred feet in length. By the time the fire department arrived, the fire was in a smoldering state. It also achieved its intended goal of an early dismissal for the students.

Schools and Other Active Shooter Incidents

Schools are locations where children and young adults should be exposed to positive learning experiences and relationships that last a lifetime. It is a place where they should feel completely safe and where good things happen. Sadly, in recent years there have been too many occasions where terrorist events have caused the exact opposite to occur. In addition to the increased occurrences of school violence imposed by one student upon another or the instances of gang activity involving shootings at schools, there are many random attacks against students and teachers. The incidents of school shootings have changed some elementary school, high school, and college campuses from docile settings to shooting fields in what are now called **active shooter incidents**.

Due to these factors and the need for full preparation, responders must gain a better understanding of how to deal with these incidents. A thorough hazard analysis should be made of case study reviews to identify problems that occurred in incidents at other schools and other venues, to anticipate the risk assessment needed if an incident occurs, and to define the possible need for multi-agency procedural changes to handle any potential problems.

active shooter incidents ■ Violent acts perpetrated on the student body at schools, or occupants at other venues, with the intent of killing and maiming.

These low-frequency, high-risk, massacre-type shootings have been committed at schools and other venues by students, employees, and intruders, and are performed for the sheer goal of seeing how much carnage can be inflicted.

Some instances of school violence are:

January 17, 1989; Stockton, California A 24-year-old intruder fired more than 100 rounds into a school playground, killing five students and wounding 29 other students and one teacher before killing himself.

April 20, 1999; Littleton, Colorado Fourteen students (including the killers) and one teacher were killed, and 23 others were wounded by 17- and 18-year-old students at Columbine High School in one of the nation's more deadly school shootings.

October 3, 2006; Nickel Mines, Pennsylvania A 32-year-old civilian entered the one-room West Nickel Mines Amish School, and shot 10 schoolgirls, ranging in age from 6 to 13 years old, and then killed himself. Five of the girls died.

April 16, 2007; Blacksburg, Virginia A 23-year-old Virginia Tech student killed two students in a dorm. Two hours later he killed 30 more in a classroom building.

His suicide brought the death toll to 33. Fifteen others were wounded. This was the deadliest school shooting rampage in US history.

February 14, 2008; DeKalb, Illinois   A gunman, who was a former Northern Illinois University student, killed five students and then himself and wounded 18 others when he opened fire on a college classroom.

February 23, 2010; Littleton, Colorado   A 32-year-old gunman shot and wounded two students outside a middle school before being subdued by a teacher as he tried to reload his rifle.

December 8, 2012; Newtown, Connecticut   A gunman at Sandy Hook Elementary School killed 27, including 20 children, and injured two others before committing suicide.

October 1, 2015; Roseburg, Oregon   A gunman at Umpqua Community College shot and killed 10 people including himself, and another seven were injured when he opened fire in a classroom.

September 1, 2004; Beslan, Russia   The Beslan school hostage crisis occurred in the Russian Federation and lasted for three days as the intruders made political demands. Thirty masked and heavily armed Chechen terrorists forced their way into the school, shooting and killing 12 adults, and took hostages of more than 1,000 students and parents. Some of the terrorists were wearing bomb belts. Police cordoned off the school as the intruders threatened to blow it up. Hostages were chained to explosives and shots were heard at various times coming from the building. Threats were made to kill hostages. On the third day, Russian security forces stormed the building using tanks, rockets, and other heavy weapons. Explosions rocked the school, followed by heavy fire involvement, engulfing the building as a gun battle ensued. The siege left 334 hostages dead, including 186 children, and 704 injured.

In addition to school violence, there have been many active shooter incidents at other venues. The FBI compiled "A Study of Active Shooter Incidents, 2000–2013" (Blair and Schweit, 2014). It reviews the carnage from all types of active shooter incidents and shows a total of 1,043 casualties that stemmed from 160 offenses—a sum that includes 486 deaths. Following are some key points from the report.

The agreed-upon definition of an active shooter by US government agencies—including the White House, US Department of Justice/FBI, US Department of Education, and US Department of Homeland Security/Federal Emergency Management Agency (FEMA)—is "an individual actively engaged in killing or attempting to kill people in a confined and populated area." Implicit in this definition is that the subject's criminal actions involve the use of firearms.

The FBI identified incident location categories, seeking to identify the primary locations where the public was most at risk during an incident. These included:

- Malls
- Businesses open to pedestrian traffic
- Businesses closed to pedestrian traffic
- Health care facilities
- Houses of worship
- Schools
- Institutions of higher learning
- Other (non-military) government properties
- Open spaces
- Military properties

A random look at some of the active shooter incidents cited shows a variety of venues. The FBI's full report is available at www.fbi.gov (search for "active shooter report 2000 to 2013").

Churches

June 22, 1980.   In a church in Daingerfield, Texas, a gunman entered and opened fire, shooting 15 people and killing five.

September 15, 1999.   In a church in Fort Worth, Texas, seven people were killed and seven more injured by a lone gunman who then took his own life.

June 17, 2015.   In a church in Charleston, South Carolina, a gunman killed nine people.

Movie Theaters

July 20, 2012.   In a movie theater in Aurora, Colorado, 12 were killed and 58 were wounded.

July 23, 2015.   In a movie theater in Lafayette, Louisiana, three people died and nine were wounded after a gunman opened fire; the gunman was among the dead.

Government and Military Installations

January 30, 2006.   At a postal distribution center in Goleta, California, a gunman killed six people. The shooter committed suicide before police arrived.

November 5, 2009.   In Fort Hood, Texas, a gunman killed 13 people and wounded 32 others; the shooter was wounded and taken into custody.

September 16, 2013.   In Washington, DC, at the Navy Yard, a gunman killed 12 and injured three before police took his life.

Miscellaneous Locations

July 2, 2004.   At a plant in Kansas City, Kansas, a gunman killed six people and wounded two others. The shooter committed suicide before police arrived.

August 4, 2009.   At a fitness center in Collier Township, Pennsylvania, a gunman killed three people and wounded nine others. The shooter committed suicide before police arrived.

November 29, 2009.   In a coffee shop in Pierce County, Washington, four police officers were killed and the gunman was killed by gunfire after a two-day manhunt.

September 6, 2011.   At a restaurant in Carson City, Nevada, three members of the US Air National Guard were killed, and two were wounded. In total, four people were killed and seven were wounded. The shooter committed suicide before the police arrived.

DEVELOPING AN EMERGENCY OPERATING PLAN

Terrorism can affect schools on a local, regional, national, and international level. Statistically, school violence is increasing, and fire departments must be prepared for this type of low-frequency but high-risk incident. In an era of instant national news coverage, cell phones, and social networking, it is essential that agency leaders develop a workable all-hazard-all-risk EOP. The contemporary plan must address interagency relationships, organizational partnerships, and clear communication between school, business, or government representatives and first responders.

An effective EOP is a necessary tool that can be used as a blueprint for the actions required of all agencies should they be confronted with an active shooter event. A well-written document provides a diagram for incident scene decision making.

The EOP needs to anticipate that the attacks may be over in a short period of time, or they can be spread over hours. The assault at Columbine High School in Littleton, Colorado, lasted approximately one hour, while the Virginia Tech rampage started in a dorm room and ended two hours later in a classroom building.

Once a plan is developed, it needs to be tested by using tabletops and full-scale exercises to fully examine each component and discover areas that can be improved upon. Improving the plan must be an ongoing concern. The EOP should include specific considerations for the school facility personnel, police, fire, and EMS. Concerns are listed under the specific agencies, yet some general considerations are the following:

- There is a need for interaction between the school facility officials, police, fire, and EMS.
- Restrict the use of lights and sirens within two blocks of the reported active shooter location for all emergency responders.
- Seal off the area and set up a safe perimeter to prevent the entry of unsuspecting civilians and to stop the shooter(s) from escaping and causing damage elsewhere.

- Create zones on the exterior similar to hazardous materials responses. The red or hot zone would be the area of the active shooter and the immediate vicinity. The yellow or warm zone would be a safe area located outside of the red zone where standby units are gathered and equipment is ready for deployment. Triage can be located in the yellow/warm zone to assure rapid medical treatment for the injured. Once an area is deemed safe, EMS and fire personnel can assume the role of medical care and sorting of victims. When establishing the yellow/warm zone, consideration must be given to the range of high-powered rifles and the line of sight from the reported shots-fired area to ensure responder safety. The green or cold zone will be outside the danger area and can be used for the Incident Command Post (ICP), staging of resources, transportation requirements, and logistical drop-points. The managing of these zones will ensure that everyone who leaves the affected building is searched for weapons and that occupants can be safely escorted through a safe corridor away from the building.
- Decide upon radio procedures and any interoperability issues and have designated frequencies that can be used for the incident.
- Have a withdrawal plan should it be necessary to remove police personnel prior to securing the active shooter if it is discovered that a Beslan type of incident is in progress.
- Generate contact information for mutual aid requests, which can include: police, fire, EMS, SWAT, bomb squads, hostage negotiation teams, hazardous materials units, mass casualty units, etc.
- Helicopters can be used for overhead surveillance, though shots fired from high-powered rifles should be considered. Some agencies may utilize drones for surveillance.
- Have and use a mass casualty procedure for dealing with multiple injuries.
- Consider the hospitals and trauma centers within a reasonable response distance and include them in your preincident planning process.
- Have the coroner or medical examiner involved in the training to enable them to express their concerns.
- Recognize the need for establishing a nearby holding area/safe haven with police presence for everyone who leaves or is evacuated from the affected building. If a school is involved, then teachers should have their roll book to ensure the accountability of every student in their class. No one should be permitted to leave the scene unless the police have released them. If a school is involved, ensure that all students and staff of the school are accounted for. If buses are needed to transport the occupants away from the school, have the school district secure buses.
- Anticipate that helicopters from television stations can congregate in the air space above the incident and may interfere with ground and air operations. If this concern exists, the air space should be restricted to police, fire, and MEDEVAC helicopters.

ACTIVE SHOOTER INCIDENTS OCCURRING AT LOCATIONS OTHER THAN SCHOOLS

Active shooter incidents at locations other than schools can possibly be preplanned and training given to employees at locations like government installations, manufacturing plants, or office buildings. Proactive planning on the part of managers can recognize the need for active shooter training. Yet there are many public places where training would not be feasible, nor occur. For example, with the exception of the employees, it would be impossible to plan for the shootings that occurred in recent years at movie theaters in Aurora, Colorado, and Lafayette, Louisiana.

The Department of Homeland Security has published a guide of good practices if confronted with an active shooter incident. It states the following:

- Be aware of your environment and any possible dangers.
- Take note of the two nearest exits in any facility you visit.
- If you are in an office, stay there and secure the door.

- If you are in a hallway, get into a room and secure the door.
- As a last resort, attempt to take the active shooter down. When the shooter is at close range and you cannot flee, your chance of survival is much greater if you try to incapacitate him or her.

This guide further discusses evacuating, hiding, and as a last resort taking action against the active shooter. It is available at: http://www.dhs.gov/xlibrary/assets/active_shooter_booklet.pdf.

Another guide prepared by FEMA, "Guide for Developing High-Quality Emergency Operations Plans for Houses of Worship," can assist houses of worship in preparing for a variety of emergency operations. This guide can be found at: https://www.whitehouse.gov/sites/default/files/docs/developing_eops_for_houses_of_worship_final.pdf.

INTERACTION OF FIREFIGHTERS

In order for firefighters to assist the police at active shooter incidents, they need to understand exactly how police will react when faced with an intruder who starts to randomly shoot people. This is critical knowledge for any interagency incident with police, but even more critical should firefighters or medics find themselves in an affected building as a shooting situation evolves.

Similar to the FBI definition, the police refer to an active shooter as "an armed person who has used deadly physical force on other persons and continues to do so while having unrestricted access to additional victims."

POLICE PROCEDURES

In the past, when confronted with a shooting situation, procedures called for the police to seal off the area and await the arrival of the Special Weapons and Tactics (SWAT) unit or Special Response Team (SRT). Once these special units arrived on the scene, they would handle the situation. At Columbine High School in Littleton, Colorado, the first arriving police officers followed this protocol and the perpetrators continued their killing spree. It is now realized that these are dynamic situations that evolve very rapidly. To combat this, an immediate action rapid deployment tactic is needed. This tactic is an aggressive police action, and it is the most effective countermeasure where the police move toward the sound of gunfire to terminate the life-threatening situation in dealing with an active shooter.

When an active shooter begins an attack, it is imperative that the initial police responders immediately pursue and establish contact with the shooter at the earliest opportunity. In addition to hand guns, rifles, and automatic weapons, explosives must also be anticipated. Law enforcement's goal is to stop the active shooter at all costs. The sooner the shooter can be contained, captured, or neutralized, the greater likelihood that fewer casualties will be incurred.

The first arriving police officer(s) must get whatever information they can from occupants, building security, the principal, teachers, or anyone who meets them. It should include:

- The number of shooters
- Their age, description, clothing worn
- Actions taken by the perpetrator(s) (shootings, threats, hostages, etc.)
- Their last reported location
- Types of weapons or explosives seen or spoken about
- Reports of how many people are shot and their location
- What threats have been made
- What, if any, demands have been made

If an active shooter is currently harming or killing individuals, the threat must be neutralized in one of three ways:

1. The responding police officers confront, contain, subdue, and apprehend the active shooter(s) with the use of less than lethal force.
2. The responding police officers are able to contain the subject(s) to a location so that no additional deadly harm is actively being inflicted upon members of the community.

3. The subject(s) is confronted; the police officers are unable to subdue, apprehend, or contain the subject; and the subject continues to pose a threat of death or serious physical injury. Appropriate use of deadly physical force is then used to subdue the suspect.

The ultimate goal is to neutralize the threat. In any circumstances, time is very important and the responding police officer(s) must take quick and decisive action and stop the violence as soon as possible. Any delay could mean additional deaths and injury to the community by the hostile intruder(s). The police should be aware that:

- They should not sound the fire alarm to evacuate the building. Persons may be placed in harm's way when they are attempting to evacuate, or shooters may be located on the exterior waiting for them as targets.
- EOPs may have occupants locking themselves in classrooms and offices.
- If the intruder can be contained to a room or area, the violence stops, and the situation has stabilized, then the police officer(s) will have two options:
  1. Continue to contain the situation if possible and await further assistance
  2. Apprehend the subject utilizing maximum officer safety.

INTERIOR OF THE BUILDING

Police should anticipate finding a situation of chaos reigning inside an affected building where an active shooter has been operating. During the interior pursuit, police officers must be disciplined to move through unsecured areas, and bypass wounded and panicked occupants while approaching the perpetrator(s). It is important for law enforcement personnel to survive the encounter to end a massacre, rather than become additional victims. Wounded and frightened occupants will be imploring the police to assist them. Anyone threatened will seek a hiding place that provides them some protection. A major problem for the police is trying to distinguish between the victims and a shooter(s) who could be hiding among the occupants and acting as a victim.

Police should possess building plans and gain familiarity with any security camera systems. These cameras may be utilized by police to assess the location(s) of the active shooter(s) and ensure a thorough search of the affected building during the pursuit of an active shooter. Utilization of these security systems requires close coordination between the police monitoring the security system and those seeking the active shooter. The dedication of a radio channel for this tactic should be considered.

The affected building(s) may have multiple escape routes for the victims and for the shooters. Most shooters have the mindset to kill and injure as many as possible. Typically, they are not looking to escape, but they may attempt to take their shooting spree to another location. Many of these active shooters have planned out the scenario and are prepared to engage the police in a highly visible shootout situation. Automatic weapons with clips holding many rounds of ammunition can be quickly loaded and cause tremendous damage. Even explosives have been set by intruders to commit suicide rather than be captured alive. In reality, many police officers initially only have handguns and can be outgunned.

ACTIVE SHOOTER OUTSIDE OF THE AFFECTED BUILDING

In addition to an active shooter within a building, a hostile person could be operating on the grounds outside, actively causing the imminent threat of deadly harm. In this type of scenario, the initial arriving police officer(s) must confront the individual(s) and use the appropriate level of force, which includes, if necessary, deadly force. The police officer should be prepared for the active shooter(s) to attempt to flee during the containment process. If this occurs, the police officer(s) will try to prevent the shooter from entering other buildings.

POLICE FRONT-LINE SUPERVISORS

Command will need to be assumed very early into an active shooter incident. The IC's first duty is to conduct a situation assessment and sort through the potentially confusing flood of initial intelligence. The IC needs to determine the number of shooters, and analyze the situation based upon risk assessment to determine how to neutralize the active shooter(s). In most cases, immediate action must be taken to stop further injuries. However, if the IC identifies a Beslan-type situation, with numerous, heavily armed attackers, a delaying action may be the best that can be achieved with limited resources. Adding responder bodies to the pile is a noble, yet futile gesture that will only compound an already complex problem.

Depending upon departmental protocols, the first-arriving police supervisor may be part of the initial attack team, or if sufficient police officers are already on scene, then he/she should assume command of the incident and establish a stationary CP in a safe location.

Other considerations include:

- Immediately request sufficient multi-agency resources, including special units to address heavy weapons, SWAT units, hostage negotiators, bomb squad, search dogs, firefighting forces, medical care, etc.
- Establish divisions and groups to ensure a thorough search pattern is completed in the designated areas.
- Establish liaison with the school principal, security personnel, building managers, church representatives, maintenance personnel, or anyone who can provide basic information on the affected building and what has occurred.
- When sufficient operational personnel are on scene, establish an Intelligence Officer and a Planning Section for gathering and collating all incident-related data, tracking all on-scene personnel, and maintaining strict accountability for personnel operating within the red/hot zone.

FIRE DEPARTMENT AND EMS RESPONSIBILITIES

Though the fire department and EMS will play an integral part in assisting the police at incidents involving an active shooter, it is primarily a police department function. A firefighter's role in any activity is the preservation of life and not performing law enforcement duties. If fire department units find themselves on the scene of an active shooter, they must immediately notify Command or Dispatch and withdraw to an area of safety until the police can stabilize the scene. The sounding of a Mayday should be used judiciously since the reason for the Mayday could be misinterpreted. Other firefighters hearing the Mayday may not realize the situation and place themselves into harm's way trying to assist. Any radio report transmitted by a unit(s) that is in danger should be as specific as possible of the situation to allow Command or Dispatch to send the proper assistance. Once Dispatch learns of a situation where fire department personnel are endangered, they should consider dedicating the radio channel that the report was received on to that incident. Incoming units on an active shooter response should set up in a staging area in a safe location as dictated by the IC.

The initial arriving fire department officer should report to the police IC and if Unified Command is to be established, he/she can become part of Unified Command. If the system in place does not utilize the Unified Command concept, then a fire department representative should report to the Liaison Officer. Some EOPs, in lieu of using Unified Command, state that once the active shooter is neutralized the police IC will transfer command to the fire department to treat the injured and suppress any fires. Once all injuries are stabilized and any fires extinguished, Command is transferred back to the police department to enable them to preserve the crime scene.

The police will need to fulfill many roles to assist the firefighters, including a thorough search of all victims to check for concealed weapons and bombs before treatment by firefighters or medics. If police find it best to evacuate victims or other occupants from hostile areas, they can bring them to areas of safety for fire and EMS personnel to treat. They will need to provide security to ensure the safety of medical personnel if the active shooters are arrested and require medical attention.

Realize that police will not treat the injured when initially entering the affected building; their main objective is to defuse the situation, which means neutralizing the shooter(s). As additional police are deployed within the affected building, initial triage may be given by these additional teams of police officers to the injured occupants. This can be accomplished in unsecured areas by police officers that are EMTs. Once an area is deemed safe, or if injured are removed to a safe location, EMS and fire personnel can assume the role of triage.

The fire department will also need to:

- Anticipate the need for mutual aid and possibly the recall of off-duty members.
- Stage all personnel and assign a Staging Area Manager.
- Anticipate a MCI and request an MCI response.
- Set up triage, treatment, and transportation areas.
- Ensure that triage tags are filled out to identify the victim and their injuries. This will also assist in accountability of occupants.
- If preliminary reports indicate multiple injuries, have EMS identify available trauma beds at area hospitals, medical centers, and availability of emergency room staff.
- Consider MEDEVAC helicopters and determine a landing zone. An engine company with foam capabilities should be dispatched for monitoring the landings and departures.
- If extreme weather conditions exist, consider utilizing nearby buildings for some operations, if approved by the IC.
- Firefighters should not be involved in searches for secondary devices.
- Some fire departments and EMS agencies are proponents of their personnel wearing bulletproof vests. This is a departmental decision, but in any case, these personnel should not be allowed to operate in any area that has not been deemed safe by the police.
- Fire department members must recognize that active shooter incidents are criminal investigations and evidence must be preserved.

DEPLOYMENT OF FIRE DEPARTMENT PERSONNEL

Firefighters should not be deployed in unsecured areas during any police activity until the situation is stabilized. It is unacceptable for firefighters to be operating in a hostile police environment. Fire department personnel should not be used in any law enforcement functions that could negate the perception that the fire department is neutral during police activity. This includes the use of fire department apparatus as shields to allow law enforcement personnel to approach an active incident. Police officers should not be allowed to utilize any fire department clothing to act as a decoy. Police may require the use of fire department tools for forcible entry, first-aid kits, backboards, or stretchers to remove injured from areas that have not yet been deemed safe for firefighters to operate in. It is important that firefighters and emergency medical personnel wear distinctive clothing to maintain this designation. A fire helmet is a very specific way of identifying a firefighter.

 ON SCENE

During the Columbine school shootings, a fire department training officer wearing a dirty turnout coat and holding a portable radio was targeted by the police as an active shooter. The fire officer at the time was in the process of assisting a helicopter coming in for a landing. He was not wearing a fire helmet. The police mistook his turnout coat for a trench coat and the portable radio in his hand for a handgun. They approached him with guns drawn and forced him to lie down; he was immediately surrounded by police officers before they realized their mistake.

Firefighters are not normally members of police SWAT teams. If a fire department has personnel cross-trained to operate on a SWAT team as a paramedic or an EMT, that member should be subject to orders from the police during SWAT operations. These personnel should operate in safe environments and not be allowed to enter areas of danger until the police have secured the scene. A much better arrangement is provided by police departments who require EMT training for police SWAT team members to allow them to be active members of the team and also treat the injured in areas of danger.

In 2013, the USFA produced a guide, "Fire/Emergency Medical Services Department Operational Considerations and Guide for Active Shooter and Mass Casualty Incidents." It discusses maximizing survival, coordinated/integrated planning and response, incident command, unified command, interagency, operational practices, and post-incident/demobilization. It also provides a checklist. The report can be found at: http://www.usfa.fema.gov/downloads/pdf/publications/active_shooter_guide.pdf.

SCHOOL PERSONNEL RESPONSIBILITIES

Active shooter incidents at schools require that the school personnel must follow the EOP in the event of an active or a suspected shooter. An announcement over a PAS can alert teachers and students of the situation and allow them to seek an area of refuge. At one school shooting that was in progress, an announcement was made over the PAS declaring a "Code Blue" to alert the teachers and students to lock down all classrooms. Some students took refuge in closets and bathrooms, while others fled the area. Some school districts do not use codes and only use clear text to avoid any confusion on the part of a student or a substitute teacher not understanding the message. A common message could be, "Attention all teachers, staff, and students. The school is in lockdown due to an intruder. Lock all classroom doors and move all students away from the doors."

Schools and universities that encompass multiple buildings will need a notification system in place for students and faculty for releasing timely information about the perpetrators. This informational system must anticipate the need for a hotline for the many calls that will be received in regards to students' safety. What actions need to be taken initially? What actions are necessary once the situation is stabilized?

The school EOP must spell out the responsibilities of the principal, teachers, security guards, students, and other school personnel. There needs to be personnel training so that backups will be available for all positions. During an active shooter incident, classrooms need to be locked down and students moved away from the visibility of the classroom doors. Hallways and common areas should be avoided. The EOP should address actions to be taken by teachers when students are in the auditorium, cafeteria, and gymnasium since they are difficult areas to secure. Since these areas may have doors that exit to the exterior, a consideration may be to evacuate the students.

There will be a need to establish a notification system for the families of anyone injured or killed. This should consider including social services, faith leaders, counselors from the school/school district, police, coroner, and hospital personnel.

Another factor that could assist public safety responders is the numbering or color coding of the school's doors to assist in identifying locations.

NORTHERN ILLINOIS UNIVERSITY

What can communities do to address a disaster waiting to happen? A comprehensive report on the school shootings at Northern Illinois University (NIU) was compiled by the USFA (Stambaugh, USFA Technical Report 167). This report examines the actions of the university and the local emergency responders. The report found that NIU developed a basic emergency operations plan, and showed how training can lead to a better outcome. NIU, in developing its plan, took the progressive steps of contacting Virginia Tech to find out firsthand the problems that it was faced with and how it handled them in April 2007 when 32 people were killed on its campus. NIU took those recommendations and integrated the lessons learned from that report into the EOP for the university and the city of DeKalb as well as their fire and police departments. They implemented those plans on February 14, 2008, when a gunman opened fire on a college classroom, killing five students and himself, and wounding 18 others.

Lessons Learned and Lessons Reinforced at NIU
- The benefit of developing an EOP with input from school officials, police, fire, and EMS personnel.
- The ability of the Kishwaukee Community Hospital to be well prepared to handle the surge of the injured in the emergency room because they had early notification from the prehospital providers about the number and condition of the victims.

- The university quickly activated its campus alert system on the university's website and simultaneously sent emails, voicemails, and recorded a hotline message.
- The use of triage tags saved lives and facilitated medical treatment while greatly assisting with patient identification.
- Knowledge of the guns used in the shootings was helpful to the hospital staff treating gunshot wounds because knowing the ballistics of the guns can assist in establishing the type of internal damage that can be anticipated.
- It was realized that vital signs in young victims can change quite rapidly; they can be stable and maintain that condition for a long time and then suddenly drop to critical levels.
- That HIPPA laws prevented hospitals from releasing information over the phone about those admitted, being treated, or those released. This can cause frustration and anger on the part of family members trying to obtain information about their loved ones. The hospital staff realizes the importance of the information to the relatives, but cannot disregard the strict regulations that are imposed upon them.
- A debriefing must be conducted for all personnel who respond to these incidents. This should include dispatchers and staff personnel who were involved in receiving phone calls during the incident. Counseling sessions should be offered for the students and teachers to assist them in getting past the incident and hopefully avoiding psychological problems in the future. (See References at the end of this chapter to access the full report.)

INCIDENT COMMAND SYSTEM

There is a distinct demand for the implementation of a unified command system at an active shooter incident. All disciplines need to be working together toward common incident objectives under a unified command structure, although each incident must be judged on its own merits. It may be more expedient to initially have a single police IC until the active shooter is neutralized. At the point that the IC declares the scene stabilized, Unified Command can be established. The implementation of Unified Command should include police, fire, EMS, and possibly school, facility, or military installation officials, depending upon the location of the incident and their need to be part of Unified Command. Unified Command can help to ensure that the correct incident objectives are identified and accomplished. They also should:

- Have all agencies adopt and train on the ICS and implementation of Unified Command.
- The appointment of a Safety Officer (SO) must be an individual versed in police operations, or an Assistant Safety Officer with training in active shooter incidents. As the incident progresses to the point where the perpetrators have been neutralized and the situation warrants fire and EMS activities, then any qualified Incident Safety Officer can be assigned.
- The Public Information Officer (PIO) can release timely information on approval of the IC. Consider the establishing of a Joint Information Center (JIC) to ensure the coordination of all information.
- Consider basic locations for safely setting up a CP and staging in areas of safety.
- Consider the activation and use of an emergency operating center (EOC).
- During ICS training, ensure that the police and fire departments understand the sectoring system that will be used. The system utilized by fire departments is alphabetical starting with the front of the building being A or Alpha and working clockwise around the building. Many police departments may still be using numerical sectoring systems, which start with the front of the building being Side 1 and also work clockwise around the building. On interior operations some departments break down the interior floors of a building into quadrants to allow better sectoring/visualization. If either agency uses a quadrant system, that system must be included in the EOP.

SIZE-UP FACTORS FOR ACTIVE SHOOTER INCIDENTS
Water
- Once the shooter has been neutralized, any fires that are burning require an adequate water supply for an offensive and, if necessary, a defensive attack.
- Fires in buildings in rural areas may have a limited water supply available.

- Once cleared by the police, it may be necessary to wash down an area to remove blood and other bodily fluids at the scene after an active shooter situation has occurred and victims have been transported.

Area

- Buildings and facilities can vary drastically in size. If a fire exists in an affected building and the situation has been deemed safe, then the common denominator is the life hazard of anyone remaining in those structures.
- Fire doors installed and functioning properly can assist in compartmentalizing a fire situation.
- If explosives or Molotov cocktails have caused any fires, the area involved may be large.

Life Hazard

- Varies, depending if the facility is still occupied.
- Fast-moving fires will necessitate a prompt response and attack on the fire once deemed safe by the police.
- Occupants will usually attempt to exit the building the same way they entered, but with an active shooter incident they may try to exit by extreme means, such as jumping from windows.
- Exit doors may be locked or chained shut in some facilities.
- In a school or other facility involving children, parents will immediately respond to the scene looking for their children and possibly interfering with police or fire department operations. Someone must be prepared to handle this influx of parents, answer their questions, and reunite them with their children when possible.
- Occupants may be hiding in various locations and be afraid to come out. TICs can assist in the primary and secondary searches.
- A fire or violent acts may initiate an MCI and place numerous patients in danger. At an active shooter incident, triage and treatment cannot be started until the incident scene has been declared safe by the police, or victims have been removed to a safe location for treatment.
- An active shooter incident will threaten the safety of fire and EMS personnel.

Location, Extent

- HVAC systems can spread smoke or tear gas throughout a building. If these systems are not assisting the fire department's operations, they should be shut down.
- Active shooter violence may be accompanied by numerous fires intentionally set. Fire control cannot start until the scene has been declared safe by the police.

Apparatus, Personnel

- Call for help early and keep units at Base. A large number of resources may be needed to handle an MCI, control fires, achieve adequate ventilation, and accomplish search, rescue, and evacuation once the police have deemed the affected building safe.
- Determine the best method of utilizing the building's staff to assist in the accountability of the occupants.
- Consider requesting MEDEVACs for patient transport in the event of an MCI.

Construction/Collapse

- All types of building construction can be found including fire-resistive, ordinary, frame, and mixed construction.
- See if preplans exist for the affected buildings to assist in the overall operation.
- Active shooters may place explosives with the intent of injuring or killing of occupants or as an attempt to cause building collapse.
- Active shooters may initiate numerous small fires throughout the school.
- Active shooters may sound the fire alarm in an attempt to move the occupants to the exterior to use them as targets.

Exposures

- Interconnected buildings can cause concern in an active shooter incident.
- Adjacent buildings could be threatened by an active shooter.

Weather

- Cold weather will impact occupants evacuated from the affected buildings; a nearby indoor area will need to be secured to protect them.
- Triage and treatment areas, if established, will be affected by cold and inclement weather.

Auxiliary Appliances

- Sprinklers and standpipes may be present. These systems should be pressurized by the fire department to ensure a continuous and adequate supply of water if the systems are activated and it can be accomplished safely.
- Intentionally set fires can activate sprinkler systems for fire control.

Special Matters

- Surveillance camera systems may be utilized by police to assist in identifying the location of active shooters and possibly any fire conditions that exist in the affected buildings.
- Ensure that the police have searched and cleared everyone leaving an affected building during an active shooter incident prior to fire or EMS personnel treating them.
- If multiple victims exist, develop a list of hospital availability for the transport of patients.

Height

- One-story buildings will allow evacuation through hallways and via windows if necessary.
- Multistory buildings may compound rescue/evacuation efforts.

Occupancy

- Exit doors can become congested with occupants trying to leave the building, restricting egress and possibly panicking others.

Time

- The critical time period is when the active shooter is still operating freely within the affected building.

CONSIDERATIONS FOR ACTIVE SHOOTER INCIDENTS

Strategic Goals and Tactical Priorities for an Offensive Attack During an Active Shooter Incident

No offensive firefighting attack can commence until the affected building is deemed safe by the police. At that time, the incident will be handled as any other incident.

- Evacuate and initiate search and rescue for occupants.
- Ensure that all exits are unlocked and unobstructed.
- Recognize the potential for an MCI.
- Call for a sufficient amount of resources (fire and EMS).
- Ensure that hose-lines are not stretched through doorways being used as exits by occupants.
- Ventilation can act as a lifesaving tool.
- If any improvised explosive devices are found, immediately withdraw personnel and have the bomb squad secure the device.
- Locate and confine the fire to the area of origin if possible while attempting to protect avenues of egress for the students.
- Depending on the location and size of the fire, 1½-inch or 1¾-inch hose-lines may be ineffective; 2½-inch hose-line may be needed for large meeting rooms or auditoriums.
- Horizontal ventilation should be sufficient for a room and contents fire. For fires involving large areas, vertical ventilation may be necessary.
- Try to utilize the HVAC system to assist in removing smoke from the building.
- Roof vents may be present over common areas; ensure that they are open.
- To use PPV, you need to know the exact location of the fire, the location for smoke to exit, and whether any occupants would be endangered. This

information will be difficult to obtain in the early stages of a fire. It will also be difficult to control building openings since the exit doors should all be opened to permit a safe and timely evacuation. These factors impact PPV, making it not a good initial choice.

Incident Management System Considerations/Solutions for an Offensive Attack

- Incident Commander
- Safety Officer
- Liaison Officer (to interact with building personnel on accountability, and assist in communicating information to concerned family members)
- Rapid Intervention Crew(s)
- Operations (if needed)
- Staging
- Ventilation Group
- Search and Rescue/Evacuation Group
- Division 1, 2, 3, etc. (for multistoried building can be implemented for any floor where deemed necessary)
- Exposure Divisions A, B, C, D (if needed)
- Logistics (if needed to assist in water supply and/or sheltering of occupants)
- Medical Group or Branch (to set up triage, treatment, transportation, morgue, etc.)

Strategic Goals and Tactical Priorities for a Defensive Attack

- If changing from an offensive to a defensive attack, ensure that PAR is taken for accountability purposes.
- Set up and maintain a collapse/safety zone.
- Utilize master streams.
- Protect exposures if necessary.

Incident Management System Considerations/Solutions for a Defensive Attack

- Incident Commander
- Safety Officer
- Liaison Officer (to interact with facility and police personnel on accountability, and assist in communicating information to concerned family members)
- Public Information Officer
- Operations (if needed)
- Staging
- Logistics (for water supply and/or sheltering of occupants if needed)
- Divisions on the exterior on Sides A, B, C, D
- Exposure Divisions (if needed)
- Medical Group or Branch

Summary

Health care and high-risk populations include facilities with occupants who oftentimes are unable to assist themselves. Hospitals, nursing homes, assisted living facilities, penal institutions, and schools include elderly, young, sick, and incarcerated populations. These groups will require special attention when a fire or emergency occurs. Since their occupants are so diverse, specific preplans will be needed to ensure that all facets are considered and addressed when you are called to an incident. Additionally, responding to a mass casualty incident will tax every response agency. There will be a need for timely and appropriate care to minimize injuries and deliver the injured to medical facilities for continued professional treatment.

Active shooter incidents, though initially a police activity, will usually involve firefighters in the operation.

Review Questions

1. Visit a local hospital and describe the hospital codes in its emergency operating plan.
2. Do all hospitals in your jurisdiction use the same hospital codes? If not, what differences exist?
3. What dangers exist with magnetic resonance imaging? Are there any MRI facilities in your jurisdiction in addition to hospital facilities?
4. What code changes were made to correct common problems found in past hospital fires?
5. What basic information should be contained in a preplan for a nursing home or assisted living facility?
6. What are the types of building construction found in nursing homes and assisted living facilities? List any benefits or problems with each type of construction.
7. What considerations will firefighters need to make when deciding whether to evacuate residents or to protect them in place at a nursing home or assisted living facility?
8. Define a mass casualty incident. What constitutes a mass casualty incident in your jurisdiction?
9. What actions will a jurisdiction need to take to successfully handle a mass casualty incident?
10. List the various levels of trauma centers in the United States and their capabilities. List the trauma centers in your area and their level.
11. When simulating a mass casualty incident, certain components should be tested. List them.
12. What are the responsibilities of the Medical Group Supervisor at a mass casualty incident?
13. Triage tags should be employed at a mass casualty incident. Describe the tag colors and what type of injury is typically assigned to each color tag.
14. What are the most common causes of fires in houses of worship?
15. Why is it so difficult to control and extinguish fires in Gothic-style churches?
16. What are the dangers to firefighters associated with fires involving the hanging ceilings in houses of worship?

17. In addition to religious services, what programs do the various houses of worship in your community offer? What problems, if any, will these programs create for firefighters responding to a fire in these buildings?
18. Discuss the lessons learned/lessons reinforced considerations with fires in houses of worship. How would you attempt to utilize these factors if confronted with such a fire?
19. What are the three generations of confinement found in a penal institution? List the types of supervision typically in each type and the level of security that can be expected.
20. What is a prison's emergency operating plan and what information should be contained in that plan that will assist firefighters should an incident occur?
21. What information should be contained in a fire department's preplan for a penal institution?
22. What procedures should EMS follow when involved in medical calls at a penal institution?
23. Select at least three public assembly buildings in your local community or response district and list the various size-up factors that would apply.
24. What problems would be found in evacuating various public assembly buildings in your response district?
25. Select one of the special occupancies presented in this chapter, and in reviewing the size-up factors, add five additional potential problems that could be present.
26. What are the key factors in a school evacuation plan?
27. If a school fire forces an early dismissal of the students, what points need to be considered?
28. List the actions and decisions that need to be made by the first-arriving units at a school fire and what those decisions should be based on.
29. What are a fire department's considerations at an active shooter incident?
30. What are the benefits and considerations with implementing Unified Command at an active shooter incident?

Suggested Reading, References, or Standards for Additional Information

Blair, J. P., and Schweit, K. W. 2014. "A Study of Active Shooter Incidents, 2000-2013." Texas State University and Federal Bureau of Investigation (FBI). Available at: http://www.scribd.com/doc/240970283/FBI-Active-Shooter-Report.

Federal Emergency Management Agency (FEMA). 2013. "Guide for Developing High-Quality Emergency Operations Plans for Houses of Worship." Available at: http://www.fema.gov/media-library-data/20130726-1919-25045-2833/developing_eops_for_houses_of_worship_final.pdf.

US Department of Homeland Security. 2008. "Active Shooter: How to Respond." Available at: http://www.dhs.gov/xlibrary/assets/active_shooter_booklet.pdf.

US Fire Administration (USFA)

Chubb, M. "Ten-Fatality Board and Care Facility Fire, Detroit, Michigan." Technical Report 066.

Copeland, T. D. "Industrial Plastics Fire: Major Triage Operation, Flint Township, Michigan." Technical Report 025.

"Fire/Emergency Medical Services Department Operational Considerations and Guide for Active Shooter and Mass Casualty Incidents." (September 2013): https://www.usfa.fema.gov/downloads/pdf/publications/active_shooter_guide.pdf.

Kirby, R. E. "Twelve-Fatality Nursing Home Fire, Norfolk, Virginia." Technical Report 034.

Routley, J. G., and R. Bush. "Hospital Fire Kills Four Patients, Southside Regional Medical Center, Petersburg, Virginia." Technical Report 080.

"School Building Fires (2009-2011)." *Topical Fire Report Series* 14, no. 14 (April 2014): http://www.usfa.fema.gov/downloads/pdf/statistics/v14i14.pdf.

Stambaugh, H. "Northern Illinois University Shooting." Technical Report 167. Available at: https://www.usfa.fema.gov/downloads/pdf/publications/tr_167.pdf.

Yates, J. "Ten Elderly Victims from Intermediate Care Facility Fire, Colorado Springs, Colorado." Technical Report 050.

National Fire Protection Association (NFPA)

Ahrens, M. 2012. "Fires in Health Care Facilities." Available at: http://www.nfpa.org/research/reports-and-statistics/fires-by-property-type/health-care-facilities/fires-in-health-care-facilities.

"Fire Safety at Your Worship Center." Church Mutual Protection Series, Church Mutual Insurance Company.

National Fire Protection Association (NFPA) Standards:

13E: Recommended Practice for Fire Department Operations in Properties Protected by Sprinkler and Standpipe Systems

221: Standard for High Challenge Fire Walls, Fire Walls, and Fire Barrier Walls

909: Code for the Protection of Cultural Resources Properties—Museums, Libraries, and Places of Worship

914: Code for Fire Protection of Historic Structures

Related Courses Presented by the National Fire Academy, Emmitsburg, Maryland

Management of Emergency Medical Services
Emergency Medical Services: Special Operations
Emergency Medical Services: Quality Management
Advanced Leadership Issues in Emergency Medical Services
Hot Topics Research in Emergency Medical Services
EMS Functions in ICS

Related Courses Presented by the Emergency Management Institute, Emmitsburg, Maryland

Multihazard Emergency Planning for Schools
Planning for the Needs of Children in Disasters

9

Commercial and Industrial

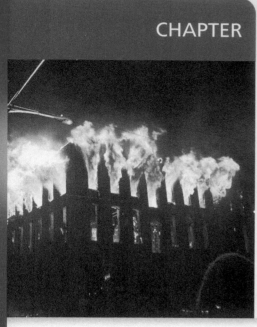

> ❝ A person's success is in direct proportion to the quality of commitment. ❞

—Vince Lombardi, NFL Hall of Fame coach

Heavy fire conditions on arrival. *Used with permission of Marty Griffin.*

KEY TERMS

commercial buildings and warehouses, *p. 395*
strip mall, *p. 403*

enclosed shopping mall, *p. 409*
supermarkets, *p. 415*

lumberyard, *p. 423*
high-rise building, *p. 428*

OBJECTIVES

Upon completion of this chapter, the reader should be able to:

- Understand the occupancies discussed and the types of fires that may confront firefighters.
- Identify pertinent characteristics of these occupancies.
- Recognize the 13 points of size-up that pertain to these occupancies.
- Recognize the strategic considerations for these occupancies.
- Understand the incident management considerations of these occupancies.
- Be able to identify the basic components needed in an operational guideline for high-rise fires.
- Have a basic understanding of high-rise building systems.
- Identify the specific components of a high-rise command system.

Chapter 9 discusses commercial buildings, warehouses, strip and enclosed malls, supermarkets, lumberyards, and high-rise buildings. These occupancies can present specific dangers to firefighters.

At the end of each section are size-up factors as well as strategic goals and tactical priorities that need to be taken into account when confronted with fires in these occupancies. These lists are not meant to be all inclusive. Other factors can and will exist that also must be considered.

The strategic goals and tactical priorities are for both offensive and defensive attacks, with suggested considerations for implementing an incident management system to meet the potential problems. The strategic goals and tactical priorities that are noted are in response to some of the potential problems that may occur. The suggested incident management positions do not mean that they must be used or that using them is the only correct method. These are initial positions that should be considered. Depending on the problems presented by each individual situation, the Incident Commander can implement these or other incident management positions to deal with the problems.

Commercial Buildings and Warehouses

Commercial buildings can be found in many cities and towns, from small manufacturing plants to large industrial factories and **warehouses**. The height can range from one to eight stories. Building sizes can range from 25 feet wide to structures that spread over hundreds of feet or city blocks. Some of the older structures were constructed near the turn of the twentieth century utilizing cast-iron columns. Newer structures and renovated buildings typically replace the cast iron with steel I-beams. Though these beams should be protected against fire, some installations utilize only a wooden enclosure.

commercial buildings and warehouses ▪ Can range from one to eight stories spread over hundreds of feet or a city block(s).

Commercial and industrial buildings are often grouped together in a cluster forming a commercial district. These buildings can be L-shaped, fronting on two streets, and they can be deep with limited access on the sides, preventing firefighters from reaching a fire burning in the middle of the building. When confronted with L-shaped buildings, confusion can occur if a fire officer describing the rear of the fire building is referring to side D and not side C. Other irregular-shaped buildings can cause similar confusion when you are attempting to identify specific areas utilizing an alphabetic sectoring system.

BUILDING RENOVATIONS

The conversion of an industrial building to a retail or factory outlet store can increase the life-safety concerns. There may be limited exits and many people unfamiliar with the building layout. (See Chapter 7, Special Situations and Occupancies, for additional information.)

Renovations can create concealed spaces. Changes could include multiple ceilings, lightweight wood paneling, and other alterations that could impact firefighting operations. The installation of central air-conditioning systems on the roof for offices and retail areas can include ductwork that penetrates the floors and becomes a readily available means for fire extension from floor to floor.

Adjoining structures converted into one building may not be discernible from the exterior. Fire-stops that once existed between the two buildings may be removed and fire doors may be nonexistent or kept permanently in an open position.

FIGURE 9-1 Large-diameter hose-line is deployed and master streams are operating at a heavily involved commercial building. *Used with permission of Deputy Chief Thomas Lyons.*

Chapter 9 Commercial and Industrial   **395**

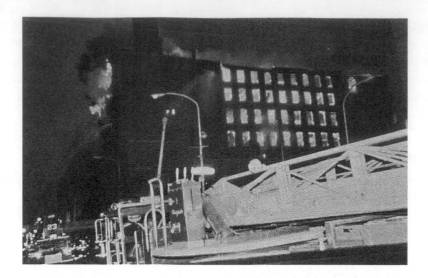

FIGURE 9-2 Large mill buildings create tremendous amounts of heat and threaten nearby structures.
Used with permission of Joseph Hoffman.

NIOSH FIREFIGHTER FATALITY REPORT F2013-16

On May 31, 2013, a 35-year-old career captain, a 41-year-old career engineer operator, a 29-year-old career firefighter, and a 24-year-old career firefighter were killed when the roof of a restaurant collapsed on them during firefighting operations. The captain was assigned to Engine 51 (E51). The engineer/operator was assigned to Ladder 51, but was detailed to E51 and assigned to the left jumpseat (E51B). The two firefighters were assigned to Engine 68 (E68). Upon arrival, the captain of E51 (E51A) radioed his size-up, stating they had a working fire in the restaurant with heavy smoke showing, plus a temperature reading from his thermal imager. E51 made an offensive attack from Side Alpha with a 2½-inch preconnected hose-line in the restaurant. District Chief 68 (D68) arrived on scene and established Command. He ordered E51 out of the building because the engine operator of E51 (E51D) advised that E51 was down to a quarter tank of water. Engine 68 had arrived on scene and had laid two 4-inch supply lines from E51 to a hydrant east of the fire building on the feeder road. Once E51 had an established water supply, E51's crew re-entered the building. Engine 68 (E68) was ordered to back-up E51 on the 2½-inch hose-line. Engine 82 (E82) (fourth-due engine company) was pulling a 1¾-inch hose-line to the front doorway that E51 had entered when the collapse occurred. The roof collapsed 12 minutes after E51 had arrived on scene and 15 minutes, 29 seconds after the initial dispatch. The firefighter from E51 (E51C) was at the front doorway and was pushed out of the building by the collapse. The captain from E82 called a Mayday and Rapid Intervention Team (RIT) operations were initiated by Engine 60. During the RIT operations, a secondary wall collapse occurred, injuring several members of the rescue group. Due to the tremendous efforts of the Rescue Group, a successful RIT operation was conducted. The captain of E68 was located and removed from the structure by the Rescue Group and transported to a local hospital. The engineer operator from E51 (E51B) was removed from the structure by the Rescue Group and later died at a local hospital. A search continued for the captain of E51 and the two firefighters from E68. Approximately two hours after the collapse, the body of the captain from E51 was located on top of the restaurant roof debris. The two firefighters from E68 (E68B and E68C) were discovered underneath the restaurant roof debris. The officer and two firefighters were pronounced dead at the scene.

Contributing factors include fire burning unreported for three hours; delayed notification of the fire department; building construction; wind-impacted fire; scene size-up; personnel accountability; fireground communications; and lack of a fire sprinkler system.

See Appendix B: NIOSH Reports in www.pearsonhighered.com/bradyresources to read the complete report and recommendations.

FIREFIGHTING

The basic aim to locate, confine, and extinguish remains the same when dealing with fires in commercial buildings or warehouses. The problem of location can be made easier if a fire occurs during the building's normal operating hours because employees can give the location, what is involved in the fire, and other particulars. When the location is known, the responding fire forces must determine the best approach to take.

Fire control can be achieved through aggressive firefighting methods when assisted by sound building construction practices. If proper fire stops are built into a structure, the fire

FIGURE 9-3 Fire on multiple floors on the arrival of the first-due units demands an immediate request for assistance. Remember, it is better to have the resources and not need them than to need them and not have them. *Used with permission of Marty Griffin.*

department will have a fighting chance of controlling the fire. Large, wide-open areas, as well as vertical openings within buildings, allow the spread of fire.

The Incident Commander (IC) may have a difficult time doing a walk-around of the fire building due to the large area and surrounding buildings that may restrict this effort. If unable to complete the walk-around, the IC must rely on reports from the divisions and groups to ascertain the problems.

When faced with fires in commercial buildings, firefighters must realize they are different from residential structures. Firefighters operating on the interior of a typical residential property are usually no more than 20 feet from a window or a doorway to the exterior. A rapid exit or at least access to a window can be accomplished if they are in danger. Commercial buildings can have limited access to windows, and long distances may need to be traversed to reach the exterior. There is a real possibility of getting disoriented or lost. Firefighters not operating a hose-line (which can be followed to exit the building) should stretch ropes or guidelines to ensure a safe method of egress.

The residential fire is a routine operation for firefighters. Commercial buildings are different. Storage areas can involve a tremendous fire load of stock piled to the ceiling that will restrict the operation of the sprinklers and limit the movement and reach of hose-lines.

Though 1½-inch or 1¾-inch hose is the workhorse of today's fire service, the initial use of 2½-inch mobile hand lines may be required. A second hose-line of equal or greater size must be stretched to back up the first hose-line. The ability of 2½-inch hose-line to knock down a large volume of fire while cooling the area is tremendous. In commercial buildings containing large open areas, maneuverability may not be a problem. In restricted areas containing shelving and narrow aisles, a sufficient number of firefighters will be needed for the advancement of the hose-line and will require companies to work together.

If there is difficulty in locating a fire in a heavily smoke-charged building, the basement or subbasements should be checked. These basement or cellar fires can be punishing fires to fight. Ventilation is difficult. Many establishments utilize this area for storage of records, and housekeeping is haphazard. Checking for upward extension of fire is critical. Venting the floors above the fire and the roof area can create flow paths and pull the fire upward; crews must be assigned to stop any upward extension of fire.

Ceilings may not be finished. The wooden beams supporting the floor above the fire area may be subjected to heavy fire conditions. Should there be penetrations between the floors, fire can easily extend to the floor above.

While the initial lines are being placed into operation from the front of the fire building, a hose-line should be advanced up the rear. These structures can have fire towers or fire escapes on the perimeter of the building and an interior stairway(s). The floors

FIGURE 9-4 A major building fire requiring numerous master streams. *Used with permission of Joseph Hoffman.*

above the fire can sometimes be accessed from the fire tower or fire escape. If firefighters are operating from fire escapes, secondary means of egress should be considered in the event that the fire escape becomes untenable due to heavy fire pushing from windows or doors on lower floors. This can be accomplished by raising ground ladders or main ladders to their working location on the fire escape.

If it is necessary to use the interior stairway, units operating on the fire floor must be aware of your location and realize that your safety depends upon their protection of that stairway. Firefighters operating hose-lines on the floor above the fire will check for extension of fire via chases, floor openings, or conduction through the floor. An examination of the floor above the fire may reveal minor fire extension that can be quickly extinguished. If heavy involvement is found, this should be immediately reported to the IC, who may need to assign additional units to combat the fire.

Once sufficient hose-lines are in place and an interior attack has commenced, it is critical that the hose-lines advance. If they are stalled and cannot move forward, conditions will only worsen.

Deteriorating conditions that could signal the need for withdrawal of firefighters and a switch to a defensive operation include:

- A large body of fire that is not diminishing with attacking hose-lines
- Water from the hose-lines turning to steam with no obvious effect on the fire
- Continued or worsening high heat and/or heavy smoke conditions

SPRINKLERS

If firefighters find that the sprinklers operating at a building fire involving rack storage are controlling the fire, they should stretch hose-lines to contain any fire burning low on the storage racks or spreading near the floor.

A sprinkler system can affect the fire in a variety of ways: immediate extinguishment, fire control, or hampering, with continued spread at floor level. This continued spread of fire can be due to warehousing methods and storage configurations that restrict sprinkler operation. This can occur in warehouses with rack storage and very high ceilings and only ceiling sprinklers present.

If sprinklers are supplied by the same water main feeding the immediate hydrant system in the area, hooking up pumpers to these hydrants could rob the sprinklers of water.

Because sprinkler systems are placing water directly onto the fire area, a large amount of steam and smoke will be produced and will restrict visibility. The temptation to immediately shut down the sprinkler system to reduce water damage and improve visibility should be resisted. Shutdown should occur after sufficient handheld hose-lines have been stretched to the fire area and the fire is under control. The handheld hose-lines can then be used to wet down any remaining hot spots during overhauling.

When shutting down a sprinkler system, it should be determined whether the system contains sectional stop valves, which allow a portion of the system to be shut down without disrupting sprinkler protection in other areas of the building. When the order is given to shut down, the firefighter assigned to do so must remain at the valve location, maintain communications, and be prepared to open the valve if the fire flares up. The fire area must be monitored for hot spots. This can be especially difficult in buildings with high ceilings, as fire can flare up in overhead areas and, due to reduced visibility, may not be seen. This is another reason why sprinklers should not be shut down prematurely.

Warehouses with high rack storage may contain sprinkler systems within the rack storage area. This in-rack sprinkler system gives greater protection than that provided with only ceiling sprinklers. The activation of these sprinkler heads will be placing water directly onto the fire.

FIRE WALLS

Compartmentation accomplished through the use of fire walls and properly functioning fire doors will assist in the confinement of fire. Masonry fire walls are self-supporting and should maintain their structural integrity in case of collapse on either side of them. They are commonly thicker than ordinary walls, and if located in a structure with a combustible roof, they must extend above the roof. That portion of the wall above the roof is classified as a parapet wall, and its purpose is to prevent the spread of a roof fire.

Horizontal openings can be protected by fire doors, but caution should be exercised by firefighters. Fire doors are often blocked open or weights may be removed, defeating the mechanism of automatically closing doors. Forklifts can damage doors, making them inoperable. Vertically sliding or overhead rolling doors can close automatically when activated by heat, injuring firefighters operating beneath them or passing through as doors are closing. Doors closing behind firefighters may

FIGURE 9-5 A warehouse fire gives off a tremendous amount of radiant heat. *Used with permission of Deputy Chief Thomas Lyons.*

disorient them, and they may be unable to find their way out. When entering an area to fight a fire, any fire doors that are passed through should be blocked open to ensure a ready path of escape. A standby team should be stationed at these locations. Fire inspectors must be aware of the importance of fire doors and inspect them for damage, checking that they are not blocked open or prevented from closing by stock piled against them. Door control must be used if the open doors lead into the fire area, creating low pressure areas that the fire can draw oxygen from.

Fire doors can be utilized for controlling a fire through compartmentation of the fire area. This segregation of the fire building is one of the best methods of containment available. In this case the firefighters should ensure that the fire doors are in a closed position and that fire has not passed through these openings to involve other parts of the building. The fire side of the fire doors would be considered lost and a defensive attack can be initiated. The unburned side of the fire doors would be protected by interior hose-lines. In multistoried buildings, the fire doors must be closed at the same fire wall location on each floor; this takes coordination of companies on all floors.

RAPID FIRE SPREAD

A relatively stable situation can rapidly become out of control with little forewarning. Older manufacturing buildings with oil-soaked wooden floors, once ignited, are difficult to extinguish. This adds to the fuel load and increases the speed with which fire can race through a structure. The escalation of fire spread throughout a building, causing rapidly deteriorating conditions, will necessitate the withdrawal of firefighters. If a fire is not controlled quickly, it can reach the stage where the only strategy left is a strictly defensive one. This poses the possibility of fire spreading to exposed buildings. A specific exposure concern should be for structures higher than the original fire building. Convected heat can impinge on the upper floors of these buildings, even when located a distance from the fire building, and tremendous radiant heat will allow rapid fire spread.

VENTILATION

Horizontal and vertical ventilation is a must, and close coordination must be maintained between line advancement and ventilation. For hose-lines to advance into the fire area sufficient openings must be made ahead of these hose-lines. This often can be accomplished by breaking windows on the fire floor and the floor above. Additionally, roof scuttles, skylights, or hatches can be opened, as directed by the crews operating on the interior. Wire glass in skylights and hatch covers should be pulled back onto the roof

FIGURE 9-6 Interviewing bystanders can often lead to valuable information about the building and occurrences at the scene prior to the fire department's arrival. *Used with permission of Deputy Chief Thomas Lyons.*

and not driven into the opening where they could strike a firefighter, causing an injury. Firefighters operating on a roof who find the presence of a scuttle cover or broken wire glass should recognize it as a warning that a roof opening exists and that it should be avoided. In addition to scuttles and skylights, roof openings may need to be created to provide adequate ventilation.

BLITZ ATTACK

Consider utilizing a blitz attack if initially encountering a large volume of fire pushing from multiple windows. A blitz attack is accomplished through the use of master streams directed onto the fire from the exterior. With luck, this initial deluge of water will have a positive effect on the fire. If the blitz proves successful and it appears that control can be accomplished by utilizing an interior attack, then handheld hose-lines can be stretched in preparation for that attack. Once the large body of fire has been knocked down, the master streams can be shut down and the hand lines advanced into the fire area. The timing is critical. If the exterior lines are shut down too soon, a large body of fire may remain, and may be beyond the control efforts of handheld hose-lines. Attempting to utilize both exterior and interior attacks in the same area simultaneously must be avoided. A safety measure that allows rapid control over master stream devices can be achieved by placing a gate valve in the hose-line at least one length back from the master stream device.

DEFENSIVE ATTACK

The volume of fire, building stability, and safety of firefighters may require a continued use of master streams. This tactic may keep the fire contained and allow knockdown before major fire involvement destroys the building. The extinguishment of a fire utilizing a defensive attack depends on the number of windows through which the fire can be fought and the potential for fire spread through them. In areas that have a high density of buildings, master streams may be deployed from windows of exposed buildings. Interior partitions and high-piled stock will restrict the penetration of exterior streams; the size of the structure itself must be considered. Knowing the effective reach of a hose stream is important. If the stream reach is approximately 50 feet and the master stream appliances are placed directly at wall or window openings, exterior streams cannot reach a fire in the center of a building more than 100 feet deep.

There have been innovative changes made in master stream appliances that allow a flow of a large volume of water with considerably greater reach than in the past. Field tests have shown flows of 1,850 gallons per minute with a reach of 200 feet from a master stream with a 2½-inch nozzle. This is one option available to combat fires in buildings of considerable depth.

EXPOSURES

Hose-lines operated from the windows of threatened exposures will have a twofold effect: They will place water on the fire and prevent the extension of fire through that window. Water for these streams can often be supplied by the standpipe system within the exposed building. Too often, the standpipe is overlooked and companies stretch many additional lengths of hose-line.

FIGURE 9-7 A heavily involved warehouse requires large-diameter hose-line and master streams. *Used with permission of Brian Feeney.*

Exposed buildings subjected to heavy fire conditions must be checked for fire extension. When adjacent buildings share common bearing walls, floor and roof supports may be contained in the same wall socket. This can allow fire in one building to extend to the adjacent structures via conduction of fire through the wooden joist.

WATER SUPPLY

Problems will be compounded if the water supply is insufficient. This can happen when buildings are located in remote areas that have no hydrant system and insufficient drafting sites. To be considered adequate, a hydrant system must be supplied by water mains of a size sufficient to meet the needed fire flow. Older water mains often have tuberculation lessening the expected flow.

> Tuberculation is the reduction of the inside diameter of a pipe caused by deposits of chemicals or growth of organisms on the inside of iron pipe. These tubercules roughen the inside of the pipe, increasing its resistance to water flow.

SIZE-UP FACTORS FOR COMMERCIAL BUILDINGS AND WAREHOUSES

Water
- Interior attacks will demand 2½-inch mobile hose-lines.
- Firefighters must supply water to the sprinkler intakes to supplement the system.
- Ensure that if private hydrants are provided, utilizing them will not affect the supply of water for the sprinklers.
- Building fires past the incipient stage will require large volumes of water. Hydrant systems or drafting sites may be inadequate for a major fire.
- Defensive attacks will need to supply numerous master streams.
- Water may be required for wetting down exposed buildings.

Area
- Building layout may be irregular and, if surrounded by other buildings, difficult to ascertain without a preplan.
- Large open areas within these buildings are common.
- Fire doors may be secured in an open position.
- Closing fire doors set in fire walls can segregate a fire and make control possible.
- Buildings may be deeper than the reach of exterior master streams.
- The building may be overstocked, restricting access and overwhelming the sprinkler system.

Life Hazard
- Fast-moving fires can threaten employees by reducing visibility and causing confusion.
- Accountability for the occupants in these large buildings will be difficult.
- Firefighters can easily get lost within these buildings. The use of hose-lines, guidelines, or search lines for easy exit capability should be mandatory.
- Thermal imaging cameras (TICs) can assist in the primary and secondary searches.

Location, Extent
- Smoke conditions may impede firefighters in locating the seat of the fire.
- Determining the size of the fire and what is burning will be difficult. Operating sprinklers can drastically reduce visibility by creating a large amount of steam that will mask the fire area. Utilize TICs.
- Narrow aisles, smoke-filled rooms, and stock wetted by hose-lines and falling into aisles will reduce mobility.
- Fast-moving fires can occur.
- High ceilings can mask heavy fire conditions and allow fire spread through the overhead, threatening firefighters operating on the interior.

Apparatus, Personnel
- Operating 2½-inch hose-lines will require the response of a larger number of personnel.
- Fires past the incipient stage will require special apparatus for master stream operations and a sufficient number of firefighters to stretch hose-lines to implement relay or water shuttle operations.
- Salvage operations on floors below the fire may be required.

Construction/Collapse
- Ordinary, heavy timber, frame, and noncombustible construction can be found.
- Heavy timber buildings will contain a large amount of wooden structural members that can add fuel to the fire.
- Noncombustible buildings will not supply fuel to the fire, but early collapse must be considered.
- Older frame buildings may be of balloon-frame construction.
- Renovations are commonplace.

Exposures
- These fires can produce high heat conditions and create a firestorm effect, threatening nearby buildings with radiant heat.

Weather
- Flying brands may be spread downwind, starting numerous other fires.
- A firestorm can occur, which will create its own wind and magnify the firefighters' problems.

Auxiliary Appliances
- Sprinklers, standpipes, or special extinguishing systems may be present.
- Buildings containing high ceilings and rack storage may need in-rack sprinkler systems.

Special Matters
- Stock may include material that could create a hot, fast-moving fire.
- Hazardous materials may be present.
- Forklift motors may be fueled by propane gas.

Height
- These buildings are typically one to eight stories.
- High ceilings in one-story buildings can exceed 50 feet. Ceiling sprinklers alone may be ineffective in controlling and extinguishing a fire on shelving near the floor level.

Occupancy
- Large quantities of stock may be found in these structures. Determine what the stock is and how it will affect firefighting operations.
- Baled paper and other absorbent materials can retain large volumes of water and add a tremendous live load to the structure.

Time
- An occupied building may permit the early discovery of a fire.
- A fire brigade may respond during business hours and control a fire in its incipient stage.
- A business may have a greater amount of stock being stored in anticipation of a holiday season or other strong sales period.

CONSIDERATIONS FOR FIRES IN COMMERCIAL BUILDINGS AND WAREHOUSES

Strategic Goals and Tactical Priorities for an Offensive Attack
- Perform search and rescue for occupants.
- Call for a sufficient amount of resources.
- Check exposures if they are threatened for extension of fire.
- Check that fire doors are closed to contain the fire.

- The most effective hose-line for a fire of any consequence in these structures is 2½ inches in diameter. After knocking down the fire, overhauling can be accomplished with 1½-inch or 1¾-inch hose-line.
- Depending on the conditions found, both vertical and horizontal ventilation might be needed. The roof may contain scuttles or skylights that can be removed.
- Positive pressure ventilation (PPV) may be effective if the building openings can be controlled.
- Anticipate the use of steel bar joists supporting the roof of a noncombustible building and an early failure if attacked by fire.
- Recognize the need for a sufficient water supply.

Incident Management System Considerations/Solutions for an Offensive Attack

- Incident Commander
- Operations (if needed)
- Safety Officer
- Rapid Intervention Crew(s)
- Staging
- Ventilation Group
- Search and Rescue/Evacuation Group
- Division 1, 2, 3, etc. (a division can be implemented for any floor where deemed necessary)
- Medical Group (if needed)

Strategic Goals and Tactical Priorities for a Defensive Attack

- If changing from an offensive to a defensive attack, ensure that a personnel accountability report (PAR) is taken.
- Set up and maintain a collapse zone.
- Determine the amount of water needed and a source for its supply.
- Utilize master streams.
- Protect exposures, if necessary. This may involve interior exposures in which fire doors can be closed. Master streams are set up at those locations to prevent the extension of fire into uninvolved areas.

Incident Management System Considerations/Solutions for a Defensive Attack

- Incident Commander
- Operations (if needed)
- Safety Officer
- Logistics (for water supply, if needed)
- Rapid Intervention Crew(s) (if needed for exposures)
- Staging
- Divisions on Sides A, B, C, D

Strip Malls

A **strip mall** contains a row of stores with a parking lot directly in front of the stores. Strip malls are also referred to as *strip stores* or *taxpayers*. The stores in these buildings are interconnected in a row.

strip mall ■ Row of stores with parking in front.

> The term *taxpayer* originated when cheaply built stores were erected to generate revenue to pay the taxes on the ground until it could be developed with higher-priced structures.

Successful operations at a strip mall fire will depend on a number of factors. The building's fire protection (if any) will rely on the codes in force at the time the building was built. Though newer codes stress the installation of sprinkler systems, it is more common to find no fire protection in most strip malls.

FIGURE 9-8 Strip malls may contain two levels. The photo shows the parking lot at the front of the strip mall at one level and the contour of the land allowing an additional level on the C (rear side) and the D sides. *Used with permission of Jim Smith, Jr.*

A height of one story is predominant with strip malls, yet some two-story structures exist. Older structures may have a truss roof covering the entire assembly. The second floor may contain one large room over a number of stores and be used for social functions, or each store may be separate on the first and second floors, with the second floor utilized for storage or as a residence.

CONSTRUCTION

Strip malls can be built on a concrete slab or may contain a basement. Common basements can be found in older strip malls. These common basements are usually separated by flimsy partitions and become gathering places for combustibles. Poor housekeeping seems to be the norm rather than the exception in these areas.

Strip malls may be constructed so that the first floor shopping area is at street level, and the basement is at a lower level and accessible by vehicles from the rear via a street or driveway, or the basement may be at a lower grade level created by landscaping the terrain. (See Figure 9-8.)

Strip malls are commonly of frame, ordinary, or non-combustible construction. They can be built with masonry walls between each building that extend through the roof and serve as fire-stops between each store. Another method of construction creates large store areas that are then subdivided into smaller stores. This permits flexibility to increase or decrease the size of stores. This division of one large area into smaller stores can allow a fire to spread and encompass other stores because separation is normally achieved by installing partition walls. There are usually large plate glass windows covering the front of the stores.

In addition to masonry fire walls, walls may consist of fire-rated dry wall constructed between the individual stores. The problem created with this type of wall construction is that it can be violated to run utilities through the attic or cocklofts by creating a poke-through in the drywall. These seemingly harmless openings can allow a fire to spread from the original area to encompass a much greater area.

Roof construction in newer strip malls utilizes lightweight components. In a number of instances, a heavy snowfall caused the collapse of these roofs during the construction stage. Dead loads on the roof, such as large air conditioners or refrigeration units, can accelerate collapse.

Suspended ceilings may be installed, or the underside of the roof may be painted and left exposed. Partition walls may end at the underside of the roof or the underside of the ceiling. A close inspection should be made. If the partition wall ends on the underside of the ceiling, a fire that extends above the suspended ceiling can spread to adjacent stores and attack the entire roof area.

Strip malls may contain storage areas in the rear of the first floor, on the second floor of two-story buildings, or in the basement. If no storage area is provided, then stock will be kept in the customer areas. (See Figure 9-9.) It is often piled high to the underside of the ceiling on large shelving units. Ceiling heights of more than 25 feet are not uncommon.

CANOPIES

Strip malls may be built with canopies or large decorative false fronts with signs attached to the fronts of the buildings. These canopies are usually erected with lumber and covered

FIGURE 9-9 If no storage area is provided in the mall, then stock will be kept in the customer areas. This can reduce aisle space and supply more fuel to a fire.

with siding. They provide shelter from the rain for customers as they window-shop. Heavy fire involvement of the canopies can cause them to collapse. They are rarely fire-stopped and often contain poke-throughs in the front walls that allow fire attacking the canopies to enter the stores. These void spaces contain electrical wiring that has caused fires that are often difficult to locate and extinguish. (See Figure 9-10.)

FIREFIGHTING

An initial visible inspection through the front windows of the fire building and the adjoining properties can detect the presence of smoke or fire, giving an indication of the severity and involvement of the fire. The objective is to keep firefighting forces ahead of the fire spreading to the exposed stores.

The fire must be attacked with hose streams of sufficient volume to contain the fire. If there are no civilians in the store, the attack on the fire is almost

FIGURE 9-10 Canopies at strip malls are often built of wood framing and are not fire-stopped. There are interconnections between the canopies and the stores through which fire can travel.

always initiated from the front of the store. This occurs because rear entry to strip mall stores is a major problem due to the numerous methods of security that are employed.

Information from the roof team of the presence of a parapet wall between the structures will indicate that a fire barrier exists. Units must still check for spread of fire in the adjoining exposures since the fire wall may have been breached, or roof collapse could crack the wall, allowing fire to spread.

A fast-moving fire can quickly involve more than one store. If rescue is not a problem and the fire has gained control of an exposed building, it will be necessary to consider that store lost and move to the next building. A decision on which exposure to defend initially should consider protecting the greatest number of stores or the long side of the mall. (The *long side* refers to the side with the greatest length of stores threatened on the B or D sides.) This decision must consider where a fire can be stopped by the fire wall locations and the end of the mall. The employment of a blitz attack with a master stream by the first-arriving unit can supply a large volume of water, providing good penetration and heat absorption to knock down the fire. This quick attack must be accompanied by ventilation and can often be the difference between fire containment and an expanding fire.

REAR ENTRY

Because rear doors are often padlocked, customers or workers seeking refuge in the rear of the store can find themselves trapped. Under these circumstances, the initial attack on the fire and the rescue of those trapped inside must occur through the rear of the store.

Access to the rear may be difficult. Narrow streets or alleyways can contain trucks making deliveries or trash containers that can impede fire department operations.

For security reasons, the rear area may not display the names of the individual stores. If a large complex is involved, it may be difficult finding the exact location in the rear area.

Entry into the rear of the store can be challenging. Because theft and vandalism are a constant problem, rear doors are fortified to resist illegal entry. The various security measures employed compound our dilemma in gaining entry. In addition to

FIGURE 9-11 Heavy smoke conditions are found on arrival of the fire department in a strip mall. Initial decisions will need to be made about rescue, ventilation, exposures, and confinement. *Used with permission of Brian Feeney.*

multiple locks, brackets are welded or bolted onto the doors, and drop-in wooden beams (restraints) are used. These beams prevent the door from being opened, even if the door locks and door hinges have been removed. In buildings of frame construction, it is sometimes easier to gain rear entry through the rear wall alongside the door.

ON SCENE

Though not involving a strip mall, a collapse in Seattle occurred after renovations in the fire building included the installation of a concrete floor on the first floor above the fire. A support beam in the basement was destroyed by fire, causing the first floor to collapse into the basement, and killing four Seattle firefighters.

RENOVATIONS

Renovations affecting a building's stability can drastically change a structure from when it was first erected. New facades on buildings may include brick veneer attached to a frame building. The brick will not give the structure any additional support and will add to the building's dead load.

Suspended ceilings are commonly found and multiple ceilings are not unusual. A fire in a relatively newer strip mall in which two firefighters died in Chesapeake, Virginia, contained two suspended ceilings.

As businesses prosper, they may expand, encompassing several adjacent stores. To accommodate their expansion, they may breach walls between the stores to create one large store, though the individual storefronts may remain. Arriving firefighters anticipating fire in one store may find multiple stores involved due to the breached walls.

Another problem with renovations is that they may occur without the proper permits being secured and may not comply with the applicable codes.

ON SCENE

In a strip mall in Philadelphia, the wooden floor was changed to a terrazzo floor set on a concrete base. This drastically increased the dead load and caused the early collapse of the floor, killing four firefighters. The building was located in a row of strip stores and was being used as a delicatessen. The first-floor front exited onto a parking lot at an elevated grade level. The basement was accessed from the rear, which was at street level. Hose-lines had been stretched into the basement from the rear and members were attacking the fire. There was little smoke or heat discernible at the first floor. When smoke started showing around the seating booths, firefighters started opening these areas to check for fire spread. With no warning, the floor canted, causing four firefighters on the first floor to fall into the burning basement to a fiery grave. The masonry floor had given little indication of the fire below. It was found later that the fire had burned through the wooden floor supports.

VENTILATION

Common cocklofts and attics should be anticipated. A fire entering a non–fire-stopped space above the ceiling can spread horizontally. A strip mall fire that has progressed past the incipient stage demands that the roof be opened to draw the fire to the exterior rather than permitting it to mushroom under the roof and spread to the adjacent stores via common roof spaces. The aim is to open the roof directly above the fire area to relieve pressure and vent the fire to the outside. A problem for firefighters is that the use of steel bar joist or lightweight truss roofs can make roof ventilation a dangerous operation. Units going to the roof must be cognizant of the roof structure that they will be operating on, and safety must be a prime consideration in their operation. Safety measures that can be employed when operating on a roof containing lightweight components can include:

- Working from the platform of a tower ladder
- Operating from an adjoining roof separated by a masonry fire wall
- Spreading a ground ladder across the roof that a firefighter can work on to spread his/her weight

Access into the stores and ventilation can be achieved by breaking large plate glass windows in the front of the store. This may be requested by the units operating on the interior. Though it will allow access and may alleviate smoke conditions, it will allow a tremendous influx of air that can rapidly accelerate a fire and generate high heat conditions. If the IC gives the order to break out the windows, the units operating on the interior must receive warning of these ventilation actions. This will allow them to reach areas of safety to see the reaction caused by the vented window openings.

The employment of a trench cut can sometimes contain a fire involving a large non–fire-stopped cockloft. This tactic can be advantageous, but will involve a large commitment of resources. When considering a trench cut, it should initially be placed on the side of the fire where extension of the fire would cause the most damage. This typically is the side where the greatest number of stores can be damaged or destroyed should the fire spread. This decision will be made in conjunction with firefighting strategies on which stores are being protected initially. A fire located in the center of a strip mall could require a trench cut on both sides of the fire area. A factor to consider is the presence of fire walls and their exact location. This knowledge will allow the IC to determine the number and location of a trench cut(s). (Trench cut is discussed in Chapter 4.)

SIZE-UP FACTORS FOR STRIP MALLS

Water
- If there is no compartmentation, a fire can quickly involve multiple stores, and a large water supply will be required.
- A blitz attack can be successful if there are no occupants.

Area
- One large open area will permit a fire to involve the entire store in a short period of time.
- There may be a stockroom in the rear of the store, in a basement, or on the second floor.

Life Hazard
- Severe during business hours.
- There may be living quarters, if a second floor exists.
- It is not unusual to find bedding in basements or rear storage areas that are being used as living quarters by an employee.
- TICs can assist in the primary and secondary searches.

Location, Extent
- A fire can quickly spread to involve other stores.
- A quick determination as to the extent of the fire must be made.
- Ventilation directly over the fire area can relieve the pent-up smoke and heat, and assist in fire control.
- Rear entry is difficult, and most operations will need to commence from the front of the store.
- Narrow driveways, delivery trucks, and the presence of large trash containers may restrict rear access.
- TICs can assist in checking walls, ceilings, and floors for hidden fire.

Apparatus, Personnel
- Initial operations will involve handheld hose-lines.
- Fires involving multiple stores will require a large commitment of personnel and often master stream operations.

Construction/Collapse
- Roof assemblies may be triangular, flat, or bowstring. Components may be wood, lightweight wood assemblies, steel bar joists, or heavy timber.

- Roof loads may include air conditioners, signs, or refrigeration units.
- Common basements, attics, and cocklofts may exist.
- Canopies or overhangs can have interconnections to the various stores. A fire can spread either from the store to the canopy or from the canopy to the store.

Exposures
- There may be some form of fire-stopping between the stores.
- There may be masonry walls between the individual stores that extend through the roof assembly.
- There may be partitions between the stores that offer little fire resistance to the adjoining stores.
- There may be a common attic or cockloft above the stores separated by a suspended ceiling and ceiling tiles that offer no fire resistance.
- Exposure protection will be an immediate concern.

Weather
- There are no weather concerns out of the ordinary.

Auxiliary Appliances
- Sprinklers may be found in new construction.

Special Matters
- Building renovations can drastically change the structure. Floors may be converted to terrazzo or concrete, adding a tremendous dead load to the floor.
- Multiple ceilings may exist.

Height
- Strip mall buildings are usually one or two stories.

Occupancy
- A store may contain stock that will add substantially to a fire, such as household goods, large amounts of plastics, and other combustibles.
- Heavy stock found in a hardware store, auto parts store, or plumbing supply store, for example, can indicate the potential for an early collapse if a basement is present.

Time
- The critical time period is when the store is occupied.
- Delayed discovery of the fire may occur when the store is closed.

CONSIDERATIONS FOR FIRES IN STRIP MALLS

Strategic Goals and Tactical Priorities for an Offensive Attack
- Perform search and rescue for occupants.
- Call for a sufficient amount of resources.
- Stretch hose-lines into exposures and check for extension of fire in ceiling and walls.
- Check for common cockloft and attic area above ceilings.
- Depending on the size of the building and the amount of fire, 1½-inch or 1¾-inch hose-lines may be effective. Larger stores may require the use of 2½-inch hose-lines.
- Horizontal ventilation will be needed. Roof ventilation may be required if fire heavily involves the area under the roof or if a common attic or cockloft exists and fire has entered those areas. (If the roof can be opened directly above the fire area and vented to the exterior, it will minimize mushrooming of the fire and the threat to the adjacent stores.)
- PPV may be effective if the rear door can be opened in a reasonable amount of time and fire has not entered the void space above the ceiling.
- Anticipate the use of lightweight building components in the roof and an early failure if they are attacked by fire.

Incident Management System Considerations/Solutions for an Offensive Attack
- Incident Commander
- Safety Officer
- Rapid Intervention Crew(s)

- Staging
- Ventilation Group
- Search and Rescue/Evacuation Group
- Divisions 1, 2 in original fire building
- Divisions in Exposures B and D
- Medical Group (if needed)

Strategic Goals and Tactical Priorities for a Defensive Attack

- If changing from an offensive to a defensive attack, ensure that a PAR is taken.
- Set up and maintain a collapse zone.
- Utilize master streams.
- Protect exposures. This may involve interior operations within the exposed buildings.

Incident Management System Considerations/Solutions for a Defensive Attack

- Incident Commander
- Safety Officer
- Rapid Intervention Crew(s) (if needed for exposures)
- Operations (if needed)
- Staging
- Logistics (for water supply, if needed)
- Divisions on Sides A, B, C, D
- Medical Group

Enclosed Shopping Malls

The **enclosed shopping mall** concept has resulted in the building of larger regional malls that can contain tens of thousands of shoppers. Enclosed malls allow a controlled, indoor atmosphere with piped-in music that is conducive to shopping. Open areas within the mall are used for shows as added attractions. (See Figure 9-12.)

enclosed shopping mall ■ Found in urban and suburban areas. Everyday merchandise consists of many flammable hazards.

The malls are designed for customer convenience. There usually is carpeting installed within the individual stores. The interconnecting areas vary. Wood flooring, which is less tiring on shoppers' legs, may be used instead of concrete and allows for longer shopping trips. When concrete, vinyl, or ceramic tiles are used, they are highly polished and become very slippery when wet.

To attract people to these malls, bus trips are arranged with incentives to lure prospective customers, which prompts increased congestion and people who are unfamiliar with the mall and its exits, adding a severe life hazard in case of fire.

The stores that occupy these malls are similar to those in any shopping district. They range from restaurants to auto parts stores or paint stores. Their everyday merchandise consists of many flammable hazards. A linen store may cover an area of 100 by 125 feet and be located in a noncombustible structure, with a 25-foot or higher roof of exposed steel bar joists. If sprinklered, the sprinklers are located at the roof area. They would have to protect wooden shelves up to 20 feet high, loaded with linens, plastic products, bedding, and anything found in an average bedroom. There are no in-rack sprinkler systems as found in a commercial warehouse of similar design and loading. The fire load will require an upgraded response from the fire department.

High activity time for enclosed malls is similar to other kinds of shopping areas. The exception is that because malls are enclosed, rainy weather seems to increase crowd size.

FIGURE 9-12 The enclosed shopping mall concept allows a controlled, inside atmosphere, with piped-in music that is conducive to shopping.

EMPLOYEE TRAINING

Store employees rarely are trained in fire safety procedures. They are paid minimum wages, and there is a rapid turnover in workers. Many are unfamiliar with the store itself, let alone the mall in general. This is often true of the security staff, as well. Fire safety and prevention, if taught initially, are not stressed. The emphasis is on security.

The use of fire drills is looked upon as an annoyance and disruption of business. Because profit is the primary purpose of the malls, understaffing occurs, and response to fire alarms may become a secondary consideration after security problems are handled. With poorly trained personnel and no practical application of fire safety, a fire can bring chaos and panic.

TYPES OF MALLS

Enclosed mall complexes can be divided into two general areas: the downtown areas of older cities and those built in suburban areas. The trend toward establishing shopping malls in the older downtown areas is an attempt to compete with the large suburban malls.

The downtown malls are often multistoried, due to the high cost of real estate. The multistory configuration frequently is centered around a stairway opening with escalators and glass elevators for ease of movement to all floor levels, permitting a large open area. (See Figure 9-13.)

A problem associated with multistoried structures is the presence of trash chutes, which terminate at the first floor or basement locations where the trash is unloaded into metal trash containers or a compactor system. These areas are trouble spots for firefighters because fires often occur within the systems. A smoldering fire can fill large areas with smoke and set off smoke alarm systems before enough heat is given off to activate a sprinkler head that can extinguish the fire.

PARKING AREAS

The need to attract shoppers with plenty of parking is dealt with by constructing large, multistoried parking garages adjacent to or beneath the downtown malls. These parking structures usually are built to accommodate cars only and restrict entry of fire apparatus. Firefighters responding to a vehicle fire or an accident with resultant injuries must decide the best location to gain entry to facilitate their operation. These parking garages are usually sprinklered and equipped with standpipes.

Above-grade, open-air garages can be accessed from street level via main or portable ladders. Below-grade garages can be a problem if an attempt to evacuate all the cars occurs at one time. A buildup of carbon monoxide can develop. Built-in carbon monoxide detectors should indicate the presence of high levels of carbon monoxide. If the exhaust fans cannot remove the carbon monoxide fast enough, it may be necessary to have the vehicles shut down their car engines. In any case, firefighters should utilize their carbon monoxide monitoring equipment to verify safe levels.

ON SCENE

A July 4th celebration at Independence Hall in Philadelphia climaxed with late-night fireworks. The large crowd dispersed at the same time. When leaving the below-grade parking garage, a few stalled cars caused a traffic jam for a number of levels. Due to the heat of summer, the motorists kept their car engines running to use their air conditioners. People became sickened by the carbon monoxide, leading firefighters to evacuate the lot quickly and tend to those who were ill. Monitoring carbon monoxide in such a situation is necessary so that firefighters will know when the proper oxygen level is restored, and it is safe to remove their self-contained breathing apparatus.

MAGNET STORES

Large urban malls may build adjacent to an existing store and incorporate it into construction. The existing store usually is a large, established institution that will act as a magnet or anchor store. The use of the well-known name will assist in drawing shoppers to the new mall.

In the construction of suburban malls, magnet or anchor stores also are secured to act as a customer draw. These stores usually are located at the ends of the mall, with specialty stores located in between. Mall construction often occurs prior to the leasing of all the stores. The rentals of these stores occur after construction, with the number of square feet for each store approximated.

Sprawling suburban malls can cover many acres. If built on an uneven landscape, ramps or steps are necessary to accommodate the terrain. This is usually accomplished in increments of two or three steps at a time. If both ramps and stairs are provided alongside each other, they may be separated by handrails. The railings can impede customers exiting a fire in a panic situation.

STORES

Large, spreading malls containing hundreds of stores need a system of identification for mall areas. One that works quite well is the use of specific colors, both on the exterior and common interior walls of the mall walkways to identify a particular section. This allows easy identification of the reported fire area. The use of red, yellow, and blue sections, for example, makes it easy for a person reporting a fire to indicate the store name, the color of the zone, plus any additional information. Firefighters can respond to and position their apparatus by the exterior color.

Walls between mall stores are often partition walls, a flexibility that permits increasing or decreasing the size of stores. The fire resistance of these walls could be questionable, allowing a fire to spread and encompass other stores. Suspended ceilings may be installed, or the underside of the roof may be painted and left exposed. If a suspended ceiling is present, partition walls often end at the underside of the suspended ceiling. This feature allows fire to move above the suspended ceiling to extend to other stores. The front of a mall store is often exposed during business hours. Security is maintained after hours by open-web metal security gates, permitting security guards to see into the store, but allowing a fire to extend from one store to another. (See Figure 9-14.)

FIGURE 9-14 The front of the stores in the enclosed mall have no doors. Security is maintained by open-web metal security gates, allowing security personnel to see into the store. This can allow a fire to extend from one store to another.

FIGURE 9-15 Sprawling malls overcome differences in terrain by adding ramps and steps. Under smoke conditions, these areas that have railings and other obstructions can become bottlenecks as the occupants attempt to evacuate.

Malls have a high turnover of businesses, which prompts constant renovations within the mall. These renovations include the presence of construction material that adds to the fire load. Renovations may necessitate shutting down the sprinkler system for specific periods. There must be a procedure in place to notify the fire department when the sprinkler is shut down, what areas are affected, and when the system is placed back in service.

OCCUPANTS

New malls cater to the physically impaired. There is an abundance of handicapped parking spaces near each entrance. Malls that are totally enclosed offer the convenience of a wide variety of stores, ramps for ease of movement, and escalators and elevators in multistoried structures. In emergency situations, one firefighter can safely lead 20 or more ambulatory people to safety. However, when rescuing the physically impaired, it may be necessary to assign rescuers on a one-to-one basis, impacting the resources required.

Bowling alleys and arcades attract children. Parents frequently leave them in these areas while they shop. If a fire occurs, the first concern of parents is for the safety of their offspring. They will attempt to enter unsafe areas, endangering themselves to find their children. A child may not evacuate a fire area because a parent previously admonished him or her with a phrase such as, "Don't you dare take one step out of this area."

Baby carriages, wheelchairs, and people on crutches or with walkers can complicate evacuation. They will have difficulty maneuvering in even the slightest panic situation. Electric carts used by elderly and physically challenged individuals, as well as people pushing, shoving, and tripping over these mechanical aids, add to the confusion.

RESPONSE

The ability of a fire department to respond effectively in parking lots that are jammed with cars will be the first hurdle. Accurate dispatch information will enable the firefighters to know the location of the fire, but their ability to gain access to that area can be tested severely if hordes of people are attempting to leave the mall as firefighters try to enter. The use of service entrances can provide a ready access for the fire department.

The preplan should show all the access points that are available, and the first-alarm companies should respond so coverage is possible from more than one entrance. Even in stores with many square feet of store space, the store layout usually funnels down near the front of the store, providing a narrow exit space and cash register area. This enables security personnel to monitor those leaving the store.

Large-area or department stores may contain specialty departments within the main store. These may be shoe or camera sections or jewelry areas. In heavy smoke conditions, a customer or firefighter entering these areas could become disoriented and trapped.

Some malls are designed with separate stockrooms. If no stock area is provided, there is a tendency to overload display areas, congesting aisle space and reducing exit ways. Shelving may be arranged to cordon off an area for storage, which creates problems for firefighters trying to ensure total evacuation or locate a fire within this space under heavy smoke conditions.

RESOURCES

The resource needs for complete evacuation can overtax the responding firefighters. Use of police and security, along with available firefighters, allows other important fire operations to continue. An aggressive attack on the fire is a prime way to ensure life safety.

Hose-lines must be brought into the interior mall area to contain the fire and protect the mall walkways. This hose-line placement also protects the interior exposures and avoids pushing a fire from one store to another.

An early request for ambulances must be anticipated due to the large number of people present. First-aid stations may need to be set up at more than one location.

Urban areas usually have sufficient water supplies. Suburban malls may be located in areas where water supply can be a problem. The suburban mall may necessitate long hose-line stretches, relay, or water tender operations. The use of large-diameter hose-line allows large amounts of water to be delivered, but it also restricts vehicle movement when stretched past driveways and through parking lots.

If the mall is sprinklered, lines should be connected into the system's intakes, if necessary, to supplement the system. If the system is not operating, someone must be assigned to investigate why and, if possible, correct the problem. Malls that cover large areas will have zoned sprinkler and standpipe systems. There may be more than one intake that needs to be pressurized.

Long hose-line stretches from the apparatus to the fire must be anticipated. As in any structure, care must be used in stretching these lines. They cannot be stretched into an exit and impede the evacuation process.

Ventilation may be difficult. Windows are nonexistent in most enclosed malls. Rear doors, if provided, are difficult to open from the exterior. However, ventilation must be afforded to allow the line to be advanced into the fire area. This can be accomplished in a number of ways. There may be service doors or truck-receiving areas that can be opened. There may be automatic roof vents that the fire department will need to verify are operational. Roof ventilation may require opening skylights, scuttles, or the roof itself. The mall ventilation system, if reversible, can be used to vent the toxic smoke to the exterior.

SIZE-UP FACTORS FOR ENCLOSED MALLS

Water

- Long hose-line stretches must be anticipated.
- Sprinkler systems will need to be supplemented.
- There may be a standpipe system that can be utilized.
- Large-diameter hose-line may be needed in areas of limited water supply.
- Cars in the parking lots may impede the stretching of hose-lines.

Area

- Large square footage areas may contain a substantial amount of stock.
- Stores can exceed 15,000 square feet, with a high open ceiling and stock on shelving reaching from floor to ceiling.
- Preplans containing plot plans are critical to ensure that responders arrive at the correct location. Callers may identify the location of the fire by the store name, the color of the zone, or the store number as identified on a plot plan.

Life Hazard

- Life hazard is severe during hours of operation; chaos and panic must be anticipated.
- TICs can assist in the primary and secondary searches.
- Handicapped individuals may require assistance to evacuate.
- Restaurants may operate past normal shopping hours.
- People may walk for exercise in the open areas of the mall during hours when the stores are closed.
- Many people unfamiliar with the mall will attempt to exit the same way they entered.
- Large numbers of shoppers can have an impact on firefighters attempting to evacuate the area.
- Maintenance personnel and security staff should be available in the mall at all hours to assist the firefighters.
- Hose-lines cannot be stretched into the building through areas that people are using for exits; this will impede their evacuation.

Location, Extent

- Determining the location of the fire can be difficult and time-consuming. These malls can be sprawling. Multiple entrances will exist, and occupants will be exiting if a fire exists.
- Utilize fire alarm panels to determine the location from which alarms are being received. One company should be assigned to this location in the preplan.
- TICs can assist in checking walls, ceilings, and floors for hidden fire.

Apparatus, Personnel

- Due to the severe life hazard, an adequate number of personnel should be initially dispatched.
- If the sprinkler system functions as designed, the fire should be controlled in the incipient stage. If the system fails, a major fire can ensue.
- Seek the assistance of security personnel and police in evacuating the fire area.

Construction/Collapse

- Newer suburban malls are built of noncombustible construction. Older malls located in the downtown areas of cities can be fire-resistive, ordinary, or noncombustible construction.
- Renovated malls must comply with the codes in effect at the time of renovation.

Exposures

- Fire exposures are usually internal and severe. Adjacent stores will need to be checked for fire spread.

Weather

- Normally, weather has no appreciable effect, though rainy days may increase the number of occupants.

Auxiliary Appliances

- Sprinklers, standpipes, detection systems, and heating, ventilation, and air-conditioning (HVAC) systems may be present.
- Stock may be piled up to the sprinkler heads and, in some cases, above them, negating their effectiveness.
- Sprinklers may be nonfunctional in areas where renovations are occurring.
- Utilizing the HVAC systems for ventilation purposes should be considered.

Special Matters

- Fire drills are nonexistent.
- Security is normally not too helpful with fire-protective devices.
- Parking garages may be located below or above grade.
- Large, sprawling parking lots surrounding the suburban malls can contain numerous vehicles attempting to exit, which can impede the fire department's response.

Height

- Enclosed shopping malls are generally one to four stories in height.

Occupancy

- Large quantities of stock will feed a fire.
- Limited storage space can cause clutter in shopping areas.
- There may be store renovations underway with construction material involved in fire.

Time

- The life hazard is most severe during hours of operation, but the building can be occupied at any hour.

CONSIDERATIONS FOR FIRES IN ENCLOSED MALLS

Strategic Goals and Tactical Priorities for an Offensive Attack

- Evacuate and perform search and rescue for occupants.
- Call for a sufficient amount of resources.
- Hose-lines cannot be stretched through doorways being used as exits by occupants.
- Confine the fire to the store of origin, if possible. This can best be accomplished by a simultaneous attack on the fire and an immediate check of adjacent stores on Sides B and D and the floor(s) above.

- Hose-lines of 1½ inches or 1¾ inches may be ineffective. Instead, 2½-inch hose-line may be needed.
- Ventilation will be difficult. Roof vents may be present; see if they are open. Opening the roof may be possible. Try to utilize the HVAC system to remove the smoke. With the numerous openings that are difficult to control, the effective use of PPV is doubtful.
- Overhaul will be extensive and resource intensive.
- Salvage can be tremendous. In a multistoried mall, the floors below should be considered.

Incident Management System Considerations/Solutions for an Offensive Attack
- Incident Commander
- Safety Officer
- Rapid Intervention Crew(s)
- Operations (if needed)
- Staging
- Ventilation Group
- Search and Rescue/Evacuation Group
- Divisions 1, 2, 3, etc. (for multistoried mall, a division can be implemented for any floor where deemed necessary)
- Divisions B and D (to check the adjacent exposed stores)
- Medical Group or Branch

Strategic Goals and Tactical Priorities for a Defensive Attack
- If changing from an offensive to a defensive attack, ensure that a PAR is taken.
- Set up and maintain a collapse zone.
- Utilize master streams.
- Protect exposures, if necessary.

Incident Management System Considerations/Solutions for a Defensive Attack
- Incident Commander
- Safety Officer
- Operations (if needed)
- Staging
- Logistics (for water supply, if needed)
- Divisions on the exterior on Sides A, B, C, D
- Exposure Divisions (if needed)

Supermarkets

Supermarkets have undergone radical changes in merchandise and services since the first supermarket, King Kullen, was opened in Long Island, New York, on August 4, 1930. Today, in addition to food products, the variety of merchandise found in supermarkets is endless. It can include household products that could be hazardous, kerosene for lamps, motor oil, large concentrations of aerosol cans, and other hazardous and combustible materials. The mixing of products to form dangerous combinations can occur quite easily.

Many supermarkets today prepare food in kitchens that operate the same as large restaurants. This change in store contents and operations creates a situation where a responding firefighter may encounter a fire in a supermarket involving a vast majority of products. It should also change the perception of a supermarket as only having shoppers walking down aisles to one with restaurant areas filled with customers.

supermarkets ■ Typically one-story self-service stores that contain food supplies and household goods for customer purchase arranged in long aisles, and may include commercial kitchens and restaurant areas as well.

GENERAL INFORMATION

The supermarket will vary widely in design and construction depending upon the area of the country in which it is located and when it was built. It may be a standalone structure or part of a strip mall complex. Store configurations usually consist of a large rectangular box, though in urban settings they may be configured in an irregular shape to fit the available lot size.

The presence of narrow aisle spacing is another challenge. The stretching of hose-lines through supermarket aisles can cause merchandise stacked at the end of the aisles to topple, clogging the aisles as occupants try to exit the store.

Parking is an essential component, and in urban settings parking lots can become congested due to limited capacity. Crowded parking areas can create a problem for emergency vehicles trying to gain access to the supermarket. Underground parking garages can also cause congestion.

Loading dock areas can be filled with trucks waiting to unload their contents and the presence of trash containers can restrict access and placement of fire department vehicles.

NIOSH FATALITY INVESTIGATION REPORT #98-F04

A supermarket fire took the life of a firefighter due to smoke inhalation. The fire started at approximately 1030 hours. Evacuation of the store was completed and a captain and firefighter entered to make an internal evaluation. They did not take a hose-line or a safety line. Initially the smoke conditions were light and no heat was felt; however, conditions drastically changed to heavy smoke and heat conditions. With zero visibility, the firefighters became disoriented and could not find their way out. The firefighter ran out of air and passed out. The captain started crawling and was able to crawl toward the outside light and to safety. NIOSH investigated the incident and recommended that firefighters entering a structure should have a hose-line or safety line that they can use to exit a building, and that at least two firefighters should be standing by outside ready to provide assistance or rescue if needed. **See Appendix B: NIOSH Reports in www.pearsonhighered.com/bradyresources to read the complete report and recommendations.**

REFRIGERATION

There is a need for large refrigeration and freezer units, which require the use of a refrigerant. The threat to the ozone layer has led to a ban in the United States of chlorofluorocarbons (CFCs) and to the use of hydrochlorofluorocarbons (HCFCs), though they are scheduled for phase-out by 2030, as well as the use of hydrofluorocarbons (HFCs), which will also be phased out. These two products and mixtures of the two are the current alternative cooling agents being used while research continues on future coolant alternatives including hydrofluoroethers.

Many firefighters equate the term Freon with CFCs. Realize that Freon is a trademark used by DuPont for marketing a wide variety of coolants, including CFCs and the current substitutes. Substitute coolant gases are marketed by other manufacturers.

A 35,000-square-foot modern supermarket will require 3,000 to 5,000 pounds of refrigerant. Industry-wide expectations project that leakage can range from 20 percent to 50 percent annually during the operation of supermarket freezers, display cases, and walk-in coolers. Responders must anticipate that if a catastrophic system failure occurs, it is possible for the store area to be flooded with coolant. The coolant acts as an asphyxiant, displacing oxygen or creating a toxic atmosphere. Coolant release can hydrogenate an area, presenting some flammability concerns.

CFCs may still be found in use at facilities that have not yet been converted to alternative coolants. CFCs present hazards to those who have heart conditions, causing cardiac arrhythmias, though others exposed usually have no serious long-term effects.

Ammonia is used in large refrigeration systems. It will not be found in the supermarket environment, but will be found at supermarket warehouse operations where large quantities of food products must be kept cold. Central refrigeration centers for supermarkets can be quite massive and contain tremendous amounts of ammonia.

ON SCENE

Supermarket fires have proven deadly for firefighters as noted under Timber Truss in Chapter 5, where six firefighters died in 1978 in Brooklyn, New York, while fighting a fire and operating on a bowstring truss roof that collapsed.

CONSTRUCTION

The type of building construction in a supermarket can allow responders to anticipate how a fire can spread and attack the building, and the impact it can have on the life safety of the occupants.

Prefire planning should include a close review of the building's construction features. Successful firefighting operations need to recognize that fire behavior has a direct relationship to the building's construction. The knowledge that can be gained during prefire planning can be critical in terms of firefighting, rescue operations, and personnel safety.

Common supermarket construction today consists of one-story noncombustible buildings, including masonry walls and steel bar joist roof assembly with metal Q decking. There are also supermarkets of ordinary and frame construction. These structures may contain roofs constructed with wooden parallel chord or triangular trusses, or steel or wood bowstring timber truss roof assemblies. A common trait is the height of these one-story structures that can be the equivalent of a three-story building.

The roof surface may be a poured concrete surface or a built-up roof design. A common roofing material is a rubber membrane. The type of roof assembly can dictate the stability of a structure when attacked by fire. Supermarkets usually have large loads on their roofs that can increase the potential of an early roof collapse. This mandates the need for the type of roof construction to be noted on the building's preplan. If responding to a fire in a supermarket without knowledge of the type of roof assembly, it should be presumed that it is lightweight and subject to an early collapse.

STORAGE AREAS

Older structures in urban areas may contain basements that are used for storage. These stores may contain access from two different levels. A loading dock in the rear servicing the basement or lower level can be at street level while the store proper is accessed from the parking lot level. Freight elevators will be utilized for moving stock from the storage level to the store level.

Supermarkets require large storage areas adjacent to the shopping areas. The storage areas can consist of high rack storage and be serviced by fork lift trucks. If space is limited in the storage area, additional stock may be placed in the store proper. This additional stock can reduce the aisle spacing and cause congestion, especially during peak shopping hours. Large quantities of merchandise may be stored in trailers either adjacent to the store or left at loading docks and unloaded as the stock is needed. This depends upon the size and number of loading docks available. Trailers at loading docks can limit access and ventilation capabilities.

REAR ENTRY

Access to the rear of supermarkets can be limited. The only windows are usually the plate glass windows contained in the store fronts. The rear and sides of the store may contain loading docks at which trailers may be situated, restricting fire department access at those locations. Due to security concerns, doors located at the rear of the supermarket may not contain any hardware from the exterior and must be opened from the interior. This will slow operations if access is attempted from the rear. Preincident planning can determine hours when the rear doors are accessible. The fire department's delay in opening the rear doors can limit horizontal ventilation until access can be achieved. Once rear doors are opened, they should be blocked open so they don't close and relock.

Apparatus access to the rear can be restricted by illegally parked vehicles, tractor trailers making deliveries, and trash dumpsters.

LIFE SAFETY

Due to the likelihood of a high occupancy load, a supermarket fire has the potential for injuries to customers and employees. The life safety problem for customers is usually limited due to self-evacuation; however, elderly or disabled customers and employees may need assistance. It should be noted that these facilities may operate continuously or have restocking and/or custodial personnel after normal shopping hours.

Most search and rescue operations are accomplished from the front of the store. Though exit signs are posted in these structures, most people rarely observe them and will try to exit the store the way they entered.

Life safety for employees will vary depending on the time of day and ability to reach an exit. In most cases a primary search should be conducted. Realize that light smoke conditions can rapidly change to zero visibility and firefighters without the aid of a hose-line or search rope can easily become disoriented due to smoke conditions. A primary search can be an extensive undertaking in large stores. Large stores demand that firefighters maintain contact with a hose-line or stretch safety ropes to ensure that crews can safely find their way to the exterior. Primary searches must include walk-in freezers since store employees may seek shelter within them during a fire.

Firefighter safety must be constantly evaluated. Early collapse of the roof assembly with its dead load of air-conditioning units and other mechanical components is always a potential hazard. The IC must evaluate risk versus gain when deciding upon the type of fire attack to be employed. The early establishment of a Rapid Intervention Crew(s) (RIC) with sufficient personnel is critical to ensure that downed firefighters can be rescued in a timely manner.

> Studies performed by the Phoenix Fire Department, following the line-of-duty death of a firefighter in a supermarket fire, indicate that up to 12 personnel may be required to rescue one downed firefighter.

FIRE CONSIDERATIONS

Understanding the problems that can be encountered during a supermarket fire requires that the fire officer understand the critical signs or cues that the incident presents. The solving of those problems can be achieved through the implementation of the correct strategies, tactics, and tasks. This will allow firefighters to control these fires effectively, while protecting civilian lives and minimizing the risk to firefighting personnel. The failure to base decisions on good information or observations can result not only in inefficiency of firefighting operations and increased fire loss, but the possibility of injury or death to responders.

Determining what is burning and its location can assist in the decision-making process on how to fight the fire, while ensuring life safety. The IC needs to identify the various problems that may be encountered. Safety considerations for response personnel and effective operations are directly related to the recognition of these problems.

PROBLEM IDENTIFICATION

The IC needs to review what was observed. What smoke conditions were noticeable during the initial 360-degree walk-around? Was a TIC used during the walk-around to detect heat within the building? With minimal windows, any smoke showing from the exterior could indicate a serious fire burning within. In performing size-up, were any heavy loads visible on the roof such as HVAC units that may accelerate the collapse of the roof? Realize that high parapet walls can hide these units from ground view. Were any wall cracks present?

The rapid extinguishment of the fire will solve many problems. Yet the IC must be constantly sizing-up fireground conditions for any indicators that an interior attack should be abandoned and an exterior attack employed. This will include:

- Are the rescues completed?
- What is the type of building construction?
- What is burning or supplying fuel to the fire?
- What is the burn time?
- Is the building sprinklered, and are the sprinklers operating, and is the water supply to the sprinkler system sufficient?

The possibility of fire extension should be checked with TICs. This is especially important if a fire that initiated in storerooms has spread to involve higher storage, the roof area, or laterally to storage on adjacent racks. The ability to observe fire conditions in the roof assembly from the interior will be minimal, if any, due to smoke conditions.

The determination of the fire being in the roof structure, in addition to the TICs, may be made by observing wall cracks on the exterior near the roof line indicating that steel in the roof assembly is expanding. If fire has involved the roof structure, it must be decided whether it can be knocked down with interior hose-lines.

The IC needs to continue to assess the fireground conditions:

- How much time has elapsed from the time the fire was discovered?
- How long has the fire department been operating on the fireground?
- Has adequate ventilation been accomplished?

If firefighting has been going on for a long period of time, the IC needs to assess if an offensive attack is controlling the fire or merely chasing it. If fire control cannot be accomplished in a timely manner, the IC should strongly consider changing to a defensive attack. This decision should include evaluating reports from division and group supervisors. Are they making progress? Do they feel that they can gain fire control shortly? Do they have a good handle on the exact problems they are faced with and will the plan solve those problems?

Once the IC has gathered sufficient information, he/she must decide whether to continue to employ an interior attack or go defensive. Factors that will guide this decision are:

- Have all occupants been removed?
- What is the risk versus gain to the firefighters?
- Is it possible to safely continue the interior attack?

 ON SCENE

One supermarket fire that I responded to was started by an employee's carelessly discarded cigarette in the storage area. Once ignited, the smoke from the fire burning in rack storage started to bank down. The employees opened a door to the exterior to ventilate the area, while propping open another door that entered the supermarket. The wind drove into the storage room, forcing the smoke into the supermarket and threatening the shoppers. The fire was slowed down by the sprinklers on the ceiling, but the fire continued to burn on the lower rack storage, fueled by combustible products. Narrow aisles allowed the fire to spread at the floor level unimpeded by the sprinklers. Though the fire department quickly extinguished the fire, the smoke damage to the store and some smoke inhalation injuries to shoppers and workers could have been avoided by proper fire prevention procedures.

STORAGE AREAS

Storage space is at a premium in supermarkets and high rack storage is often utilized. These racks can reach heights of 25 feet. A fire originating on a lower rack can spread upward and laterally. If the storage area is sprinklered, are the sprinklers operating? Has the sprinkler system been supplied by the fire department? Rack storage with only ceiling sprinklers can allow a fire to burn at the lower levels without being extinguished by the sprinklers. If storage aisles are narrow, or stock is left in aisles, the fire can extend from one rack to another, causing the opening of numerous sprinkler heads, while the fire continues to spread. The steam and smoke created by the burning material and the partial extinguishment by the sprinklers will reduce visibility to zero.

Fires burning in unsprinklered storage areas can become well-involved and develop into fast-moving fires that will be difficult to control. The use of 2½-inch hose-lines will be needed. Stock falling from the storage racks into the aisles will impede firefighters attempting to enter the fire area and will bury hose-lines as it slows down their advancement.

VENTILATION CONSIDERATIONS

Ventilation will need to be accomplished to assist in store evacuation and for firefighting. When confronted with smoke conditions, and when there are occupants still in the supermarket, there will be an immediate need for ventilation to assist in evacuation and rescue to minimize the number or severity of injured occupants. The exact method of ventilation could involve utilizing the built-in features such as operable hatches, skylights, the exhaust capability of the HVAC system, etc. The lack of windows may be overcome by attempting cross-ventilation by opening rear and front doors. Opening the roof is also an option, if no roof hatches or skylights exist. The IC must recognize that ventilation openings will create flow paths, and if water is not being placed onto the fire, it will accelerate.

Ventilation may be accomplished by removing trailers that are docked at the loading platforms. This can sometimes be done by employing the truck drivers waiting to unload their trailers, in cooperation with the police. Emergency ventilation can sometimes be achieved by the utilization of large tow trucks to remove the trailers a sufficient distance to effect ventilation. Removal of the trucks can also remove material that could add fuel to the fire.

Removal of large plate glass windows in the front of the stores will allow a tremendous influx of air. Though it can alleviate smoke conditions, it can also rapidly accelerate a fire and generate high heat conditions. Wind conditions also need to be considered. If the wind is driving toward the windows to be broken, it would not be advisable to break the windows since it could produce a wind-driven fire. Prior to taking out plate glass windows, units operating on the interior must give approval for these actions; if they approve, they need to receive warning of when these ventilation actions are occurring. This information will allow them to reach areas of safety and to see the reaction caused by the vented window openings. All of the windows should not be broken at one time; instead, you should break one out and check with the interior crews on the effect it has on the fire. If it has a positive effect on ventilation, additional windows can then be removed; if a negative effect, then no more windows should be broken.

Firefighters reaching the roof should give a report to the IC on conditions found there. Are the roof vents open, and are they pushing heavy smoke or fire from them? Has the fire vented through the roof?

OVERHAUL AND SALVAGE

Prior to a supermarket being permitted to resume sales after a fire, the store must be inspected by the authority having jurisdiction (AHJ) and adjudged that the fire and smoke did not contaminate the store's food products. The AHJ may require that some products be destroyed due to possible contamination.

SIZE-UP FACTORS FOR SUPERMARKETS

Water
- There must be an adequate water supply for both an offensive and, if necessary, a defensive attack.
- Supermarkets located in rural areas may have a limited water supply available.

Area
- Supermarkets are usually rectangular but can vary in size.
- Smoke conditions on the interior can confuse some occupants. They will most often attempt to exit the same way they entered.
- Large open areas will permit a fire to spread quickly in the store proper and storage areas.

Life Hazard
- Life hazard can be severe if occupied. It is difficult to know who the occupants are and if anyone is still in the building.
- Though exit signs may indicate exits through storage areas, many occupants will not notice them or will ignore them.
- TICs can assist in the primary and secondary searches.
- Fast-moving fires will necessitate a prompt response and attack on the fire.
- Smoke may reduce visibility and confuse occupants. Early ventilation is critical.
- Elderly and handicapped may need assistance in exiting.
- Shopping carts full of goods will be abandoned in aisles, adding to the congestion as occupants flee the building.

Location, Extent
- Locating the fire can be difficult.
- Light smoke conditions can worsen in a short period of time, and panic is possible.

- TICs can assist in checking walls, ceilings, floors, and storage areas for hidden fire.
- HVAC systems can spread smoke throughout a building. If it cannot be determined if the system is assisting the fire department's operations, it should be shut down.

Apparatus, Personnel
- Call for help early. A well-advanced fire will require a large number of resources in an attempt to control these fires; achieve adequate ventilation; and accomplish search, rescue, and evacuation.
- A decision on whether to fight the fire or assist in the evacuation may have to be made if resources are limited.

Construction/Collapse
- These buildings only contain windows in the front, which can impact ventilation operations. Loading docks can be used for horizontal ventilation if no trailers are occupying them.
- It is difficult to gain access to rear entrances even during hours of store operations.
- A decorative façade can hide roof-mounted equipment and can allow a sizeable drop to the roof assembly for a firefighter stepping from a ladder onto the roof.
- Large supermarkets may contain multiple front entrances.

Exposures
- External exposure problems can exist if the supermarket is part of a mall.

Weather
- Nothing out of the ordinary.

Auxiliary Appliances
- Sprinklers will be present in some of these occupancies.
- Kitchens may have special extinguishing systems over deep fryers.
- If set back from the street, it may be a long stretch to supply sprinkler system intakes.

Special Matters
- Employees trained in evacuation procedures can assist the fire department.
- Shopping carts can impede occupants evacuating and firefighters attempting to enter the building to fight the fire.

Height
- One story is the norm, though some multistory supermarkets can be found. Some buildings have a mezzanine area.
- There may be a basement that is accessed at street level and the supermarket at the parking lot level.

Occupancy
- Overcrowding can be a problem during prime shopping times.
- Exit doors can become congested with people trying to leave the building, restricting egress and leading to panic. Ensure that those exiting the store under emergency conditions do not attempt to take shopping carts with them, restricting egress for others.

Time
- The critical time period is when the supermarket is open for business.

CONSIDERATIONS FOR FIRES IN SUPERMARKETS
Strategic Goals and Tactical Priorities for an Offensive Attack
- Evacuate and perform search and rescue for occupants.
- Ensure that all exits are open.
- Recognize the possibility of a mass casualty incident.
- Call for a sufficient amount of resources [fire and emergency medical services (EMS)].
- Hose-lines cannot be stretched through doorways being used as exits by occupants.
- Locate and confine the fire to the area of origin, if possible, while attempting to protect avenues of egress for the occupants.
- Depending on the size of the supermarket, 1½-inch or 1¾-inch hose-lines may be ineffective; 2½-inch hose-lines may be needed.
- Storage room fires should be fought with 2½-inch hose-line.

- Ventilation can act as a life-saving tool. Horizontal ventilation should be an early consideration. Roof vents may be present; see if they are open. Try to utilize the HVAC unit to assist in removing the smoke. Large plate glass windows in the front of the stores can be broken out if necessary.
- To use PPV, you need to know the exact location of the fire, the location of where the smoke would exit, and whether any occupants would be endangered. This information will be difficult to obtain in the early stages of a fire, making PPV a poor initial choice.

Incident Management System Considerations/Solutions for an Offensive Attack
- Incident Commander
- Safety Officer
- Rapid Intervention Crew(s)
- Operations (if needed)
- Staging
- Ventilation Group
- Search and Rescue/Evacuation Group
- Division 1 (for multistoried building, a division can be implemented for any floor where deemed necessary)
- Medical Group or Branch (to set up triage, treatment, transportation, morgue, etc.)

Strategic Goals and Tactical Priorities for a Defensive Attack
- If changing from an offensive to a defensive attack, ensure that a PAR is taken.
- Set up and maintain a collapse zone.
- Utilize master streams.
- Protect exposures, if necessary.

Incident Management System Considerations/Solutions for a Defensive Attack
- Incident Commander
- Safety Officer
- Public Information Officer
- Operations (if needed)
- Staging
- Logistics (for water supply, if needed)
- Divisions on the exterior on Sides A, B, C, D
- Exposure Divisions (if needed)
- Medical Group or Branch

NIOSH FIREFIGHTER FATALITY REPORT F2008-07

On March 7, 2008, two male career firefighters, aged 40 and 19 (Victims #1 and #2 respectively) were killed when they were trapped by rapidly deteriorating fire conditions inside a millwork facility in North Carolina. The captain of the hose-line crew was also injured, receiving serious burns. The victims were members of a crew of four firefighters operating a hose-line protecting a fire wall in an attempt to contain the fire to the burning office area and keep it from spreading into the production and warehouse areas. The captain attempted to radio for assistance as the conditions deteriorated but firefighters on the outside did not initially hear his Mayday. Once it was realized that the crew was in trouble, multiple rescue attempts were made into the burning warehouse in an effort to reach the trapped crew as conditions deteriorated further. Three members of a RIC were hurt rescuing the injured captain. Victim #1 was located and removed during the fifth rescue attempt. Victim #2 could not be reached until the fire was brought under control. The fourth crew member had safely exited the burning warehouse prior to the deteriorating conditions that trapped his fellow crew members.

Key contributing factors identified in this investigation include radio communication problems (unintelligible transmissions in and out of the fire structure that may have led to misunderstanding of operational fireground communications); inadequate size-up and incomplete preplan information; a deep-seated fire burning within the floor of the office area that was able to spread into the production and warehouse facility; the procedures used in which operational modes were repeatedly changed from offensive to defensive; lack of crew integrity at a critical moment in the event; and weather that restricted fireground visibility. **See Appendix B: NIOSH Reports in www.pearsonhighered.com/bradyresources to read the complete report and recommendations.**

Lumberyards

The vast amount of combustible materials heightens the problems associated with **lumberyard** fires. (See Figure 9-16.) Heat generation can make exposure protection critical. Lumberyards can contain sawmills, millwork shops, woodworking factories, and kilns. Sheds are often of frame construction, not sprinklered, and built only to provide protection for the lumber. They are open toward the yard area to facilitate the loading and unloading of materials and are located close to other piles of lumber.

Lumberyards are often located adjacent to railroad sidings or locations accessible to ships. If situated on a waterway, a lumberyard blaze will require an attack from the waterside. Otherwise, the fire department will be restricted to attack from only three sides. One benefit of having a lumberyard located on a waterway is an adequate supply of water for fighting fires.

Railways alongside a lumberyard fire must be shut down to ensure firefighter safety. Firefighters can dig under the rails to place a hose-line in operation until the shutdown can be accomplished. (Hose-lines can be fed under the tracks; however, the use of large-diameter lines can restrict this practice.) This tactic does cause a delay in getting hose-lines into operation. If confronted with life endangerment, hose-lines will have to be stretched over the rails to provide immediate protection. It is important to remember that railroad tracks can mean dead-end water mains on a hydrant system, reducing the available water supply.

FIGURE 9-16 A large amount of stock, combined with narrow aisles, can set the stage for numerous problems for firefighters responding to a lumberyard fire.

lumberyard ■ May contain sawmills, millwork shops, and a large quantity of highly combustible lumber in exposed exterior piles.

PILING OF LUMBER

There are two distinct methods of piling lumber. The first allows for air to flow through the material by placing furring strips between the boards. When a lumberyard operates its own sawmill, the piles of lumber often have intervening wood strips that facilitate air-drying of the boards. With this piling method in use, fire can attack a much greater surface area of lumber. The second method piles the boards one atop another, exposing only the surface and sides of the pile. This is the preferred method of piling lumber and can help firefighters control a lumberyard blaze. (See Figure 9-17.)

Piles of lumber are often found adjacent to roadways throughout a lumberyard to ensure easy loading of lumber onto and off of trucks. These piles can exceed 20 feet in height. Though the piles are evenly stacked facing the roadway, the rear of the pile can be uneven if various board lengths protrude against a pile from the opposite roadway. Uneven board lengths allow increased surface area for a fire to attack.

In yards that manufacture lightweight building components, stacked truss or wooden I-beams create natural air spaces. Fire reaching these products will have sufficient fuel and air available to culminate in a large, fast-spreading, and difficult-to-control fire. As lumber businesses expand, the scarcity of yard space often reduces the spacing between piles. This increases immediate exposures, and piles can increase in height.

FIGURE 9-17 Lumber piled unevenly with furring strips separating the boards will be subject to attack over a much greater surface area.

FIGURE 9-18 Lumberyard fires generate a tremendous amount of heat and can ignite structures and piles of lumber a distance away. *Used with permission of Joseph Hoffman.*

DIVERSIFIED PRODUCTS

In addition to lumber, lumberyards carry other product lines. These can range from roofing materials to hot tubs to swimming pool supplies. Roofing materials may include propane tanks. Swimming pool supplies almost always include chemicals for filtration systems. Hot tubs are often constructed of fiberglass or plastic, which will burn furiously. Firefighters battling lumberyard fires should also expect to find plastic pipe and flammable adhesives.

INITIAL ACTIONS

First-arriving units at a fire involving a few piles or an involved shed of lumber must make an aggressive attack on the fire from the leeward side. They should wet down any immediate exposures in an attempt to contain the fire. A 2½-inch hose-line should be the minimum size used. A blitz attack with a master stream can be most effective. The object is to knock down the fire and prevent extension.

The large quantity of highly combustible material in exposed exterior piles of lumber and material stored in combustible buildings create a severe fire hazard. Lumberyard fires develop quickly, with intense heat and rapid fire spread. (See Figure 9-18.) A fire past the incipient stages has enormous potential to escalate to major proportions and will necessitate a defensive attack on arrival. The spread of fire will be determined by the direction of the wind. The exposures on the leeward side will need immediate protection if the fire is to be contained. The IC's initial assessment must predict how long it will take to place effective water streams into operation. This prediction of time frames will allow the IC to know how far the fire will extend in the interim and what practical assumptions can be made on containment. It seems that fire always extends in the direction that will tax a fire department the most.

The ability to maintain a position in front of a moving fire requires courage, experience, and common sense. Courage needs no explanation. Common sense and experience require monitoring of the positions taken by firefighters. If positions become untenable, relocation to the flank, or side, will be required to control the fire. The Division Supervisors and the Safety Officer (SO) must constantly monitor the fire in relation to firefighter safety.

A fast-moving fire must be fought from the flanks of the fire. The aim when flanking a fire is to prevent further fire extension. Lines operated from elevated heights (roofs, elevated platforms, or ladder pipes) can sweep a large area above, around, and between burning piles. Water applied to the tops of burning piles will extinguish fire on the sides and bottom as it runs off.

Wind is unpredictable and can change direction without warning. This should be anticipated, and positions must not be so close that a wind shift will endanger firefighters. A wind shift can occur for only a short time, and then the wind can revert to its previous direction. If shifting winds are encountered, exposures on more than one front will require protection.

SPREAD OF FIRE

Fire extension can occur in a combination of ways. The large volume of radiated heat can ignite structures and piles of lumber a distance away. The convected hot gases can take fiery brands on upward currents of air a distance from the fire and, as the currents cool, allow the brands to settle on exposed structures or wildland areas, igniting independent fires remote from the original burning site. Brand patrols must be organized with the responsibility of checking areas downwind for fire extension. Checking nearby roofs can consume a great deal of resources. Urban lumberyards can be surrounded by dwellings or commercial properties that restrict access and firefighting capabilities, setting the stage for a possible conflagration. (See Figure 9-19.)

FIGURE 9-19 An urban lumber-yard surrounded by dwellings and commercial properties can restrict access and firefighting capabilities. *Used with permission of Joseph Hoffman.*

A fast-moving fire may require the relocation of apparatus. Shutting down pumpers for relocation purposes will require additional hose-lines stretched to piece out the existing stretches or connecting the already-stretched hose-lines to available outlets on other apparatus or outlets on large-diameter hose-lines.

ON SCENE

A lumberyard situated in a residential neighborhood had a fire that occurred late at night and led to a delay in notifying the fire department. On arrival there was no visible fire, but heavy smoke hung over the area. A serious fire was burning, but it was difficult to discern its exact location. The IC struck four alarms and had ordered some companies to abandon the fire hydrants that were located close to the lumberyard to which they were attached. He realized that once the fire broke out it would cause major problems. When the fire broke out, it quickly overwhelmed firefighters in their attempts to control it and remove the fire apparatus. Master streams, which had already been set up, had little effect on the fire. Two pumpers were destroyed and a few others received damage. The fast-moving fire also destroyed or damaged more than 60 homes.

WATER

Large quantities of water will be required if the fire is past the incipient stage. The water will be needed to reduce radiant heat, knock down the spreading fire, and reduce flying brands. Exposed buildings and piles of lumber will need to be wetted down. Knowledge of the water supply, including hydrant locations, water main sizes, and drafting sites will be needed.

Water runoff can be massive. Hose-lines within the lumberyard can be completely covered with runoff water, negating the ability of firefighters operating in heavy smoke conditions and reduced visibility to retrace their steps by following the hose-line out of the area.

When the fire is under control, handheld hose-lines can be employed for overhauling. This will be a time-consuming venture, but it can be facilitated by the use of forklift trucks. They can raise the lumber while firefighters drive water into and beneath the piles. Water should

FIGURE 9-20 Large quantities of water will be needed at lumberyard fires. This fire boat can pump 6,000 gallons of water per minute. *Used with permission of Joseph Hoffman.*

be used abundantly until no doubt is left that the fire is out. During overhaul, burned piles of lumber can become unstable; shifting piles must be monitored.

FULLY ENCLOSED LUMBERYARDS

There has been a change from traditional outside lumberyards to a one-stop shopping concept. Businesses such as Home Depot and Lowe's are contained in large buildings with high ceilings. The problems normally associated with exterior lumberyards spread over many acres are now contained within one building. This places many hazards under one roof, including pesticides, plastics, and combustible liquids. The ability of a fire department to fight a fire successfully in these structures is dependent upon a properly installed and operating sprinkler system. This must include in-rack sprinklers. These large enclosed structures can be handled with the assistance of the sprinkler system. Should the sprinklers fail to operate properly, there is potential for a disastrous fire.

These occupancies could require an upgraded response due to the large number of occupants during hours of operation. There could be problems of ventilation and evacuation. Many of the problems associated with these structures are described in an earlier section of this chapter, "Commercial Buildings and Warehouses."

SIZE-UP FACTORS FOR LUMBERYARDS

Water
- Large volumes of water will be needed.
- Drafting sites may exist along the banks of a river or waterway.
- Lumberyards situated alongside railroads or bodies of water may be protected by hydrants supplied from dead-end water mains.

Area
- Access may be limited by railroad sidings or by a river or waterway.
- Narrow roadways within the lumberyard leave minimum spacing between piles of lumber.

Life Hazard
- Life hazard is severe if a lumberyard is occupied at the time of the fire. The lumber products and other goods will generate high heat conditions.
- TICs can assist in the primary and secondary searches.
- Firefighters' safety must be a prime concern due to the rapidly changing conditions found at these fires.
- Fully enclosed lumberyards can contain a large number of occupants during hours of operation.

Location, Extent
- The location and intensity of the fire will dictate the mode of attack. Anticipate a fast-moving, extremely hot fire.
- TICs can assist in the latter stages of a lumberyard fire in checking walls, ceilings, floors, and lumber piles for hidden fire during overhauling.

Apparatus, Personnel
- Call for help early. A large number of resources will be needed to control these fires.
- Apparatus placement on the leeward side is dangerous. A fast-developing lumberyard fire can damage and destroy apparatus parked too close to the fire.

Construction/Collapse
- Buildings may be flimsy and can collapse early in a fire.
- High-piled lumber can shift and fall.
- Newer buildings may have bar joist roofs that can fail if subjected to the extreme heat of the fire.

Exposures
- Large flying brands are common. Fires must be anticipated downwind for great distances.
- Exposures should be wetted down to cool them and prevent extension of fire.

Weather
- Wind will drive this fire. Operating into the wind while fighting this fire is dangerous. In the initial stages, the fire can be fought on the leeward side to prevent extension. As the fire involves a larger area, it must be fought from the windward side or with master streams flanking the fire on the leeward side.
- Anticipate changes in wind direction that could endanger firefighters and exposures.

Auxiliary Appliances
- Normally no auxiliary appliances are present, though the newer fully enclosed structures are sprinklered.

Special Matters
- Stock can include plastics, roofing supplies, propane tanks, pool supplies, and other hazards.
- The newer fully enclosed structures will contain anything found in a household.

Height
- Lumberyard buildings are usually one story with high ceilings.
- Stacked lumber in yard areas can reach up to 20 feet.
- Racks in enclosed buildings can reach the ceiling.

Occupancy
- Occupancy includes large areas with stacks of lumber, frame buildings, sawmills, woodwork shops, and kilns.

Time
- Delayed discovery can be a problem if a lumberyard is closed for business.

CONSIDERATIONS FOR FIRES IN LUMBERYARDS
Strategic Goals and Tactical Priorities for an Offensive Attack
- Perform search and rescue for occupants of buildings.
- Call for a sufficient amount of resources.
- Stretch 2½-inch hose-lines. The use of 1½-inch or 1¾-inch hose-lines is ineffective. The larger hose-line will provide a sufficient water supply to knock down a large volume of fire.
- If piles of outside lumber are involved, attempt a blitz attack with a master stream to contain the fire.
- Horizontal and roof ventilation may be required if fire is within a building. PPV may be effective if the building openings can be controlled.
- In newer buildings or renovated buildings, anticipate the use of lightweight building components in the roof and an early failure of the roof.
- Support the sprinkler system, if one is present.

Incident Management System Considerations/Solutions for an Offensive Attack
- Incident Commander
- Safety Officer
- Rapid Intervention Crew(s)
- Staging
- Ventilation Group
- Search and Rescue/Evacuation Group
- Divisions A, B, C, D for an outside lumberyard, or Division 1 for an interior attack
- Medical Group (if needed)

Strategic Goals and Tactical Priorities for a Defensive Attack
- If changing from an offensive to a defensive attack, ensure that a PAR is taken.
- Request additional personnel because these fires are resource-intensive.
- Operate on the windward side. If it is necessary to operate on the leeward side, flank the fire with master streams.
- Set up and maintain a collapse zone.
- Utilize master streams.

- Protect exposures from the tremendous radiant heat generated by these fires.
- Consider an ember patrol to control the fires from flying brands downwind.
- Overhauling will be labor-intensive.

Incident Management System Considerations/Solutions for a Defensive Attack
- Incident Commander
- Safety Officer
- Rapid Intervention Crew(s) (if needed for exposures)
- Operations (if needed)
- Staging
- Logistics (for water supply, if needed)
- Divisions on Sides A, B, C, D
- Exposure Division or Branch
- Medical Group

High-Rise Buildings

high-rise building ■ This type of structure is more than 75 feet tall and presents numerous life safety problems to firefighters fighting a fire.

High-rise building fires can present different firefighting challenges than low-rise structure fires and have proven to be deadly to both civilians and firefighters. A number of fires have led to major changes in life safety features in high-rise buildings.

- The Triangle Shirtwaist Company fire in New York City in 1911 took the lives of 146 workers and led to a public outcry. A major factor was a locked door leading to one of the stairs and a fire escape that collapsed under the weight of fleeing occupants. This fire brought awareness to the general public of the dangerous conditions that many American workers were subjected to and the need for life-safety changes in the workplace.
- The Hotel LaSalle fire in Chicago in 1946 brought a similar public outcry after which city officials made it mandatory to install fire alarm systems and to post instructions in all hotel rooms on actions to be taken by occupants should a fire occur.
- The Winecoff Hotel fire in Atlanta in 1946 involved the greatest loss of life in a hotel fire in US history. The hotel had only one wooden stairway in the center of the building in which lightweight doors had been propped open and transoms above the doors to the hotel rooms were open for air circulation. This allowed fire and smoke to spread to the upper floors, killing many of the occupants. In the aftermath of this fire, many southern cities were forced to comply with the national codes, including the installation of automatic alarm systems.
- The MGM Grand Hotel fire in 1980 and the Hilton Hotel fire in 1981, both in Las Vegas, reinforced the need for sprinklers in high-rise buildings.
- The Interstate Bank fire in Los Angeles in 1988 and the One Meridian Plaza fire in Philadelphia in 1991 both destroyed multiple floors of high-rise office buildings. The Interstate Bank fire destroyed five floors of a 62-story high-rise office building that was in the process of being retrofitted for sprinklers. The One Meridian Plaza fire destroyed eight unsprinklered floors of a 38-story partially sprinklered high-rise office building and took the lives of three Philadelphia firefighters. That fire caused the withdrawal of the firefighters from the fire building due to the potential of a pancake collapse of the building. The fire, which had started on the 22nd floor, destroyed eight floors and was finally extinguished by nine sprinkler heads that were activated when the fire reached the 30th floor, showing the excellent performance of a properly installed sprinkler system and the need for sprinklers in all high-rise buildings. (See Appendix C: Case Studies in www.pearsonhighered.com/bradyresources for more information on the One Meridian Plaza fire.)

High-rise fires have proven to be very deadly. High-rise fires that caused large losses of life include:

- March 26, 1911—Triangle Shirtwaist Co. fire in New York City (146 dead)
- June 5, 1946—Hotel LaSalle fire in Chicago (61 dead)

- December 7, 1946—Winecoff Hotel fire in Atlanta (119 dead)
- December 25, 1971—Tae Yon Kak Hotel fire in Seoul, Korea (163 dead)
- February 1, 1974—Joelma Building (Crefisul Bank Building) in Sao Paulo, Brazil (179 dead)
- June 21, 1980—MGM Hotel fire in Las Vegas (85 dead)
- December 31, 1986—Dupont Plaza Hotel in San Juan, Puerto Rico (96 dead)
- September 11, 2001—World Trade Center Complex, New York City (2,749 dead)

In addition to civilian deaths, many firefighters have lost their lives while fighting high-rise fires. These incidents include:

- February 23, 1991—One Meridian Plaza office building fire in Philadelphia (three firefighters killed)
- February 5, 1992—Indianapolis Athletic Club fire in Indianapolis (two firefighters killed)
- August 14, 1993—Geneva Tower apartment house fire in San Francisco (one firefighter killed)
- April 12, 1994—Regis Towers apartment building fire in Memphis (two firefighters killed)
- December 18, 1998—Vandalia Avenue apartment building fire in New York City (three firefighters killed)
- September 11, 2001—World Trade Center fire and collapse in New York City (343 firefighters killed)
- August 18, 2007—Deutsche Bank Building fire, building under demolition, New York City (two firefighters killed)
- July 5, 2014—Wilson Street apartment building fire in New York City (one firefighter killed)

 ON SCENE

High-rise buildings have become targets for terrorist activities, as seen at the World Trade Center in New York City on February 26, 1993, where a car bomb killed six and injured more than 1,000 people. The resulting smoke migration into the high-rise towers and the loss of electricity created a mammoth evacuation problem. Elevator extrications and EMS were required throughout the building.

On September 11, 2001, two hijacked commercial Boeing 767 airliners heavily loaded with fuel crashed into the 110-story twin towers of the New York World Trade Center. The damage from the crash and the ensuing fire caused the collapse of both towers and the deaths of 2,749 people, including occupants of the towers and the planes, as well as firefighters.

At the Alfred P. Murrah Federal Building in Oklahoma City on April 19, 1995, a truck bomb destroyed the high-rise building and inflicted 168 deaths and hundreds of injuries.

HIGH-RISE CONSIDERATIONS

A high-rise fire can have many problems occurring simultaneously. This presents a tremendous challenge to even the most experienced fire officer. The basis for success will depend upon a number of factors, including organization, preplanning, knowledge of the building's construction, and its internal systems, and applying experience and training to solve the many predicaments that undoubtedly will occur.

DEFINING A HIGH-RISE BUILDING

A high-rise building has the following characteristics:

a. More than 75 feet tall and constructed for human habitation
b. A dependency on the building's systems
c. Part of the building is beyond the reach of the fire department's longest ladder (See Figure 9-21.)
d. An unreasonable evacuation time

FIGURE 9-21 One definition of a high-rise building is if any part of the building is beyond the reach of the fire department's longest ladder.

The high-rises built in the first half of the twentieth century contained large masses of masonry material. The second half of the twentieth century and continuing into the twenty-first century saw a drastic change to lighter-weight core-constructed buildings. The steel supports are often protected by encapsulation in fire-rated drywall or sprayed-on cementatious mixtures.

CORE CONSTRUCTION

The core area is the location through which the utilities, shaftways, and elevators reach up the building. The core may be located in the center, front, rear, or side of the building.

A core-constructed building built today can weigh less than eight pounds per square foot. Compare that to the Empire State Building in New York City, which was completed in 1931 and weighs more than 24 pounds per square foot.

This reduction in weight is due to the changes in the material used in building construction. For example, metal studding and drywall have replaced plaster and concrete materials in many applications. Exterior walls in the past were constructed of marble and granite; today, they often consist of glass and other lightweight materials. These changes in material create problems for firefighters. The older material absorbed the heat of the fire, and the fire tended to be more localized. The newer construction materials, along with the change in buildings' contents to a large dependence on plastics, allow a fire to generate much higher temperatures that the building materials do not absorb. This causes the fire to spread to larger areas, making containment difficult.

KNOWLEDGE OF THE BUILDING AND ITS SYSTEMS

The best way to protect life and control or extinguish high-rise fires is through the installation and maintenance of an automatic sprinkler system. Many high-rise buildings, though, are not protected by sprinkler systems. Attacking and controlling fires in these buildings demands that firefighters have knowledge of the structure and the built-in protective systems. Thorough inspections and familiarization tours can accomplish this. These visits allow firefighters to identify problem areas and to interact with the building engineer and security personnel in nonemergency situations. This interaction builds trust that carries over to emergency incidents.

FIGURE 9-22 A curtain wall being installed on a high-rise building.

SPRINKLERS AND STANDPIPES

An adequate and continuous water supply needs to be delivered to the standpipe and sprinkler systems. (See Figure 9-23.) The fire department should know the size and capacity of the building's standpipe systems and the pumps supplying these systems. The building's fire pumps may directly feed these systems, or water tanks may be used. Should the building contain water tanks, their capacities and locations should be known.

A dry standpipe system may be provided in fire towers. If there are multiple stairways in the building, it should be noted on the preplan whether the dry standpipes are interconnected (looped system) or each standpipe riser is independent. Fire department pumpers can pressurize an interconnected system at any fire department standpipe connection.

If independent risers are installed, the specific standpipe being used must be known so that the correct riser can be pressurized. The preplan should contain the location of the system's intakes in an easily recognizable form. The department should preassign units to pressurize these systems. (Additional information on sprinklers and standpipe systems can be found in Chapter 4, Engine Company Operations.)

FIGURE 9-23 Fire department pumpers can pressurize standpipe and sprinkler systems to augment their water supplies. *Used with permission of Joseph Hoffman.*

DEDICATED WATER SUPPLY

Water tanks supply the domestic and fire protective systems in some buildings. These systems may have a dedicated water supply for the standpipes and sprinklers. This can be accomplished by having outlet points on the water tank for domestic supplies and fire protection at different levels. This ensures a dedicated water supply in an emergency should the building's fire pumps fail. (See Figure 9-24.)

High-rise buildings may contain pressure-reducing valves, which reduce the pressure and the volume of water. (See Figure 9-25.) They may be adjustable or nonadjustable. If adjustable, the tool to adjust them must be readily accessible on the premises. Fire departments should require that these tools be kept at a designated location for their use. Familiarity with the type of pressure-reducing valves in a high-rise building should be incorporated into a fire department's routine building inspection program.

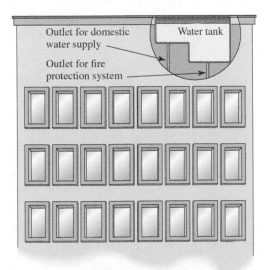

FIGURE 9-24 A cutout view of a water tank showing how water is dedicated to supply fire-protection systems in a high-rise building. *Used with permission of Pearson Education.*

FIGURE 9-25 Pressure-reducing valves reduce both pressure and volume. This type of valve was used in the deadly One Meridian Plaza fire.

During fire responses, in buildings with dry standpipes, units should be assigned to check that standpipe valves in the fire stairs are closed. This is the responsibility of the engine company from the ground level to the fire floor and the truck company above the fire floor.

HIGH-RISE STAIRWAYS

There are different construction types of stair systems in high-rise buildings. (See Figures 9-26 A-D.)

- U-return–type stairs consist of two sets of stairs connected with landings at each floor level and at the halfway point between each floor level, that allow entry to, and exit from, the stairway to be made from relatively the same location on each floor level.
- Return stairs consist of a straight flight of stairs that has the access and egress points from each set of stairs at a slightly different location on adjacent floors. This means that a person entering the stairway in the north corridor of a center core building would exit closer to the south corridor on the floor above or below.
- Scissors-type stairs consist of two sets of straight flight stairs in a common stair shaft. Scissors stairs can alternate floors with each set of stairs in the stair shaft (one set may serve odd-numbered floors, the other set serves even-numbered floors).
- Access stairs, convenience stairs, or private stairs are stairways that interconnect floors when the floors are used by the same company. These stairways may be of various configurations, including circular. In fully sprinklered buildings they may be unobstructed and contain no doors to segregate the floors and virtually interconnect multiple floors into one large area.

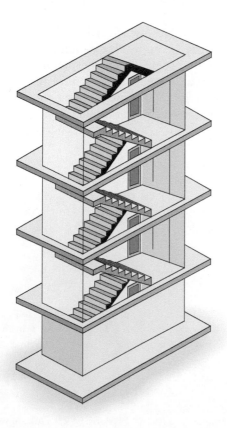

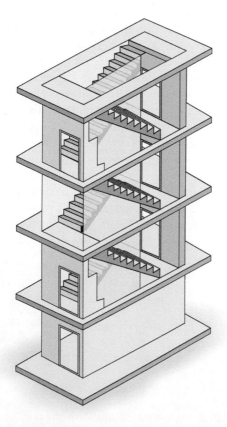

FIGURE 9-26 A-D U-return stairs, return stairs, scissors-type stairs, and circular stairway. *Used with permission of Pearson Education.*

A

B

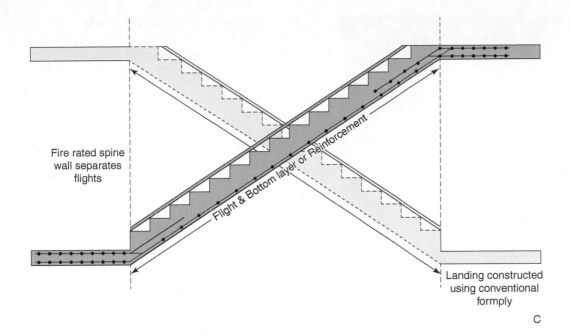

Fire rated spine
wall separates
flights

Flight & Bottom layer or Reinforcement

Landing constructed
using conventional
formply

C

Though there may be many sets of stairways in high-rise buildings, they are not designed for total building evacuation. It is not practical to build a sufficient number of stairways to fully evacuate a high-rise building. Firefighters should have a working knowledge of the different types of fire protection systems found in high-rise stairways:

- The fire tower is a stairway that has access to outside air. An open-air balcony protected by self-closing fire doors separates the stairs from the hallways on each floor. Any smoke entering this hallway is ventilated to the exterior, preventing smoke from entering the stairs.
- There are stair shafts that contain a vestibule between the hallway and the stairs; this will exhaust any smoke that enters into a vent shaft, preventing the contamination of the stairs. These stairways may be referred to as a *smoke-proof stairways*.
- The enclosed stairway may not provide for smoke removal unless pressurized.
- There may be unprotected open stairs between floors in older buildings. A common example is a large open stairway connecting a lobby area with a mezzanine floor.
- Access stairs connect floors of individual business concerns. These private stairs permit an employee access to multiple floors without entering public stairs or elevators. This allows a degree of safety and convenience.
- Some stair shafts may be pressurized. This prevents smoke migration into the stairways and is excellent for firefighting. In buildings that contain multiple stair shafts, all may not be pressurized.

D

FIGURE 9-26 A-D (*Continued*)

STAIRWELL PRESSURIZATION

Stairwell pressurization can be accomplished through various methods. The basic premise is to introduce outside air through fans to create pressure within the stairway. This pressure will prevent the smoke on a fire floor from entering the stairway. The outside air can be introduced at the top, bottom, or at various levels throughout the stair shaft. (See Figure 9-27.)

The system can function in a variety of ways. One method has an exhaust fan that allows for constant air changes, removing any smoke that may enter the stairs. This occurs at a rate to maintain pressurization. Another system works similar to a relief valve on a pumper. When the pressurization reaches a certain level, some air is exhausted to maintain the correct amount of pressure in the stair shaft.

FIGURE 9-27 Part of the stairwell pressurization system is shown in this staircase photo in a high-rise fire tower.

Too much pressurization in a stairway would make it difficult to open doors. Likewise, if too many doors are propped open to fight the fire or evacuate floors, the stairwell pressurization can be negated.

There has been success in setting up positive pressure fans at the base of enclosed stairways to pressurize them. NIST examined the use of PPV in high-rise firefighting. Tests were performed to pressurize the stairs in high-rise buildings by NIST and the Chicago and Toledo fire departments. This testing proved that properly sized and placed PPV fans can pressurize the stairways in a high-rise building and clear them of heat and smoke.

COMMUNICATION SYSTEMS

The large amount of concrete and steel in a high-rise building can hinder portable radio communications. To counteract these problems, fire departments can use other communication systems. These include:

- Stationary telephones
- Cellular telephones
- Hardwired systems
- Public address systems
- Elevator intercoms
- Built-in emergency telephone systems (both stationary and plug-in phones)

Some fire departments automatically set up a secondary phone link between the Operations Section Chief and the IC. One method of accomplishing this is by Operations locating a phone in the vicinity of the Staging Area and assigning a firefighter to stand by at that location. The firefighter then contacts the lobby fire Command Post (CP), relaying the phone number. Should the IC wish to discuss sensitive material with Operations, that phone can be used.

Cell phones can also be used. The cell phone allows more freedom on the part of Operations. Before ascending in the building, the Operations Section Chief makes note of the cell or stationary phone number at the lobby CP and leaves his or her cell phone number there. Some fire officers attach medical tape directly onto the phone and write the lobby CP phone number on the tape. By attaching another piece of tape in the vicinity of the lobby phone, they can write the cell phone number on the tape.

Hardwired systems (portable phones with dedicated wiring) can be utilized by an exterior CP by stretching a line to the lobby area. This provides an avenue of communications between these vital positions and keeps the IC up-to-date on relevant information without time-consuming radio communications.

Public address systems keep the occupants informed of emergencies occurring or actions expected of them. Firefighters may take advantage of these systems to facilitate their operations.

Many elevators have state-of-the-art intercoms. These systems can be accessed from the lobby CP to keep units apprised of conditions on various floors and to assign them specific duties.

FIGURE 9-28 Built-in emergency phones can be utilized by firefighters as a secondary means of communications during high-rise fires.

Built-in phone systems, both stationary and plug-in, can prove beneficial in a large-scale operation. The main drawback to using a built-in system occurs when it is also accessible to the building occupants. A frantic occupant can cause numerous interruptions, disrupting communications.

The use of multiple means of communications provides versatility, reduces radio communications, and facilitates the overall operation.

COMPUTER SYSTEMS

High-rise buildings constructed today have a strong dependency on computers. Computers are programmed to control fire protection systems, security systems, elevators, and other building systems. These systems can be quite sophisticated. They can:

FIGURE 9-29 Lobby command centers can access the building's communication systems and the built-in computers that monitor the building systems.

- Identify the locations of activated heat detectors, smoke detectors, or water flow alarms
- Indicate doors that are being opened
- Unlock stairway doors when a fire alarm is activated
- Pinpoint elevator locations and specific functions that each elevator is performing
- Identify the activation of any component in the HVAC system

The computer assists the fire department operation by monitoring the protective systems in the building and providing a printout or history of what has occurred in the building. The IC can assign a firefighter to review the printout to determine what happened prior to the fire department's arrival and what is currently happening.

At one particular fire, it was difficult to locate a smoldering fire. The ventilation system had spread smoke throughout numerous floors. By reading the printout, the suspected fire area was narrowed down and valuable time was saved in locating the fire.

Because each detector, fire or security door, or other building feature must be programmed individually, there is the possibility of human error. Should they be improperly programmed, false data will result. To minimize mistakes, a process to certify the building's systems must be in place.

FLOOR SEPARATIONS

There are two basic design concepts for horizontal floor separations in high-rise buildings:

1. Compartmentation
2. Open area or open space

The compartmentation concept divides each floor into small units. An example is an apartment building in which each apartment is segregated from the others. To be truly compartmentalized, the walls must extend from concrete floor to concrete floor and not end under the suspended ceiling. This construction feature provides fire confinement and gives the fire department time to attack and control a fire.

The open-space concept is utilized in today's modern high-rise office buildings. Each floor is basically wide open. Movable partitions may be used to separate the floor area into workstations. These partitions are 5–6 feet high and provide a modicum of privacy. Under fire conditions, this wide-open concept lets a fire spread rapidly throughout the entire floor. With few barriers and no sprinklers, the fire will be difficult to contain and extinguish. The partitions restrict firefighter movement and reduce the reach of fire streams while allowing a fire to spread unimpeded.

HVAC SYSTEMS

Air is supplied to high-rise buildings through air intakes that service a bank of floors. Outside air is brought into the building, filtered, and sent to various floors. At each floor, the air is heated or cooled and dispensed as indicated by individual thermostats. The air is then recirculated from the occupied areas back to the HVAC unit.

The finished ceiling in a high-rise building is lowered from the concrete above by installing a suspended ceiling. The space created between the concrete and the suspended ceiling is called a *plenum*. The plenum is used for wiring, ducts, and other utilities. It is often used as an air return to recirculate air to the HVAC system. No material should be located in the plenum that can supply fuel to a fire. Realize that many high-rise buildings still exist that were constructed before this code requirement was in effect. The plenum areas in many older high-rise buildings may contain electrical wiring and other combustible material that will add fuel to a fire.

A plenum is an area created by lowering a suspended ceiling from the original concrete ceiling.

The recirculation of air can spread smoke throughout a building. Control of the HVAC system during a high-rise fire is critical. The system may be used to assist in ventilation of the building. Some systems can be utilized to purge smoke from a fire floor. It may be possible to pressurize the floor above and below the fire floor to prevent spread of smoke to these floors. Each system is different. The building engineer's knowledge of the system will be necessary to assist the fire department in the proper use of the system.

Should it be decided to use the HVAC to purge or exhaust smoke from the fire floor, the IC or Operations Section Chief will need to know the exact location of the fire on the fire floor, the location of the exhaust fans, and the location of the stairs from which the fire attack is being made. It then must be determined whether the fans will intensify or spread the fire. Will the exhaust fans pull the fire toward or away from the firefighters making the attack? Could the heat of the fire force the dampers to close? If all of this information cannot be determined or if this operation cannot be accomplished safely, it may be best to not use the exhaust fans while firefighters are attacking the fire on the fire floor.

OPERATIONAL GUIDELINE

A high-rise operational procedure or guideline is an integral plan that is needed for successful high-rise operations. It must be created by the fire department to serve as a blueprint for handling high-rise incidents. This guideline should spell out initial assignments and how the incident management system can be implemented.

BUILDING ENGINEER

No one knows more about a building than the building engineer. This makes locating him or her an initial concern. There should be an understanding that the engineer will await the arrival of the fire department in the lobby area or other preassigned location and remain at that location for the duration of the fire department's involvement. The first-arriving fire officer can glean the needed information from the engineer and then proceed to the fire location.

There is a tendency on the part of the first-arriving officer to ask the engineer to "show me" the fire location. The engineer will be only too happy to accommodate. This, however, causes a problem—when the chief officer arrives, no one is available to answer his or her questions and accomplish tasks that the chief deems necessary.

USE OF ELEVATORS

The only practical method of movement in a high-rise building is by using elevators. During a fire, this can be dangerous. Many fire departments forbid the use of elevators until units have reached the fire floor and can ensure their safe use. Some fire departments dictate when an elevator may be used.

A common rule is that the elevator should not be used by firefighters if the fire is located within the first seven floors. This rule recognizes the danger involved in the use of elevators and that firefighters can safely arrive at these fire locations in a reasonable amount of time by climbing stairs. If a fire is located above the seventh floor, the use of the elevators is at the discretion of the IC.

If the decision is made to utilize an elevator, the elevator should have firefighter's service and the firefighters must be familiar with its emergency operation. This will enable them to have control over the elevator car. A factor to consider is whether the elevator services the fire floor. If it does not service the fire floor, it will be much safer to use.

In taller buildings, there are often split banks of elevators using different shafts. A typical example would be a 30-story building with three banks of elevators. The lower bank would service the first 10 floors, the middle bank would service the first floor and then floors 11 to 20, and the upper bank would service the first floor and then floors 21 to 30. A fire on the 23rd floor could be reached by using the middle bank to the 20th floor and walking up from that point. This would save a tremendous amount of time while providing a safety factor for the firefighters.

FIGURE 9-30 A bumper on which the elevator counterbalance rested when in the down position was located in the base of an elevator shaft. It was struck by a 3,400-pound counterbalance that had free-fallen from the 14th floor after its cables were attacked by a fire that destroyed an ornate elevator car. The bumper was driven through more than 12 inches of concrete and the basement floor into the subbasement offices.

ELEVATOR SAFETY

Safety when using elevators under emergency conditions requires members to have their self-contained breathing apparatus (SCBA) ready for immediate use. Firefighters should not pack everyone onto the elevator. There should be sufficient room to operate should trouble develop with the elevator car. Ensure that hand tools are carried by firefighters should a problem arise.

When ascending in an elevator, it should be stopped at random floors to check that the controls are operating properly. Firefighters should exit the elevator at least two floors below the fire floor and then climb the stairs to the fire floor. Portable radio transmissions can affect electronic controls on some elevators, causing them to shut down and requiring maintenance personnel. If erratic operation occurs, the elevator should be abandoned and the stairs used. Realize that in extremely high buildings on windy days, the elevator may be programmed with a safety feature of built-in intermittent stops to realign the elevator car due to building sway.

If any doubt about the safe use of the elevator exists, climb the stairs until the safe operation of the elevators can be ascertained.

 ON SCENE

Firefighters should ensure that they do not crowd into an elevator car leaving little room to maneuver. In one fire, the elevator became disabled at the fire floor. The door would not open, and the firefighters had no room in the elevator to use their hand tools. Another company had to extricate them. Fortunately, the fire was located a distance from the elevator.

FREIGHT ELEVATOR USAGE

Some fire departments forbid the use of freight elevators. There have been incidents in which firefighters were confronted with heavy fire conditions upon exiting the elevator because maintenance personnel often store trash for removal near the freight elevator doors.

FIGURE 9-31 Fire departments should consider utilizing 2½-inch hose-line with smooth-bore nozzles to increase the water's reach and the cooling effect at working high-rise fires. *Used with permission of Joseph Hoffman.*

Many departments, however, rely on the freight elevators. They are larger and able to carry more firefighters and equipment than passenger elevators. Many freight elevator cars have no ceilings, permitting firefighters to observe whether there are smoke conditions above in the elevator shaft as the elevator ascends. As with passenger elevators, firefighters should exit the elevator at least two floors below the fire floor and then climb the stairs to the fire floor.

WATER SUPPLY

To gain control of a high-rise fire requires an aggressive interior attack with hose-lines of sufficient size to achieve a quick knockdown. In noncompartmentalized high-rise buildings, 1¾-inch hose may be unable to accomplish this. Fire departments should consider utilizing 2½-inch hose with a smooth-bore nozzle to increase the reach and cooling effect of the hose stream. (See Figure 9-31.) Though current standards require a minimum of 100 psi at the standpipe outlets, the use of automatic nozzles that require 100-pound nozzle pressure may be ineffective if reduced standpipe pressures are encountered.

If necessary, supplemental water supplies can be secured by stretching hose-lines via the stairs, or by attaching to outlets from tower ladders or other elevated devices. However, either of these can present problems. The use of apparatus as exterior standpipes places firefighters operating the equipment in danger from falling glass. The stretching of hose-lines via the interior stairs will be labor-intensive and time-consuming.

 **ON SCENE**

The One Meridian Plaza fire in Philadelphia took the lives of three firefighters. A major problem at this fire was that the water pressure on the fire floor was drastically affected by the improper settings of the pressure-reducing valves. Since adequate pressure and supply was not forthcoming from the standpipe, it was decided to stretch five-inch large-diameter hose-line via the stairways to fight the fire. It took approximately one hour and 40 minutes for one stretch of five-inch hose-line to reach the 20th floor. Three alarms were struck for 12 additional companies to assist in this operation. Three separate stretches via three towers were accomplished. It was an arduous and dangerous task.

FIREFIGHTING CONSIDERATIONS

When deciding how to attack a high-rise fire, we must recognize that some normal firefighting procedures may not apply. There will be a strong reliance on the building's systems. We may not be able to take any actions on the exterior due to the height of the fire.

Working high-rise fires may show few signs from the exterior. In fact, smoke showing on the exterior may be like looking at the tip of an iceberg. The real problems may not be visible and can be much greater than could be estimated from outside the building.

Upon entering the building, the first-arriving units must determine the location and scope of the fire. Information can be gleaned from fire indicator panels in the lobby showing detection equipment that has been activated, and verbally from those in the lobby and on the upper floors. A computer printout can indicate the alarms that have been received and their location and can be taken by the first-arriving company.

Time plays a critical role. How long will it take to reach the fire area? What will be the reflex or lag time (the time lapse between receiving an order and accomplishing it)? It is not unusual for units to take 15–20 minutes or longer to reach the fire area in a high-rise building. How much additional damage will occur from a free-burning fire due to lag time? How will time be factored into the overall operation?

As firefighters ascend the building, they should take note of how many sets of stairs are present in the building, their location in relation to the fire area, their type, and which stairs contain a standpipe. With this information, they should proceed to the floor below the fire floor to the identical location found on the fire indicator panel or printout and decide on the stairs that would allow a safe evacuation of the building's occupants and permit units to perform a search of the floors above the fire floor. This information will allow them to decide on the evacuation/search and rescue stairway and the stairway from which to attack the fire. (Is the fire located close to a stairway? Is the fire being pushed toward a stairway, which would make entry very difficult? Are all the stairways pressurized?) Deciding on the stairways to be used for evacuation/search and rescue and firefighting should consider:

a. A stairway containing a vestibule that exhausts smoke, or a fire tower that has access to outside air are ideal for use as the evacuation/search and rescue stairway.

b. The selection of a firefighting stairway should consider a pressurized stairway if available. This will minimize the amount of smoke on the fire floor from migrating into the stairs above the fire floor since the hose-line will keep the door from closing. The use of a stairway that contains a vestibule with a fan exhausting the smoke should be avoided if possible since it will tend to pull the fire and smoke towards the stairs and onto the firefighters as they attempt to reach the fire area.

This information must then be transmitted to the IC and Lobby Control (if it has been established) so that units entering the building can be apprised of the proper stairs to utilize.

FIGHTING THE FIRE

The first unit to arrive at the fire area must:

1. Do a quick size-up and decide on the best way to attack the fire. (Should the door to the fire room remain closed until evacuations are effected? Should a direct attack on the fire be started immediately?)
2. Determine the need for specific operations (rescue, containment, ventilation, protect occupants in place, and others, as needed).
3. Estimate the number of personnel that will be needed. (This can often be determined by the number and types of assignments required and the personnel needed to accomplish them.) The solution to many problems is often an adequate number of resources. A high-rise fire will demand a great number of personnel. If the on-duty strength is minimal, there will be a need to request mutual aid and probably recall off-duty personnel in career departments.
4. Give a comprehensive report to the IC and request additional units, if needed. An initial status report to the IC will permit him or her to prioritize the situation and direct units to solve the problems that have been identified.
5. Initiate an attack on the fire.

Firefighters should attach their hose-line to the standpipe on the floor below the fire and stretch it up the stairs to the fire floor. This will allow the hose-line to back down the stairs if confronted with heavy fire. Hose-lines stretched to back up the initial line can be attached to the standpipe connection at the fire floor. Before opening the door to the fire floor to initiate fire attack, the stairs above the fire floor must be clear of firefighters and occupants. If there is anyone on those stairs, the fire attack must wait until they can either pass the fire floor as they are descending, or they can be directed to enter another floor above to seek protection.

Firefighters must be prepared to encounter high heat and heavy smoke. They may experience difficulty in advancing hose-lines down hallways and through offices or apartments. Furniture, fallen ceiling tiles, and collapsing metal grids from suspended ceilings will become obstacles.

The plenum area should be checked as firefighters move along the fire floor. Suspended ceiling tiles can be pushed up with the butt end of a pike pole or other tool to observe if fire has involved the immediate area. The ceiling tile should then be allowed to drop back down to its previous position.

After the initial hose-lines are in place, other hose-lines can be stretched to check adjacent areas for fire spread. Stretching more than two hose-lines through the same doorway will cause the lines to become entangled, and the forward movement of all lines will stop. Secondary locations from which the lines can be stretched should be found. This can include other fire towers not being used for evacuation, access stairs, or first-aid systems located on the fire floor.

It can be difficult to assess whether progress is being made in controlling a fire in a commercial high-rise. Hose-lines may be unable to advance and the effect the hose-lines are having on the fire may not be known. The inability to advance hose-lines can be caused by:

1. An intense fire that has had a long preburn
2. Large amounts of combustible material feeding the fire
3. Hose-lines too small to cool and extinguish the fire
4. Opposing hose-lines
5. An insufficient water supply
6. Wind blowing into the fire area through failed windows, pushing heat and fire onto the hose-line crews
7. Ventilation systems pulling a fire toward the ventilators and onto the firefighters

 ON SCENE

A fire on the 12th floor of the 37-story Cook County Administration Building in Chicago took the lives of six occupants who, while evacuating via the southeast stairway, were unable to pass the fire floor once firefighting operations had begun. Occupants reported that when they reached the 12th floor, they were instructed by a firefighter to go back up the stairway. In compliance with the firefighter's instruction, these tenants reversed course. They attempted to reenter floors above the 12th floor, but all stairway doors were locked. An investigative report was highly critical of the actions of the Chicago Fire Department.

An executive summary of the report can be found at: http://writer-tech.com/pages/ccab/CCAB%20Exec%20Summary%2093004%201550%205.0.pdf.

FIRE SPREAD

Predicting how a fire will spread in a high-rise building will assist in attempting to control it. In addition to horizontal extension, fire spreading vertically to the floors above must be anticipated. (See Figure 9-32.) The most common ways that this occurs are through:

- Auto-extension via the windows
- Lack of fire-stopping where the curtain walls meet the concrete floor
- Poke-throughs in the concrete floors

Auto-extension or auto-exposure of fire occurs when fire laps out of windows of the fire floor and attacks the window glass on the floor above. (See Figure 9-33.) The possibility of fire spreading via this means is increased if the vertical spacing between the windows on each floor is narrow (this area is often referred to as the *spandrel space*). Spread of fire through this means has been sensational at times.

 ON SCENE

At the Las Vegas Hilton Hotel fire, an arson fire that started on the eighth floor extended via the exterior windows to the 28th floor, causing eight fire deaths, numerous injuries, and extensive damage to the fire floor and the floors above.

Fire can extend to the floor above through spaces between the concrete floor and the exterior curtain wall. This space is called the *safing gap*. In older buildings, this space may not be fully protected. In the past, it was common practice in many areas to fill this space with fiberglass insulation. In time, the insulation becomes wet from moisture and provides little protection in stopping a fire from extending to the floor above. A notable case in which fire breached this space was the Interstate Bank fire in Los Angeles. The fire spread from the 12th floor through the 16th floor, causing massive damage. Today's codes specify adequate fire-stopping material for this space.

Fire can also spread vertically through poke-throughs. These are holes created between floors for utilities to pass through. These floor openings are usually found in the core area. The space around the piping, wiring, and ducts must be properly fire-stopped. Fire-stopping in this area is often overlooked, especially during building renovations. Tests performed on poke-throughs have shown that high temperatures in the fire area are quickly equated on the other side via poke-through openings.

Fire that has taken control of a floor area in excess of 10,000 square feet is beyond the control of handheld hose-lines. Lightweight deluge guns should be set up to attack these fires. These appliances will provide a greater water flow to absorb the heat of the fire. They will also create water supply problems. Where will the water come from to feed these appliances?

Most standpipe systems cannot supply multiple appliances, even with fire department pumpers supplementing the system. The situation will be further complicated because the reach of these streams will be reduced by movable partitions set between workstations.

A difficult situation that the IC must recognize is whether a fire is past the point of control by an interior attack and whether there is the threat of auto-extension of fire to the floor above. If the fire is located on a lower floor of the building, an exterior attack may be feasible.

Interior firefighting forces can withdraw from the fire floor. Through close coordination with interior units, outside streams can be utilized to knock down the fire. After the fire has been knocked down from the exterior, interior hose-lines can move onto the fire floor and extinguish smoldering fires. Units must be sent to the floors above to extinguish any points of fire extension before they can gain a foothold.

These actions can mean the difference between control of a fire or total loss of the floors above. Should the fire start to auto-extend from the fire floor and gain control of the floor above, it may be too late to stop the spread to additional floors.

The highest floor on which outside streams will be effective depends on the reach and height of the exterior apparatus. If elevated streams are not available, then the highest practical height would probably be the third or fourth floor. Though streams from windows of nearby buildings have been used in the past, they are usually too distant to penetrate into the fire building effectively and can only attempt to knock down some extension of fire via windows to the floor above. [An excellent case study on the effective use of exterior streams at a high-rise fire is "New York City Bank Building Fire: Compartmentation vs. Sprinklers" (Routley, Technical Report 071).]

WIND DANGERS

High-rise buildings are built to withstand high winds. The amount of wind shear that can be withstood depends on a building's location and height. It is not unusual for a 60-story building to be built to withstand winds of 110 miles per hour.

FIGURE 9-32 Fire can be seen lapping up the exterior to the floor above. *Used with permission of Joseph Hoffman.*

FIGURE 9-33 Fire lapped up the building's exterior, cracking windows on the floors above and necessitating the assignment of firefighters to those floors to check for fire extension.

ON SCENE

Wind can be a dangerous component at a high-rise fire. We must be concerned with how wind will affect a fire on an upper floor. At one particular fire, we had wind speeds of less than five miles per hour at ground level, yet at the 14th floor, the wind was driving into the building at more than 40 miles per hour. This created a blowtorch effect that caused a rapid breakdown of fire barriers throughout the entire floor.

There was a problem attempting to approach the fire area. Heat and smoke were being driven onto the advancing firefighters, forcing them to take a great deal of punishment. The wind's intensity attacked structural steel beams, causing the floor above to sag more than 15 inches. This was unrecognizable on the fire floor due to zero visibility. Companies sent to the floor above noticed the severe sagging and notified the Operations Section Chief. In the aftermath, repairs necessitated the replacement of structural steel members in a 25-by-125-foot section of the 15th floor.

BREAKING WINDOWS FOR VENTILATION

Before attempting to ventilate a fire area by breaking a window in a high-rise, one must consider wind direction and speed. It will be detrimental to break out a window in a fire area if the wind is being driven into that opening. The driving wind will push the fire toward the firefighters, preventing them from advancing hose-lines.

The best method for determining wind direction is to proceed to the floor below the fire floor and remove a window directly beneath the intended one on the fire floor. If it is found that the window faces the leeward side of the building, removing the window will be beneficial. However, if the wind is blowing into the building, then removal of the window in the fire area will be detrimental to the fire attack. Another method of ventilation should be found.

Breaking glass in a high-rise building presents many dangers. Falling glass can travel up to 200 feet from a high-rise building, causing injuries and property damage on the street below. Additionally, actually breaking out a window in a fire area in a high-rise building can be a problem. It may be possible to effect this from the exterior via a main ladder or tower ladder if the building is within its reach. A firefighter leaning out an adjacent window may be successful. It may be possible to attach a tool to a rope and break the window from above. Windows made of polycarbonate materials (Lexan is one example) will not be broken using a hand tool.

NIST conducted wind-driven testing in a laboratory structure, and field testing in high-rise structures in conjunction with numerous agencies including the Fire Department of New York (FDNY) and Polytechnic University. These experiments at high-rise fires were intended to improve the safety of firefighters and civilians by developing a better understanding of wind-driven fires and the firefighting tactics needed to address these fires, including ventilation and suppression. The testing showed how wind driven into the window of a high-rise apartment fire (generated naturally or with fans from the exterior) can drastically impact fire conditions.

Numerous studies were conducted including the conditions created when the door to the apartment and the public corridor were left open. The conditions in the public corridor and the stairwell area were measured when exposed to a wind-driven, post-flashover apartment fire. It was found that with winds of 20–25 miles per hour extreme thermal conditions existed in the corridor and stairwell that were not tenable, even for a firefighter in full protective gear. The conditions created have been described by firefighters as a "blow torch" flame.

The study noted that these fires were initiated when window glass failed and wind was driven into the fire apartment. If the door to the apartment was left open, or opened by firefighters, the flow of air and fire would be driven into the corridor and also the stairwell if that door was also open.

The study looked at ways that firefighters could identify the potential for a wind-driven fire. One method was door control, where the firefighters should check the fire stair door before opening it for heat or hot gases flowing around the edges. The door should only be opened a few inches to allow firefighters to observe if there are any rapid changes in smoke volume, or velocity and/or thermal conditions. If the firefighters recognize that the thermal environment is changing quickly, the door must be closed to interrupt the flow path. The same method of investigation should also be used when opening the door to the fire apartment. Under heavy smoke conditions, firefighters should use a thermal imager.

TACTICS USED AT WIND-DRIVEN FIRES

Blankets (made of a silica fabric aluminized with a foil material or of a treated fiberglass material) were used as wind control devices (WCDs). The WCDs were secured and suspended from the windows on the floor above the fire floor and secured at the windows on the floor below the fire floor. The WCDs covered the failed windows of the fire apartment through which the wind had been driven, thus preventing the wind from fanning the fire. This worked well. Another method tested was by applying water from the floor below in a variety of ways, including fog and solid streams. In all cases the water flows suppressed the fires, thereby causing reductions in temperature in the corridor and stairwell.

PPV was tested to see how it would perform at wind-driven fires and it was found that fans alone could not overcome the effects of a wind-driven condition. However, when used in conjunction with wind-control devices and water lines from the window on the floor below, fans were able to maintain tenable and clear conditions in the stairwell.

This excellent study can be found at: http://fire.nist.gov/bfrlpubs/fire09/PDF/f09015.pdf.

TIME FACTORS

Climbing stairs in a high-rise is a time-consuming task. Studies have shown that firefighters starting at ground level and carrying equipment above the 15th floor can take up to two minutes per floor. Firefighters report that after climbing heights over 15 floors, they need an initial rehab before engaging in a firefighting assignment.

Naturally, time frames are dependent on the physical condition of the firefighters. National competitions have shown fantastic times in accomplishing physical feats, but company unity is critical for firefighter safety and accountability. Firefighters must stay together as a team and are only as fast as the slowest member.

STAFFING HOSE-LINES FOR EXTENDED PERIODS

If a hose-line will be operating for an extended period of time, a sufficient number of units should be assigned to maintain a continuous and uninterrupted water flow. The minimum number of crews or companies to accomplish this is three units per hose-line. The units are deployed as follows:

a. One crew will operate the line.
b. The second crew will be standing by in the fire tower, waiting to relieve the first crew.
c. The third crew will restore its SCBA and then move up to a standby position.

Other factors could require that the minimum be four or five crews. Hose-line size and available resources will dictate the number of personnel in each crew.

AIR CYLINDERS OF LONGER DURATION

The use of 45- or 60-minute air cylinders should be considered. The additional time afforded will permit a concentrated initial attack and a longer primary search to be made.

A drawback to the use of air cylinders of extended duration is the physically draining effect it can have on firefighters. Firefighters base their stamina on the number of depleted air cylinders they normally use. A firefighter operating under this premise and using air cylinders of longer duration will most likely require rehab sooner. The tendency of most firefighters is immediately to return to firefighting after replenishing an expended air cylinder. It is important to have medical screening adjacent to the Staging Area and enforce rehab rules to ensure timely rest periods.

A rule that should be strictly enforced is that when crews are operating on floors above the fire floor, each firefighter should carry a spare SCBA cylinder, and the company officer and the individual firefighter must closely monitor the air usage.

ON SCENE

REMOTE CASCADE

The replenishment of SCBA cylinders will place an immediate drain on the personnel assigned to Logistics. Consideration should be given to establishing a remote cascade system adjacent to Staging. Air cylinders can be quickly refilled at this location.

The remote cascade is accomplished by stretching a reinforced air line (one that can withstand sufficient outside pressure) up the face of the building. It can be supplied by an air unit or mobile air cascade system. By attaching valves and shutoffs to the air line, an engine or truck company can fill one to three air cylinders at a time. The cylinders are refilled at the same speed as they would be at the air unit.

The air unit can be located a distance from the building by adding more lengths of air line. The use of a remote cascade system will permit air cylinders to be refilled with a minimum number of personnel in a timely manner.

NIOSH FIREFIGHTER FATALITY REPORT F2007-37

On August 18, 2007, a 53-year-old male career firefighter (Victim #1) and a 33-year-old male career firefighter (Victim #2) became trapped in the maze-like conditions of a high-rise building undergoing deconstruction. The building's standpipe system had been disconnected during the deconstruction and the partitions constructed for asbestos abatement prohibited firefighters from getting water to the seat of the fire. An hour into the incident, the fire department was able to supply water by running an external hose-line up the side of the structure. Soon after the victims began to operate their hose-line, they ran out of air. The victims suffered severe smoke inhalation and were transported to a metropolitan hospital in cardiac arrest where they succumbed to their injuries. By the time the fire was extinguished, 115 firefighters had suffered a variety of injuries.

Key contributing factors to this incident include delayed notification of the fire by building construction personnel; inoperable standpipe and sprinkler system; delay in establishing water supply; inaccurate information about standpipe; unique building conditions with both asbestos abatement and deconstruction occurring simultaneously; extreme fire behavior; uncontrolled fire rapidly progressing and extending below the fire floor; blocked stairwells preventing firefighter access and egress; maze-like interior conditions from partitions and construction debris; heavy smoke conditions causing numerous firefighters to become lost or disoriented; failure of firefighters to always don SCBAs inside structure and to replenish air cylinders; communications overwhelmed with numerous Mayday and urgent radio transmissions; and lack of crew integrity. **See Appendix B: NIOSH Reports in www.pearsonhighered. com/bradyresources to read the complete report and recommendations.**

LIFE SAFETY

What is the best way of protecting the occupants of a high-rise building when a fire occurs? Should they be evacuated or protected in place?

Every fire has unique circumstances. Firefighters should be familiar with the building's evacuation plan. Evacuation should neither endanger occupants nor interfere with control efforts. Coordination should permit both of these efforts to function simultaneously. If this is not practical, life safety of the occupants takes precedence. For instance, if a fire in an apartment can be quickly controlled, yet it has created heavy smoke conditions in the hallway, it may be better to let the other residents of that floor remain in their apartments rather than attempt to evacuate them into the toxic smoke conditions in the hallway.

A fire that has control of a large floor area will necessitate some evacuation. Initial evacuation considerations should target the fire floor and one or two floors above. These occupants can be moved a few floors below the fire floor. Other occupants can be protected in place. Firefighters can be assigned to stay with people who feel threatened. These firefighters must be authoritative to reduce panic and keep the occupants calm. They should constantly reassess their situation. If conditions change and evacuation becomes necessary, they should recommend to the Operations Section Chief the need to evacuate.

 ON SCENE

Prior to the collapse of the World Trade Center towers, occupants would typically heed the direction given to them by firefighters. In the aftermath of the World Trade Center collapse in which some civilians were told to remain and not to evacuate, I have found that many occupants will self-evacuate despite being told otherwise.

SEARCH

Primary and secondary searches consume time and energy. A list of building occupants is helpful. Obtaining master keys prior to ascending the building will facilitate the search. If keys are not available, forcible entry will be required. The use of hydraulic devices will enhance this operation. Should hand tools be used, a greater number of firefighters will be required, and fatigue will set in quickly.

When performing a search in hotels, hospitals, and other occupied buildings, firefighters operating in hallways under heavy smoke conditions should recognize that doors with hinges protruding indicate that the door will open into the hallway. This usually indicates a nonliving area; most likely it is a closet or a utility room. Do not waste time and energy on forcible entry to perform a search.

A marking system must be in place to avoid duplication. One method uses chalk to mark doors. A single line indicates that the primary search has been performed in the marked room. A second line forming a cross or an X would show that the secondary search has been completed.

HIGH-RISE COMMAND STRUCTURE

High-rise building fires present problems that require implementation of specific functions or duties in a command system. The ability to expand from a routine fire to one presenting unusual types of hazards can be accomplished only through a system designed for this type of growth. The basic incident command system does not change; it expands to include additional positions or functions.

The utilization of an incident management system and a well-conceived high-rise operational guideline sets the foundation for a successful operation. The main components of Command, Operations, and Logistics are always important, but success will also depend heavily on functions specifically associated with high-rise fires. These include Lobby Control, Systems Control, Ground Support Unit, Stairwell Support, Elevator Control, Base, and remote cascade systems.

COMMAND

The function of the IC changes little at a working high-rise fire. The assignment of an Operations Section Chief will allow the IC the time to focus on the many challenging and complex demands associated with these fires, including building an organization that will address the problems found. The IC must consider the early assignment of an SO and establish a CP from which to manage the incident.

FIGURE 9-34 A working high-rise fire will test any fire department.

FIGURE 9-35 Stretching hose-lines and performing search and rescue in long narrow hallways can be punishing work for firefighters. Note the extreme heat that was present in this hallway. *Used with permission of Joseph Hoffman.*

A comprehensive command system is a must, yet you should only implement those parts of the system that are needed and not the whole system just because of the large size of the building. As important as it is to implement a command system, there is an urgent demand for personnel to quickly respond to the reported fire location and attempt quick intervention to resolve the problem. Do not prioritize implementing command system functions at the expense of a rapid assessment and attack on the fire. A fast attack on a fire in the early stages permits control and extinguishment with one or two handheld hose-lines.

SAFETY OFFICER

The assignment of an SO is necessary on all working high-rise fires. In reality, it is impossible for one person to accomplish the many areas of responsibility. There will be a need to assign additional personnel or companies to assist the SO in tracking units. Assistant Safety Officer(s) (ASOs) or a safety company (an engine or truck company assigned to assist the SO) should be assigned to the lobby area to continue the tracking of personnel entering the building that was started by Lobby Control. If multiple fire towers are being utilized for fire attack, this will increase the problem of accountability. (See Figure 9-36.) ASOs may be needed to assess hazardous situations on the fire floor and the floors above.

COMMAND POST

The CP is the nerve center of every operation. Its implementation should be a priority during a high-rise fire. It can be established in one of two locations. Some departments utilize the lobby of the fire building. Others situate their CP a minimum of 200 feet from the building with a view of the fire building, when possible.

Both locations have benefits and drawbacks. The benefits of a lobby CP are:

1. Communications are improved. Handheld radios operate better and the IC, Operations, or Division or Group Supervisors can use building phone systems as a primary or secondary means of communication with the lobby.
2. The IC has firsthand knowledge of the events as they occur. He or she can have face-to-face contact with the Lobby Control Unit Leader.

The drawbacks of the lobby CP are:

1. A fire on the first floor or basement levels or one in which smoke migrates downward to the lobby can make this position untenable.
2. Some high-rise buildings contain only an elevator lobby area, which can be too small to operate a CP.
3. Access to an interior CP can be impeded if entry into the building is hampered by falling glass.

The benefits of an exterior CP are:

1. It may permit a view of the fire building.
2. The location of the fire will not interfere with CP operations.
3. Access is easily obtained.

The drawbacks of the exterior CP are:

1. Communications with interior units are often not as good.
2. Face-to-face communications must be replaced with radio messages, hardwired systems, or cellular phones instead of the building's communication systems.
3. Problems can arise if no feasible locations are available for a CP. The exterior CP can be restrictive in cities that have a high density of high-rise buildings and few locations to set up a functional CP with a good view of the building.
4. Severe weather conditions can impact the efficiency of the CP functions.

FIGURE 9-36 Safety is a critical element in every fire department operation. Its importance is magnified at a high-rise fire.

My personal preference is the lobby CP. This is probably due to the fact that it is the system that I learned and with which I am familiar. I feel that more information is at my fingertips and I have a constant feel for the situation. A lobby CP is a readily available place out of the weather with sufficient lighting and comfort.

OPERATIONS

The function of Operations on most assignments is at the discretion of the IC. A working high-rise fire, however, demands the immediate implementation of this critical position. Strong emphasis is placed on the Operations Section Chief, who will be directing suppression activities in the vicinity of the fire floor. This will permit him or her to talk to Division and Group Supervisors to keep abreast of changing developments. Some fire officers feel that operating in the fire tower achieves a greater measure of control over the situation. They are able to get firsthand information from units returning from upper floors to replenish their air cylinders. This allows Operations to find out exactly what is happening and, if necessary, adjust the tactics to meet changing conditions (see Figure 9-37).

To prevent themselves from becoming overloaded, individuals assigned to positions of authority within the incident command structure must delegate assignments, ensuring that unity of command is not violated by too many people reporting to any one person. The complexity of these fires demands that Operations develop an organization that will address life safety and incident stabilization in a timely manner. This organization must be manageable and ensure timely reports and the assigning of companies to address the problems encountered. The implementation of divisions and groups enhances control of units operating at a high-rise fire. As previously stated, a sufficient number of resources need to be sent to the immediate fire area. Operations should designate these units to operate in a Division on that floor. Units then need to be sent to the floor above to check for extension of fire, search and rescue and evacuation, and to set that floor up as a Division.

A working high-rise fire would also need to address all of the floors above the fire floors. An example would be a 30-story occupied high-rise building with a working fire on the 10th floor. The fire has extended to the 11th floor via auto-exposure. The 30th floor contains a restaurant that is heavily occupied at the time of the fire. Smoke is migrating to the floors above the fire floor, and there are numerous calls for assistance to the front desk in the lobby. Operations must create an organization to address these problems while assuring that he/she is not overwhelmed. The fire on the 10th and 11th floors can be addressed by assigning Division 10 and Division 11 with sufficient personnel to accomplish the needed tasks. Progress reports state that the fire is now threatening the 12th floor via auto-exposure and a Division 12 can be assigned. There are reports of smoke migrating to the restaurant on the 30th floor and the numerous phone calls from the 13th to the 30th floor. It would not be practical to create a division for each floor of the building. Operations can address the floors above the 12th floor by assigning a chief or company officer as an Exposure Branch Director. This officer would be responsible for the 13th through the 30th floors and report to Operations. As the Exposure Branch Director, he/she would develop the organization (divisions and groups) needed to address the problems on the floors above the fire floors. The 30th floor with the tremendous life safety problems could be assigned as Division 30. The numerous phone calls from multiple floors could be handled by assigning a Search and Rescue Group to investigate the calls. Smoke in the towers could be addressed by assigning a Ventilation Group to vent the towers by opening doors at the roof top and possibly setting up positive pressure fans to pressurize the stairs. This organization should be manageable having Division 10, Division 11, Division 12, and the Exposure Branch reporting to Operations. Realize that Staging and RIC would also report to Operations, but they are support roles and not counted against the rule of five for a manageable span of control.

RAPID INTERVENTION CREWS

In a high-rise fire, the RICs report to the Operations Section Chief and will generally operate on the floor below the fire. Should more than one stairway be used for fire attack, more than one RIC may be needed. The RICs should study the layout on the floor

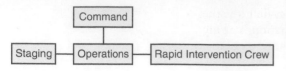

FIGURE 9-38 Staging is usually located at least two floors below the fire floor in an attempt to turn the high-rise fire into a two-story fire. The Rapid Intervention Crew will also report to the Operations Section Chief once Operations is established.

below the fire; if possible, RIC members can review floor plans of the fire floor(s). They should monitor radio messages for information and have a handle on conditions and operations that are occurring. After an RIC is committed because a firefighter is down, another crew must be immediately put in place. They will assume the duties of the RIC should other emergencies occur, or they may be needed to relieve or assist the original crew in removing the fallen firefighter.

Units that are sent above the fire floor must maintain communications and keep Operations apprised of their current location.

STAGING

Because time factors play such a critical role in controlling a fire, the plan should be to move personnel and equipment to the staging floor to have them available for immediate deployment on the fire floor and the floors above. (See Figure 9-38.) The Staging Area in a high-rise building should be a minimum of two floors below the fire floor. Under conditions where a reverse stack effect occurs and causes smoke to bank down more than two floors below the fire floor, Staging will need to be located a sufficient number of floors below the fire floor so that it is in a smoke-free environment. Staging is an assembly location for personnel and equipment that allows their rapid deployment.

To control these personnel, a Staging Area Manager must be assigned. This person will track all units as they ascend into the fire building, receiving information from Lobby Control on the disposition of units as they enter the building. Except for the first-arriving companies, or companies directed differently by Command, all other companies will proceed to Staging for assignment. As units are needed in the various divisions and groups, the Operations Section Chief will direct Staging on where to send the companies. After units have returned to the staging floor for replenishment of their SCBAs and are ready for reassignment, they will report to the Staging Area Manager.

When selecting a location for the Staging Area, it should be large enough to handle the large number of personnel, with adjacent areas for rehab, a first-aid station, SCBA refilling if a remote cascade is established, and changing of air cylinders. If the space available for the Staging Area is small, rehab and the first-aid station can be located on the floor below staging.

LOGISTICS

Success of a high-rise operation will be determined by the ability of the fire department to move resources and equipment up into the building in a timely manner. (See Figure 9-39.) This becomes the responsibility of the Logistics Section Chief.

Due to the nature of high-rise fires, it is not unusual to utilize two logistics companies for every company engaged in suppression activities. This high demand for resources can quickly strip all but the largest fire departments. Logistics will be responsible for most activities occurring on the exterior of the building. This could include the utilization of an observer to keep the IC and Operations informed of developments that are out of their sight (e.g., visible signs of fire, fire lapping from floor to floor, or window failure). Depending on the size and complexity of the incident, Logistics must develop an

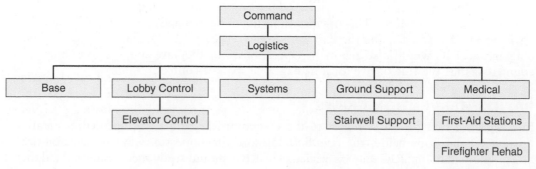

FIGURE 9-39 Logistics can be the key to success at a high-rise fire.

organization to ensure proper support for the firefighting efforts. Logistics will be responsible for Lobby Control, Elevator Support, Systems, Ground Support Unit, Stairwell Support, Base, Medical, Rehab, and First-Aid Stations.

BASE

At a high-rise fire, Base is established on the exterior a minimum of 200 feet from the fire building to prevent injury due to falling glass. Base can be viewed as an exterior Staging Area, and it is utilized for the marshaling of apparatus, resources, and equipment in preparation to move it to the upper floors. Since the aim of a high-rise fire is to turn it into a two-story fire, personnel are typically not kept at Base; they leave their apparatus there, find out what specific equipment they should bring with them, and then proceed to Staging. Base should be considered a secure area, and entry by unauthorized persons should be prevented. Unattended apparatus and equipment can be secured by a minimum number of personnel, and drivers/engineers can enter the fire building with their companies to fight the fire. (This can vary in areas that may require the engineers to remain with the apparatus.) Units entering the building need to take equipment with them to the upper floors.

LOBBY CONTROL

The lobby area demands fire department control throughout the operation. For this reason a Lobby Control Unit Leader must be assigned. The importance of this position is obvious in that some incident management systems elevate this function to a general staff position reporting directly to the IC. Under most systems, Lobby Control reports to the Logistics Section Chief unless the Support Branch Director position is established.

Lobby Control is responsible for:

- Controlling all building entry and exit
- Controlling all building access points
- Directing personnel to the correct stair/elevator
- Controlling and operating elevator cars
- Directing building occupants and exiting personnel to proper ground-level safe areas or routes

They may be assigned the duties of the Systems Control Unit in the early stages of an incident, or on less complex incidents where the position of Systems is not initiated.

An immediate priority of Lobby Control is clearing the area of civilians. Because fire department resources are limited initially, police and building security can be utilized. A secure area should be found for relocation. Leaving site selection to a ranking police officer is a method of delegation that will assist the fire department while building interagency support. This safe area can be a location within the fire building (if approved by the fire department) or a nearby building. Keeping those evacuated at one location until a list of names can be secured will assist in finding anyone reported missing.

Lobby Control must determine the status of all elevators within a building. Have they all been recalled to the lobby? Has each elevator car been checked to ensure its location? Do they contain any injured occupants? If elevators have not returned to the lobby, Lobby Control must contact the building engineer to establish their location and relay this information to the Operations Section Chief.

The movement of firefighters and equipment to the upper floors is the shared responsibility of Lobby Control. This task will be accomplished through the use of elevators (for which Lobby Control is responsible), or stairwell support, which is the responsibility of the Ground Support Unit Leader. Lobby Control directs firefighters to the proper stairwell or elevator to be used for ascent within the building.

Lobby Control must handle many assignments. Initial tracking of all units entering the building and their destination will be required. This task will become the responsibility of the SO, but early documentation will assist the accountability effort. Initial handling of the HVAC will come under Lobby Control. They will need to gather information on people still in the building who are calling to the front desk seeking assistance. Lobby Control must note the location of each caller and the conditions being reported, and then assess whether an immediate danger exists and whether physical disabilities could hinder the caller's escape. If the occupant is in danger, the Division or Group Supervisor operating in the area where the individual is located should promptly receive that individual's location and all pertinent information. These phone calls can provide information on fire and smoke spread that can be relayed to the units operating in the affected areas. Lobby Control must also anticipate numerous calls from concerned relatives of the people who live or work in the building.

ELEVATOR CONTROL

When the safe use of elevators has been established, Elevator Control can be initiated and assigned by the Lobby Control Unit Leader. This position oversees the movement of all elevators within the building to move civilians, firefighters, and equipment. The elevator car should be in the firefighters' emergency mode under manual operation and controlled by a firefighter with an SCBA, hand tool, and portable radio. A call letter should be assigned to each car. Elevator Control should handle the dispatching of elevators.

SYSTEMS

The Systems Control Unit Leader (Systems) monitors and maintains built-in fire control, life safety, environmental control, communications, and elevator systems. The Systems Control Unit Leader reports to the Support Branch Director, if established, or to the Logistics Section Chief. Systems must establish close liaison with building/facility engineering staff, utility company representatives, and other appropriate technical specialists.

Systems will work closely with Lobby Control. Normally, Lobby Control will be one of the first functions established at a high-rise fire. Their sharing of information that has already been acquired will facilitate Systems in addressing problems.

Systems will:

- Monitor and operate the building's display and control panels in conjunction with the building engineer.
- Evaluate the building's domestic water pumps and water supply. This includes knowledge of areas that are protected by sprinklers and the location of sectional stop valves to permit the shutting down of a sprinkler zone while maintaining protection for other areas of the building.
- Determine the presence and location of any pressure-reducing devices on standpipes and class and types of standpipes (whether a looped system or individual risers, etc.)
- Evaluate and operate the HVAC along with any smoke removal systems that may be located in the various stairways and elevator shafts. The location of air returns on the HVAC should be determined should there be a need to utilize the HVAC to exhaust smoke from an area.
- Determine the locations for shutting off electricity and natural gas on the individual floors and relay this data to units operating on those floors.
- Evaluate all communications systems within the building (public address, telephones, intercom, emergency phones, etc.) to assist Command in determining which systems will be used.
- Work closely with Operations, which will often involve numerous communications between these two functions.

GROUND SUPPORT UNIT

A Ground Support Unit can be assigned at a working high-rise fire. The Ground Support Unit Leader (GSUL) will be responsible for refilling SCBA cylinders and their maintenance, and for providing tools and assisting Lobby Control in evaluating and ascertaining safe entry and exit routes for the fire building. The GSUL will report to the Support Branch Director, if established, or the Logistics Section Chief.

The GSUL will work closely with the SO and the police to ensure that a safety zone is established and maintained on the exterior of the fire building.

Working in conjunction with Operations and Base, the GSUL will see that needed tools and equipment are brought to the Staging Area. The use of elevators will assist in this movement of personnel and equipment. If this is not possible, then stairwell support must be initiated by the GSUL.

STAIRWELL SUPPORT

Stairwell Support enables the movement of equipment from the lobby to the Staging Area via those stairwells selected for firefighting by the Operations Section Chief. This function is initiated by the GSUL if the use of elevators is questionable. Stairwell Support can be accomplished by stationing a firefighter every two floors (one at the first floor, another at the third floor, etc.). These individuals will carry equipment a maximum of two floors and give it to the next firefighter to do likewise. This is a laborious and tiring assignment, yet a very important one. Discipline is needed to maintain this vital link. A company officer should oversee the operation. Frequent relief will be needed if this function is utilized for an extended period of time.

FIRST-AID STATIONS

A minimum of two first-aid stations should be established. One should be adjacent to Staging to monitor firefighters as they return from upper floors to replenish their air supply and to monitor those in rehab. (Rehab is typically set up adjacent to the first-aid station.) This medical treatment area can attend to occupants removed from the fire floor and those above. If necessary, a triage area can be established for occupants prior to their being moved to the lobby.

Another first-aid station should be established in the vicinity of the lobby to handle building occupants. If sufficient resources are available, and a need exists, another first-aid station can be established at the occupants' relocation site.

SIZE-UP FACTORS FOR HIGH-RISE BUILDINGS

Water
- Dependency on the building's systems will be needed. Pressurizing the sprinkler and standpipe by fire department pumpers may be necessary.
- Pressure-reducing or pressure-regulating valves can reduce water flow and pressure.
- If additional water supply is needed on the lower floors of a building, hose-lines can be stretched from the exterior. For fires on higher floors, stretching hose-lines via the stairs will be demanding and labor-intensive. Hose-lines can be attached to outlets on raised aerial platforms and be stretched onto the floors within reach of the aerial devices.

Area
- Large floor areas can be found in commercial high-rises and those locations used as places of assembly.
- Stationary or movable partitions may be the only separation within some areas.
- Residential high-rises often are compartmentalized.
- Fire involving areas over 10,000 square feet is beyond the control of handheld hose-lines.

Life Hazard
- The life hazard in an occupied high-rise building will most probably be severe.
- TICs can assist in the primary and secondary searches.
- Buildings are not designed for total evacuation; consider protecting occupants in place or evacuating only those on the fire floor and the two floors immediately above the fire floor.
- Those being evacuated can be taken to floors directly below the Staging Area, and when elevators are available, they can then be taken to safety via the elevators.

Location, Extent
- Smoke may be spread via the HVAC system to other floors, making it hard to determine the fire's location.
- It can be difficult to ascertain the location of the fire floor from the exterior. No signs may be showing or smoke may be showing on multiple floors. Floor numbers may not coincide with the count of the floors via the exterior.
- Use the building's fire indicator panels for information on fire location and areas from which alarms have been triggered.
- Obtain master keys for access to locked areas.
- TICs can assist in checking walls and ceilings for hidden fire.

Apparatus, Personnel
- High-rise fires are resource-intensive. Ensure that there is an adequate initial response, and if a working fire exists, an immediate call for additional resources should be made.
- Contact the building engineer and utilize his or her knowledge of the building and its systems.

Construction/Collapse
- Changes in high-rise buildings that remove fire-resistant protection from structural steel allow it to be exposed to the high heat of a fire and can lead to failure of structural components.

Exposures
- Curtain walls attached to the building can create a space between the wall and the concrete floor through which fire can spread to the floor above.
- Fire can extend to the floor above by lapping out of windows on the floor below (auto-exposure).
- Poke-throughs can allow a fire to extend to the floor above.

Weather
- Higher winds will be found as you ascend within a high-rise. Wind-driven fires will need to be recognized to ensure firefighter safety.

Auxiliary Appliances
- Sprinklers will exist in newer and some older buildings. Standpipes should exist in all buildings.
- Special extinguishing systems may exist in selected areas, such as computer rooms and kitchens in restaurants.
- The HVAC can spread smoke to uninvolved areas. It can possibly be utilized by the fire department to remove smoke from the building and to pressurize areas to prevent the spread of smoke into those areas.

Special Matters
- The plenum area in older high-rise buildings can contain combustibles that can add fuel to a fire. A plenum fire can spread to other areas of the fire floor by dropping down through missing ceiling tiles. A plenum fire can also spread smoke to other areas.

Height
- The higher the fire is in the building, the longer it will take for firefighters to reach the fire area to assess and mitigate the problem.

Occupancy
- High-rise buildings contain residential, office, mercantile, places of assembly, and other occupancies.
- Vacant high-rise buildings can be a problem. A major concern is whether the building's systems are being maintained.

Time
- The impact of time can vary, depending upon the type of occupancy. The critical time is when the building is occupied.
- Reflex time or lag time will impact the overall operations.

CONSIDERATIONS FOR FIRES IN HIGH-RISE BUILDINGS

Strategic Goals and Tactical Priorities for an Offensive Attack

- Obtain an assessment of the fire and fire area as soon as possible.
- Ensure that an adequate number of firefighters are sent to the fire area to control a fire.
- Call for a sufficient amount of resources immediately if a working fire is found.
- Fire towers should be designated for firefighting and evacuation.
- Confine the fire to the area of origin, if possible. Hose-line of 1½-inch or 1¾-inch diameter may be ineffective; 2½-inch hose-line with smooth-bore nozzles may be needed.
- Floors above the fire floor must be checked for fire extension.
- Evacuate occupants from necessary areas. Consider protecting other occupants in place.
- Perform search and rescue.
- Ventilation will be difficult. Opening the fire tower doors at the roof may ventilate the smoke in the stairs. Try to utilize the HVAC system to remove the smoke. Breaking windows can be dangerous. Wind conditions should be considered.
- Positive pressure can be effective under the right conditions.
- A high-rise fire will be labor-intensive. Logistics will need a sufficient number of resources to move tools, equipment, and air cylinders to the Staging Area.
- Overhaul can be extensive and resource-intensive.
- Salvage can be tremendous in a commercial occupancy, especially if a computer room is involved. It may involve water removal from lower floors due to water runoff.

Incident Management System Considerations/Solutions for an Offensive Attack

- Incident Commander
- Safety Officer/Safety Company
- Rapid Intervention Crew(s)
- Operations
- Logistics
- Base
- Staging
- Lobby Control
- Systems Control
- Elevator Control
- Stairwell Support
- Ground Support Unit
- Ventilation Group
- Search and Rescue/Evacuation Group
- Divisions 10, 11, 12, etc. (a division can be implemented for any floor where deemed necessary)
- Exposure Branch [for addressing multiple floors above the immediate fire floor(s)]
- Medical Group or Branch
- First-aid stations

Strategic Goals and Tactical Priorities for a Defensive Attack (though a rare occurrence, it is possible)

- If possible, complete a primary and secondary search of the entire building.
- If changing from an offensive to a defensive attack, ensure that a PAR is taken.
- Get technical advice from structural engineers and set up and maintain a collapse zone.
- Utilize master streams, possibly from the windows of adjacent buildings.
- Continue supplying sprinkler systems if sprinklers protect the fire floor or any floors above.
- Protect exposures, if necessary.

Realize that a fire on a lower floor that cannot be controlled by an interior attack can be knocked down with a blitz attack using exterior streams. The goal is to operate in a defensive/offensive mode of attack. Once the fire is knocked down, the exterior streams can be shut down, and if deemed safe by the SO, an offensive attack can commence. This tactic will be dependent upon:

- The fire floor: The lower the floor, the better chance of utilizing a variety of apparatus and equipment to deliver the master streams. Ground-level master streams are ineffective above the third floor.
- A sufficient amount of special apparatus on the scene properly positioned to accomplish the blitz attack on fires located above the third floor (aerial ladders and platforms to deliver the elevated master streams).
- The switching to the exterior attack in a timely manner before the fire is too large to be controlled.
- The assurance that all firefighters on the interior have been withdrawn to locations of safety and that fire doors leading onto the fire floor are closed.

Incident Management System Considerations/Solutions for a Defensive Attack

- Incident Commander
- Safety Officer
- Operations
- Staging
- Logistics
- Base
- Divisions on the exterior on Sides A, B, C, D
- Exposure Branch/Divisions (if needed)
- Medical Group, or Branch

Summary

Commercial and industrial buildings include warehouses, strip malls, enclosed malls, supermarkets, lumberyards, and high-rise buildings. These structures will typically be larger and thus more challenging than the more common residential building fires, which all fire departments are adept at handling. They can contain large amounts of stock and/or a variety of contents that once ignited will support a fire.

The high-rise building is of particular concern for fire departments because we can be faced with a fire on floors higher than our ladders can reach. The reliance on the building systems becomes a must and that reliability will be the difference between success and failure.

Review Questions

1. What procedures should be followed when shutting down a sprinkler system?
2. What problems do overhanging canopies attached to the front of strip malls create for firefighters?
3. What problems are associated with entry into the rear doors of a strip mall store?
4. Discuss the types of malls in your response district and any unique problems they present.
5. What are the benefits and drawbacks to using a color-coded system for identification purposes at an enclosed mall?
6. What potential problems would need to be considered when attempting to gain exterior access to supermarkets?
7. What would be your ventilation considerations for a fire in a supermarket?
8. What initial actions would you take at a lumberyard fire involving a few piles of lumber burning alongside a frame storage building?
9. If operating at a lumberyard fire, what actions would you take if flying brands were starting numerous fires throughout the lumberyard and the surrounding community?
10. Select one of the special occupancies presented in this chapter, and in reviewing the size-up factors, add five additional potential problems that could be present.
11. How do the building and fire codes in your jurisdiction define a high-rise building?
12. Describe a core-constructed high-rise building. Identify one in your jurisdiction.
13. What are the responsibilities of the first-arriving unit at the fire area in a high-rise building?
14. List the different types of stairways that can be found in high-rise buildings.

15. List the rules that should be followed for safe operation of elevators. What are the rules of your department pertaining to elevator operation? What changes could be made to improve those guidelines?
16. Discuss the benefits and risks of having firefighters utilize freight elevators.
17. What are the various types of alternative communications that can be used at a high-rise fire?
18. Describe the difference in floor separations between compartmentation and open area.
19. Discuss the common ways that fire extends to the floor above in a high-rise building.
20. List the causes that could prevent the advance of hose-lines in a high-rise fire.
21. How would you break a window on an upper floor to effect ventilation in a high-rise building? What factors would negate proper ventilation through these broken windows?
22. Discuss a scenario in which a hose-line would be staffed for a long period of time at a high-rise fire. How many crews or companies would be required?
23. List the duties and the location of operation of a Rapid Intervention Crew at a high-rise fire.
24. What is the ideal location for the Operations Section Chief when fighting a high-rise fire? Explain.
25. Discuss the many and varied duties that could be assigned to the Logistics Section at a working high-rise fire. Then, decide on the approximate number of personnel needed to accomplish those assignments and how those resources would be obtained.
26. Draw up a high-rise command system to handle a working fire in the highest building in your response district. Assume that the fire involves the third floor below the roof, and the building is occupied.

Suggested Reading, References, or Standards for Additional Information

US Fire Administration (USFA)

Copeland, T. D., and P. Schaenman. "Sherwin-Williams Paint Warehouse Fire, Dayton, Ohio." Technical Report 009.

Howell, S. M., investigator, and J. G. Routley, ed. "Manufacturing Mill Fire, Methuen, Massachusetts." Technical Report 110.

Jennings, C. "Five-Fatality Highrise Office Building Fire, Atlanta, Georgia." Technical Report 033.

Miller, T. H. "Logan Valley Mall Fire, Altoona, Pennsylvania." Technical Report 085.

Routley, J. G., and J. Stern. "Two Firefighter Deaths in Auto Parts Store Fire, Chesapeake, Virginia." Technical Report 087.

Routley, J. G. "Four Firefighters Die in Seattle Warehouse Fire, Seattle, Washington." Technical Report 077.

Routley, J. G. "Interstate Bank Building Fire, Los Angeles, California." Technical Report 022.

Routley, J. G. "New York City Bank Building Fire: Compartmentation vs. Sprinklers, New York, New York." Technical Report 071.

Rush, R., and J. G. Routley. "Special Report: Operational Considerations for Highrise Firefighting." Technical Report 082.

Schaenman, P. "Schomburg Plaza Fire, New York City (Harlem)." Technical Report 004.

National Fire Protection Association (NFPA)

Standards

92A: Standard for Smoke Control Systems Utilizing Barriers and Pressure Differences

92B: Standard for Smoke Management Systems in Malls, Atria, and Large Spaces

Related Courses Presented by the National Fire Academy, Emmitsburg, Maryland

Executive Analysis of Fire Service Operations in Emergency Management

Incident Command System and Resource Management for the Fire Service

Incident Command for High-Rise Operations

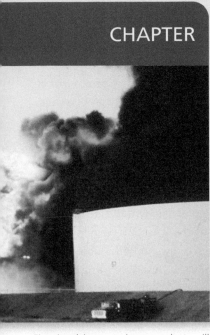

“ I've got a theory that if you give 100 percent all of the time, somehow things will work out in the end. ”

—Larry Bird, NBA Hall of Fame player

Fires involving petroleum products will develop a tremendous amount of heat. *Used with permission of Joseph Hoffman.*

KEY TERMS

hazardous materials incident, *p. 458*

tank farm/refinery fires, *p. 473*

terrorism incidents, *p. 484*

dirty bomb, *p. 494*

clandestine drug labs, *p. 501*

natural disasters, *p. 511*

OBJECTIVES

Upon completion of this chapter, the reader should be able to:

- Discuss the initial concerns at a hazardous materials incident.
- Set up initial exclusion zones at a suspected hazardous materials incident.
- Identify the difference between evacuation and protecting in place.
- Discuss types of storage tanks at tank farms and refineries.
- Discuss fire extinguishment concerns at tank farms and refineries.
- Discuss boilover, slopover, and frothover at a burning storage tank.
- Discuss the weapons of terrorism.
- Discuss the actions of the first-arriving officer at a suspected terrorism event.
- Understand the stages of response at a suspected or confirmed terrorism event.
- Discuss strategic considerations if confronted with a dirty bomb incident.
- Describe the protection provided by a firefighter's personal protective gear at a suspected dirty bomb incident.
- Discuss radiation exposure considerations for firefighters at dirty bomb incidents.
- Discuss strategic considerations at an incident involving a clandestine drug lab.
- Discuss the problems associated with natural disasters.
- Discuss the strategic considerations needed for a natural disaster incident.

There are many special types of operations that can confront responding firefighters. These incidents often require the expertise of specially trained personnel to mitigate the problems found. However, the initial responders will often face difficult decisions that will need to be made prior to the arrival of the specialists or technicians. For this reason, this chapter looks at those special areas: hazardous materials incidents, tank farm or refinery fires, terrorist incidents, dirty bombs, clandestine drug labs, and natural disasters. Each of these subjects is complex enough to fill a book of its own. This chapter presents an overview of these important subjects. The initial actions at a scene are the keystone of an operation. When performed correctly, the incident usually runs smoothly, and the problems that arise can be addressed.

Hazardous materials incidents can occur anywhere and anytime. The response time for a hazardous materials team can often be too long to await their arrival without attempting to mitigate a minor situation or one that involves life safety of civilians. This demands that firefighters have a basic understanding of hazardous materials. They should be able to reference the Department of Transportation guidebook to identify the potential of hazardous materials and initiate actions to mitigate the situation. This is especially important when dealing with tank farm or refinery fires.

Incidents involving terrorism are rare. Yet their occurrence will have a tremendous potential for a large loss of life and injury. We must recognize the signs that identify these events and act accordingly. This is especially true with the possibility of the detonation of a dirty bomb. Terrorism incidents will place firefighters at the leading edge. Clandestine drug labs pose not only the potential for explosion injuring emergency responders, but the hazardous waste created can affect anyone who comes into contact with it.

Natural disasters can strike anywhere and anytime. Preparation is a key element in attempting to address the many and varied problems that these disasters pose.

We as initial responders must ensure that our actions do not compound a situation that will be difficult at best. Our actions should be directed to minimizing death or injury to civilians and firefighters alike.

At the end of each section are size-up factors as well as strategic goals and tactical priorities that need to be taken into account when confronted with fires or incidents in these occupancies. These lists are not meant to be all-inclusive. Other factors can and will exist that must also be considered.

The strategic goals and tactical priorities are for both offensive and defensive attacks, as well as nonintervention, with suggested considerations for implementing an incident management system to meet the potential problems. The strategic goals and tactical priorities that are noted are in response to some of the potential problems that may occur. The suggested incident management positions do not mean that they must be used or that using them is the only correct method. These are initial positions that should be considered. Depending on the problems presented by each individual situation, the Incident Commander can implement these or other incident management positions to deal with the problems.

Hazardous Materials and the Initial Responder

With hundreds of millions of metric tons of hazardous materials (hazmat) used in the United States yearly, firefighters will routinely be faced with a variety of dangerous situations. (See Figure 10-1.) A seemingly minor incident involving hazmat can quickly become a major problem if not handled expeditiously. A **hazardous materials incident** requires a more cautious and deliberate size-up than most incident scenes. One undeniable fact is that under fire conditions hazmat can be totally unpredictable. A situation that seems under control one minute can become a raging inferno the next minute.

hazardous materials incident ■ Can expose firefighters to uncontrolled situations that present a more complex set of occupational health and safety concerns than structural firefighting.

As initial responders, firefighters must react if confronted with life-threatening situations. If there is no threat to life, they should control the site and await the arrival of the Hazmat Unit.

Each firefighter must know his or her limitations and that of the equipment. With changing technology, the level of expertise of the first responder must increase accordingly to handle the many varied situations that may be encountered.

Hazmat scenes can expose firefighters to uncontrolled situations. They may be confronted with a wide range of hazards that present a more complex set of occupational health and safety concerns than traditional structural firefighting.

There is the potential for serious complications because of the presence of unknown chemicals, their interaction with other products, or their reaction under fire situations. A person exposed to hazmat may feel the effects immediately or may experience delayed effects. There is also the possibility of developing chronic ailments.

The best method for firefighters to prevent contamination is by avoiding contact. They can

FIGURE 10-1 Hazardous materials scenes can expose firefighters to uncontrolled and dangerous situations. This refinery fire destroyed a tank and threatened overhead electrical wires before being controlled. *Used with permission of Joseph Hoffman.*

reduce their chances of contact by being methodical in their operations. This means early detection and avoidance of hazmat after discovering their presence. One problem is that contact with these substances is often difficult to detect.

INITIAL CONCERNS

A thoughtful, cautious approach, together with a basic knowledge of what hazards may exist, is needed to ensure the safety of all personnel. The first-arriving unit at a suspected hazmat incident must be prepared to initiate specific actions based upon the crew's training level:

■ Make a comprehensive size-up.
■ Assume command.
■ Request sufficient resources.
■ Identify the products involved.
■ Determine the immediate need for rescue and evacuation.
■ Establish scene control by isolating the area and denying access.
■ Consider diking or diverting escaping liquids.
■ Eliminate sources of ignition.
■ If there is ignition, take actions to control or minimize the problem.

First-arriving units at a suspected hazmat incident have the difficult task of defining the problem. Before the dangers of a product can be established, the product must be recognized and identified. Firefighters must be alert for signs of hazmat, and they should seek information from all available sources.

Analysis of an incident scene is enhanced by the amount of knowledge gained by the first-arriving units. The more information that is secured, the higher the probability that appropriate actions will be taken to mitigate the problems found, thereby reducing the probability of injuries to emergency responders.

Securing this information is critical and forms the basis for making incident scene decisions. Identification includes determining:

1. The product and amount of it involved
2. Whether any product has been or is being released and, if so, at what rate
3. Whether it can be contained or controlled
4. Whether a fire hazard exists
5. Whether there is a health hazard

GLOBALLY HARMONIZED SYSTEM

Changes have occurred in hazmat handling. The Globally Harmonized System of Classification and Labeling of Chemicals (otherwise known as the Globally Harmonized System or simply GHS) is a worldwide standardized approach to hazard communication. It was developed by the United Nations, and the United States was represented in its creation by the Occupational Safety and Health Administration (OSHA), the US Department of Transportation (USDOT), the Environmental Protection Agency (EPA), and the Consumer Product Safety Commission (CPSC).

GHS was created because internationally there were different methods of both labeling and Safety Data Sheets (SDSs). Countries and organizations had established laws and regulations that could be similar or significantly different enough to cause confusion, and some countries had no systems at all. These inconsistencies between national and international laws created regulatory and compliance problems that could interfere with commerce and compromise safety.

GHS is designed to provide clear, consistent label messages to chemical handlers and users, emergency first responders, and the public. Signal words, pictograms, and hazard statements have the same meaning in all settings, domestically and internationally. As a standardized system, GHS can protect workers, emergency response personnel, the environment, and the public by identifying chemical hazards.

The EPA states, "The primary goal of GHS is better protection of human health and the environment by providing chemical users and handlers, emergency first responders, and the public with enhanced and consistent information on chemical hazards" (EPA, 2015).

What Impact Does GHS Have on the United States?

OSHA adopted GHS through the Hazard Communication Standard (HCS) in what is referred to as HCS 2012 or HazCom 2012. It became mandatory as of June 1, 2015, and can be summarized into three main components:

1. Standardized hazard definitions and symbols
2. Standardized MSDS format and content, which will be called SDSs
3. Standardized hazard warnings and symbols on container labels

What Do These Standardized Areas Use?

- Pictography: There are nine pictograms under GHS. HazCom 2012 requires eight of these pictograms, the exception being the environmental pictogram. More than one pictogram may be displayed. The pictograms are used as a primary element to communicate important information about the hazards of a chemical. If hazardous chemicals are being shipped or transported from a manufacturer, importer, or distributor, labels must have hazard pictograms framed in a red square frame set at a point (like a diamond).
- Labeling: Responders may find both National Fire Protection Association (NFPA) and GHS labeling. Only the GHS labeling will be regulated.
- Hazard Ratings: Like the NFPA, the GHS standard assigns numerical ranking to hazard levels. However, the two systems are inverse: NFPA has five levels and the scale ranges from 0-4 and ranks 0 as the lowest hazard level. GHS has four levels, which range from 1-4 and ranks 1 as the highest hazard level. This conflicting approach does not affect labelingbecause the hazard ranking by number does not appear on a GHS-aligned HazCom 2012 label. However, the rankings could cause confusion when reading the SDSs.
- Signal Words: A single word used to indicate the relative level of severity of hazard and alert the reader to a potential hazard on the label. The signal words used are *danger* and *warning*. *Danger* is used for the more severe hazards, while *warning* is used for less severe hazards.
- SDSs: They will replace MSDS sheets. The format of the 16-section SDS should include the following sections:
  - Section 1. Identification
  - Section 2. Hazard(s) identification
  - Section 3. Composition/information on ingredients

- Section 4. First-aid measures
- Section 5. Firefighting measures
- Section 6. Accidental release measures
- Section 7. Handling and storage
- Section 8. Exposure controls/personal protection
- Section 9. Physical and chemical properties
- Section 10. Stability and reactivity
- Section 11. Toxicological information
- Section 12. Ecological information
- Section 13. Disposal considerations
- Section 14. Transport information
- Section 15. Regulatory information
- Section 16. Other information, including date of preparation or last revision

How Does OSHA/HazCom 2012 Affect Firefighters?

OSHA (through HazCom 2012) and the NFPA are two separate and distinct systems. HazCom 2012 looks at hazmat from the perspective of informing workers about the hazards of chemicals in the workplace under normal condition of use and foreseeable emergencies. The NFPA provides basic information for emergency personnel responding to fire or spills and planning for emergency response. Responders may find HazCom 2012 labeling, pictograms, and "NFPA 704: Standard System for the Identification of the Hazards of Materials for Emergency Response"(NFPA 704) symbols identifying hazmat; they need to realize that there are some differences between the two systems.

Responders need to understand that GHS uses only two signal words: *danger* and *warning*.

Transportation vehicles are regulated by the USDOT and there have been some changes to US placards to make them more consistent with international markings. The 4-digit identification (ID) number will still be displayed on vehicles carrying hazmat and can still be referenced in the Emergency Response Guide (ERG). HazCom 2012 has helped to align OSHA with the USDOT labeling requirements.

To help understand the differences, the NFPA and OSHA provide a "Quick Card" that compares NFPA 704 and HazCom 2012 labels. It is a free download at www.nfpa.org/704.

OSHA states that, through GHS, "The standard that gave workers the right to know, now gives them the right to understand." This statement refers to the standard requiring that workers understand the pictography, labels, and signal words, not just have certificates stating that they had training.

HAZARD COMMUNICATION STANDARD PICTOGRAM

The Hazard Communication Standard (HCS) requires pictograms on labels to alert users of the chemical hazards to which they may be exposed. Each pictogram consists of a symbol on a white background framed within a red border and represents a distinct hazard(s). The pictogram on the label is determined by the chemical hazard classification. (See Figure 10-2.)

HAZARDOUS MATERIALS INDICATORS

Hazmat indicators are discussed for occupancy, transportation, fixed facilities, and laboratories.

Occupancy

Occupancy can include businesses using industrial processes involving hazmat and commercial establishments using or selling potentially hazardous products. Among these are paint stores, exterminators, auto parts stores, pool suppliers, lumberyards, or lawn service stores selling pesticides and herbicides.

Facilities are required to have SDSs on the premises for use by responding firefighters. The manufacturer of the material provides these sheets. They identify a product and contain information on its physical properties and procedures for handling it during an emergency.

HCS Pictograms and Hazards

Health Hazard

- Carcinogen
- Mutagenicity
- Reproductive Toxicity
- Respiratory Sensitizer
- Target Organ Toxicity
- Aspiration Toxicity

Flame

- Flammables
- Pyrophorics
- Self-Heating
- Emits Flammable Gas
- Self-Reactives
- Organic Peroxides

Exclamation Mark

- Irritant (Skin and Eye)
- Skin Sensitizer
- Acute Toxicity
- Narcotic Effects
- Respiratory Tract Irritant
- Hazardous to Ozone Layer (Non-Mandatory)

Gas Cylinder

- Gases under Pressure

Corrosion

- Skin Corrosion/Burns
- Eye Damage
- Corrosive to Metals

Exploding Bomb

- Explosives
- Self-Reactives
- Organic Peroxides

Flame over Circle

- Oxidizers

Environment (Non-Mandatory)

- Aquatic Toxicity

Skull and Crossbones

- Acute Toxicity (fatal or toxic)

FIGURE 10-2 HCS pictograms and hazards.

Source: OSHA Quick Card, US Department of Labor

Facilities may have numerous hazmat that would require thousands of pages of SDSs sheets. In some jurisdictions, these facilities are permitted to store their SDSs data on a CD for use by the responding fire department. Realize that the responders will need a computer with a CD drive to access the information and a printer to produce a hard copy. There should also be a hard copy available in case there is a problem accessing the SDSs on the computer.

NFPA 704 Marking System   The NFPA marking system (NFPA 704) is required at facilities storing hazmat where referenced by fire code or other documents. The symbol is diamond shaped and is divided into four color-coded quadrants, each with a rating number inside it. These placards provide emergency responders with information about the type of danger caused by the stored material. The left quadrant is blue and denotes health hazard. The top quadrant is red and denotes flammability hazard. The right quadrant is yellow and denotes reactivity hazard. The bottom quadrant carries special designations, such as "OX" for oxidizers, or a "W" (with a line struck through it) for water reactive materials. Each quadrant carries a number from 0 to 4 that indicates the relative degree of hazard, with 4 being the most severe hazard and 0 presenting the least danger. (See Figure 10-3.) Box 10.1 shows the numbering system used in the 704 symbol.

NFPA 704 NUMERICAL RATING SYSTEM

HEALTH HAZARD—CODE COLOR BLUE

4—Materials that under emergency conditions can be lethal.

3—Materials that under emergency conditions can cause serious or permanent injury.

2—Materials that under emergency conditions can cause temporary incapacitation or residual injury.

1—Materials that under emergency conditions can cause severe irritation.

0—Materials that under emergency conditions would be no hazard, beyond that of ordinary combustible materials.

FLAMMABILITY HAZARD—CODE COLOR RED

4—Materials that will rapidly or completely vaporize at atmospheric pressure and normal ambient temperature or that are readily dispersed in air and that burn readily.

3—Liquids and solids that can be ignited under almost all atmosphere and temperature conditions. Materials in this degree produce hazardous atmospheres with air under almost all ambient temperatures or, though unaffected by ambient temperatures, are readily ignited under almost all conditions.

2—Materials that must be moderately heated or exposed to relatively high ambient temperatures before ignition can occur. Materials in this degree would not under normal conditions form hazardous atmospheres with air but, under high ambient temperatures or under moderate heating, might release vapor in sufficient quantities to produce hazardous atmospheres with air.

1—Materials that must be preheated before ignition can occur. Materials in this degree require considerable preheating under all ambient temperature conditions before ignition and combustion can occur.

0—Materials that will not burn. This includes any material that will not burn in air when exposed to a temperature of 1500 degrees Fahrenheit (F) for a period of five minutes.

INSTABILITY HAZARD—CODE COLOR YELLOW

4—Materials that by themselves are readily capable of detonation, explosive decomposition, or explosive reaction at normal temperatures and pressures. This includes those that are sensitive to localized thermal or mechanical shock at normal temperatures and pressures.

3—Materials that by themselves are capable of detonation, explosive decomposition, or explosive reaction but that require a strong initiating source or that must be heated under confinement before initiation.

2—Materials that by themselves undergo violent chemical change at elevated temperatures and pressures.

1—Materials that by themselves are normally stable but that can become unstable at elevated temperatures and pressures.

0—Materials that by themselves are normally stable even under fire conditions.

Source: Reprinted with permission from NFPA 704-2017, *System for the Identification of the Hazards of Materials for Emergency Response,* Copyright © 2016, National Fire Protection Association. This reprinted material is not the complete and official position of the NFPA on the referenced subject, which is represented solely by the standard in its entirety. The classification of any particular material within this system is the sole responsibility of the user and not the NFPA. NFPA bears no responsibility for any determinations of any values for any particular material classified or represented using this system.

FIGURE 10-3 NFPA 704 System of marking is required whenever another standard or fire code requires such marking. Examples include the storage, handling and use of hazardous materials listed in NFPA 400 or NFPA 1. Reprinted with permission from NFPA 704-2017, *System for the Identification of the Hazards of Materials for Emergency Response,* Copyright © 2016, National Fire Protection Association. This reprinted material is not the complete and official position of the NFPA on the referenced subject, which is represented solely by the standard in its entirety. The classification of any particular material within this system is the sole responsibility of the user and not the NFPA. NFPA bears no responsibility for any determinations of any values for any particular material classified or represented using this system.

Transportation

A vehicle carrying hazmat should have placards that contain a 4-digit identification number (ID) identifying the product it is transporting. All modes of transportation are required to have shipping papers identifying their cargo. These papers are referred to differently, depending on the mode of transportation: highway vehicles carry bills of lading; trains have waybills, wheel reports, and consists; ships carry cargo manifests; and airplanes have air bills. If shipping papers cannot be located or if more information is needed, the Chemical Transportation Emergency Center (CHEMTREC) can be called. This is a service provided by the chemical industry that can assist responders at potential hazmat incidents by providing information and technical assistance to emergency responders (phone number: 1-800-424-9300).

After providing CHEMTREC with the name of a chemical, responders can receive immediate advice on the nature of the product and steps to be taken in handling the early stages of an incident. CHEMTREC can also contact the shipper of the product for more detailed information.

The driver of a vehicle may have information on what has happened (breakage or spill). If a mixed load is on the vehicle, some deliveries may have been made. This will help in determining which contents still remain on the truck.

There are various methods of marking hazmat. The USDOT regulates labels and placards that must be attached to vehicles during transportation. Placards on vehicles can indicate specific hazards. The major drawback with placards is that too often vehicles are improperly labeled or no placards exist.

A ready resource for assisting firefighters in identifying hazmat is provided by the USDOT in its *Emergency Response Guidebook (ERG)*. A copy of this publication should be carried on every emergency response apparatus and be readily accessible. By referencing the 4-digit ID number in the ERG, valuable response information can be found on the products you are dealing with. The ERG gives vital initial information on placarding systems, chemicals, and initial evacuation distances. It assists in identifying products within a facility or aboard a mode of transportation.

ON SCENE

Firefighters cannot rely only on placards. Facilities must be routinely visited by the local fire department to achieve a first-hand look at the facility and interact with the plant personnel. At a major building fire that was heavily involved on our arrival, a 704 symbol indicated material in the building that posed a health, flammability, and instability hazard. Units were removed from the immediate exterior areas until additional information could be found. In checking with updated SDSs, it was found that a previous business had occupied the building and failed to remove the 704 symbol when it moved to another facility.

Fixed Facilities

A fixed facility can offer some benefits to initial responders. The material stored in tanks can be marked with NFPA 704 identification numbers. (These placards permit identification with a numerical system that can be referenced through the use of the *ERG* or other hazmat textbooks.) SDSs sheets should also be available for quick reference of the products. There may be fire-protective systems in place to assist in mitigation. They can include sprinklers, special extinguishing systems, private hydrants, or ventilation systems.

There is also the likelihood that someone knowledgeable about the emergency will be on location or readily available, if requested.

A benefit of a fixed facility is the ability of the local engine company to visit and inspect it on an annual or a semi-annual basis. These visits should be used to discuss with the plant manager hazmat within the plant, systems that have been installed to mitigate the hazards associated with these materials, and actions to take in an emergency situation.

Some fire departments include plant managers in their response plans. They have issued portable radios to the managers, along with designated call letters. This ensures that the transfer of information can readily occur between the fire department and plant manager.

Laboratories

Many responses to incidents occur in laboratories within hospitals, research facilities, or industrial plants. These incidents normally involve a small

FIGURE 10-4 Fire department during a response checking SDS sheets at a bulk storage facility.

amount of a product yet can encompass a wide range of dangers, including fire, explosion, or health hazards. Elaborate ventilation systems exchange the air within these laboratories every few minutes or less. There is usually no recirculation of air within the building. This prevents spreading a problem within the building should an incident occur. One downside is that rapid ventilation to the exterior can spread the situation from the interior to the exterior. Weather conditions may exist that restrict the dissipation of noxious fumes. Because relatively small amounts of a product are used in experiments, the release of fumes to the exterior may be completed by the time the fire department arrives. In many instances, those working within the labs are unaware of the situation created and that can add confusion to the fire department's investigation. Another problem has occurred in systems where floor drains in labs interconnect with the drains at each lab station. Materials disposed of go to a holding tank to prevent contamination of the central sewer system. The floor drains have trap systems (similar to any drain) that prevent fumes within the system from entering living spaces. A common occurrence is one in which the liquid within these drains has evaporated, permitting fumes to back up into the labs after technicians have properly disposed of a product down the lab station drains. Many labs have addressed this problem by placing antifreeze in the drains to minimize evaporation.

Fixed facilities can handle most problems that occur. They will request assistance when dealing with hazmat when someone is injured or the spill, leak, or fire is out of control.

NIOSH FIREFIGHTER FATALITY REPORT F2009-31

In December 2009, a 33-year-old male firefighter died, and eight firefighters, including a lieutenant and a junior firefighter, were injured in a dumpster explosion at a foundry in Wisconsin. At 1933 hours, Dispatch reported a dumpster fire at a foundry in a rural area. Eight minutes later, the initial responding crews and the Incident Commander (IC) arrived on scene to find a dumpster emitting approximately two-foot high bluish-green flames from the open top and having a 10-inch reddish-orange glow in the middle of the dumpster's south side near the bottom. The IC used an attic ladder to examine the contents of the dumpster: aluminum shavings, foundry floor sweepings, and a 55-gallon drum. Approximately 700 gallons of water was put on the fire with no effect. Approximately 100 gallons of foam solution, starting at 1 percent and increased to 3 percent, was then put on the fire, and again there was no noticeable effect. Just over 12 minutes on scene, the contents of the dumpster started sparking, then exploded, sending shrapnel and barrels into the air. The explosion killed one firefighter and injured eight other firefighters, all from the same volunteer department.

See Appendix R: NIOSH Reports in www.pearsonhighered.com/bradyresources to read the complete report and recommendations.

HAZARDOUS MATERIALS SCENE CONDITIONS

Each incident must be handled on its own merits. An overview of the scene, combined with a rational consideration of potential problems, must be made.

- What do we see?
- What do we know?
- What is actually involved?
- What type of containers are involved?
- What is their condition?
- What are the properties of the involved material?
- What are the potential hazards?
- Could there be a synergistic effect between the involved products?
- Is there a spill or fire involved?
- What life hazard exists?
- Where is the problem located?
- Is it an urban or rural setting?
- What decisions need to be made on evacuation?

The time of year, day of the week, and time of day can all affect the incident. Rain could spread a leaking material, compounding containment problems. On suspected hazmat responses, Dispatch should transmit wind direction and speed as noted by the

FIGURE 10-5 This fire at a petroleum packing plant endangered surrounding structures and an adjacent elevated interstate highway. *Used with permission of Joseph Hoffman.*

National Weather Service or local weather stations. These data will prepare initial responders to approach the site from an upwind location.

The terrain must be considered when approaching a suspected hazmat site. When possible, we should use a downhill approach. Personnel positioning apparatus must consider the possibility of apparatus being exposed to products that will require them to be decontaminated. This could be costly and time-consuming.

High temperatures can cause some products to vaporize readily, compounding a precarious situation and possibly endangering a larger area. We should determine the most likely method of how the problem material can spread. Through sewers? Vapors spreading downwind? From water runoff as a result of firefighting efforts? When the problems are known, plans can be made to address them.

A hazmat incident requires a more cautious and deliberate size-up than most other types of responses. One consideration is, "What if the fire department does nothing?" Will this nonaction change or affect the overall situation?

The IC must decide what actions, if any, can be performed by the initial responders. What does their level of training permit them to do? He or she must consider the reality of a delayed arrival and then a prolonged period of time before the hazmat team can set up and get into operation.

If life safety of civilians is a threat, then an assessment of how best to protect and rescue those involved must be made. The IC may choose to remove those trapped and then withdraw personnel and await the arrival of the Hazmat Unit. (Firefighters should not be put at risk to remove dead bodies.) He or she may decide to utilize initial personnel to set up diking and seal off storm sewer drains in an attempt to contain the product.

Although there is a demand for scene safety and control, we must exercise some common sense. A truck carrying only two 55-gallon drums of kerosene with a minor leak and no fire in one drum should not require evacuations and shutting down of all traffic for 2,000 feet. The total amount of product that could be involved would not exceed 110 gallons. Yet a truck with numerous chemicals that could drastically affect a community and is involved in fire could demand exclusion from the area, evacuations, and withdrawal of firefighters, allowing the product to burn.

SETTING UP ZONES

Setting up hot, warm, and cold zones can assist firefighter safety. Size consideration for the zones should be based upon wind direction, terrain, accessibility, vapor clouds that exist, and any explosion potential.

The data in the USDOT *ERG* can be utilized by the first-arriving fire department units for setting up the initial zones and for the evacuation of civilians. When the Hazmat Unit arrives, they can adjust the zones accordingly to meet the current hazards.

The actual size of the zones and the distances between each zone will be dependent on:

a. Safe operating distances
b. Practical work considerations
c. Physical/topographic barriers
d. Explosion/exposure potential
e. Specific characteristics of the contaminants

Hot Zone

The immediate danger area is the hot zone. This is an exclusion area and should be considered contaminated. Entry into the hot zone must be strictly monitored. Those entering this area will require a high level of personal protection. The exact type of protection will vary, based upon the product involved.

Warm Zone

The warm zone is adjacent to the hot zone and allows an area for decontamination of personnel and equipment. The warm zone prevents or minimizes the spread of contamination. Similar to entry into the hot zone, access to the warm zone also must be controlled. One method is through the use of access points that are staffed to record those entering and leaving the area. This will assist in monitoring and decontamination of personnel exposed.

Cold Zone

The cold zone is the outermost area where support personnel and equipment are located. Contaminated personnel are not permitted in this zone.

The size of the cold zone can be enlarged or reduced as the incident stabilizes or expands. Once the Hazmat Unit arrives, they can utilize their monitoring devices to document any changes required in the already established zones. The zones can be designated through the use of banner tape or ropes. Street locations or other recognizable landmarks may denote extremely large areas.

Once established, these safety zones must be strictly enforced. It should be spelled out in the operational guidelines who can enter and operate in a zone.

Fire Conditions

An incident in which the fire involves hazmat compounds an already dangerous situation. The factors that must be considered are:

- Is the product readily identifiable and can the immediate dangers be assessed?
- In what type of vessel is the product contained?
- Is fire impinging on other vessels, compounding the situation?
- Is a boiling liquid expanding vapor explosion (BLEVE) a possibility?
- Can an attack on the fire or the immediate exposures be implemented?
- When these factors are known, the decisions need to be based upon the following:
  - Are sufficient resources available?
  - What is the proper extinguishing agent?
  - If water is used, will runoff cause a worse problem than if the material was allowed to burn?

Certain pesticides when burning at elevated temperatures become less harmful than if the fire were extinguished and the pesticide allowed to vaporize into the air or run off into the soil.

If fire is in or impinging upon a vessel or tank, cooling of that container should be an immediate concern. This action will reduce the possibility of a BLEVE.

BLEVE—Boiling Liquid Expanding Vapor Explosion: a container failure with a release of energy, often rapid and violent, accompanied by a release of gas to the atmosphere, followed by ignition (fireball) and propulsion of the container or container pieces.

STRATEGIC MODES FOR HAZARDOUS MATERIALS

There is a distinct difference between structural firefighting and hazmat incidents. An additional mode titled *nonintervention* is added to assist the IC in handling incidents involving these materials:

- *Offensive:* An aggressive attack to either extinguish a fire, stop a leak, or control a spill.
- *Defensive:* A holding action where a defensive stand is taken. Firefighters are kept a safe distance from the threat of exposure to the product or potential explosion.
  Transitional is a combination of offensive and defensive modes:
- *Offensive/Defensive:* Making an aggressive attack with either hose-lines or foam lines to rescue someone trapped, or attempting a quick shutdown of valves to

minimize or mitigate the problem. Once the rescues are accomplished or the valves are shut down, a defensive attack is then initiated.

- *Defensive/Offensive:* An initial holding action is utilized until a fire reaches the state where it can be controlled, or until sufficient resources arrive, such as foam or Hazmat Units. The Safety Officer (SO) can then make an assessment of the situation and recommend to the IC if an offensive attack can commence.

- *Nonintervention:* This mode means that no action will be taken. This can be due to a number of considerations: the IC has decided that no civilian life is at stake and that any course of action would be too dangerous to the firefighters attempting to extinguish the fire or mitigate the problem, or that intervention would have little or no effect on the incident.

SAFETY

Safety dictates that we approach hazmat incidents in a cautious manner. The tendency to rush in and solve the problem as quickly as possible must be resisted. First responders must not attempt to mitigate an incident if they are not qualified. They should avoid touching or walking in spilled material or inhaling the fumes. Some products, though odorless, may be harmful.

Hazmat incidents require that an Assistant Safety Officer (ASO) be assigned to work with the hazmat team. Competencies for this position include training as a responder at the awareness, operational, and technician levels in addition to the required training as a SO. The hazmat ASO will report to the SO and coordinate with the Hazmat Group Supervisor or Hazmat Branch Director, depending upon the incident organization. Complex hazmat incidents may require the use of multiple ASOs to ensure the safety of responders. There may be a need for air monitoring, runoff monitoring, checking on hazmat personnel's vital signs, decontamination, and so on.

When hazmat personnel are operating in a hot zone, there needs to be a Rapid Intervention Crew (RIC) in place in the warm zone who can be immediately deployed. This RIC needs to be equipped with at least the same level of protection as those operating in the hot zone.

Firefighters must realize the limitations of their personal protective equipment (PPE). Structural firefighting gear is designed to protect a firefighter from heat and flame. If it comes in contact with hazmat, it can be saturated or penetrated by them. If the gear is not decontaminated, the toxic chemicals can cause chronic exposure to the firefighter each time it is worn.

Firefighters must document hazmat exposures. These data will be useful in the future in determining specific exposures during an individual's career. They may assist in diagnosing a later ailment. Accurate health records must be maintained on all responders. These records should list any chemicals to which the firefighters have been exposed. They should also list any effects that these substances have been known to cause. This information should be available for review by the firefighters.

The fact that in the past many incidents were taken care of without a hassle does not mean that it is okay to handle them that way today. Specialists in the public and private sectors have received training and certification for these complex incidents. Initial responders who are not trained to this level should not allow a brief and seemingly innocuous contact with hazmat to result in a lifetime of misery.

INCIDENT MANAGEMENT SYSTEM

Federal law mandates utilization of an incident management system at a hazmat incident. The complexity of these incidents may require the implementation of Unified Command (see Chapter 2). The initial IC should set up a Command Post (CP) in the cold zone and assume Command. The location of the CP must anticipate the expansion of the incident. If the signs of a potentially dangerous and expanding situation are present initially, then the CP should be placed a sufficient distance from the hot zone. It is a difficult task to move the CP after it has been established. (See Figure 10-6.)

The IC must ensure that initial responding units operate within the realm of their training and capabilities. The SO can assist in confirming that this occurs.

The IC or the Liaison Officer can coordinate with police to establish site control. This includes limiting access to the scene and police assistance in the evacuation of areas where it can be accomplished safely.

A Staging Area should be designated, considering topography, wind, and accessibility. A large-scale incident may require the establishment of a Base for the marshaling of equipment and resources a distance from the scene. A good location for such a Base is a school or a large fire department facility. Units then can be rotated to the Staging Area from the Base in anticipation of their utilization at the incident.

A first-aid area should be established where paramedics can monitor vital signs of the hazmat team members prior to the donning of their protective suits. A method that facilitates this monitoring is the use of a preprinted check-off form. This will ensure that all vital signs are taken, and the check-off sheet can be used as a baseline for comparison when the hazmat team member is again monitored after completing the assignment and removing the suit. After the incident, the sheet can become part of the member's permanent record.

FIGURE 10-6 Interagency assistance at hazardous materials incidents will aid the overall operation.

Be prepared to expand the incident management system to handle the many and varied problems associated with these types of incidents.

Assistance at the Scene

Request assistance from federal, state, local, and private agencies that can help at the incident, such as providing sand trucks for diking, heavy equipment, absorbents, firefighting foam, dry powder, etc. This request for assistance could include a representative from the EPA to assist in the cleanup efforts. (See Figure 10-7.)

To adequately recoup funds and equipment that are expended, fire departments must be meticulous in their record keeping. At the emergency scene, this is usually not a prime consideration, but it must be. If funds spent are not replenished, it will cause a shortage in the department's budget, requiring cutbacks in other areas.

EVACUATE OR PROTECT IN PLACE

After it is established what civilian dangers exist, a plan on how to protect them can be developed and implemented. The weather conditions must be considered along with the current situation. Might they change? For the better or for the worse?

We must determine the population that is threatened, including how many people are involved and their exact location. A prime consideration would include places of public assembly or institutions, such as hospitals, schools, nursing homes, and prisons. We then must decide how long it will take to evacuate or whether it would be best to protect in place. Deciding which method to employ or whether to use both methods to address particular situations will be difficult.

Evacuation means removing everyone from the area of concern to a safe location. This method can be employed if it can be controlled and there is sufficient time and resources to accomplish it.

In-place protection means moving or keeping people indoors in the affected area and having them remain there until the danger passes. This method is used when an evacuation cannot be performed or when there would be a greater risk involved in removing the people rather than letting them remain. Protecting in place requires that all doors and windows be closed. Heating and cooling systems have to be shut down. Communications should be maintained between those remaining and emergency response personnel to keep them abreast of changing conditions.

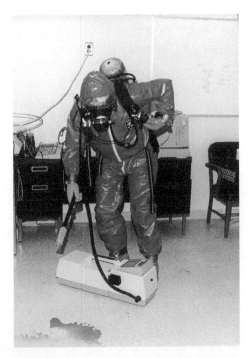

FIGURE 10-7 Initial responders will need the assistance of a hazardous materials reconnaissance team to identify hazards associated with spills. *Used with permission of Philadelphia VA Medical Center.*

DECONTAMINATION

A decontamination plan should be in place prior to firefighters' entering an area where an exposure to hazmat may occur.

Civilians and firefighters exposed to harmful products will need to be decontaminated. Contamination can be caused by chemical, biological, or radiological exposures. Gross decontamination is typically a washing down of an individual with either a handheld hose-line or a water spray through which people can walk. Secondary decontamination can then be accomplished by having those involved remove their clothing. The removal of a person's clothing will typically remove 80 percent of the contamination.

When confronted with chemical contamination, it may be visible on the individual or it may be detectable only through monitoring. If monitoring indicates that the individual has not been fully decontaminated, then secondary or tertiary decontamination can be initiated. This may include showering until suitable readings are reached.

Biological or radiological contamination will not be visible to the human eye. Biological contamination is impossible to detect since it will not manifest itself until an incubation period has passed.

Radiation can be detected through monitoring, and if detected, decontamination can be initiated. Inhalation of radiation may indicate radiation readings that will not be lowered by decontamination procedures.

Water that is used for decontamination should be captured if possible to minimize contamination of soil and sewers. If the water used for decontamination seeps into the ground, that area should be cordoned off and mitigation can be accomplished at a later time. If contaminated water enters the sewer system, the water treatment plant must be notified. These types of contamination can occur if interior showers are utilized for decontamination.

Secondary decontamination of individuals will be performed when needed to ensure the complete removal of the contaminated products. When possible, this can be achieved by showering. It should be realized that life safety of those contaminated should be addressed first; consideration for the environment is secondary.

When dealing with possible weapons of mass destruction events, the tracking of those who have been decontaminated will allow the monitoring for any symptoms exhibited by those individuals so that proper medical treatment can occur.

Medical personnel must protect themselves to prevent their being contaminated when treating a patient who has been exposed. Prior to treatment and transport to a hospital, decontamination must occur to protect the medical unit and the personnel.

Hazardous Materials Task Forces

Hazardous materials task forces (HMTFs) have been created to respond to incidents involving hazmat. Typically their responsibility includes handling the hazmat aspects at an incident, including decontamination. Some fire departments have purchased trailers that can be towed to an incident scene for mass decontamination incidents.

Before the HMTF is deployed, there needs to be an Incident Action Plan (IAP) in place (see Chapter 3 for Incident Action Plans). This plan must specify the projected actions of the HMTF. Each member of the team needs to be briefed on the plan prior to engaging in the incident.

In addition to the HMTFs, a Squad Company concept has been adopted in many fire departments. Their training can include being hazmat technicians as well as acquiring the skills utilized by heavy rescue personnel.

SIZE-UP FACTORS FOR HAZARDOUS MATERIALS INCIDENTS
Water
- Can water be used to fight a fire or for decontamination?
- Is there a sufficient water supply to fight a fire?
- Can water be used to disperse fumes or knock down plumes of chemicals being released from a spill?
- Will water compound the problem or produce another hazardous material? (For example, water and chlorine could produce hydrochloric acid.)

Area

- It may be difficult to define the size of the area being threatened.
- Hazmat incidents can involve large areas.
- Determine how large of an area is threatened by the fire or spill.
- If a vessel is leaking product, is the product readily identifiable? If involved in fire, can the dangers be assessed? Are other vessels exposed? Is a BLEVE possible?

Life Hazard

- Individuals can be exposed to harmful products that can threaten their health and safety.
- No one should touch or walk in spilled product. Eating or smoking in the area must be prohibited. Full protective personal equipment must be utilized.
- Decisions on evacuating or protecting in place may need to be made. Evacuation can involve areas of up to one mile or more from the incident scene.
- Are there residential or commercial buildings nearby that could be threatened by either the fire or toxic fumes? Should evacuation orders be given?
- Relocation centers and transportation may be needed if mass evacuations are required.
- Have all civilians been accounted for?
- Are firefighters operating in safe areas?
- Is all PPEbeing fully utilized?

Location, Extent

- When possible, approach the site from an upwind location and downhill.
- Hot, warm, and cold zones must be defined, marked, and enforced.
- Extensive decontamination and cleanup may be required.

Apparatus, Personnel

- Specialized knowledge is required of hazmat personnel. A decision must be made if initial responders can act to mitigate the situation or if they must wait for the Hazmat Unit personnel.
- Plant managers should be utilized for their expertise when dealing with a hazmat problem at a fixed facility.
- Decontamination must be set up prior to firefighters' entering contaminated areas.
- Sand, absorbents, diking material, firefighting foams, or special extinguishing equipment may be needed.

Construction/Collapse

- Buildings and vessels must be assessed for stability.

Exposures

- These incidents can affect buildings a great distance away. They can involve extension of fire or the spread of toxic fumes or gases.
- Transportation accidents can threaten any type of structure.

Weather

- Dispatch should transmit the current weather conditions to the responders.
- Rain can react with a spilled material, flush it into sewers or streams, or assist in diluting it. Wind can spread a plume of toxic material or assist in dispersing it. Cold weather can keep the temperature below a product's flashpoint and minimize vaporization. Hot weather can cause a material to reach or exceed its flashpoint, causing ready vaporization.

Auxiliary Appliances

- Sprinklers may be used to mitigate a spill or fire. Sprinklers may cause a runoff problem that will need to be addressed.
- Special extinguishing systems may be utilized where dangerous processes are in use.
- There may be special ventilation systems for hazardous processes or research areas.

Special Matters

- Check to see if SDSs sheets are available at the scene.
- Check shipping papers to identify suspected hazmat cargo.

- Check with CHEMTREC for data on hazmat being transported.
- Nonintervention needs to be a consideration. What would the outcome or result be if the fire department did nothing?
- If operating at a facility, consider supplying a portable radio to the plant supervisor to enhance communications during an emergency.

Height
- Hazmat incidents are possible in any building.

Occupancy
- Various occupancies can cause different problems for responders (e.g., paint stores, exterminators, auto parts stores, pool suppliers, lumberyards, lawn service stores, refineries, chemical plants, and tank farms). Know the occupancy and recognize the specific problems related to those occupancies.

Time
- When confronted with an occupied structure, a hazmat incident will affect the fire department's ability to protect the occupants.
- Time of day may determine if a plant coordinator or engineer at a commercial plant or facility is on duty and can provide fire department personnel with critical data about hazmat incidents occurring on their premises. This could include the product involved and possible solutions to mitigate the problem.

CONSIDERATIONS FOR HAZARDOUS MATERIALS INCIDENTS

Strategic Goals and Tactical Priorities for an Offensive Attack
- Obtain an assessment of the fire or incident area as soon as possible.
- Confine the spill or fire to the area of origin, if possible. Consider unattended master streams if a threat to firefighters exists.
- Ensure that the firefighters are qualified to handle the problem or whether hazmat technicians are needed.
- After a plan is established, ensure that an adequate number of firefighters are available to handle the incident, or call for additional resources immediately.
- Evacuate occupants if necessary and consider protecting other occupants in place if evacuation is not possible.
- Perform search and rescue, if needed.
- Ventilation of a building may be required. Try to utilize the heating, ventilation, and air-conditioning (HVAC) unit to exhaust toxic fumes or smoke. Wind direction and velocity should be considered.
- Positive pressure ventilation may be effective under the right conditions.
- Runoff water may need to be captured and disposed of properly.

INCIDENT MANAGEMENT SYSTEM CONSIDERATIONS/SOLUTIONS FOR AN OFFENSIVE ATTACK
- Incident Commander
- Unified Command (if needed)
- Safety Officer(s)
- Rapid Intervention Crew(s) (The level of protection of the crews must be equal to that of those operating within the hazardous areas.)
- Operations (if needed)
- Hazardous Materials Group or Branch
- Logistics
- Base
- Staging
- Ventilation Group
- Search and Rescue/Evacuation Group
- Division(s) (This could involve interior or exterior areas.)
- Medical Group or Branch
- First-Aid Stations

Strategic Goals and Tactical Priorities for a Defensive Attack

- If changing from an offensive to a defensive attack, ensure that a personnel accountability report (PAR) is taken.
- Reassess the hot, warm, and cold zones and adjust if necessary.
- Set up and maintain a collapse zone, if necessary.
- If using water, utilize unmanned master streams.
- Protect exposures, if necessary.

Strategic Goals for Nonintervention

- If it is too dangerous to attempt to mitigate the incident in a defensive mode, then all responders must be moved a safe distance from the incident. The scene can be monitored from a distance via binoculars or unmanned monitors, and the fire, spill, or leak can run its course before emergency responders are allowed to return to the scene. This can take an extended period of time and should not be rushed.

INCIDENT MANAGEMENT SYSTEM CONSIDERATIONS/SOLUTIONS FOR A DEFENSIVE ATTACK AND NONINTERVENTION

- Incident Commander
- Unified Command (if needed)
- Safety Officer(s)
- Operations
- Staging
- Logistics
- Base
- Hazardous Materials Group or Branch
- Rapid Intervention Crew(s) (The level of protection of the crews must be equal to that of those operating within the hazardous areas.)
- Divisions on the exterior on Sides A, B, C, D
- Exposure Divisions or Branches (if needed)
- Medical Group or Branch
- First-Aid Stations

Tank Farm/Refinery Fires

Refineries and tank farms will contain numerous hazards for firefighters. The units that respond to these facilities should have more than a basic understanding of the inherent dangers they could encounter. **Tank farm and refinery fires** need to consider all the hazardous factors previously covered, yet there are other factors that also must be assessed. One aspect in favor of the initial responders is the time element. Arriving in the incipient stages of a fire can be beneficial if the strategies and tactics, which were considered in the preplanning stages, can be initiated.

A quick response to the exact location of the emergency requires that the responding units have a plot plan of the facility. The installation should have an adequate numbering system on the tanks, along with well-marked streets. The preplan should indicate the size and capacity of the individual tanks as well as the product normally stored in each. A common problem associated with these complexes is the presence of narrow streets that restrict movement and proper location of apparatus, limiting the fire department's ability to bring special apparatus close to the fire scene.

At tank farms located in pier areas, tanker ships are unloaded and the contents placed in holding

tank farm/refinery fires ■ Involve highly flammable products and there is the ever-present danger of rapid incident scene changes.

FIGURE 10-8 Hot, cold, and warm zones must be established and controlled. Firefighters tend to let their guard down when fighting a fire of extended duration. This tank fire burned for days before final extinguishment. *Used with permission of Robert T. Burns.*

tanks. The contents are then downloaded into over-the-road tank trucks or rail cars. The product in the tanks at these facilities, as well as at refineries, can change on a routine basis. An up-to-date list of the product contained within each tank, and SDSs for each product, should be available. Obtaining this information is a priority of the first-arriving unit, as is obtaining information on firefighting actions that are underway.

FACILITY FAMILIARIZATION

Fire departments should pay annual visits to tank farms and refineries to gain familiarity with the facility and personnel. The inclusion of an annual drill improves the interaction between emergency responders and fire brigades, as well as communications between fire department and plant officials and will carry over to the emergency scene.

> The product handled at tank farms and refineries is normally referred to in barrel quantities. A barrel contains 42 gallons. A typical tank, 140 feet in diameter and 48 feet high, could contain 121,000 barrels. A full tank would mean more than 5 million gallons of product.

STORAGE TANKS

Recognition of the type of storage tank assists in determining the type of attack to be made on a tank fire. There are three common types of flammable liquid storage tanks:

- Open-top floating roof
- Covered floating roof
- Cone roof

Open-Top Floating Roof Tank

The open-top floating roof tank has a roof that floats on the surface of the product in the tank (normally crude oil). The roof has a set of legs approximately six to seven feet long that allow it to rest on the bottom. The tank design prevents a buildup of vapors between the surface of the liquid and the underside of the roof, which can occur with a fixed roof installation. There is a seal, protected by a weather shield, between the floating roof and the sidewalls of the tank to prevent any release of product. One way of identifying this kind of tank is by the wind girder encircling it at the top. (See Figure 10-9.)

Most fires in floating roof tanks are located in the seal area. Small fires can often be handled with extinguishers by directing agents so that they will flow between the shield and the seal. Fires that involve large portions of the seal area should be fought with foam streams, working from opposite directions, to confine and extinguish the fire. In situations where a large amount of foam is used, the roof must be monitored for accumulation of foam that could sink the floating roof. Roof drains exist and should be opened to prevent this from happening, though they should be monitored for any product release.

Covered Floating Roof Tank

The covered floating roof tank has a floating roof and a permanent solid cone roof. (See Figure 10-10.) The cone roof is permanently attached to the sidewalls with a weak shell joint. This weak shell attachment is meant to separate the roof from the sides if there is any pressure buildup within the tank. These tanks can be distinguished by vents near the top of the tank on the sidewalls, which allow the release of vapors. The tanks are usually free from ignitable vapors except during filling and for a period of approximately one day thereafter, depending on the volatility of the product.

Fires in this type of tank are caused by an ignition source (such as lightning) coming in contact with flammable vapors between the floating roof and the cone roof and causing an

FIGURE 10-9 Open-top floating roof tank. One way of identifying a floating roof tank is by the wind girder encircling it at the top. *Used with permission of Mike McCool.*

explosion. The explosion can cause the cone roof to break its attachment from the sidewalls and create a number of situations:

- It can create openings between the sidewalls and the cone roof, as the cone roof is forcibly pushed upward by internal pressure from the explosion.
- The roof can fully separate from the sidewalls and be blown clear of the tank. Depending upon the amount of force from the explosion, I have seen these roofs travel intact more than 200 feet from the tank and land on another tank.
- The cone roof can be blown straight upward, fully detaching from the sidewall and then collapsing downward into the tank onto the floating roof, displacing product from the tank into the dike area and spreading the fire.
- The roof can remain hinged to the sidewall and be blown clear of the top of the tank, hanging onto the sidewall outside of the tank.

The explosion in these tanks will also affect the floating roof. The pressure of the explosion can push downward on the floating roof, displacing product in the tank and causing the product to push past the seals and be exposed above the floating roof, which provides additional fuel to the fire. There is also the possibility of the floating roof sinking to the bottom of the tank.

The fire record for this type of tank is excellent; however, fires that do occur are extremely difficult to extinguish if the tank is not equipped with a fixed extinguishing system or if the extinguishing system is damaged.

Extinguishment can occur by placing foam onto the burning surface directly above the floating roof. If the cone roof is completely blown away, access to the burning surface will be viable. If the roof has only peeled back in some areas, then foam streams will need to be placed, taking advantage of those openings to deliver foam to the burning surface under the cone roof. If the roof is somewhat intact, the foam chambers in the extinguishing system may have been damaged by the explosion, or they may still be operable, in which case they can be supplied to place foam directly onto the burning surface of the tank.

Subsurface injection is not recommended for covered floating roof tanks, since the roof is either afloat and will prevent or restrict the passage of foam, or may be in a sunken state and could obstruct foam travel to the tank surface.

Cone Roof Tank

The cone roof tank has a permanent solid cone roof attached to the sidewalls with a weak shell joint, allowing for separation of the roof and sides should pressure buildup occur within the tank. (See Figure 10-11.) The solid roof allows a vapor space between the surface of the liquid and the underside of the roof. If an ignition source comes in contact with the vapors on the liquid surface while they are in the flammable range, an explosion will occur. An explosion involving the cone roof will cause it to react in ways similar to the cone roof on a covered floating roof tank. The roof will separate from the sidewalls and may remain in one piece, or it can fragment. In some instances the roof has been known to separate from the shell in a few places and at times has been propelled a distance from the tank. On other occasions the roof has been lifted into the air by the explosion and fallen back.

Fires in these tanks usually involve the entire surface area of the liquid. Strategies must be based on containment and control of the burning product, while recognizing the burning characteristics of flammable and combustible liquids.

FIGURE 10-10 Covered floating roof tank. These tanks can be distinguished by vents near the top of the tank on the side walls, which allow the release of vapors. *Used with permission of Mike McCool.*

FIGURE 10-11 Cone roof tank. The cone roof tank has a permanent solid cone roof attached to the side walls with a weak roof to shell joint. *Used with permission of Mike McCool.*

FIRE EXTINGUISHMENT

Water is the most common firefighting agent used on tank fires. It is usually the most readily available, has the greatest heat-absorbing quality of any common material, and is the prime ingredient in foam. Water can be applied as a spray to extinguish fires involving hydrocarbons where the flash point of the fuel is more than 100 degrees F. It can also be used to control fires, though not extinguish them, when flash points of less than 100 degrees F are encountered. Water can also be used to disperse flammable vapors or fumes, or as a cooling stream to protect firefighters operating at flammable liquid fires.

Firefighting foams are the principal agents used in controlling these kinds of fires. Hydrocarbon fires can be fought with either a topside application of foam, or subsurface injection. If a polar solvent is involved, topside application is required when using alcohol-resistant foam.

Extinguishment of the burning product without prior planning as to how to keep it extinguished is useless. It is necessary to determine the type of foam required, how much is immediately available at the site or nearby, how it will be transported to the scene, what the expected flow of any built-in foam system is, and what actions the responding fire forces can take.

After gathering information, it should be decided whether any fire department intervention will indeed be able to control and extinguish the fire. The size of the area to be covered, the amount of foam needed initially, the amount needed to keep the area covered, as well as the appropriate method to employ, must be established. Often, maintaining a blanket of foam for a period of 30 to 60 minutes is necessary. If the demands of the job cannot be met (insufficient foam, insufficient foam-producing apparatus or equipment to properly apply the foam, insufficient resources or water supply), then a defensive posture must be taken. This will involve letting the fire burn and protecting surrounding structures, tanks, and piping until sufficient resources are available to start and continue an offensive attack, or possibly a complete withdrawal from the area. A strategy that merits consideration is the removal of fuel from a burning tank by pumping it to other tanks in the complex through interconnected piping.

WATER SUPPLY

Tank farms and refineries may have independent hydrant systems to supply water for cooling, as well as foam lines. If demand exceeds hydrant supply, supplemental lines will have to be stretched. If units rely solely on water tender operations, their ability to handle large-scale tank fires will be limited.

Open-top floating roof and covered floating roof tanks typically are equipped with a stationary or fixed fire-protection system. These systems allow the delivery of foam over the top of the tank and place it directly onto the top surface area of the tank. They

FIGURE 10-12 When arriving at a fire involving hazardous materials, we must determine what is burning. Is fire impinging on other vessels? Are there sufficient resources? Size-up must be a continuous process. This fire involved piping and threatened numerous tanks containing petroleum products. *Used with permission of Robert T. Burns.*

may be attached to pump houses and adjacent foam tanks for immediate delivery of foam. They can also be supplemented by foam pumpers. One problem that can be encountered is an explosion, damaging the foam chambers on these systems and reducing their foam delivery capability or making them completely inoperative. The foam chambers can also become clogged if untreated water (e.g., river water) containing foreign objects is used.

GROUND FIRES

Ground fires around tanks must be controlled quickly. Fires involving flammable liquids may be extinguished with dry chemical, water spray, or foam. Exposed metal should be cooled with water lines to prevent reignition. If the product feeding the fire is coming from a vessel or tank, the supply should be shut off as soon as possible. It may be necessary to use cooling streams to protect firefighters or plant personnel as they shut down the valves. If a spill fire occurs within a dike area, the dike's drain valves should be shut to prevent fire extension to other areas.

FIGURE 10-13 Firefighters and fire brigade members were operating in pooled water with a foam covering. A short time later, the foam broke down, and product that was floating on the surface of the water ignited when it came in contact with the muffler on this foam unit. This action took the lives of eight Philadelphia firefighters. *Used with permission of Robert T. Burns.*

These facilities have a separator system to prevent products that enter the plant sewer system from being introduced into the municipal sewer system. These systems separate the product for future refining or proper disposal and allow the water to be sent to the sewage treatment plant.

Firefighters should not operate within a dike area that contains either a burning tank or a ground fire. There is a potential for a pooling of water and foam. If there is any product leaking from the tank and floating on the surface of the water beneath the foam blanket, it can ignite, seriously endangering firefighters. Firefighters should not operate in any area where there is a buildup of water because of this possibility—eight firefighters lost their lives at a refinery fire in Philadelphia, Pennsylvania, under similar circumstances. (See Figure 10-13.) See Appendix G: Case Studies, Gulf Oil Refinery Fire at www.pearsonhighered.com/bradyresources.

Storage tanks are built to collapse inward when under severe attack by fire. The sidewalls fold in toward the center of the tank, above the product line. There is always the possibility of seam failure, which would expel the burning contents of the tank into the dike area, exposing all tanks in that dike to impingement.

PROTECTING EXPOSURES

The protection of exposed tanks will be necessary if there is direct impingement of heat or flame on these tanks. The deluging of all tanks with water that are in the vicinity of a burning tank is not needed and will increase the demand on water supplies, depleting the volume of water available, and overtaxing sewer and product separation systems. The excessive water can also erode dike areas. To determine the need for protection of an exposed tank, a hose-line stream can be directed onto the exposed surface to see if steam is created, signifying the need for continued protection. Tanks that are exposed to radiant heat and flame are better able to withstand the heat if the impingement occurs at a level where there is product in the tank to absorb and distribute the heat. Heat impingement above the liquid level can severely distort the tank. Endangered LPG or LNG tanks must be protected by water streams.

Overhead electrical wiring can be threatened by flame impingement. The shutting down of the electricity in these lines can impact on the overall operation. In many instances, the pumps that would be used to pump product from the involved tank to other tanks will become inoperable. Operation of pumping stations and water pumps used to move water from dikes or low-lying areas within the tank farm itself may also be affected.

BOILOVER, SLOPOVER, AND FROTHOVER

Problems associated with tanks containing various types of oil are:

- Boilover
- Slopover
- Frothover

Boilover

Boilover is a phenomenon that can occur in an open-top tank containing crude oils that have a wide range of components. This occurrence can develop in a tank where the roof has been blown off, or in an interior floating tank where the top has sunk. For boilover to occur, the tank must contain water at the tank bottom. The crude oil in the tank releases its lighter components, which have been distilled off and are burned by the fire. This releases the heavier residue, which is heated to temperatures of 300 degrees F. The heat then generates downward at a rate that does not allow it to dissipate quickly enough and creates a heat wave within the product in the tank. (These storage tanks typically have some water residue at the bottom of the tank.) When this heat reaches the water in the bottom of the tank, it causes the water to boil, resulting in a sudden ejection of some of the residual oil in the tank. This phenomenon can cause the crude oil to be dispelled a distance from the tank, threatening workers and firefighters.

Slopover

Slopover can occur if a fire stream is applied to the hot surface of burning oil, provided the oil is viscous and its temperature exceeds 212 degrees F (the boiling point of water). Since this involves only the surface oil, a slopover is a relatively minor occurrence.

> Viscous refers to the resistance to flow of a liquid. It typically is used in reference to heavy liquids that do not flow easily.

Frothover

Frothover is the overflowing of a container that is not involved in fire. Frothover occurs when the water in the container boils under the surface of a viscous hot oil. An example is a tank car that contains water residue and is filled with hot asphalt. The initial asphalt that enters the tank is cooled by the metal of the container, but as more asphalt is added, the temperature of the water is raised and starts to boil. The asphalt may then overflow the tank car.

SAFETY OFFICER

The incident scene SO should designate an ASO who is qualified in dealing with hazmat. This individual will be responsible for monitoring the hot and warm zones. When confronted with a tank fire, this ASO will monitor for product release from piping and flanges as well as for airborne contaminants. He or she can recommend draining dike areas and using sand to channel runoff water.

Ear protection should not be overlooked. Many installations use high-pressure steam lines. If these lines fail, the noise level is deafening, and permanent or temporary loss of hearing can result. Additionally, communications are destroyed, and it can take a long period of time to isolate and shut down the broken line.

When dealing with highly flammable products, the possibility of rapid incident scene changes is ever-present. Constant monitoring is vital to be prepared for all possibilities. Due to the unpredictability of these incidents, there is a distinct need for command and control of personnel. Safety zones must be stipulated by the SO to ensure protection of firefighters. As with many hazmat incidents, operations should not be rushed. The decision to declare the fire under control and release units should be made only after careful assessment. The situation differs from the basic structure fire because of the volatility of the products involved. What appears to be a stable incident scene one moment can become a fire scene of major proportions the next.

ON SCENE

Plan B, Plan C, Plan D, etc.

A fire at a refinery or tank farm demands that alternative plans be in place. At an eight-alarm fire that I commanded, a covered floating roof tank containing more than one million gallons of a gasoline-like product was struck by lightning, causing an explosion and ignition of the product. The cone roof was peeled back in two places, creating openings in the roof of 35 feet in the front and 20 feet in the rear of the tank.

The initial plan (Plan A) was to engage the pumps at the stationary foam pump station to supply foam to the fixed system on the tank. As this operation was initiated, a flange in the piping within the foam house blew apart, putting the pump out of service.

The next actions (Plan B) involved the delivery of foam from fire department units, which was pumped into the piping to deliver a supply of foam to the fixed system. As this operation was started, it was found that one of the foam chambers on the tank had been damaged by the initial explosion and the utilization of the fixed system was proving ineffective.

The firefighters then attempted to aid the fixed system by setting up foam lines (Plan C) that were directed onto the burning liquid through openings created by the lightning in the cone roof. This also failed to extinguish the fire, as additional foam chambers on the fixed system broke down.

Subsurface injection of foam (Plan D) was tried and though it originally seemed to be effective it, too, proved unsuccessful.

Transfer of the product to another tank (Plan E) was initiated, but fire impinging on the overhead electric lines necessitated the shutting down of the electricity to the area, causing the transfer operation to cease.

Finally, protective water lines (Plan F) were placed on the exposed tanks and the remaining product was allowed to burn itself out. But planning was still underway in case the tank failed catastrophically, which could have dumped burning product into the common dike area, threatening other tanks.

It was decided that certain master streams that were supplying water to the cooling operation could be utilized to supply foam (Plan G) to extinguish any ensuing dike fire.

Firefighters must strive to find the safest and most effective methods of handling fires at these facilities. If the initial strategies are not successful, they must continue to seek alternative strategies to control the incident scene. Fire department responses to tank farms and refineries typically find fires that can be controlled with an extinguisher or a foam line, or they can also be confronted with time-consuming, challenging, and dangerous situations. If ICs are not anticipating what *may* occur, the situation can catch them off guard. By being proactive and anticipating problems, the IC can have a sufficient number of resources at the scene and be prepared to handle any problems that arise.

SIZE-UP FACTORS FOR TANK FARM/REFINERY FIRES

(The size-up factors for hazmat incidents and tank farm/refinery fires are very similar.)

Water

- Can water be used to fight a fire or for decontamination?
- Is there a sufficient water supply to fight a fire?
- Can water be used to disperse fumes or knock down plumes of chemicals being released from a spill?
- Is there an independent water source supplied by pumping stations? If so, what is the system's capacity?
- If fireboats are available, can they be utilized to supply or supplement the water supply?
- What is the capacity of the separator systems? Will they be overwhelmed by a large flow of water used for fire extinguishment? If so, will any released product be fed into local sewer systems?

Area

- It may be difficult to define the size of the area being threatened.
- If a vessel is leaking product, is the product readily identifiable? If involved in fire, can the dangers be assessed? Are other vessels exposed? Is a BLEVE possible?
- Are any surrounding tanks threatened?
- If a tank is on fire, is it in a common dike area with other tanks? Do the other tanks need to be protected?
- How large an area is threatened by the fire?

Life Hazard

- Individuals can be exposed to products that can threaten their health and safety.
- No one should touch or walk in spilled product. Eating or smoking in the area must be prohibited.
- Decisions on evacuating or protecting in place may need to be made. Evacuation can involve areas of up to one mile or more from the incident scene.
- Relocation centers and transportation may be needed if mass evacuations are required.
- Have all civilians at the facility been accounted for?
- Are firefighters operating in safe areas?
- Is PPE being fully utilized?
- Are there residential or commercial buildings nearby that could be threatened by either the fire or toxic fumes? Should evacuation orders be given?
- Administrative buildings may need to be evacuated to protect employees from the threat of fire and toxic releases.

Location, Extent

- When possible, approach the site from an upwind location and downhill.
- Hot, warm, and cold zones must be defined, marked, and enforced.
- Extensive decontamination and cleanup may be required.

Apparatus, Personnel

- Specialized knowledge is required of hazmat personnel. A decision must be made if initial responders can act to mitigate the situation or if they must wait for the Hazmat Unit.
- Plant managers should be utilized for their expertise when dealing with a hazmat problem at a fixed facility.
- Decontamination must be set up prior to firefighters entering contaminated areas.
- Sand, absorbents, diking material, firefighting foams, or special extinguishing equipment may be needed.
- These facilities typically have narrow streets. Apparatus need to park along the side of the street, and hose-lines should be stretched alongside roadways to facilitate apparatus access.

Construction/Collapse

- Buildings and vessels must be assessed for stability.
- Storage tanks are built to collapse inward and resist complete tank failure that would spill their contents out of the tank.
- Refineries may contain cooling towers made of redwood that can reach heights of four to five stories. These towers are prone to ignition if exposed to fire and can collapse under heavy fire conditions.

Exposures

- These incidents can affect buildings a great distance away. They can involve extension of fire or the spread of toxic fumes or gases.
- Transportation accidents can threaten any type of structure.
- Explosions within a storage tank can threaten nearby tanks.
- Explosions in cone roof tanks have been known to project the roof a distance from the tank, causing damage to nearby tanks, piping, and buildings.

Weather

- Dispatch should transmit the current weather conditions to the responders.
- Rain can react with a spilled material, flush it into sewers or streams, or assist in diluting it.
- Wind can spread a plume of toxic material or assist in dispersing it.
- Cold weather can keep the temperature below a product's flash point and minimize vaporization.
- Hot weather can cause a material to reach or exceed its flashpoint, causing ready vaporization.

Auxiliary Appliances

- Sprinklers may be used to protect tanks. Sprinklers may cause a runoff problem that will need to be addressed.
- Special extinguishing systems may be utilized where dangerous processes are in use.

Special Matters

- Check to see if SDSs sheets are available at the scene.
- Nonintervention needs to be a consideration. What would the outcome be if the fire department did nothing?
- Check with the plant supervisor or engineer to confirm what product is in the involved tanks and any nearby threatened tanks.
- Certain areas within a refinery can be extremely dangerous—e.g., hydrofluoric acid is utilized in some refining processes, exposure to which can be fatal.
- Consider supplying a portable radio to the plant supervisor to have ongoing contact during an emergency.

Height

- Storage tanks and cooling towers can reach many stories in height.

Occupancy

- These facilities contain areas of varying degrees of danger. Communications with the plant supervisor will allow firefighters to be cognizant of the more dangerous locations.
- Tank farms and refineries have many similar and many different problems that responders will be faced with.

Time

- Time of day may determine if a plant coordinator or engineer at a commercial plant or facility is on duty and can provide fire department personnel with critical data. This could include the protective systems available, product involved, and possible solutions to mitigate the problem.
- Refinery processes are typically 24/7 operations with full crews at most times.
- Tank farm hours of operations can vary widely, leaving only a watchperson during some time periods. This individual may provide little or no information for responding firefighters.

CONSIDERATIONS FOR TANK FARM/REFINERY FIRE INCIDENTS

Strategic Goals and Tactical Priorities for an Offensive Attack

- Obtain an assessment of the fire or incident area as soon as possible.
- Define the hot, warm, and cold zones.
- Confine the fire to the area of origin, if possible. Consider unattended master streams if a threat to firefighters exists.
- Ensure that the firefighters are qualified to handle the problem or whether hazmat technicians are needed.
- After a plan is established, ensure that an adequate number of firefighters are available to handle the incident, or call for additional resources immediately.
- Evacuate employees and nearby residents if necessary and consider protecting other occupants in place if evacuation is not possible.
- Ventilation systems may need to be shut down in nearby buildings. If toxic fumes or smoke have entered buildings, consider utilizing the HVAC unit to exhaust them.
- Consider wind direction and velocity.
- Runoff water may need to be captured and disposed of properly.
- Monitor separator systems to ensure that they are functioning properly and that they can handle the volume of water being used to fight the fire.
- If utilizing apparatus foam units, ensure that they are located in an area safe for firefighters to refill foam to the unit without being threatened by the fire. This may require utilizing relay pumping.

Incident Management System Considerations/Solutions for an Offensive Attack

- Incident Commander
- Safety Officer(s)
- Rapid Intervention Crew(s) (The level of protection of the crews must be equal to that of those operating within the hazardous areas.)
- Operations (if needed)
- Hazardous Materials Group or Branch
- Logistics
- Base
- Staging
- Ventilation Group (if needed for exposed buildings)
- Search and Rescue/Evacuation Group
- Division(s) (This could involve interior or exterior areas.)
- Medical Group or Branch
- First-Aid Stations

Strategic Goals and Tactical Priorities for a Defensive Attack

- If changing from an offensive to a defensive attack, ensure that a PAR is taken.
- Reassess the hot, warm, and cold zones and adjust, if necessary.
- Set up and maintain a collapse/safety zone, if necessary.
- If using water or foam, consider utilizing unmanned master streams.
- Protect exposures, if necessary.

Strategic Goals for Nonintervention

- If it is too dangerous to attempt to mitigate the incident in a defensive mode, all responders must be moved a safe distance from the incident. The scene can be monitored from a distance via binoculars or unmanned monitors, and the fire, spill, or leak can run its course before emergency responders are allowed to return to the scene. This can take an extended period of time and should not be rushed.

Incident Management System Considerations/Solutions for a Defensive Attack and Nonintervention

- Incident Commander
- Unified Command (if needed)
- Safety Officer
- Operations
- Staging
- Logistics
- Base
- Hazardous Materials Group or Branch
- Rapid Intervention Crew(s) (The level of protection of the crews must be equal to that of those operating within the hazardous areas.)
- Divisions on the exterior on Sides A, B, C, D
- Exposure Divisions (if needed)
- Medical Group or Branch
- First-Aid Stations

Terrorism Incidents

A threat that now looms for civilians and responders is that of terrorism. Terrorists strike unsuspecting, innocent people and then claim victory over dead and maimed men, women, and children. The terrorist acts at the World Trade Center in New York City, the Pentagon in Arlington, Virginia, and the Murrah Federal Building in Oklahoma City are prime examples.

On September 11, 2001, concerted acts of terrorism were committed against the United States of America. Religious zealots acting as suicide bombers hijacked commercial airliners. (See Figure 10-14.)

- 0846 hours: A Boeing 767, American Airlines Flight 11, with 92 people aboard hit between floors 93 and 99 of the 110-story One World Trade Center, the North Tower, in New York City. The impact of the crash and the ensuing fire fed by more than 20,000 gallons of jet fuel weakened the steel supports, causing the building to collapse at 1028 hours.
- 0903 hours: A Boeing 767, United Airlines Flight 175, with 65 people aboard struck between floors 77 and 85 of the 110-story Two World Trade Center, the South Tower. The impact of the crash and the ensuing fire fed by more than 20,000 gallons of jet fuel weakened the steel supports, causing the building to collapse at 0959 hours.
- 0937 hours: A Boeing 757, American Airlines Flight 77, with 64 people aboard struck the northwest side of the Pentagon in Arlington, Virginia. The burning fuel-fed fire caused the eventual collapse of a section of the five-story building, taking the lives of 125 occupants.
- 1003 hours: A Boeing 757, United Airlines Flight 93, with 45 people aboard crashed into a field in Shanksville, Pennsylvania, when hijackers were foiled by passengers in their attempt to inflict more damage.

The collapse of the World Trade Center towers occurred quickly. Within a span of 1 hour and 42 minutes, the 220 stories of the twin towers became debris in the street laden with bodies of occupants and would-be rescuers.

At 1725 hours, the rescue problem was complicated by the collapse of Seven World Trade Center. The 47-story office building collapsed due to the fire that had spread from the towers and collateral damage caused by the collapsing towers.

All seven buildings in the World Trade Center complex totally or partially collapsed. (See Figure 10-15.)

The death toll from the attacks on the World Trade Center was 2,749, which included a staggering 343 firefighters—the largest single loss of life of firefighters in history. The attack on the Pentagon caused the deaths of 125 occupants and 64 passengers aboard the airplane. The crash of Flight 93 in Shanksville took the lives of the 45 people aboard the airplane. This one-day act of terrorism took 2,983 lives on American soil.

In the aftermath of the terrorist attacks on the United States of America, the president and Congress created the Department of Homeland Security (DHS). It became the third-largest cabinet-level federal

department by merging 22 disparate agencies. Its mission is to protect the citizens of the United States from terrorism. There are many facets to its operation, one of which is to analyze threat information that is received by the various federal agencies in regard to terrorism. In utilizing this information, when the DHS receives information that a credible threat has developed that could impact the public, it will provide information to keep families and communities safe.

TERRORISM TODAY

Terrorism continues to be a problem both worldwide and in the United States. Federal agencies are vigilant, and many plots have been uncovered, while some incidents have occurred, including cyber terrorism.

- May 1, 2010: New York, New York. Times Square car bomb attempt and plot: An attempted evening car bombing in crowded Times Square in New York City failed when a street vendor saw smoke emanating from an SUV and called police. The White House has blamed Tehrik-e-Taliban, the Pakistani Taliban, for the failed attack and an American of Pakistani origin who was arrested in relation to the incident was working for the group.
- December 2014: The Guardians of Peace, linked by the United States to North Korea, launched a cyberattack against SONY pictures. Embarrassing private emails were published and the organization threatened attacks against theaters that showed "The Interview," a satire movie that depicted the assassination of North Korean leader Kim Jong Un. North Korea denied responsibility for the attack and proposed a joint investigation with the United States.
- January 15, 2015: Washington, DC. A US Capitol terror attack was stopped by the FBI. Investigators said a 20-year-old Ohio man wanted to set off pipe bombs at the US Capitol as a way of supporting the terrorist group ISIS.
- April 15, 2013: Boston, Massachusetts: Two bombs detonated within seconds of each other near the finish line of the Boston Marathon, killing three and injuring more than 180 people. The terrorists surfaced late in the evening of April 18, in Cambridge, Massachusetts. They shot and killed an MIT campus police officer while he was sitting in his squad car. Police killed one suspect and wounded the other. The suspect was arraigned on federal terrorism charges and later sentenced to death. A note that was found stated that the bombings were in retaliation for US actions in Iraq and Afghanistan against Muslims.
- May 3, 2015: Garland, Texas. Outside the Curtis Culwell Center, two gunmen opened fire during an art exhibit hosted by an anti-Muslim group called the American Freedom Defense Initiative. The center was hosting a contest for cartoons depicting the Muslim prophet Muhammad. Both gunmen were killed by police. The attackers were motivated by the Charlie Hebdo shooting in France and the 2015 Copenhagen shooting in Denmark earlier in the year. ISIS claimed responsibility for the attack through a Twitter post.
- December 2, 2015: San Bernardino, California. An all-staff meeting and training event of the San Bernardino County Department of Public Health was in the process of transitioning to a department holiday/luncheon when a coworker and his wife opened gunfire on the participants. Three homemade explosive devices were placed by the perpetrators but failed to explode. Police pursued the fleeing terrorists and a gun battle ensued, leaving the two terrorists dead. In addition to the terrorists, 14 civilians were killed and 22 others were seriously injured.

TERRORISM INCIDENTS VERSUS HAZARDOUS MATERIALS INCIDENT

terrorism incidents ■ Violent acts, or acts dangerous to human life, in violation of criminal laws of the United States. The Federal Bureau of Investigation (FBI) recognizes two categories of terrorism: domestic and international.

There is a distinct difference between a hazmat incident and a **terrorism incident**. In the terrorism incident, the release has been done intentionally, and there is a strong possibility of the presence of secondary devices that are meant to kill or injure responders. At hazmat incidents, first responders often have the time to identify the involved material without immediately having to address mass casualties or mass decontamination. A terrorist response demands immediate actions to save lives and to try to limit the spread of a harmful agent.

The US Department of Justice describes terrorism in part as "a violent act or an act dangerous to human life, in violation of criminal laws of the United States." The Federal Bureau of Investigation (FBI) recognizes two categories of terrorism:

a. Domestic
b. International

Domestic terrorism is classified as actions by those groups or individuals whose terrorist acts are directed at elements of our government or population without foreign direction.

International terrorism involves groups or individuals whose terrorist activities are foreign-based and directed by countries or groups outside the United States.

A terrorist incident will almost always include a criminal activity and technological hazards. The motives of the perpetrators must be determined. The amount of actual damage is a secondary concern of the terrorists. Their primary concern is the psychological impact of the attack. The claiming of responsibility for a terrorist act can give credibility to an extremist group by providing it autonomy through the media.

WEAPONS OF TERRORISM

The weapons of a terrorist attack include armed attack; biological, nuclear, incendiary, chemical, and explosive weapons (B-NICE); and weapons of mass destruction (WMD).

Biological

Biological weapons include bacteria, such as anthrax and plague, and viruses. These toxins can enter the body through inhalation or skin absorption. EMS and hospital personnel will often be the first to recognize the fact that biological weapons were used. They must immediately notify Command of this finding.

FIGURE 10-16 Firefighters suspecting a terrorist incident should don all protective equipment to conduct their investigation. *Used with permission of Philadelphia VA Medical Center.*

Nuclear

A nuclear weapons attack is doubtful due to the enormous expense. A potential problem could exist if an attack occurred on a fixed nuclear facility with the intention of a nuclear release. Because the security is high at these facilities, this is probably a remote possibility.

Incendiary

Incendiary weapons are economical and easily acquired. They are a favorite of some terrorist groups. They can be thrown or triggered remotely by chemical, electronic, or mechanical means. They can include flammable liquids, gases, combustible liquids, or a mixture of chemicals that will ignite on contact with each other. Unfortunately, this is also an instrument used in drug and gang violence and is something with which the fire service has experience. Differentiating between terrorism and other types of violence when incendiary devices are used will not be easy.

Chemical

Chemical weapons include nerve agents, blister agents, blood agents, and others. With the exception of hydrogen cyanide (which is lighter than air), these agents are heavier than air and will tend to seek the lowest level. They will pool in basements and other low-lying areas. These chemicals attack the nervous system, skin, eyes, mucous membranes, and gastrointestinal tract. They include sarin, mustard gases, cyanides, arsenics, chlorine, and phosgene, among others.

Explosives

Explosives are the choice for the greatest number of terrorist acts. They produce a pyrotechnic event that causes damage and frightens those exposed. The bomb can be quite sophisticated or as basic as a pipe bomb. Explosives can also be used to disperse other agents, such as biological, chemical, or incendiary weapons. If a bomb is used to disperse a chemical, it must contain a low-level explosive so as not to destroy the chemical agent.

Airplanes

As witnessed on September 11, 2001, commercial airplanes were used as WMD. Of the four hijacked airplanes, three reached important targets, resulting in many deaths and injuries. The DHS drastically changed the rules for air flight in the United States. In addition to increased security for boarding flights and increased baggage checks, air marshals have been trained and routinely accompany both national and international flights to guard against potential hijacking and other acts of terrorism.

DISPATCHERS AND EMERGENCY OPERATORS

The roles of the emergency operator and dispatcher are significant for early recognition and identification of a terrorist event. Key information received can be critical. This would include verbal indications of:

- Multiple seizures in a public place
- Many people sickened in a public building
- An explosion in a public building
- A group claiming responsibility for a terrorist act at a location to which units have been dispatched but have not yet arrived

Dispatch centers must develop a list of questions to assist in identification of a terrorist event. Training for dispatchers should include signs and symptoms that could indicate the possibility of terrorism.

By developing a target hazard analysis, a list of locations or various types of occupancies could trigger a warning. These could include historic or governmental targets, airports, airplanes, places of assembly, or controversial occupancies (e.g., family planning clinics).

Special events during religious or governmental holidays could trigger assassination attempts.

If Dispatch receives additional information about an incident, it is critical that it is immediately transmitted to the responding units.

ACTIONS OF FIRST-ARRIVING FIRE OFFICER

As firefighters, we are put at high risk in responding to a terrorist act. The initial units will not be aware of what has occurred until a size-up has been completed. They must recognize the signs and symptoms of a chemical or biological incident. Unfortunately, the time it takes to complete the size-up may be sufficient to have already exposed some firefighters to harmful chemicals or secondary devices. Similar to radiation exposure, the best methods of protection are time, distance, and shielding. Akin to hazmat responses, time plays a major factor. Unlike hazmat incidents, the terrorist often targets the civilian population, compounding the firefighter's job because there will be a need for medical intervention, and the resources on scene will be overwhelmed.

The first-arriving fire officer must read the cues presented by the incident. The earlier the recognition of terrorist involvement, the faster safeguards can be initiated. As firefighters, our priority of life safety and wanting to make an immediate impact on saving lives cannot override the need to recognize obvious signs of terrorism:

- An unexplained number of sick, injured, or dead
- Extensive damage to a structure, property, or vehicles
- Unexplained damage over a wide area
- Mass hysteria due to nerve agents
- The presence of clouds or vapors
- Unusual packages or devices that could contain a secondary device
- Multiple victims with seemingly serious afflictions with no apparent cause
- Seemingly unusual conditions, things out of the ordinary, chaos
- Dead animals
- Unusual odors

Approach from uphill and upwind of the suspected area. Don all protective equipment. Attempt to cover all exposed skin. Set up a hose-line for immediate decontamination of firefighters after they exit the contaminated area. If any runoff is occurring, be aware of its direction.

Minimize the number of personnel in suspected areas. The fewer people exposed, the fewer problems with exposure and decontamination. Accountability of personnel must be in place, and they must be monitored for signs or symptoms of exposure.

Without advance information, if the terrorist act involves a biological weapon, it is highly likely that initial responders will become part of the problem. Fire units responding as a first-responder on a medical call can be affected. Not being clothed in PPE and without self-contained breathing apparatus (SCBA), they can be severely exposed. This exposure may involve delayed symptoms.

Large coastal cities may be more likely places for international terrorism, but the attack in Oklahoma City must act as a warning that domestic terrorism can happen anywhere.

The most effective actions are protection and decontamination of civilians and responders. The scene will be complicated if the initial responders become contaminated and sickened. The normal reaction of the next-arriving emergency personnel is to rush in and rescue everyone. This will compound the problem if the rescuers are not properly protected and trained to recognize the signs and symptoms of a terrorist attack.

Hot, warm, and cold zones must be established. Once zones are established, access to these areas must be controlled. Police play a vital role in supporting the scene. They should deny entry to unauthorized personnel. The cross-training of both Hazmat Units and bomb squads allows interaction and cooperation on joint operations.

IDENTIFY THE TYPE OF AGENT

Is the agent chemical, biological, or an explosive? Identification of the agent should occur as early as possible. Initially, this may be aided by the signs and symptoms of those injured. The Hazmat Unit will need to determine the specific agents and concentrations. This will assist in the decontamination and emergency medical operations in treating patients. Identification will also assist in the search for evidence.

SCENE CONTROL

Controlling the scene will be difficult. There is the potential for a mass casualty incident. Depending upon the type of attack, there can be massive injuries due to a bomb, for example, or complex medical injuries due to a chemical attack. A secondary device may exist that is meant to injure responders. Secondary devices may be another improvised explosive device (IED) hidden near the initial detonation, or a car bomb may be used as the initial IED and secondary or tertiary car bombs may be located near the initial blast to kill and disable responders. Police may have bomb-detecting dogs that can assist in finding a secondary device.

Medical personnel can be confronted with numerous people injured and pleading for assistance. Based on signs and symptoms, decontamination of victims may be required before medical personnel can intervene.

See References at the end of this chapter to download the reference document "Homemade Explosives Recognition Guide."

 ON SCENE

In January 1997, a bomb exploded at the rear of the Atlanta Northside Family Planning Service. A second bomb exploded about an hour later as emergency workers, including firefighters, were attempting to secure the scene and evacuate the area. The secondary bomb was placed to kill and injure paramedics, firefighters, and police.

REQUEST ASSISTANCE

Determine the type and amount of assistance that may be needed. This includes:

- Hazmat teams
- Replacement protective clothing
- Air supply for SCBA cylinders
- Medical teams
- Decontamination units
- Clothing for those who have been decontaminated
- Communication equipment

Because local and regional resources will be overwhelmed, an early request for federal assistance will be needed. Realize that some governmental agencies require at least 12 hours to activate and then be transported to the scene. Even military units may require six hours of activation time. Recognize the resources available through mutual aid. Consider assistance from:

- Public health departments
- Public works departments
- Department of Homeland Security
- Federal Emergency Management Agency
- Federal Bureau of Investigation
- Bureau of Alcohol, Tobacco, Firearms, and Explosives
- Other federal, state, local, and private agencies

UNIFIED COMMAND

Operations will succeed if all agencies are working toward a common goal. This cooperation can occur only if established prior to the incident by conducting training and exercises that employ a unified command system. Each major agency will have a stake in the success of the operation. When the incident occurs, each agency's role should be identified. Without prior commitments and Unified Command, there will be turf battles and duplication of effort. Evidence may also be destroyed that could affect the arrest and conviction of the perpetrators. (See Unified Command in Chapter 2.)

INTERACTION WITH POLICE

Incident scene safety and scene security are a must. If an armed attack has occurred, the police must assure scene safety by having the terrorists in custody or confirming that they have left the scene before firefighters enter the area. The scene should be thoroughly checked for secondary devices by the bomb squad. The scene will need to be treated as a federal crime scene. The preservation of evidence can become a monumental task. At the Murrah Federal Building in Oklahoma City, each bucketful of debris was removed for examination. The level of evidence preservation must be discussed with the police prior to the event, and a plan must be put in place.

There have been instances in which police have allowed fire personnel to enter incidents where the police suspected foul play without notifying the firefighters of the potential danger. In at least one incident, the firefighters were pinned down by heavy

FIGURE 10-17 A tremendous number of resources responded to the World Trade Center bombing in 1993. It included multi-agency response from local, state, and federal agencies. The collapse of the towers in 2001 destroyed 89 fire department vehicles. *Used with permission of Steve Spak.*

gunfire. An exchange of information between the police officers and the firefighters can help to avoid these occurrences. Firefighters should not be subjected to hostile situations. Likewise, police officers should not be placed into areas where atmospheres have not been deemed safe to operate without proper protective clothing. Multi-agency exercises and training will help in eliminating these problems.

THEFT OF FIRST RESPONDER GEAR LINKED TO TERRORISM

According to the Emergency Management and Response-Information Sharing and Analysis Center (EMR-ISAC), several incidents show terrorist organizations' interest in impersonating first responders to make it easier to commit acts of terrorism. In several cases, the actors either planned on stealing equipment, uniforms, and badges or had already done so:

April 2015: France. Police find bulletproof vests, police arm bands, and emergency vehicle lights when investigating a plot against churches.

March 2015: Illinois. A member of the National Guard allegedly offered his uniform to his cousin to allow him to enter an armory and conduct an attack.

February 2015: New York City. One of several men arrested on terrorism-related charges expressed interest in shooting law enforcement officers so he could take their gear, to allow him to carry out more attacks.

January 2015: Belgium. Authorities conducting raids of people with suspected links to ISIS found police uniforms.

Strategically and tactically, being in a military or law enforcement uniform gives an attacker a distinct advantage because most civilians will assume that they are what they look like and many officers will as well, at least at a distance.

It is important to note that many EMS and fire uniforms are similar enough to law enforcement uniforms that they may also be of interest during the planning stages of an attack. It is vitally important to notify the proper authorities and chain of command as soon as you notice the theft of any gear, uniform, keys, badges, or other identification.

Source: EMR-ISAC, The InfoGram 15, no. 34, August 20, 2015.

DECONTAMINATION

Gross decontamination means removing the largest amount of contamination. This is accomplished by removing the victim's clothing, flushing the victim with water, and providing him or her with temporary clothing. The contaminated clothing will need to be placed in a bag, tagged, and kept secured to establish a chain of custody. The clothing may contain evidence needed for a criminal investigation. Secondary decontamination means a thorough rinsing with handheld hose-lines, decontamination showers, and possibly the use of a detergent or cleaning solution. The victim will then need to be clothed so that he or she can be transported to a medical facility for treatment.

Civilians who are suspected of being contaminated should be detained at the scene until decontaminated and medically cleared by EMS personnel. Allowing them to leave will spread the problem. Should they go to a hospital, the contamination will spread, and decontamination of the hospital will be necessary. If contaminated civilians come directly to the hospital, police and fire personnel should be dispatched to the hospital to set up a decontamination area.

Decontamination of protective equipment may require the replacement of that equipment. A problem that may result in the contamination of PPE, including gloves and footwear, is that residual contamination left from exposure to a liquid can result in *off-gassing* (the dispersion of a product that was absorbed by a person's clothing) that could cause injury long after the incident is resolved. A procedure must be in place to provide replacements. This could be PPE issued for temporary use.

There will be a need for separate decontamination sites for responders and civilians. (Decontamination is also discussed earlier in this chapter under Hazardous Materials.)

Antidotes for Exposures to Nerve Agents and Cyanide

Acts of terrorism can expose firefighters to nerve agents or cyanide. To counteract these exposures, it is critical that medical personnel have access to antidotes to treat them. Each jurisdiction should mandate who has the authority to carry and administer these drugs.

Pralidoxime and atropine are the initial treatments for patients with symptoms of a nerve gas exposure. These drugs may be given as individual injections, but they are conveniently packaged in autoinjectors known as *Mark 1 kits*. Generally, one Mark 1 kit is used to treat mild or moderate symptoms and three kits are used for moderate-to-severe symptoms. The autoinjector is designed to be given intramuscularly through clothing directly into the thigh or buttocks.

Diazepam is used to treat or prevent seizures in patients. Diazepam may be carried in vials or in autoinjectors. Sodium thiosulfate is used to treat patients with symptoms that may be related to exposure to cyanide.

The dosage of the drugs to be administered and any additional treatment will be decided by the medical personnel at the scene.

The authority having jurisdiction must decide who should carry these kits or the individual drugs. In major cities these drugs should be kept on front-line apparatus and be readily available at an incident scene. In areas where the perceived threat is low, they can be kept in prepackaged caches at area hospitals so they can be delivered to the scene by a responding unit or by an air ambulance.

PERSONAL PROTECTION

Protection is critical. Firefighters, when operating in the hot zone, must wear their SCBA. The wearers must allow a sufficient amount of time to exit the hot zone and be decontaminated before removing their masks. Otherwise, there may be residual contamination on the protective clothing that can be inhaled by the firefighters. After analysis of the scene, the Hazmat Unit may find that respirators with the proper filters will provide a sufficient safety factor for responders.

Firefighters must monitor coworkers and themselves for signs and symptoms of contamination. Personal protection means not becoming a victim. Having a firefighter become a victim complicates the problem. He or she will need to be treated, and this will diminish the treatment of civilians. A methodical assessment of the scene should be made. Do not touch, taste, or smell anything. These actions can lead only to problems. Firefighters should not make assumptions about what might have happened. If a problem exists, treat it. Be suspicious of events and act cautiously.

STAGES OF RESPONSE

A community must be prepared to meet the threat of terrorism. This preparedness includes all agencies that would be needed to meet the various types of threats. A written set of procedures for terrorist acts is needed to guide responders at these complex events. The different stages of readiness could include:

Stage 1 Alert

The alert stage is initiated in anticipation of a terrorist act. This phase is placed into motion because of intelligence from a police agency that has determined the credibility and the likelihood of a terrorist act occurring in the jurisdiction in the near future.

An alert stage will be implemented if:

1. A threat has been made
2. A threat has been confirmed
3. A previous threat that had been in the planning stage has been upgraded to a certain occurrence (a time frame in excess of 48 hours)

Implementation of the alert stage would set the following fire department events into motion:

- Evaluate available intelligence and develop contingency plans for stages 2 and 3.
- If available, issue body armor for each unit.
- Review station apparatus and security.
- Implement the necessary station training exercises to review the terrorism operational guidelines.

Stage 2 Warning

The warning stage would be initiated if the event is likely to occur within 48 hours. The warning stage could be used if a neighboring community has been threatened and the potential exists of the same threat occurring in your community.

Implementation of the warning stage would set the following fire department events (as mandated in the department's written procedures or guidelines on terrorism) into motion:

- Set up a meeting of all agencies to review pertinent data for accuracy.
- Assign someone to keep abreast of current intelligence information.
- Establish channels of communications with mutual-aid and police agencies.
- Ensure staffing or increase staffing on selected units and at the dispatch center.
- Keep on-duty officers informed of developments.
- Keep on-call officers informed of developments.
- Set up a system to monitor all responses for possible developments.
- Be prepared to relocate units located in the affected area.

Stage 3 Immediate Response

The immediate response stage would be implemented if there was no prior warning or intelligence indicating a forthcoming terrorist act. This stage will be the most difficult. With no prior warning, training procedures will be severely tested. If confronted with a B-NICE attack, responders will need to read the signs and symptoms of the scene.

Realize that as firefighters, the amount of information that the local and state police and federal authorities will share is minimal, if any. In many cases, their notification to local fire departments will be immediately prior to or after the start of a terrorist incident. If confronted with a B-NICE attack, EMS personnel, who will often be the first units on the scene, will need to identify the situation as a terrorist event.

Implementation of the immediate response stage would set the following events in motion:

- Confirm an actual incident and the type of attack (e.g., B-NICE).
- Determine the area of impact.
- Relocate any impacted units in the area.
- Notify all necessary on-duty and on-call personnel.
- Dispatch requested resources. (Realize that the amount of resources required will overwhelm most departments.)
- Restrict requests for other units in the impacted area unless approved by the IC.
- Alert mutual-aid and supporting agencies of implementation of stage 3.

Stage 4 Recovery Operations

A return to normal is the last phase. It involves restoration of equipment and securing replacements for equipment that has been contaminated or used during the medical treatment of those injured. The obtaining of PPE must be done. Contracting for the disposal of the material used in the decontamination process must be completed. There probably will be a need for critical incident stress debriefing.

These incidents will be quite expensive, and the need to recover costs will be an essential part of the operation. This will require proper documentation throughout the incident. If the documentation has been followed properly, then determining which agency or jurisdiction will cover the costs can be dealt with later. The federal government will strongly support the recovery phase, with help from local governments.

RESPONSE CONCERNS

Response to suspected terrorist acts demands alertness and caution. Safety of firefighters requires that we follow a judicious path. Treat each call as a true threat and wear the proper protective gear. Realize a terrorist group can be watching your actions and reactions to any false alarms to assist them in causing the most damage possible during an ensuing event.

SIZE-UP FACTORS FOR TERRORISM INCIDENTS

Water

- Terrorism incidents may require large volumes of water for decontamination or to fight a fire.
- Water can be used to disperse fumes or knock down plumes of chemicals being released.

Area

- The affected area may be difficult to discern and could be quite large.

Life Hazard

- Individuals can be exposed to explosions or harmful products that can threaten their health and safety.
- Look for indications of terrorist acts: multiple seizures, numerous sick people in one place, an explosion in a public building, or a group claiming responsibility as companies are responding to an incident.
- Responders should not touch or walk in any spilled product. Eating or smoking in the affected area could cause the product to enter the body. Full PPE must be utilized.
- Evacuation concerns will need to be considered. If a chemical was released, it may already be dispersed, or product and residue could remain.

Location, Extent

- A fire could involve an extensive area.
- Hot, warm, and cold zones must be defined, constantly reevaluated, marked, and enforced.
- Extensive decontamination and cleanup may be required if there has been a chemical release.
- Do vapor clouds exist? Is there extensive property damage? Is there widespread damage?

Apparatus, Personnel

- A multi-agency response will be needed. Additionally, the need for mutual-aid assistance could be tremendous.
- There may be an immediate need for EMS personnel for civilian casualties and emergency responders. A mass casualty situation could exist.
- Decontamination trailers will be needed to handle both civilians and responders.
- If a chemical is involved, the specialized knowledge of hazmat personnel may be required. A decision must be made whether initial responders can act to mitigate the situation or whether they must wait for the Hazmat Unit personnel.
- Decontamination must be set up before firefighters enter areas contaminated with dangerous chemicals.
- Sand, absorbents, or diking material may be needed.

Construction/Collapse

- A terrorist attack occurring within or alongside a building can compound the problem. If a bomb has already detonated, the building may be unstable. Collapse must be considered.

Exposures

- These incidents can involve the spread of toxic fumes or gases for great distances from the release site.
- Explosions may cause damage to nearby buildings.
- Acts of terrorism on public transportation (trains, subways, buses, etc.) can threaten many types of structures.

Weather

- Dispatch should transmit the current and anticipated weather conditions to the responders.
- Rain can react with a spilled material, flush it into sewers or streams, or assist in diluting it. Wind can spread a plume of toxic material or assist in dispersing it. Cold weather can keep the temperature below a product's flash point and minimize vaporization. Hot weather can cause a material to reach or exceed its flash point, allowing ready vaporization.

Auxiliary Appliances
- Sprinklers may be used to mitigate a spill or fire. Sprinklers may cause a runoff problem that will need to be addressed.

Special Matters
- There is a possibility that secondary devices were placed to harm emergency responders.
- Are there dead animals? Are there unusual odors? Are there unusual conditions? Are there unusual packages that could be bombs or secondary devices?
- Water used for decontamination needs to be contained and disposed of properly.

Height
- Acts of terrorism can occur within or outside of any building, although high-rise buildings, with their many occupants, are most vulnerable.

Occupancy
- Terrorist acts will often occur in places of assembly or on modes of transportation where they will have the greatest impact and the potential for numerous injuries exists.

Time
- The time of occurrence will most likely be when a large number of civilians can be affected, and it will cause a negative impact on emergency responders.

CONSIDERATIONS FOR TERRORISM INCIDENTS

Strategic Goals and Tactical Priorities for an Offensive Attack
- Obtain an assessment of the fire or incident area as soon as possible.
- When an assessment has been made, immediately call for additional resources, including EMS and other agencies that can assist.
- If a fire is involved, attempt to contain it to the area of origin.
- Determine whether the firefighters are qualified to handle the problem or whether Hazmat technicians are needed.
- Evacuate occupants and have decontamination set up, if necessary.
- Perform search and rescue, if needed.
- Ventilation of a building may be required. Try to utilize the HVAC unit to remove toxic fumes or smoke if you are assured that it can be sent directly to the exterior without contaminating other interior areas.
- Consider wind direction and velocity.
- Arrange for runoff water and decontamination water to be captured and disposed of properly.

Incident Management System Considerations/Solutions for an Offensive Attack
- Incident Commander
- Unified Command (if needed)
- Safety Officer(s)
- Liaison Officer
- Rapid Intervention Crew(s) (The level of protection of the crews must be equal to that of those operating within the hazardous areas.)
- Intelligence/Investigations Officer
- Operations (if needed)
- Hazardous Materials Group or Branch
- Logistics
- Base
- Staging
- Ventilation Group
- Search and Rescue/Evacuation Group
- Division(s) (This could involve interior or exterior areas.)
- Medical Group or Branch
- First-Aid Stations

Strategic Goals and Tactical Priorities for a Defensive Attack

- If changing from an offensive to a defensive attack, ensure that a PAR is taken.
- Reassess the hot, warm, and cold zones and adjust if necessary.
- Set up and maintain a collapse zone, if necessary.
- If using water streams, consider unmanned master streams.
- Protect exposures, if necessary.

Strategic Goals for Nonintervention

- If it is too dangerous to attempt to mitigate the incident in a defensive mode, then all responders must be removed to a safe distance from the incident. The scene can be monitored from a distance via binoculars or unmanned monitors, and the fire, spill, leak, or secondary device allowed to run its course prior to allowing emergency responders to return to the scene. The bomb squad may assist with reconnaissance of the site by remote-controlled robots. This can take an extended period of time and should not be rushed.

Incident Management System Considerations/Solutions for a Defensive Attack and Nonintervention

- Incident Commander
- Unified Command (if needed)
- Safety Officer
- Liaison Officer
- Public Information Officer
- Intelligence/Investigations Officer
- Operations
- Staging
- Logistics
- Base
- Hazardous Materials Group or Branch
- Rapid Intervention Crew(s) (The level of protection of the crews must be equal to that of those operating within the hazardous areas.)
- Divisions on the exterior on Sides A, B, C, D
- Exposure Divisions (if needed)
- Medical Group or Branch
- First-Aid Stations

Dirty Bombs

There are numerous types of terrorist events that could be directed against civilians and emergency workers. One serious threat is the detonation of a **dirty bomb**.

There have been no reports of a dirty bomb ever being used. An incident occurred in Moscow in 1995 where rebels buried a cache of radiological materials in Moscow's Ismailovsky Park. The press was notified and found cesium in a container partially buried. No explosion occurred and the device was removed.

In the United States, an American citizen was arrested in June 2002 in Chicago's O'Hare Airport on suspicion of planning to build and detonate a dirty bomb in an American city. The suspect was thought to have undergone training in the mechanics of dirty-bomb construction.

As with any type of emergency that visits our communities, it will be firefighters who will be on the front lines during the initial stages of an incident. This will place them in the line of fire, and their decisions and actions will assist in saving lives and in mitigating the problems found.

NUCLEAR BOMB VERSUS DIRTY BOMB

There is a distinct difference between a nuclear bomb and a dirty bomb. A nuclear bomb involves a fission reaction. A dirty bomb is a homemade bomb that uses conventional explosives and contains radioactive material that is intended to be dispersed as the bomb explodes. A dirty bomb is referred to as a *radiological dispersion device*. The concept is to blast radioactive materials into the area around the explosion. The intent is to cause

dirty bomb ■ A homemade bomb that uses conventional explosives and contains radioactive material that is intended to be dispersed as the bomb explodes.

damage from the explosive force of the bomb that can cause injury and death to those in the immediate area, and to expose people to radioactive material. The bomb's purpose is to frighten people and leave the buildings and land unusable for a long period of time.

A dirty bomb can be in the form of a pipe bomb, constructed from a paint can, or a large bomb that is assembled in a truck. The detonation device can be dynamite, plastic explosives, or ammonium nitrate. Any type of blasting cap attached to switches, timers, or cell phones can be used to trigger the bomb.

Another method of attempting to spread radioactivity may be placing a bomb on a vehicle or train containing radioactive material that when exploded will disperse the radioactive contents, contaminating the surrounding area.

RADIOACTIVE MATERIALS

The radioactive materials that can be used in a dirty bomb can be obtained from military, industrial, or medical applications. Though weapons-grade plutonium and uranium would be most deadly, they are the hardest to obtain and the most difficult to handle. The most likely radioactive materials for these devices would be cesium, cobalt, and iridium isotopes. These materials are widely used for industrial applications in labs, hospitals, and factories. Typically there is minimal, if any, security and they can be easily obtained by theft. There have been reports of missing radioactive materials in both the United States and Russia. Terrorists would be receptive to the possibility of buying these materials.

Some experts have identified cesium-137 as the most likely radioactive element to be used in a dirty bomb. It is created as a byproduct of nuclear reaction. It has a wide variety of uses from treating cancer to maintaining atomic clocks. Cesium is the most reactive metal found. It easily attaches to many materials, including roofing materials, concrete, and soil. As cesium decays to a nonradioactive form of barium, the isotope emits gamma radiation. These rays are extremely difficult to contain. Only concrete, steel, or lead can keep gamma radiation in check.

Because radiation cannot be seen, smelled, felt, or tasted by humans, anyone at the scene of an explosion will not know if radioactive materials were involved.

BOMB CONSTRUCTION

It takes little experience to assemble a dirty bomb other than knowing how to build a conventional bomb. The explosive is intended to disperse the radioactive material in the bomb. The only restrictive part is acquiring and handling the radioactive material. Improper handling of this material would seriously endanger the bombmaker. Though many terrorists are suicide bombers, high levels of radioactivity could sicken and kill anyone handling these materials.

A case that illustrates the danger of handling radioactive material occurred in Goiaina, Brazil, in September 1987. A worker found a discarded canister in a scrap yard that contained a sparkling blue powder that turned out to be radioactive cesium. Local residents found the powder very interesting and passed it from one household to another. More than 200 people were exposed to the cesium, resulting in four deaths. The radioactivity contaminated soil, businesses, and 85 homes that had to be leveled in the cleanup process.

STRATEGIC CONSIDERATIONS

Dirty bombs are not intended for mass destruction, but more intended to cause economic destruction. It has been said they target mass disruption and panic, more so than mass destruction.

Responders should approach any suspected or confirmed bomb site with caution. Attempt to position personnel upwind and uphill of the site if possible. Once a bomb is detonated, size-up should be performed to assess the potential damage and threat to civilian and response personnel.

Ensure that all personnel utilize SCBA or respirators to prevent the inhalation of radioactive dust that may be present. If a dirty bomb is suspected, establish hot, cold, and warm zones to protect firefighters and civilians. Any open wounds or cuts should be protected from radioactive contamination. Eating, drinking, or smoking should not be permitted while exposed to contaminated dust or smoke.

The immediate protection from any radiation exposure is time, distance, and shielding. The less time spent in contact with radiation, the better. The farther away people are from the radiation, the less their chances of being affected by it. Shielding, if available, will protect a person. Shielding can come in many forms including masonry, lead, and steel. Maximize the amount of shielding by using dense buildings between you and the bomb site.

When anticipating potential injuries with a dirty bomb, expect that few people will be subjected to acute doses of radiation in the short term. However, anyone remaining in the contaminated area will be subject to increased risk of developing cancer in the future. Since there is little immediate risk from the contamination, those near the dirty bomb should try to remain calm and exit the area. Panic could prove more deadly than the bomb.

Once a bomb is detonated, fire and bomb squad personnel should use radiation-monitoring devices to identify the presence or lack of radiation. These units should be kept in plastic bags to prevent their contamination. The monitoring devices should be used to identify areas of the highest dose rates. If radioactive readings are found, the IC must determine the need for personnel to remain in the area for firefighting or search and rescue. Areas of high dose rates should be avoided except to save lives, in which entry should be as brief as possible.

Firefighters can be equipped with individual alerting and monitoring devices to identify the presence of radioactivity. Some bomb squads have made it mandatory that each member wear an alerting device during investigations. Squads that utilize dogs to search for bombs have found that attaching an alerting device to their collars can reduce human exposures due to the early warning. Remote-controlled robots can also be fitted accordingly.

Once radiation is found, decontamination should be set up immediately and measures enacted to remove and replace contaminated clothing of civilians and response personnel. This will assist in eliminating the radioactive dust that may have accumulated on clothing. All clothing should be bagged for later disposal.

ON-SCENE CONSIDERATIONS

Whether a bomb is intended strictly as a destructive device or as a dirty bomb does not alter the actions of firefighters and bomb squad personnel. Firefighters encountering situations where reports of explosive devices are suspected or found fall into two basic categories: exploded and unexploded devices. This differentiation determines the actions of the fire department and the bomb squad.

An unexploded device should be viewed as a police scene and should be left to the bomb squad and not be handled by firefighters unless they are specifically trained as bomb squad members.

With an unexploded device, typical firefighter actions should be:

- Set up a safety zone around the suspected site. Maintain a clear zone of at least 500 feet from the location of the suspected device. This should be considered an exclusion zone that firefighters should not enter.
- Ready hose-lines to be placed into operation should the device detonate.
- Attach hose-lines to siamese connections that supply sprinkler and standpipe systems. Firefighters should be prepared to pressurize the systems if needed.
- Do not use cell phones and portable radios within the safety zone, since there is the possibility that radio waves could trigger the explosive devices.
- Establish communications with the dispatcher outside of the safety zone.

NEEDS ASSESSMENT

An exploded device should be handled as a fire and hazmat situation. The need for fire control and rescue must be assessed. This includes determining building damage and the potential for structural collapse, and finding the safest way possible to accomplish firefighting and search and rescue.

Considerations must be given to the possibility of any secondary devices or the presence of radioactive material that would be contained in a dirty bomb. If a dirty bomb is suspected, then hot, warm, and cold zones should be established.

PERSONAL PROTECTIVE GEAR

Typically the radiation that could be expected at the site of a dirty bomb would be reasonably low. A firefighter's PPE and SCBA would provide sufficient protection at low levels for a prolonged period.

Even at elevated levels, the firefighter's PPE will permit operating within an area for a few minutes. This should be enough time to permit a primary search and the removal of injured civilians.

FIREFIGHTERS EXPOSED TO RADIATION

Emergency exposures are usually allowed to exceed those tolerable to persons who work continuously with radioactive materials. In an emergency, such as a rescue operation, raising the exposure—within limits—for a single dosage is considered acceptable.

For a lifesaving action such as search and rescue, the removal of injured civilians, or entry to prevent conditions that would injure or kill numerous persons, the planned dose to the whole body should not exceed 100 rems.

During less stressful circumstances, where it is still desirable to enter a hazardous area to protect facilities, eliminate further escape of effluents, or control fires, the recommended planned dose to the whole body should not exceed 25 rems. These rules apply to a firefighter for a single emergency; further exposure is not recommended.

> Rem is known as a unit of dose equivalence and is an acronym for *roentgen equivalent man*. One rem involves the same risk regardless of the type of radiation, but the dose required to produce one rem may vary depending upon the type of radiation.

PROTECTING THE INJURED AND EXPOSED

If a dirty bomb has detonated in a public assembly building, firefighters must anticipate a mass casualty incident. Seriously injured people should be removed from the source of radiation, decontaminated, stabilized, and sent to hospitals. Those with lesser injuries can then be decontaminated, triaged, treated, and transported to hospitals.

After treatment of serious physical injuries, preventing the spread of the radioactive material or unnecessary exposure of other people is paramount. Carry out the following immediate response actions:

■ Establish an exclusion zone around the source. This should be of significant distance to anticipate the spread of the radioactivity. It is easier to shrink the zone at a later time than to expand it.
■ Mark the area with ropes or banner tape.
■ Utilize police to control and reroute traffic.
■ Limit entry to rescue personnel and strictly monitor each firefighter's exposure time.
■ Detain uninjured people who were near the event or who are inside the control zone until they can be checked for radioactive contamination.
■ Limit or stop the release of more radioactive material, if possible, but delay cleanup attempts until radiation protection technicians are on the scene.
■ Notify nearby hospitals of the incident, the type of injuries involved, and the possibility of the arrival of radioactively contaminated and injured people.
■ Check everyone near the scene for radioactive contamination.

Record-keeping is important for the long-term health of the victims and the emergency responders. Record contact information for all exposed people for future follow-up.

CLEANUP OF RADIATION SITES

A dirty bomb that is detonated in a populated city will probably not be the cause of immediate deaths, but the resulting cleanup problem can be tremendous. The cleanup at radiation sites consists of removing the layers of contamination for disposal. This process can involve demolition of structures, sandblasting the faces of buildings that are contaminated, and removing soil. This material then must be properly disposed of, since it will be many years before the radioactive contamination will decay. In reality, radiated sites cannot be decontaminated; the material can only be transferred to another site.

INCIDENT COMMAND ORGANIZATION

The complex situation created by a dirty bomb explosion will require that an incident management organization be initiated immediately. There will be the potential for establishing a Unified Command.

The typical response of the fire department and bomb squads will require close coordination. In addition to fire and police, many other agencies will respond and can be invaluable in the handling of the incident. These could include:

- Public works departments
- Health departments
- Companies to remove the decontamination materials
- Private waste cleanup companies
- American Red Cross
- County/state medical
- Mutual and automatic aid
- Environmental Protection Agency
- US Coast Guard
- Federal Emergency Management Agency
- Military
- Department of Homeland Security
- Federal Bureau of Investigation
- Department of Defense
- Bureau of Alcohol, Tobacco, Firearms, and Explosives

There will be numerous major functions occurring simultaneously that may dictate the need for branches. These could include Medical Branch, Suppression Branch, Hazmat Branch, and Evacuation Branch. Multiple operations in various areas can dictate the need for groups and divisions. They could include a Decontamination Group and a Rescue Group.

A major concern will be coordination between fire department units and the bomb squad members operating at the scene. Early consideration should be given to establishing a Liaison Officer. There will also be a demand for Information, Planning, Intelligence/Investigations, and a SO to assist the IC(s).

SIZE-UP FACTORS FOR A DIRTY BOMB INCIDENT

(These factors are very similar to a response to a hazardous material, yet some differences do exist.)

Water
- An exploded dirty bomb may require an adequate water supply for decontamination or to fight a fire.

Area
- The affected area may be difficult to discern and could be quite large.

Life Hazard
- Individuals can be exposed to explosions or harmful products that can threaten their health and safety.
- Responders should not eat or smoke in the affected area. It could cause the radioactive product to enter the body. Full PPE must be utilized.
- Evacuation concerns will need to be considered.
- A firefighter's protective clothing with SCBA or respirator will provide a measure of protection from radiation exposure.

Location, Extent
- Utilize radiological meters to determine the extent of the contaminated area.
- An extensive area of property damage and contamination could require massive decontamination and cleanup.
- Hot, warm, and cold zones must be defined, constantly reevaluated, marked, and enforced.

Apparatus, Personnel
- A multi-agency response will be needed.
- The need for mutual aid could be tremendous.
- There may be an immediate need for EMS personnel for civilian casualties and emergency responders.
- A mass casualty situation could exist.
- Decontamination will be needed to handle both civilians and responders.
- The Hazmat Unit personnel can assist in determining if radioactive elements were employed in the bomb.
- Decontamination must be set up before firefighters enter contaminated areas.
- Sand, absorbents, or diking material may be needed if there is a runoff problem.

Construction/Collapse
- A terrorist attack occurring within or alongside a building can be compounded if the bomb has left the building unstable.

Exposures
- There is the potential for the spread of radioactive materials to nearby buildings.
- Explosions may damage nearby buildings.

Weather
- Dispatch should transmit the current and anticipated weather conditions to the responders.
- Rain can cause radioactive materials to spread to other areas, including the sewer systems.

Auxiliary Appliances
- Sprinklers may be used to mitigate a spill or fire caused by the exploding bomb.
- Sprinklers may cause runoff problems.

Special Matters
- There is a possibility that secondary devices are placed to harm emergency responders.
- Check for unusual packages that could be a secondary device.
- Cell phone calls or radio transmissions could trigger an explosion.

Height
- A dirty bomb can be detonated in or outside of any building. Naturally, a taller building will compound the fire department's problems.

Occupancy
- Terrorist acts involving a dirty bomb can occur in places of assembly or on modes of transportation. The terrorist targets wherever it will have the greatest impact and the greatest potential for injuries and deaths.

Time
- The time of occurrence will most likely be when a large number of civilians can be affected and it will cause a negative impact on emergency responders.

CONSIDERATIONS FOR A DIRTY BOMB INCIDENT
Strategic Goals and Tactical Priorities for an Offensive Attack
- Obtain an assessment of the fire or incident area as soon as possible.
- Once an assessment has been made, immediately call for additional resources, including EMS and other agencies that can assist.
- If a fire is involved, attempt to contain it to the area of origin.
- Determine whether the firefighters are qualified to handle the problem or whether hazmat technicians are needed.
- Evacuate occupants and have decontamination set up, if necessary.
- Perform search and rescue, if needed.
- Consider wind direction and velocity.
- Arrange for runoff water and decontamination water to be captured and disposed of properly.

Incident Management System Considerations/Solutions for an Offensive Attack

- Incident Commander
- Unified Command (if needed)
- Safety Officer(s)
- Liaison Officer
- Intelligence/Investigations Officer
- Rapid Intervention Crew(s) (The level of protection of the crews must be equal to that of those operating within the hazardous areas.)
- Operations (if needed)
- Hazardous Materials Group or Branch
- Logistics
- Base
- Staging
- Ventilation Group Supervisor
- Search and Rescue/Evacuation Group
- Division(s) or Branch(es) (This could involve interior or exterior areas.)
- Medical Group or Branch
- First-Aid Stations

Strategic Goals and Tactical Priorities for a Defensive Attack

- If changing from an offensive to a defensive attack, ensure that a PAR is taken for accountability purposes.
- Reassess the hot, warm, and cold zones and adjust, if necessary.
- Set up and maintain a collapse zone, if necessary.
- If using water streams, consider unmanned master streams.
- Protect exposures, if necessary.

Strategic Goals for Nonintervention

- If it is too dangerous to attempt to mitigate the incident in a defensive mode (due to a number of factors, including the discovery of multiple unexploded devices in addition to the exploded devices), then all responders must be removed to a safe distance from the incident. The scene can be monitored via binoculars or unmanned monitors and the incident, or secondary device(s), allowed to run its course prior to allowing emergency responders to return to the scene. The bomb squad may assist with reconnaissance of the site by remote-controlled robots. This could take an extended period of time.

Incident Management System Considerations/Solutions for a Defensive Attack and Nonintervention

- Incident Commander
- Unified Command (if needed)
- Safety Officer
- Liaison Officer
- Public Information Officer
- Intelligence/Investigations Officer
- Operations
- Staging
- Logistics
- Base
- Hazardous Materials Group or Branch
- Rapid Intervention Crew(s) (For surrounding structures. The level of protection of the crews must be equal to that of those operating within the hazardous areas.)
- Divisions or Branches on the exterior on Sides A, B, C, D
- Exposure Divisions or Branches (if needed)
- Medical Group or Branch
- First-Aid Stations

Clandestine Drug Labs

The production, distribution, and use of illegal drugs have a direct impact on public safety. Law enforcement agencies spend a tremendous amount of time and taxpayers' dollars in an attempt to police these illegal activities. Fire department responses caused by drug-inflicted problems reduce the available resources for firefighting and other emergency responses. **Clandestine drug labs (CDLs)** are illegal laboratories ranging from primitive to highly sophisticated facilities for the production of illegal drugs such as methamphetamines.

Large quantities of illegal drugs are produced in clandestine drug labs (CDLs) each year in the United States. Though this includes a variety of designer drugs, in recent years 98 percent of the labs seized were manufacturing methamphetamine.

clandestine drug labs ■ Illegal laboratories ranging from primitive to highly sophisticated facilities for the production of illegal drugs such as methamphetamines.

METHAMPHETAMINE

Methamphetamine is a potent central nervous system stimulant that affects neurochemical mechanisms responsible for regulating heart rate, body temperature, blood pressure, appetite, attention, mood, and responses associated with alertness or alarm conditions. Most commonly it is found as a colorless crystalline solid. Impurities may result in a brownish or tan color. An impure form of methamphetamine is sold as a crumbly brown or off-white rock commonly referred to as *peanut butter crank*. Other nicknames for methamphetamine are numerous and vary significantly from region to region. Some common nicknames include: *ice, crystal, meth, crystal meth, crank, glass, chalk, tweak, uppers, black beauties, glass, bikers coffee, methlies, quick, poor man's cocaine, chicken feed, shabu, stove top, trash, go-fast, yaba, yellow bam,* and *speed*.

A primary source of methamphetamine has been the super labs operating in the United States and Mexico, but the smaller drug labs account for the majority of the CDLs seized in the United States. In a recent 10-year period, an average of more than 11,000 methamphetamine laboratory incidents per year were reported to the US Drug Enforcement Administration (DEA). The increase in the number of CDLs is due to the ease of securing the chemicals and the fact that the equipment used is basic and unsophisticated. These incidents include labs, dumpsites, or chemical and glassware seizures. These CDLs present a significant threat to the health and safety of firefighters, police officers, and other emergency responders.

Public safety personnel face a high potential of acute and chronic health risks when involved in the seizure and handling of the products and residues of these labs. The public is likewise threatened by the potential hazards due to the volatility of the chemicals and the ensuing pollution from the waste that is produced. It is estimated that for every pound of methamphetamine produced, five to six pounds of toxic waste is produced. This waste is discarded with no concern for environmental issues or the potential for exposure to children who may come in contact with it.

METHAMPHETAMINE PRODUCTION

CDLs are more dangerous than a legal laboratory or chemical manufacturing plant. Those facilities have safety equipment and procedures, fire-suppression measures, appropriate ventilation, and chemical-handling equipment on location. CDLs range from primitive operations to highly sophisticated laboratories similar to those found in modern testing facilities.

Methamphetamine labs can be found anywhere in the United States in both urban and rural areas: in houses, barns, apartments, trailers, campers, cabins, motel rooms, vacant buildings, and isolated wildland areas. The storage of the chemicals used in the production and the equipment may be kept in self-storage centers, and in a number of cases, the production of methamphetamine has taken place in these rented self-storage centers. CDLs may be located in remote areas to avoid scrutiny. Mobile methamphetamine labs have been found in trailers and vehicles as they drove down highways and through cities. The volatility of these processes is compounded when concocted in a moving vehicle and increases the possibility of explosions and fires. Those who manufacture methamphetamine are often harmed by toxic gases, especially when produced in the confines of a moving vehicle.

Those involved in the production of methamphetamine will go to extreme lengths to protect their territory. A serious concern is the presence of numerous guns that have been found in labs run by gangs. The guns are meant for protection from rival gangs, but criminals can easily turn them onto the firefighters to protect their identities and afford them time to escape before arrival of the police. Sophisticated monitoring systems may be in place in the labs, and the operation may be booby-trapped. Each clandestine drug lab is unique and presents a number of hazards. A large variety of chemicals could be present, most of which are highly flammable and toxic. The operator may be an experienced chemist who utilizes sophisticated equipment, or a novice who employs nothing more than primitive or makeshift devices. The operators of these illegal labs may be very astute in preparing the methamphetamine or may be operating from directions passed on to them from others, or from recipes found on the Internet.

The chemicals used in the manufacture of methamphetamine can generally be listed as containing dangerous characteristics, such as being flammable, explosive, irritants, corrosive, carcinogenic, water reactive, and causing rapid asphyxia. They can give off noxious gases and vapors. A specific concern exists if red phosphorus is used and it overheats due to improper ventilation. This condition can lead to the creation of large quantities of explosive gas.

The common methods of production of methamphetamine in the United States include:

- The "Red, White, and Blue Process," which involves red phosphorus, pseudoephedrine or ephedrine (white), and blue iodine.
- The "Birch Reduction" uses metallic sodium or lithium. The lithium is commonly extracted from non-rechargeable lithium batteries. This process is dangerous because the alkali metal and liquid anhydrous ammonia are both extremely reactive, and the temperature of liquid ammonia makes it susceptible to explosive boiling when reactants are added.
- Though less common, there are methods that use other means of hydrogenation, such as hydrogen gas in the presence of a catalyst.
- In recent years, a simplified "Shake 'n Bake" method was developed. The method uses chemicals that are easier to obtain, though no less dangerous than traditional methods. It is an easier, cheaper, and faster way to produce methamphetamine and can be produced in 30 minutes. A two-liter soda bottle is used along with crushed pseudoephedrine pills, and some household chemicals. The ingredients are placed in the bottle and shaken, resulting in a crystalline form of methamphetamine. This process can produce a powerful explosion, touch off intense fires, and release drug residue that must be handled as toxic waste. The instability of this process is evident because if there is any oxygen in the bottle, it has a propensity to make a giant fireball. Unscrewing the bottle cap too fast can result in an explosion. There have been numerous reports of flash fires and fatalities due to this process. Police have found it being made in moving vehicles. The plastic bottles, which can contain toxic, explosive, or flammable residual chemicals, are discarded.

 ON SCENE

Police in Tennessee found a driver passed out in his car at a gasoline pump. In the back seat, a batch of methamphetamine was actively cooking and going through a chemical process.

Firefighters in Texas fought a vehicle fire that was fully involved. Methamphetamine lab components were discovered in the charred remains.

Police in Georgia stopped a car for a routine traffic stop. As the officer approached the vehicle, she noticed a strong smell of chemicals. A blender and a cooler were visible in the back seat and about a dozen electrical plugs coming from the cigarette outlet. A K-9 dog was alerted to the scent of drugs and a response from the DEA found hazmat and methamphetamine in the vehicle. The driver was manufacturing methamphetamine inside the car and was arrested.

In Kentucky, police pulled over a driver and found three separate containers of methamphetamine brewing in the car, one in the backseat and two in the trunk. A vehicle accident would most likely have been followed by an explosive release of the contents of the methamphetamine containers. The containers were dismantled by the DEA.

CUES THAT CAN IDENTIFY A CLANDESTINE DRUG LABORATORY

There are cues that can assist in identifying a meth lab. Some of this information can come from size-up at an incident, while other information can come from complaints by neighbors who may have observed:

- Unusual, strong odors (like cat urine, ether, ammonia, acetone, or other chemicals)
- Excessive trash, including large amounts of items, such as antifreeze containers, lantern fuel cans, red chemically stained coffee filters, drain cleaner, and duct tape
- Unusual amounts of clear glass containers being brought into the home
- Windows blacked out or covered by aluminum foil, plywood, sheets, blankets, etc.
- Secretive/protective area surrounding the residence (like video cameras, alarm systems, guard dogs, reinforced doors, electrified fencing)
- Little traffic during the day, but high traffic at late night hours, including different vehicles arriving and staying for short periods of time

Clandestine Drug Laboratories May Be Categorized As:

- *Active laboratory (hot lab):* One in which the chemicals and equipment necessary to produce an illegal drug are present, and a chemical reaction is taking place. Cooling by water, or heating electrically or by open flame, is frequently observed in active laboratories.
- *Inactive lab:* One in which the chemicals and equipment necessary to produce illegal drugs are present and assembled, but no reaction is taking place.
- *Abandoned lab:* Materials and equipment are present and disassembled. These materials may be stored in a dangerous manner and contaminated with unknown chemicals.

PLANNING THE TACTICAL OPERATION

When a law enforcement agency receives intelligence of a CDL, it should plan a tactical operation to seize the lab. This planning process is a police responsibility, and their assessment will determine which agencies should respond. The police agencies should have training in handling the associated hazards. If they have not been trained, they should contact the DEA to assist in the seizure of the laboratory. The fire department should have a representative present during the planning process to advise the police on the best utilization of the fire department's resources. The decision on how to proceed with the seizure will be based upon the intelligence received on what the site contains. It could include active labs, inactive labs, abandoned labs, or chemical storage areas. Once this is known, the response can be based upon the degree of hazard anticipated. The factors involved will depend upon the location of the specific site, the amount and types of chemicals, and their concentrations and proximity to each other. The response to a CDL can include the bomb squad, DEA, fire department, Hazmat Unit(s), emergency medical personnel, and a hazardous waste contractor.

Safe operating procedures at a CDL for police operations would include:

- Avoiding the use of weapons or diversionary devices such as flash bangs, smoke, or tear gas canisters because they can ignite fumes present at the lab.
- Do not turn switches on or off; unplug cookers, heating elements, or cooling equipment; open refrigerators or freezers; or move containers that are in the way because they could be booby-trapped.
- Do not use matches or flames of any kind.
- Use an explosion-proof flashlight to look in dark areas.
- Do not taste, smell, or touch any substance.
- Do not eat, smoke, or drink at the site.
- Do not touch your mouth, eyes, or other mucous membranes with your hands.
- Decontaminate clothing, equipment, and personnel before leaving the laboratory site.

INITIAL ENTRY TEAM

The purpose of the initial entry is to secure the scene and apprehend any and all suspects in the lab. The initial entry team must be specially trained law enforcement personnel who will secure the site as the arrests are being made. Because of the physiological/psychological side effects of methamphetamine abuse, users are often highly paranoid and may not be rational. Proper protection is required by the team, and their time spent in the lab should be minimal. Their back-up team or RIC must be in place prior to the entry by the initial entry team. That team should be composed of DEA-trained law enforcement personnel and not fire personnel.

ASSESSMENT

The assessment is carried out once the building has been secured through the arrest and/or evacuation of all occupants. The assessment is conducted by qualified personnel who have been certified in CDL operations by the DEA. It should include law enforcement officer(s) and a forensic chemist, all who need to be in the appropriate level of PPE. By sampling the atmosphere within the lab, they will determine the explosive limits, oxygen levels, and the extent of toxins present. Their inspection will determine any immediate health and safety risks. Once readings are taken, identification of all chemicals and equipment present in the lab and the hazard that may exist will be performed. It will determine the air levels and whether air purifying respirators can be used in lieu of SCBAs.

DEACTIVATION OF THE LABORATORY

Deactivation of the lab will be accomplished by law enforcement personnel and the forensic chemist. This will include fully shutting down any active cooking processes after thorough analysis to ensure the safety of all responders. It can include closing valves on compressed gas, and ensuring that vacuum systems are properly shut down.

If necessary, they will perform ventilation of the lab and surrounding areas after ensuring that windows and doors are not booby-trapped.

In addition to analyzing for the potential of fire or explosion, they will investigate for the presence of boobytraps on the premises. They are responsible for identifying, documenting, and collecting physical evidence, including taking comprehensive photographs of the operation and the chemicals present.

The backup team or RIC(s) requires the same protection as the assessment team. They may include either fire or police personnel, or a combination of both. Fire personnel are not required to be DEA-certified becausetheir responsibilities prohibit them from being members of an initial entry or assessment team. In the event of a fire and/or explosion, fire personnel assigned to backup safety teams will protect escape routes and make rescues if needed.

BOOBY TRAPS

Booby traps can be well-disguised and are meant to counter an attack by anyone entering the lab. They can be intended for public safety personnel or for rivals intending to steal the contents of the lab. Booby traps may be set to destroy the evidence of the operation. The traps can be very basic or exotic and in the past have included:

- Attack dogs
- Poisonous snakes
- Trip lines to detonate chemicals or explosives
- Light switches that trigger flammable liquid containers
- Refrigerators or freezers triggered to ignite flammable liquids when opened
- Hydrogen cyanide gas generators
- Videotape cassettes that have been adulterated to explode when placed in a VCR
- Various explosives including contact explosives, which are made by combining potassium chlorate or red phosphorous with another chemical that becomes unstable when dried. The chemicals are rolled into a ball of aluminum foil and placed throughout the lab. If disturbed, the ball will explode.

POLICE REQUEST FOR THE FIRE DEPARTMENT AT A DRUG LAB SEIZURE

When through the course of everyday operations police discover a CDL, they will request assistance from other agencies including the fire department. The fire department response can include a chief officer, a Hazmat Unit, engine and truck companies, and emergency medical technicians, or paramedics.

Depending upon the command structure created, the fire department may become part of Unified Command, together with the police and DEA. However, because CDLs are primarily police matters, another possibility is for the law enforcement agency to assume the lead role as the IC and have the fire department act as an assisting agency. Regardless of the incident command structure, all activities of the fire department will be coordinated with the police.

FIRE DEPARTMENT CHIEF OFFICER'S ROLE AT A CLANDESTINE DRUG LAB

The first-arriving fire department chief officer will confer with the Police Incident Commander. Based upon whether the lab is active or inactive and the pertinent information available, he/she will:

a. Assist in determining the extent of the area to be evacuated.
b. Advise on the size of the boundaries between hot, warm, and cold zones.
c. Ensure that a sufficient number of hose-lines are stretched to unmanned monitors to adequately cover the property containing the lab and the exposures.
d. At the scene of an inactive lab, determine if a single engine response is sufficient for handling the incident. Although a chemical process is not taking place, a cold lab should be considered dangerous due to the possible presence of unstable, flammable, or toxic materials.

The fire department chief officer, in conjunction with the hazmat officer, can advise the Police Incident Commander of the safety considerations necessary to protect all personnel and the general public. They can confer with the chemist and assist him/her in determining the appropriate level of protection necessary for the assessment and processing teams. They will confer with the Police Incident Commander in all decision-making meetings to assure proper coordination between police and fire departments.

 ON SCENE

Heavy fire that started on the first floor of a six-story heavy timber warehouse, 175 by 475 feet, quickly went to multiple alarms and a defensive attack. Fire doors were closed to stop the fire from running the entire length of the building and the tactic proved to be successful. The fire was contained to an area approximately 175 by 150 feet. As the defensive operations were underway, all floors were heavily involved with fire and ground-level and elevated streams were operating. An explosion occurred that created a tremendous white flash. Because there were heavy smoke conditions on the scene, it was difficult to discern exactly what had occurred. The brilliance of the white flash was similar to an electric arc, and at first I thought one of the fire department's elevated platforms had come into contact with overhead electric wires. I was immediately concerned for the safety of the firefighters on the platform. It soon became apparent that the explosion was from the sixth floor, and fortunately, its force was directed upward and not toward the platforms. An ensuing fire marshal's investigation found that a methamphetamine lab had been constructed on the top floor and was the cause of the explosion.

HAZARDOUS MATERIALS UNIT

The Hazmat Unit Officer will ensure that the members of the Hazmat Unit assist the police assessment and processing teams in donning protective equipment and conducting a safety inspection of same before the teams enter the lab. Police teams are required to have training in the use of SCBAs.

The Hazmat Unit Officer should obtain data from the police entry team pertaining to the interior layout of the building. Information should be collected on the storage location(s) of chemicals, the location of the lab, stairways and exits, and any other unusual conditions. After acquiring this information, he/she will ensure that members assigned to backup safety teams are apprised of the same. The Hazmat Unit Officer will determine the staffing level, the number of back-up teams, the proper level of protective clothing for those teams, and the number of water lines needed to safeguard police personnel operating in the lab.

The hazmat team will ensure that all teams exiting the lab undergo proper decontamination procedures.

No overhauling by the fire department should take place in a drug lab. Because drug labs are crime scenes, the preservation of evidence is important to ensure the arrest and prosecution of those criminals who are operating the lab. Only an approved cleanup contractor should conduct removal operations. This can involve the removal of propane cylinders that contain anhydrous ammonia, mercury, radioactive waste, explosives, and other chemicals that could be reactive with water or may react with other chemicals.

MEDICAL RESPONSE TO A CLANDESTINE DRUG LAB

The response of medical personnel to a CDL will be dependent upon the jurisdiction. The minimum should be an ambulance with emergency medical technicians. Ideally, paramedics and an Emergency Medical Control Officer should respond if available.

They will be responsible for the pre- and post-entry vital sign monitoring for all members of the entry and backup teams. They will also assess, stabilize, and transport any injured and/or contaminated civilians and response personnel.

They will set up a triage area in the cold zone. The triage area should be in close proximity to the warm zone. Paramedics/EMTs will take vital signs of all police and fire personnel who don fully encapsulated suits. Readings will be taken prior to entry, and again after exiting, and upon completion of decontamination procedures.

FIREFIGHTERS ENCOUNTERING CLANDESTINE DRUG LABS

While CDLs pose the potential for a significant hazardous material release, they are, in fact, ongoing crime scenes. Approximately 30 percent of all methamphetamine labs discovered in the United States are found as a result of an explosion or fire. This fact alone shows the serious threat posed to firefighters. In addition to an explosion or a fire, there are other instances when firefighters may locate a CDL during:

- Routine building inspections
- Responses to investigations for unusual odors
- A request for the fire department to respond to a medical problem
- An investigation of a building for fire extension, or for occupants during a primary search while working on a fire in an adjacent building, or another part of the building in which the lab is situated
- Emergency medical responses

EMERGENCY MEDICAL RESPONSES

If a drug lab is found while on an emergency medical call, the fire department or ambulance personnel must leave the lab immediately. Victims in a drug lab may be overcome with fumes or have been burned. If possible, remove the patient if you can do so without being exposed to the hazardous chemicals that are present. Once the Hazmat Unit arrives, they will need to decontaminate the personnel, the patient, and all of the equipment that was carried into the lab. In cases where the patient needs an immediate response to the hospital, the hospital should be notified.

FIREFIGHTERS' ACTIONS ON DISCOVERING A CLANDESTINE DRUG LAB

Should firefighters in the course of their duties discover or inadvertently enter a suspected CDL, whether active or inactive, they should consider the situation to be potentially explosive and toxic and:

a. Immediately notify the IC
b. If the CDL is in a motel, hotel, or other type of structure, the firefighters should try containment and immediately evacuate around, above, and below the involved area and possibly extinguish a fire.

c. If using water, utilize nozzles on fog patterns to minimize contamination and runoff.

d. If the building has been evacuated, immediately leave the building. Egress should be through the entry point. Do not open any additional doors. Do not touch anything. Always consider the possibility of booby traps.

e. Do not turn light switches on or off.

f. Do not touch suspected drugs, appliances used to manufacture them, or open refrigerators.

g. Do not unplug cookers, or heating elements, or tamper with their controls.

h. Do not disconnect or reduce the flow of water used for cooling.

i. Do not shut off any utilities.

j. Do not use thermal imaging cameras within these labs because they are not intrinsically approved for use in environments containing certain gases. They may be used on the exterior to identify areas of heat sources.

k. Keep your gear on and use a hose-line to gross decontaminate once outside.

l. Immediately notify the dispatcher of the discovery of the CDL and request the response of the police and/or drug enforcement task force if one is available.

m. As conditions warrant, relocate apparatus a safe distance upwind from the laboratory.

n. Have all responding companies report to an upwind Staging Area.

o. Consider the possibility that criminal suspects may either still be inside the property or in the immediate vicinity. Safety of members shall be of paramount importance at all times.

p. When police have rendered the area secure (e.g., free of suspects), assist in evacuating adjacent properties. If the lab is in an active stage, ensure that all fire department members are wearing full PPE and SCBAs.

Firefighters operating on a scene who discover that the incident is the location of an illegal drug lab should immediately withdraw to defensive positions. Water should be used judiciously, and hand lines should never be directed into areas where chemicals are present. An extremely reactive chemical used in the manufacture of methamphetamines is lithium aluminum hydride, which can explode violently upon contact with the slightest amount of moisture.

Firefighters should be cognizant of the possibility that vapors emanating from the lab are usually narcotic and could be acutely toxic and that anyone operating in the hot zone must employ full protective clothing and SCBAs. When operating in the cold zone, firefighters will wear full protective clothing and SCBAs in the standby position. Once an assignment is completed, companies must check with the Hazmat Unit as to whether they need to undergo decontamination procedures.

INTERIOR SIGNS OF A CLANDESTINE LAB

Firefighters need to recognize the interior signs of a lab by being observant. Indicators could include:

- Laboratory glassware like beakers, flasks, funnels, etc.
- Distillation equipment
- Plastic or rubber tubing
- Laboratory supplies
- Heating apparatus, such as hot plates
- Filter paper, scales, thermometers
- Glass containers/bottles or drums of chemicals
- Chemistry reference books
- Books referencing illegal operation, i.e., *The Anarchist's Cookbook*
- Numerous empty pillboxes or containers
- Amber-colored gallon jugs with no labels or outside identification

FIREFIGHTING

Though drug labs are normally police operations, once a fire starts the fire department takes the lead role to control and extinguish the fire. Firefighting operations at drug labs are dangerous due to the hazards and instability of the chemicals involved. If a task force consisting of police, fire, and DEA are already operating at the scene of a suspected CDL when a fire occurs, the prime objective is to safely remove everyone from the building. Care must be taken to minimize the use of water and restrict the number of firefighters entering the building. Once evacuation is completed, a working fire should be handled by going to a defensive attack in which the fire is allowed to burn and the protection of exposures is initiated. The threat of attempting to extinguish these fires is a danger to firefighters due to the potential for explosions, toxic smoke, and contaminated runoff that will cause pollution to waterways. Firefighters should set up fixed monitors to protect the surroundings and then be withdrawn from the hot zone.

If firefighting operations are underway when it is realized that a drug lab is involved, the fire attack teams should be withdrawn immediately. Use hose-lines to gross decontaminate personnel. Set up for a defensive attack with unmanned monitors and move personnel upwind.

Some indicators of methamphetamine lab fires are:

- Unusual flame color due to the chemicals involved
- Violent reactions to water streams
- Strong chemical odors
- Empty chemical containers

A hot zone must be established and entry must be limited to those involved in suppression. Full protective equipment must be enforced in the hot zone. The warm zone should be set up for decontamination so firefighters can be immediately decontaminated prior to removing their SCBAs to prevent off-gassing of hazmat being inhaled or absorbed.

Utilities should not be turned off until the chemist gives his/her approval. Premature shutting down of the utilities can impact the cooking process in a negative way. Water for cooling purposes is an integral part of the process; shutting it down can prevent cooling water from reaching the process and cause overheating that could result in an explosion. Shutting down the electric could have a similar negative effect on a process, if electrically driven pumps are present. Disruption of the cooking process can occur by turning off the gas supply.

Prior to operating in a hot zone, members must have their vital signs checked and recorded. After being decontaminated, members must undergo medical evaluation by having their vital signs retaken and checked against those taken prior to entry. Additionally, members should be checked for any symptoms of exposure to toxic chemicals that could include nausea, vomiting, headaches, burning sensations in nose, throat, or lungs, drowsiness or being unusually tired, numbness of lips, or eyes not focusing.

CLANDESTINE DRUG LAB CLEANUP

A major factor with CDLs is that after the evidence has been seized and processed, law enforcement is responsible for the removal and disposal of the hazardous waste at the illegal laboratory. Cleanup costs for methamphetamine labs can run from $3,000 to $100,000. The waste can range from a small amount to several tons. It can include solvents, reagents, precursors, and byproducts, and it can be reactive, explosive, flammable, corrosive, and/or toxic. A variety of gases are used in CDLs, including phosphine, anhydrous ammonia, and methylamine. These gases are compressed in small cylinders or in tanks similar to those used for propane gas grills.

SIZE-UP FACTORS FOR A CLANDESTINE DRUG LAB

Water
- Water should not be used in active CDLs due to the potential for an explosion.
- Master stream devices should be placed as a precautionary measure should the CDL ignite or explode.

Area
- The affected area may be quite large since the fumes can be deadly and spread throughout a structure.
- An explosion could cause a fire to spread rapidly and involve a large area.

Life Hazard

- Individuals can be exposed to the chemicals used in the illegal labs.
- Explosions of the labs can threaten the health and safety of those involved in the illegal activities as well as others who may come in contact with the fumes.
- Responders should not eat or smoke in the affected area. Full PPE must be utilized and immediate decontamination must occur upon exiting the lab.
- Evacuation of a building housing the CDL and surrounding buildings will need to be considered.
- The Hazmat Unit in conjunction with the forensic chemist will determine the exact levels of personal protection, including SCBA or respirator, that are needed. Level B personal protective suits would normally be the minimum protection required. If a higher level of skin protection is required, then Level A protection can be used.

Location, Extent

- The assessment team will sample the atmosphere within the lab to determine the explosive limits, oxygen levels, and the extent of toxins present.
- The Hazmat Unit should secure air readings in surrounding areas to assess the dangers that exist.
- Typically an extensive area of contamination would require decontamination and cleanup.
- Hot, warm, and cold zones must be defined, constantly reevaluated, marked, and enforced.
- The extent of the area involved would expand if an explosion occurs with the cooking process.

Apparatus, Personnel

- A multi-agency response will be needed, including police, fire, Hazmat Unit, DEA, cleanup contractors, and emergency medical personnel.
- There may be an immediate need for EMS personnel for civilian casualties.
- Medical monitoring of emergency responders prior to entering and upon leaving the hot zone will be required.
- Decontamination will be needed to handle responders and any civilians that are contaminated.

Construction/Collapse

- An explosion in the lab could seriously damage a structure and collapse should be considered.

Exposures

- There is the potential for spread of fire to nearby buildings if the lab becomes heavily involved in fire due to a malfunction of the chemical process.
- Explosions may damage nearby buildings.

Weather

- Dispatch should transmit the current and anticipated weather conditions to the responders.
- Rain can spread chemicals in vessels that have been left outside, or from discarded containers used in the cooking process, contaminating areas and possibly entering well water or sewer systems.

Auxiliary Appliances

- Sprinklers, if present, may cause runoff problems.

Special Matters

- There is a possibility that booby traps or explosive devices have been placed to harm emergency responders.

Height

- A CDL can be found in a building of any height. Naturally, a taller building will compound the fire department's problems.

Occupancy

- CDLs have been found in every type of occupancy including moving vehicles.

Time

- Response to a CDL can occur at any time.

CONSIDERATIONS FOR A CLANDESTINE DRUG LAB

Strategic Goals and Tactical Priorities for an Offensive Attack

- An offensive attack would be made for rescues followed by a switch to a defensive attack once rescues are completed.
- Obtain an assessment of the fire or incident area as soon as possible.
- Once an assessment has been made, immediately call for additional resources, including EMS and other agencies that can assist. This is especially the case if the fire department is the first agency on scene.
- If a fire is involved, attempt to contain it to the area of origin.
- Evacuate occupants and have decontamination set up, if necessary.
- Set up hot, warm, and cold zones.
- Consider wind direction and velocity.
- Arrange for runoff water and decontamination water to be captured and disposed of properly.

Incident Management System Considerations/Solutions for an Offensive Attack

- Incident Commander
- Unified Command (if needed)
- Safety Officer(s)
- Liaison Officer
- Intelligence/Investigations Officer
- Rapid Intervention Crew(s) (The level of protection of the crews must be equal to that of those operating within the hazardous areas.)
- Operations (if needed)
- Hazardous Materials Group or Branch
- Logistics
- Base
- Staging
- Ventilation Group Supervisor
- Search and Rescue/Evacuation Group
- Division(s) or Branch(es) (This could involve interior or exterior areas.)
- Medical Group or Branch
- First-Aid Stations

Strategic Goals and Tactical Priorities for a Defensive Attack

- If changing from an offensive to a defensive attack, ensure that a PAR is taken for accountability purposes.
- Reassess the hot, warm, and cold zones and adjust, if necessary.
- Set up and maintain a collapse zone, if necessary.
- Evacuate surroundings structures that are threatened.
- If using water streams, consider unmanned master streams.
- Protect exposures, if necessary.

Incident Management System Considerations/Solutions for a Defensive Attack

- Incident Commander
- Unified Command (if needed)
- Safety Officer(s)
- Liaison Officer
- Public Information Officer
- Intelligence/Investigations Officer
- Operations
- Staging
- Logistics
- Base

- Hazardous Materials Group or Branch
- Rapid Intervention Crew(s) (For surrounding structures. The level of protection of the crews must be equal to that of those operating within the hazardous areas.)
- Divisions or Branches on the exterior on Sides A, B, C, D
- Exposure Divisions or Branches (if needed)
- Medical Group or Branch
- First-Aid Stations

Natural Disasters

Natural disasters are devastating events that have destroyed numerous properties and taken many lives. They can inflict tremendous damage and strike with little or no warning. These disasters can involve hurricanes, floods, earthquakes, wildland fires, winter storms, tornadoes, volcanic eruptions, and others. They will challenge firefighters and fire officers called upon to handle the problems associated with these disasters.

natural disasters ▪ Devastating natural events, such as earthquakes, tornadoes, and hurricanes, occurring with little or no warning that destroy properties and take many lives.

The types of disasters are often germane to specific parts of the country. Earthquakes will occur in areas that sit upon faults. Typhoons and tsunamis will threaten communities located adjacent to the oceans. The greatest impact of hurricanes and tropical storms is felt at locations near the oceans or the Gulf of Mexico, though heavy rains caused by hurricanes can spread great distances inland, causing flooding. Volcanic eruptions and avalanches occur in areas where their threat is known. Wildfires will happen in forested areas and can extend into communities where there is an interface with housing. The possibility of flooding, landslides, storms, and high winds are prone to some locales but may occur anywhere.

HURRICANES

Hurricanes are giant spiraling tropical storms that can pack wind speeds of more than 160 miles per hour and unleash more than 2.4 trillion gallons of rain a day. These same tropical storms are known as cyclones in the northern Indian Ocean and Bay of Bengal, and as typhoons in the western Pacific Ocean. The Atlantic Ocean's hurricane season peaks from mid-August to late October and averages five to six hurricanes per year.

Hurricanes begin as tropical disturbances in warm ocean waters. These low pressure systems are fed by energy from the warm seas. If a storm achieves wind speeds of 38 miles an hour, it becomes known as a tropical depression. A tropical depression becomes a tropical storm, and is given a name by the National Weather Service when its sustained wind speeds top 39 miles per hour. When a storm's sustained wind speeds reach 74 miles per hour, it becomes a hurricane.

Hurricanes are classified into five categories, based on their wind speeds and potential to cause damage.

Category One—Winds 74–95 miles per hour
Category Two—Winds 96–110 miles per hour
Category Three—Winds 111–130 miles per hour
Category Four—Winds 131–155 miles per hour
Category Five—Winds greater than 155 miles per hour

These storms bring destruction ashore in many different ways. When a hurricane makes landfall, it often produces a devastating storm surge that can reach 20 feet high and extend nearly 100 miles. Ninety percent of all hurricane deaths result from storm surges. A hurricane's high winds are also destructive and may spawn tornadoes. Torrential rains cause further damage by spawning floods and landslides, which may occur many miles inland.

The best defense against a hurricane is an accurate forecast that gives people time to evacuate from danger areas. The National Hurricane Center issues tropical storm and hurricane *watches* 48 hours in advance, and hurricane *warnings* 36 hours in advance for storms that may endanger communities.

ON SCENE

On August 28, 2011, Hurricane Irene was predicted to do massive damage to the state of New Jersey. Cape May County, which is on the southern tip of New Jersey, ordered a mandatory evacuation of the barrier islands. The governor declared a state of emergency. Atlantic County shore communities were placed under a mandatory evacuation. All Atlantic City casino resorts shut down on August 26 as the city faced its first mandatory evacuation in history. Part of Ocean County was under mandatory evacuation. To relieve evacuation traffic, toll operations were temporarily suspended on the Garden State Parkway south of the Raritan River, and on the Atlantic City Expressway, which traverses New Jersey from Atlantic City towards Philadelphia.

The 172-mile-long Garden State Parkway stretches the length of New Jersey. To assist in the evacuation efforts, the governor closed the southbound lanes south of Exit 98 (a distance of 98 miles), while east-west bridges and arteries were closed to eastbound shore traffic. That same day, traffic on other southbound major routes heading toward shore points was only allowed to head north.

By the time Irene hit land in New Jersey, it had been downgraded to a tropical storm. The effects of Irene in New Jersey in 2011 included about $1 billion in damage to 200,000 homes and buildings. This made it the costliest disaster in the state's history, though this was dwarfed by Hurricane/Super Storm Sandy the following year. The storm left the southern part of New Jersey relatively untouched, with little damage and flooding, but it caused serious flooding and damage in the central and northern parts of the state.

ON SCENE

Between October 22 and November 5, 2012, Hurricane/Super Storm Sandy killed at least 117 people in the United States, and 69 more in Canada and the Caribbean. As the storm struck New Jersey and New York, it was downgraded to a tropical storm, but continued to do massive damage. At least 100 homes were destroyed by fire in the Breezy Point neighborhood of Queens, New York, and 7.9 million businesses and households lost electric power. Sandy caused at least $65 billion in damage in the United States, making it the second costliest weather-related disaster in American history, behind only Hurricane Katrina.

During Hurricane Katrina in New Orleans, forest service helicopters were placed into service to perform air drops of water and fire retardant onto structures that were heavily involved in fire since there was no ground access available due to flooding.

FLOODS

A flood is an overflow or accumulation of an expanse of water that submerges land. It can occur on land adjacent to bays, rivers, lakes, and streams. It can be caused by heavy rains, melting snow, and failure of dams or levees. Flooding in uninhabited areas is not normally a problem, yet serious situations can develop when bodies of water overflow their banks and threaten populated areas. Flood waters rising rapidly can trap occupants in their homes and their only option is to reach their roof and await rescue. Homes can be swept away by the rapidly rising and fast-moving waters.

Considerations of primary damage of floods can be the physical damage to bridges, roadways, buildings, and vehicles. Casualties can include people and livestock due to drowning and sickness caused by waterborne diseases. Secondary considerations will include the loss of water supplies for drinking and sanitary use, food supplies, spread of diseases, damaged crops, and economic hardship due to business losses, business interruptions, and numerous other factors.

Firefighters needing to operate during flood conditions will be challenged with the many and varied types of emergencies they will be confronted with. There will be a need for small boats and flotation devices for responders.

Swift-water rescues are dangerous operations that require specific training and equipment. Vehicles with occupants can be taken downstream for distances by the rapidly running waters in which they are basically helpless. The many dangers involved can place firefighters in compromising situations when attempting rescues with fast-moving waters.

EARTHQUAKES

Most earthquakes occur at fault zones, where tectonic plates—giant rock slabs that make up the Earth's upper layer—collide or slide against each other. These impacts are usually gradual and unnoticeable on the surface; however, immense stress can build up between plates. When this stress is released quickly, it sends massive vibrations, called seismic waves, often hundreds of miles through the rock and up to the surface. Other quakes can occur far from fault zones when plates are stretched or squeezed.

Scientists assign a magnitude rating to earthquakes based on the strength and duration of their seismic waves. A quake measuring 3 to 5 on the Richter Scale is considered minor or light; 5 to 7 is moderate to strong; 7 to 8 is major; and 8 or more is great.

On average, a magnitude 8 earthquake strikes somewhere every year and some 10,000 people die in earthquakes annually. Collapsing buildings claim by far the majority of lives, but the destruction is often compounded by mudslides, fires, floods, or tsunamis. Smaller tremors that usually occur in the days following a large earthquake can complicate rescue efforts and cause further death and destruction.

Loss of life can be minimized through emergency planning, education, and the construction of buildings that sway rather than break under the stress of an earthquake. Building codes in earthquake-prone areas of the United States address those concerns.

FIGURE 10-18 Street flooding will impact upon citizens' safety and the fire department's ability to respond to calls.

WILDLAND FIRES

Wildland fires, or wildfires, burn millions of acres of forest, brush, and grass-covered land annually. The wildland fire is any uncontrolled fire that occurs in the countryside or a wilderness area. Though most wildfires are extinguished when smaller than one acre, large wildfires present such different challenges that they require specific training that is very different from structural firefighting. These large wildland fires can spread to involve more than 100,000 square acres, requiring a mobilization of massive resources to control and extinguish these conflagrations. While most wildfires burn in unpopulated areas, fires often threaten or involve areas of urban interface and can cause tremendous damage in these populated areas. (See Chapter 7 for Wildland Urban Interface Fires.)

The leading causes of wildland fires are lightning, arson, debris burning, careless smoking, children with matches, campfires, and railroad use.

WINTER STORMS

Winter storms can bring heavy snow and ice accumulation and can paralyze communities. A blizzard warning is given for a storm that contains large amounts of falling or blowing snow, sustained winds of at least 35 miles per hour, and is expected to last for several hours. In addition to snow, freezing rain can be crippling. Precipitation falling as rain can freeze and cause a tremendous amount of damage. The weight of the freezing rain can cause tree limbs to break and fall into streets, blocking them. Freezing rain can form an ice coating on highways, often referred to as *black ice*. It is difficult to see and can cause numerous accidents. Overhead electric lines can sag and fail due to the weight of the ice.

A common phenomenon is a nor'easter. Although these storms occur along the East Coast of the northeastern United States and Atlantic Canada, the term *nor'easter* refers to the storm's defining characteristic, which is the continuing strong wind blowing from the northeast. Nor'easters can cause flooding, coastal erosion, hurricane force winds, and heavy precipitation, especially in the coastal areas. They can occur at any time of the year, but most commonly occur during the winter season. A problem caused by a nor'easter is that tidal waters in bays and rivers can be greatly affected. It is not unusual for a high tide to enter a bay or river and the high winds of the nor'easter prevent it from receding into the ocean. The next high tide is then pushed onto the previous one causing flooding on land adjacent to the bays and rivers. This can occur for multiple cycles depending upon how long the storm lasts, raising the water levels of the bays and rivers, which causes flooding on the adjacent land.

The inherent danger of severe cold weather is that it can cause permanent damage or even death to an exposed person. Extremities, including toes, fingers, ears, and nose, are prone to frostbite. Another potential problem is hypothermia, which occurs when a person's body temperature falls below 90 degrees F. It is recognizable by uncontrollable shivering, impaired speech, memory lapses, drowsiness, and exhaustion. Frostbite and hypothermia victims should be treated by warming the person slowly and securing medical help.

DERECHO

A *derecho* (deh REY cho) is a complex of thunderstorms or a mesoscale convective system (MCS) that produces large swaths of severe, straight-line wind damage at Earth's surface. These storms are difficult to predict. Derechos occur year-round but are most common from May to August.

Specifically, for an MCS to be classified as a derecho, the following conditions must be met:

- There must be a concentrated area of convectively induced wind damage or gusts, greater than or equal to 58 mph, occurring over a path length of at least 250 miles.
- Wind reports must show a pattern of chronological progression in either a singular swath (progressive) or a series of swaths (serial).
- There must be at least three reports separated by 64 kilometers (km) or more of Enhanced Fujita 1 (EF1 damage) and/or measured convective wind gusts of 74 mph or greater.
- No more than three hours can elapse between successive wind damage/gust events.

Source: NOAA, Derecho Facts Page

TORNADOES

A tornado is a violent, dangerous, rotating column of air that, in order to be classified as a tornado, must be in contact with both the surface of the earth and a cloud. Tornadoes are typically spawned by a severe thunderstorm and almost always contain heavy precipitation. They may be accompanied by large hail and lightning.

The United States averages about 1,200 tornadoes per year and they are most common in spring and least common in winter. Spring and fall experience peaks of activity because those are the seasons when stronger winds, wind shear, and atmospheric instability are present, though favorable conditions can occur any time of the year. Tornado occurrence is highly dependent on the time of day because of solar heating, yet destructive tornadoes can occur at any time of day.

Pulse-Doppler radar is often used to detect tornadoes as well as reports from storm spotters. Tornadoes primarily occur in the area known as Tornado Alley or the central region of the United States. Firefighters searching for missing civilians after a tornado should remember that the recommended instructions for those caught in a tornado are to seek shelter in a storm cellar basement, or an interior first-floor room of a strongly constructed building to increase their chances of survival. The Enhanced Fujita (EF) scale was implemented in the United States on February 1, 2007, and can be used to obtain a rating for a tornado event. The scale is based on a three-second wind gust speed, in miles per hour.

Since tornado cleanup activities can be hazardous, NIOSH offers informational resources for minimizing the risks of work-related injury or illness from recovery activities, including disaster site management, electrical hazards, stress and fatigue, and other issues, at: http://www.cdc.gov/niosh/topics/emres/tornado.html.

Enhanced Fujita Scale Wind Speed Ranges

EF Scale	Three-Second Gust Speed (mph)
EF0	65–85
EF1	86–110
EF2	111–135
EF3	136–165
EF4	166–200
EF5	greater than 200

Source: Enhanced F Scale for Tornado Damage, NOAA Storm Prediction Center

ON SCENE

On April 14–16, 2011, one of the largest convulsions of tornado activity recorded in US history left a wake of 44 deaths and massive destruction in Oklahoma, Arkansas, Mississippi, Alabama, Georgia, South and North Carolina, and into Virginia. AccuWeather.com stated that 241 tornadoes were reported by 14 states.

VOLCANIC ERUPTIONS

A volcano is a rupture, or opening, in the surface or crust of the earth, which allows hot magma, ash, and gases trapped below the surface to rise up and escape. The damage from a volcano can impact those living not only near the volcano, but major volcanic eruptions can spread to distant communities.

The US Geological Survey monitors and predicts volcanic activity, and these predictions are based upon detecting early symptoms before an imminent eruption.

The methods used include studying the geographical area of the volcano, taking seismic readings, measuring poison gases, and using satellites. Seismic activity (earthquakes and tremors) always occurs as volcanoes awaken and prepare to erupt and are a very important link to eruptions.

ON SCENE

The natural disaster itself can cause unforeseen problems. When Mount St. Helens erupted in Washington State, volcanic ash from the eruption was blown east-northeastward, depositing ash for hundreds of miles. Major volcanic ash deposits occurred in central Montana and ash fell as far east as the Great Plains of the central United States, more than 900 miles away. It fell like snow as it covered many communities a distance from the immediate area. The volcanic ash is extremely abrasive and was drawn into carburetion systems of many vehicles, disabling them.

On April 14, 2010, the Eyjafjallajokull volcano in Iceland erupted for the first time in nearly two centuries. Danger to planes from the volcanic ash plume led most northern European countries to shut their airspace from April 15–20, 2010, grounding an estimated 10 million travelers worldwide.

POTENTIAL PROBLEMS

The impact upon the infrastructure of cities and towns caused by natural disasters can last for days, weeks, or longer. This can include the damage or destruction of rail lines, roads, bridges, and the loss of electricity, natural gas, phones, water supplies, and sewer usage. With widespread destruction of homes and businesses, the impact to the community can be overwhelming.

The disruption of electrical and phone service can isolate large numbers of citizens in both urban and rural areas, though in many areas of the country, the reliance on cell phones is minimizing this problem if the phone towers are left undamaged by the natural disaster. The ability of local jurisdictions to assist those in need can impact heavily upon emergency responders. Residents living in cities and towns can assist in assuring the safety of their neighbors who may not be able to fully care for themselves, while this responsibility can fall upon local authorities in rural areas.

Severe snow and ice storms can place heavy loads upon roof structures that were not built to support these loads, causing them to collapse into the structure. This type of structural failure occurring in remote areas can leave the occupants exposed to the elements without any timely assistance.

Earthquakes can devastate cities and regions with collapsing buildings, and destroy highway and transportation systems.

Varied problems that could accompany a natural disaster include:

- Landslides that can damage roadways, collapse buildings, bury vehicles, etc.
- Hazmat releases from residences involving household chemicals, propane tanks, and releases from commercial properties
- Failure of home heating oil storage tanks, LPG tanks, and large tanks at tank farms or refineries
- Occupants in vehicles caught in flood waters needing rescue
- Downed power lines electrifying large areas
- Escaping natural gas due to broken pipes or the extinguishment of pilot lights on appliances
- The loss of water supply to fire hydrants to fight fires
- Airport facilities may be damaged or destroyed
- Shelters may not be usable due to damage
- Critical infrastructure

ACTIONS TO ALLEVIATE PROBLEMS

Communities must determine how they can best protect their citizens and emergency responders. The severity of any disaster can vary from a minor to a major impact. The conditions that caused the damage to the community will often impact upon the ability of responders to assist those in need. Delays may occur due to the fact that it may be too dangerous for responders to immediately attempt to give aid to those in need. High winds may exceed safe levels for responders to be outside. Damaged roads and bridges, along with downed electric wires and trees, will restrict access and maneuverability of responders. Collapsing structures housing public safety personnel can trap them and damage or disable their response equipment.

What measures are needed to address these disasters? Catastrophic events can place an unusual demand on the emergency response system's ability to maintain its normal level of service, and they create a significant demand for increased coordination with other local, state, and federal resources.

Utility companies nationwide have cooperative agreements to assist each other in times of need. Even with these mutual-aid types of agreements, the disaster may be so widespread that it can be days before assistance may arrive in some areas.

There will be an immediate need for drinking water that will require the distribution of bottled water. The lack of sanitary sewage disposal can create unhealthy conditions. The loss of refrigeration will require the distribution of ice to preserve necessary food supplies.

WARNINGS OF DISASTERS

Proactive communities can prepare for the possibility of natural disasters by receiving early warnings of some disasters that can permit them to initiate emergency actions. These indicators can be associated with winter storms, heavy rains that can cause flooding in low-lying areas, hurricanes, etc.

Ocean front communities and locations that are prone to flooding should have an evacuation plan that will allow the communities to be evacuated in a timely manner. The implementation of the evacuation plan needs to occur soon enough to ensure the safety of the occupants. Evacuation plans for some communities is to change driving directions on certain roads to one way so they can be used for full evacuation from the danger areas. If the implementation of these plans to use certain state highways and bridges needs the approval of the governor of the respective states, there can be a negative impact if getting the governor's approval is delayed in any way.

PLANNING FOR NATURAL DISASTERS

Successful incident scene operations will require teamwork at all levels of local government. Interagency cooperation will be needed to effectively plan and prepare for large scale events. Planning will require identifying the potential hazards and what is needed to mitigate those hazards, including securing and effectively deploying the necessary resources at these emergencies.

A major component in the planning process is having an emergency operational plan (EOP) ready to handle the anticipated problems. Once the plan is written, training on the plan must be initiated by all involved agencies. This should include various exercises to enable all responders, dispatchers, and emergency operations center (EOC) participants to gain familiarity with the plan, including using damage assessment forms. As the exercises are conducted, any adjustments to the plan can be made.

IMMEDIATE DAMAGE ASSESSMENT

An immediate damage assessment will need to be conducted in the initial stages of the disaster to determine the degree of damage and the impact upon life and property. This immediate damage assessment performed by response units will be larger in scope than a typical size-up. It may be necessary for an engine company to assess its entire response district to enable the EOC to get the big picture. The information gathered can be used:

- As part of size-up and for initial reports
- To assist in establishing an incident management organization
- In developing incident objectives and strategy development
- To determine tactical operations
- To determine additional and specialized resource needs
- To deploy resources to critical locations

An immediate damage assessment in the initial stages of a disaster differs from size-up performed at a structure fire. The typical structural size-up is usually followed by implementation of strategies, tactics, and tasks to address the problems found. When faced with a large-scale disaster, the immediate damage assessment is critical in determining the full scope of the situation. The units performing that assessment are often compelled to complete their assessment and cannot stop for minor incidents. Should they need to intercede due to the potential for loss of life, it will be necessary to have another unit complete the immediate damage assessment as soon as possible. This is crucial since incomplete or delayed initial assessments can lead to unnecessary or improper responses to emergencies.

The immediate damage assessment needs to address:

- Estimated life loss and number of injuries
- Potential for additional loss of life and injury
- Existing or potential need for additional emergency operations
- Life safety threats to responders and the public
- Additional hazards to emergency response personnel
- Potential for additional property loss
- Obstructions to incident access
- Damage to buildings
- Damage to roadways, bridges, and tunnels
- Damage to governmental facilities
- Damage to infrastructure
- Other unsafe conditions

An immediate damage assessment is critical to determine the impact of the event on the community and on governmental services. This applies to both the immediate and long-term impact. Identifying the need for additional resources at a scene will allow a determination of whether they will come from local agencies or need to be requested from county, regional, state, or federal agencies.

CONDUCTING AN IMMEDIATE DAMAGE ASSESSMENT

An immediate damage assessment is basically a windshield survey that starts when first responders make a rapid visual assessment of conditions. A detailed assessment by first responders is difficult because of limited time and the need to deal with higher priority issues.

Immediate damage assessments for areas that have suffered extensive damage can be completed by units assigned to the area. Using this concept, each unit does a brief survey of a specific area and reports the information, through channels, to Dispatch or the EOC.

The damage assessment process is normally a shared responsibility. While the fire department surveys specific buildings, the police department can survey jails and detention facilities. EMS personnel or public health personnel can check nursing homes, medical facilities, and hospitals. Public works can survey roads, bridges, water, sewer, dams, etc.

The fastest way to conduct an immediate damage assessment is by air. Helicopters from public or private agencies, such as the news media, can be used to provide initial damage reports. Personnel conducting aerial damage assessments should be given a list of key buildings or facilities to assess in priority order. The assessment should focus on locations where large numbers of people may have been injured, where emergency response access may be obstructed, and where potential problems may exist. This includes:

- Hospitals
- Schools
- Nursing homes and assisted living facilities
- Government buildings
- Major businesses
- Jails
- Dams and rivers
- Roadways, including bridges and overpasses

STANDARDIZED DAMAGE ASSESSMENT FORMS

Agencies should develop standardized damage assessment forms to allow units to record and then transmit the information. The use of a standard form allows the receiver to utilize the same format and place the data on the standardized form. It also permits the data to be collated with that received from other units and/or agencies doing the immediate damage assessment. The information is necessary for both an initial assessment stage and can be incorporated into the post-incident damage assessment. The damage assessment information becomes the basis for requesting emergency declarations from the state and federal government. State and federal declarations demand accurate data by which decisions can be made. A practical approach is for field units to have a smaller size form (8½-inch × 11-inch) and for Dispatch or the EOC to have a larger sized form (11-inch × 17-inch) using the same format so they can insert the information. See Appendix I on www.pearsonhighered.com/bradyresources for Sample Damage Assessment forms.

COMMUNICATIONS AT NATURAL DISASTERS

Since communication is the backbone of emergency responders, it is critical that incident scene conditions are reported to the dispatch center who can relay it to the EOC. This will allow the EOC to have an accurate picture of the situation and enable them to make decisions to guide the community's response to the disaster. Likewise, the EOC will keep field commanders apprised of the scope of the disaster. This will enable the individual ICs to realize the extent of the damage to the community and that available resources may be scarce until outside assets can arrive.

A disaster may knock out normal communications. If this occurs, it will be a critical priority to ensure that communications are reestablished as quickly as possible. There needs to be a plan in place to implement a backup system. This can be quite challenging. In Florida during previous hurricanes, complete radio systems have been destroyed and makeshift systems had to be developed.

The use of simplex portable radios that operate from one radio to another can be used in a relatively small area. The use of cell phones, if they are functional, may be needed to initially maintain communications. Many fire departments maintain caches of cell phones that are not activated unless an emergency occurs and then they are distributed to ensure adequate communications. Another communications option is the use of Cells on Wheels (COWs). These COWs are satellite dishes that phone companies can set up to allow use of cell phones when a heavy load is placed on satellites due to a community disaster.

Once the EOC gets accurate reports of the damage inflicted upon the community, it can initiate the relocation of resources. This may involve moving units to positions closer to the disaster or taking them out of damaged fire stations or stations cut off from major travel routes. Whatever the reasons for relocation, supporting the relocated units may create additional logistical demands on the EOC.

ON SCENE

In areas that do not have cellular coverage or in areas where additional cell phone coverage is needed, expanded cellular network coverage can be provided by COWs. This can be useful in areas devastated by hurricanes, earthquakes, floods, or wildland fires. Through the deployment of 36 COWs in Lower Manhattan on September 14, 2001, critical phone service was made available to FEMA and rescue and recovery workers after the 9/11/2001 attacks on New York City. COWs have since been used in many other disasters including Hurricane Katrina.

RESOURCE ASSISTANCE

Natural disasters will require assistance from outside of the impacted communities. The exact type of assistance will be predicated by the type of disaster. It is not unusual to require county, state, federal, and private resource assistance. Some considerations would include:

- Building inspectors for damaged buildings
- Dump trucks and front-end loaders to clear streets and highways
- Crews to cut trees that block streets
- Public works to fix traffic lights
- Fuel trucks to supply emergency vehicles
- Tire-changing trucks to repair/replace damaged tires on emergency vehicles
- Coast Guard to assist in opening waterways and clean up hazmat that may have entered the waterways
- Technical specialists for hazmat mitigation, air monitoring, building engineers, bridge and highway engineers, and meteorologists

INCIDENT MANAGEMENT SYSTEM

Organization is required to successfully handle any emergency scene and it is magnified at a major emergency. The National Incident Management System (NIMS) is mandated by the DHS to be used by emergency responders in the United States. Mandatory training is part of the requirements for police, fire, and other emergency personnel. This training starts very basic and expands to higher disciplines when communities are faced with a natural disaster. By using NIMS, diverse agencies can work together seamlessly using common terminology and operating under the umbrella of the same incident management system.

NIMS has the ability for modular expansion to address the many and varied problems at a natural disaster. The systems are time-tested over many years at wildland fires and have been expanded to include all-hazard incidents.

Since natural disasters occur infrequently, and are accompanied by many different and diverse problems, they place demands upon an IC that require a varied thought process to handle. Though experienced in their normal duties, they can be overwhelmed when having to manage large and expanding incidents. There will be a need for immediate actions to organize the incident and achieve effective management and control. Decisions will have to be made under extreme time pressure with many developments occurring in a compressed time frame.

The complexities of these disasters will require a written IAP. Documentation will be required during the incident and afterwards if the agency hopes to recover funds from state and federal agencies based upon their actions in handling the incident. This documentation will often require the assistance of a scribe or an aide to assist not only the IC but other major role players at the incident.

The basic method of acquiring experience in managing natural disasters is through preplanning and frequent exercises to gain knowledge of the management system while addressing potential problems that may occur.

SIZE-UP FACTORS FOR NATURAL DISASTER INCIDENTS

Water
- Water supplies may be interrupted due to broken water mains or flooding.
- Frozen hydrants may be a problem with winter storms.
- Water tender operations may be needed.

Area
- Natural disasters can be widespread and involve large areas outside of local jurisdictions, which can reduce the ability to receive assistance from normal mutual-aid partners.

Life Hazard
- Life hazard can be severe with high sustained winds that accompany hurricanes, tornadoes, and winter storms.
- Areas of refuge will need to be searched after tornadoes.
- Thermal imaging cameras can assist in the primary and secondary searches.
- People can be trapped in collapsed buildings.
- Mass casualties may occur.
- Firefighters' safety must be a prime concern due to the potential for rapidly changing conditions.
- Operations may be restricted until winds lessen to safe levels during some disasters.

Location, Extent
- Natural disasters can devastate large areas.
- The location and extent of the damage incurred with natural disasters can be widespread.
- Floods can inundate large areas and wash out bridges and highways, segregating large sections of communities.
- Flash floods are common in some areas and may require swift water rescue capabilities.

Apparatus, Personnel
- Call for help early. A large number of resources will be needed from local, county, state, and federal agencies during natural disasters.
- Large wildland fires can spread rapidly and require a large contingent of resources.
- Apparatus may not be able to access certain areas and special apparatus, including boats and snowmobiles, may be needed.

Construction/Collapse
- Building collapse can be widespread with tornadoes, hurricanes, and floods.
- Buildings not built to withstand natural disasters can fail readily if faced with earthquakes, tornadoes, and hurricanes.
- Firehouses and other public safety buildings can be affected by natural disasters, and collapse of these structures can reduce the available resources needed to serve the community.

Exposures
- Wildland fires can spread to more than 100,000 acres, destroying everything in their path. Large flying brands are common. Fires must be anticipated downwind for great distances.
- With wildland fire, a tremendous amounts of radiant heat must be anticipated. Threatened structures can be wet down to cool them or Class A foam applied to them.

Weather
- Hurricanes, tornadoes, and winter storms can have high sustained winds that can be destructive.
- Wind can drive wildland fires and predictions of wind direction and speed can assist in determining the direction that a wildland fire will travel.

- Wind can carry ash from volcanoes, causing fires near the eruption as well as causing disruption of transportation, including ground vehicles and airplanes.
- Wind from nor'easters can restrict tide waters, causing flooding.
- Low humidity can mean low moisture content of the fuel and a fast-moving wildland fire.
- Heavy rain can assist in controlling wildland fires.
- Heavy rain can add to the problem in flooded areas, raising water levels in streams and rivers.
- Lightning is the leading cause of wildland fires.

Auxiliary Appliances
- Auxiliary appliances normally have no impact on natural disasters.

Special Matters
- Hazmat releases can occur with the various natural disasters. Large fuel tank storage can be damaged. Residential and commercial outdoor fuel oil or propane tanks can be damaged and leaking.
- Damaged roadways and bridges will restrict access.
- Downed trees, tree branches, overhead electric wires, and debris in streets will limit access.
- Flat tires sustained from roadway debris will place emergency response vehicles out of service.
- Homeless pets will need to be taken to shelters.
- Loss of power or flooding can reduce available services from hospitals and other care providers.
- Hospitals can become inundated with patients and have no beds available.

Height
- Buildings can be of any height.

Occupancy
- Occupancies can include everything in a community.

Time
- Long response times can be caused by the distances that must be traveled due to damaged or flooded roadways.
- Delays may occur in dispatching responders until high winds reach safe levels.
- Natural disasters are more prone to occur during certain times of the year. Winter storms are one example. Tornadoes and hurricanes have seasons in which they would be anticipated. Floods in some areas follow the spring thaw of mountain snows.

CONSIDERATIONS FOR NATURAL DISASTER INCIDENTS

These incidents will require specific actions dependent upon the life safety of residents and types of associated damage. As stated earlier, preplanning, utilization of an incident management system, and training exercises to become proficient are basic requirements. The need for an EOP that would be a blueprint for handling these incidents is a must.

- Preplanning:
  - Must include a community risk analysis.
  - Building codes to address potential problems identified in the risk analysis, i.e., earthquake-resistant buildings, clear areas surrounding structures in wildland areas, hurricane straps for roof assemblies in hurricane-prone areas, safe rooms constructed in tornado areas for life safety of the occupants, etc.
- Utilization of an incident management system.
  - Initial and ongoing training.
  - Conducting tabletop and realistic exercises to hone the participants' skills.
  - Have all response agencies participate in the training and exercises.
  - Include responsible members in authority to also participate in the training, e.g., mayor, city manager, etc.

- Develop an EOP and test it through exercises and adjust it as needed.
  - Develop generic IAPs for anticipated natural disasters. This could include a tornado touchdown in areas that are located in tornado-prone areas, and similar types of IAPs for areas that are prone to having floods, earthquakes, hurricanes, wildfires, etc. These IAPs will not be fully implementable should a disaster occur, but can initially assist in establishing benchmarks of factors that may need to be addressed.

Summary

Firefighters who respond to incidents involving hazmat will be challenged to mitigate the ensuing problems while protecting civilians and themselves. These operations are often unpredictable due to the nature of the materials involved. A seemingly stable situation one moment can rapidly change to a scene that has escalated tremendously.

Incidents of terrorism, though seemingly rare in the United States, may increase in the coming years. There is also the potential for the use of dirty bombs. This cultural change in our society will create numerous challenges for firefighters as they respond and attempt to mitigate the problems found.

Clandestine drug laboratories are a threat to all emergency responders as well as the occupants of a building in which the lab is situated and the exposed buildings.

Natural disasters can occur with no warning and can devastate a community, taking lives and destroying property. As emergency responders, we need to assist all in need at these emergencies.

Review Questions

1. What actions are required of the first-arriving officer at a hazardous materials incident?
2. What are the various terms associated with shipping papers when discussing different modes of transportation?
3. List the strategic modes that can be utilized at a hazardous materials incident.
4. Discuss the type of hazardous materials incident in which the proper action to take is no action.
5. What considerations are needed for setting up zones at a hazardous materials incident?
6. Discuss hazardous materials situations in which it would be beneficial to protect in place rather than evacuate. How would you attempt to accomplish this at an incident scene? How would you notify those affected? Who would do the notifications? How would you monitor their well-being? How would you keep them informed?
7. Which hospitals in your area have facilities for decontamination of hazardous materials on patients?
8. Define *open-top floating roof tank*, *covered floating roof tank*, and *cone roof tank*.
9. What fire protection is provided for exposed tanks at a refinery or tank farm?
10. What are the definitions of the two categories of terrorism as recognized by the FBI? Which category would be more of a concern to your community? Explain.
11. Discuss how firefighters arriving at a scene where terrorists had used incendiary weapons might not recognize it as a terrorist act.
12. How can dispatchers help to identify potential terrorist events?
13. What signs should the first-arriving officer look for that could indicate a potential terrorist act?
14. Would implementing a unified command system be a problem in your area? Explain.
15. What assistance would you need to request for a major terrorist action occurring in your area? Be specific.
16. In your area, would you anticipate any problems of interaction between the hazardous materials unit and the bomb squad? What training exercises do they typically do together? Explain.
17. Discuss the sample stages of response to a terrorist act listed in this chapter. Would this format be easily adaptable to your area? Explain.
18. Select one of the technical operations in this chapter and, after reviewing the size-up factors, add five additional potential problems that could be present.
19. Define the difference between a dirty bomb and a nuclear bomb.
20. List strategic considerations if confronted with the possibility of a dirty bomb incident.
21. Describe the protection provided by a firefighter's personal protective gear when operating at a suspected dirty bomb incident.
22. What radiation exposures are considered acceptable for firefighters when confronted with a suspected dirty bomb incident and civilians are in the endangered area?
23. What are the common methods of production of methamphetamine in the United States?
24. What actions should firefighters take upon discovering a clandestine drug lab?
25. List the types of natural disasters that could impact upon your community. What preparations does your fire department and community have in place to deal with these events?
26. List the types of common problems that could accompany a natural disaster.
27. What problems does the initial damage assessment need to address?

Suggested Reading, References, or Standards

Furgione, L. 2013. "The Historic Derecho of June 29, 2012." US Department of Commerce, National Oceanic and Atmospheric Administration. Available at: http://www.nws.noaa.gov/os/assessments/pdfs/derecho12.pdf.

Missouri Council for a Better Economy. 2010. "Homemade Explosives Recognition Guide." Available at: http://www.bettertogetherstl.com/files/better-togetherstl/Maryland%20Heights%20PD%20%20%20%20Homemade%20Explosives%20Recognition%20Guide.pdf.

National Oceanic and Atmospheric Administration (NOAA). 2015. "Derecho Facts Page." Available at: http://www.spc.noaa.gov/misc/AbtDerechos/derechofacts.htm.

NOAA. "Enhanced F Scale for Tornado Damage." *Storm Prediction Center*. Available at: http://www.spc.noaa.gov/faq/tornado/ef-scale.html.

OSHA. "Quick Card." Available at: https://www.osha.gov/Publications/HazComm_QuickCard_Pictogram.html.

US Department of Transportation. 2016. *Emergency Response Guidebook*. (New version every four years.)

The 9/11 Commission Report: Final Report of the National Commission on Terrorist Attacks Upon the United States. 2004. New York, NY: W. W. Norton & Company.

US Environmental Protection Agency (EPA). 2015. "Pesticide Labels and GHS: Comparison and Samples." Available at: http://www.epa.gov/pesticide-labels/pesticide-labels-and-ghs-comparison-and-samples.

US Fire Administration (USFA)

Emergency Management and Response-Information Sharing and Analysis Center (EMR-ISAC). August 2015. "Theft of First Responder Gear Linked to Terrorism." *The InfoGram* 15, no. 34: https://www.usfa.fema.gov/downloads/pdf/infograms/34_15.pdf.

Manning, W. A., ed. "The World Trade Center Bombing: Report and Analysis, New York City, New York." Technical Report 076.

Stern, J. "Fire Department Response to Biological Threat at B'nai B'rith Headquarters, Washington, DC." Technical Report 114.

National Fire Protection Association (NFPA)

Standards

45: Standard on Fire Protection for Laboratories Using Chemicals

471: Recommended Practice for Responding to Hazardous Materials Incidents

472: Standard for Professional Competence of Responders to Hazardous Material/Weapons of Mass Destruction Incidents

Related Courses Presented by the National Fire Academy, Emmitsburg, Maryland

Advanced Life Support Response to Hazardous Materials Incidents

Chemistry for Emergency Response

Hazardous Materials Operating Site Practices

Hazardous Materials Incident Management

Special Operations Program Management

Hazardous Materials Code Enforcement

Emergency Response to Terrorism: Strategic and Tactical Considerations for Supervisors

Fire and Emergency Services Pre-Disaster Long-Term Recovery Planning

Related Courses Presented by the Emergency Management Institute, Emmitsburg, Maryland

IEMC/ All Hazards: Preparedness and Response

IEMC/ Hurricane: Preparedness and Response

IEMC/ Earthquake: Preparedness and Response

IEMC/ Homeland Security: Preparedness and Response

IEMC/ Hazardous Materials: Preparedness and Response

Situational Awareness and Common Operating Picture

11

After the Incident

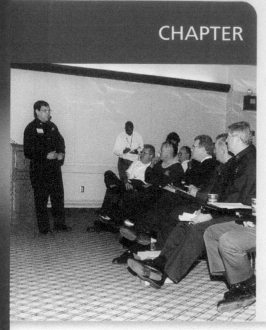

> " Life is a test; take good notes. "

—Author Unknown

A formal critique should be held after major emergencies or significant events. *Used with permission of VA Medical Center.*

KEY TERMS

incident critique, *p. 527*   critical incident stress, *p. 532*

OBJECTIVES

Upon completion of this chapter, the reader should be able to:

- Recognize the signs and symptoms of incident stress.
- Recognize the benefit of both formal and informal critiques.
- Understand how to perform a self-critique.

Chapter 11 looks at two important areas: critiques and critical incident stress. When the physical work at an incident has been accomplished, there are still tasks that must be performed. An overall review of what caused the emergency and how the fire department handled everything must be undertaken. This introspective look should be accomplished by each person individually, together as a unit, and overall by the Incident Commander. This effort to find areas for improvement by never leaving a stone unturned will ensure that our operations will constantly improve.

An incident scene review must look at the potential impact on a firefighter who has been exposed to traumatic events. Ensuring that assistance is available and compelling affected firefighters to attend a critical incident stress debriefing is a good first step. This can help in identifying whether additional help will be required.

Incident Critiques

A strong point of the fire service is the constant desire to improve operations. A basic way to do this is through critiques of previous incidents.

The **incident critique** is meant to reconstruct events and assess how the fire department performed.

- What worked well?
- Where is improvement needed?
- Do operational guidelines need revision or modification?

incident critique ■ A means to reconstruct events and assess how the fire department performed at an incident. The outcome should be the improvement of operations.

Well-run incidents can serve as benchmarks for future operations. Where many problems confronted firefighters, solutions can be shared with other department members.

In the past, critiques in some departments were a way of chastising members. They were used as whipping posts to embarrass officers, with the thinking that this form of negative discipline would improve future operations. In reality, it stifles incentive. This negative type of critique degenerated to a "cover your anatomy" philosophy. Fire officers attempted to escape the meeting unscathed and would shed no light on the operation, fearing the wrath of departmental leadership should a mistake be discovered. The overall effect destroyed the concept of openly sharing successes and failures.

Many departments have replaced the term *critique* with *post-incident analysis*. The word *critique* is defined by Webster as "a critical estimate or discussion." The success of a critique lies not in its title, but in the openness of the participants and in their willingness to conduct an honest review of the incident to improve future operations.

Hotwash is another term that is used to identify immediate after-action discussions and evaluations of performance at an incident scene. Like the critique, the hotwash is meant to identify strengths and weaknesses of a response to an event or training exercise. This identification can then lead to improved operations at future incidents. These discussion/evaluation events can lead to an After Action Review, a structured review for analyzing what happened, why it happened, how it can be done better in the future, and any changes needed in the current SOGs.

TYPES OF CRITIQUES

Critiques can range from a discussion between company members after a minor fire to a full-blown review that includes all fire officers who responded to a large-scale incident.

Informal Critique

The informal critique (or *tailgate*) is for a company or multiple companies. (See Figure 11-1.) The aim is to review the actions taken and their impact on the overall operation. This review can take place at the incident scene, the kitchen table in the firehouse, or any convenient location.

Formal Critique

A formal critique should be held after most major emergencies or significant events. A date should be selected as soon as possible after the incident. This should consider the time factors necessary to gather any technical information that may be needed. Participants should be invited and required to attend. Maps and drawings of the incident should be displayed for participants to use during their individual presentation. A facilitator should be selected to ensure that the critique proceeds without being bogged down with prolonged discussion. Time limits should not be imposed on meaningful discussion, however.

Each area of the incident must be reviewed. This includes decisions by the Incident Commander (IC), members in major assignments, and division and group supervisors.

The critique should start with the highest-ranking officer stating the anticipated benefit of the critique. He or she should stress that the purpose is not to place blame for mistakes but to decide how

FIGURE 11-1 An informal critique by company members can be accomplished at an incident scene. *Used with permission of Joseph Hoffman.*

to implement lessons learned in the future. The IC should give a brief overview of the incident and highlight the positive occurrences, as well as some problems that need review. By recognizing problems at the start, solutions can be addressed later.

A critique can be successful only if an open environment is created and the participants are encouraged to be forthcoming in their observations. Commanding officers can foster this environment by a willingness on their part to discuss problems and failures that they personally encountered. This creates an environment of sharing.

The representative from the dispatch center can start the presentations. What information was received and what equipment was dispatched? There should follow a natural progression from the first-in company.

1. What did they observe on arrival?
2. What information did they receive?
3. What personal knowledge were they able to utilize?

The discussion by the first-in company should be followed by all participants who commanded units or were responsible for a major segment of the operation. They should discuss their observations, problems encountered, orders received or given, and actions taken. The critique can conclude with the fire marshal's findings on origin and cause.

A recorder should take notes to prepare the final report. Special emphasis should be placed on problems encountered, solutions, and failed remedies.

FORMAL CRITIQUE FORMAT

Each presenter should try to touch on the following points:

- *Communications:* Discuss any problems or reports that were detrimental or beneficial to responders.
- *Size-Up:* What specifically was observed, and what information were they able to pass on to Command or other units to assist in the overall operation?
- *Incident Management System:* What was their role in managing the incident? Did they find any problems, or do they have any suggestions to improve future operations?
- *Strategy and Tactics:* Did everyone understand the incident objectives and the strategy and tactics that were being implemented? Did any problems arise that needed to be addressed? What specific orders were they given and by whom? Were they able to carry out their orders and complete their assignments?
- *Medical Assignments:* Did medical assignments enter into the operation? If so, how were they handled and what future recommendations can be made? Did they make use of Rehab or the First-aid Station?
- *Safety:* What safety issues did they encounter? Did they interact with the Safety Officer (SO)? Did they find any safety problems that were reported to either Command or the SO? If so, discuss them.
- *Apparatus and Equipment:* Did they experience any problems with apparatus placement or special equipment? Would they recommend any special apparatus or equipment that would accomplish the job in a safer or faster manner at future incidents?
- *Resources:* Were the resources sufficient to accomplish the incident objectives, strategy, and tactics that they were responsible for?
- *Outside Agencies:* How did your agency interact at the incident scene? Do you feel that you were properly utilized? Could you recommend any changes for future incidents?
- *Lessons Learned, Lessons Reinforced:* What lessons did you learn at this incident that you can share to assist everyone in future incidents? What lessons were reinforced at this incident that you can share with everyone to assist in future incidents? Do you have any recommendations for changes to operational procedures or operational guidelines that need to be made?

Self-Critique

Self-critique is an important method for firefighters seeking self-improvement. Firefighters should review the operation.

- What were their areas of responsibility?
- Was search and rescue performed in a timely manner?
- What did they attempt to accomplish with the initial hose-line?
- Why did they stretch the hose-line through the front door?
- Was that the best method of extinguishment?

- If not, what was?
- What was the best way to ventilate?
- Was the roof opened in the right location?

INDIVIDUAL DEVELOPMENT

Assigning a new recruit to an experienced firefighter is part of the development process of the recruit. In that situation, a discussion of what occurred should take place between the two after each incident. The recruit should be encouraged to ask questions about why certain actions were performed a certain way. The explanations will assist him or her in understanding the importance of assignments and the teamwork needed for safe, efficient operations. This review is a key factor in a recruit's professional development. This process allows recruits to correct mistakes and develop good fireground habits.

An officer has to review his or her areas of responsibility:

- Did I properly size up the situation?
- Were my calculations of the seriousness of the incident consistent with the overall outcome?
- Did I call for sufficient resources?
- Did I call for too many? (This is usually the exception. The error most often made is underestimating the seriousness of the situation.)
- What basic orders did I give?
- What did I hope to accomplish with those orders?
- Did I properly prioritize the incident?
- What areas needed attention that I did not initially address?
- Did I formulate an overall plan and give that information to the units performing those assignments?
- How were search and rescue, ventilation, and other strategies addressed?
- Did I overload my initial units with too many orders?
- Did I get sufficient feedback from the divisions and groups?
- Did I frustrate their efforts or assist them in accomplishing their goals?

Critical analysis starts with being truthful with yourself. Every fire that we have ever responded to has gone out. This can be as a result of a great effort by the responding firefighters or in spite of them. Supervisors must recognize exceptional effort and praise those who have earned recognition.

Firefighters want meaningful feedback. They certainly desire praise if it is earned, but also seek direction for self-improvement. If praise is given regardless of the effort expended or the outcome of the operation, a grave injustice is done.

The fact that a fire is extinguished does not mean that praise is warranted. A firefighter making a mistake and being praised for doing a good job will make the same mistake the next time. Orders must be given to rectify problems encountered. In the aftermath of the incident, either an informal or formal discussion must take place to review proper strategies and tactics and discuss any mistakes that occurred. The personnel who have made mistakes should be taken aside and have explained to them the proper way of performing the assignment.

Commanding officers should place mistakes or errors into two categories: those of commission and those of omission. A firefighter making a mistake because he or she took the initiative to attempt something is not in the same category as the one who stood by and did nothing. We must recognize those willing to attempt a solution and try to motivate those who do not attempt to intervene to mitigate the problem.

FIGURE 11-2 Utilizing simulations and then critiquing the participants is an excellent tool for officer development.

NIOSH FIREFIGHTER FATALITY REPORT F2009-23

On August 24, 2009, a 45-year-old male career lieutenant (Victim #1) died following a partial floor collapse into a basement fire, and a 34-year-old male career firefighter (Victim #2) was fatally injured while attempting to rescue Victim #1. The career fire department was dispatched for "an alarm of fire" with reported civilian(s) entrapment. Arriving units discovered a heavily secured mixed commercial/residential structure with smoke showing. Following failed initial attempts to locate an entry to the basement, crews located a door on Side 2 that provided access down a flight of stairs to a basement entry door. Repeated attempts were made to force open this basement door in order to search for trapped civilians, but crews had difficulty gaining access through this door because it was made of steel and locked and dead-bolted on both sides. Other crews on scene performed primary searches of the first and second floors and found no civilians.

Approximately 30 minutes into the basement fire, Command ordered all interior crews to exit the structure in order to regroup because crews were still unable to gain access into the basement from Side 2. Additional manpower was sent with special tools to assist in breaching the basement door on Side 2. Victim #1 and two firefighters from his crew entered into the structure from Side 1 to verify all firefighters had exited a first floor deli. Victim #1, following a hose-line into the structure, was well ahead of the other two firefighters when the first floor partially collapsed beneath him. Victim #1 fell with the floor into the basement, which exposed him to the basement fire. The other two firefighters immediately exited the deli after fire conditions quickly changed and shelving and displays fell on them. They were unaware of what had just occurred. Victim #1 made several Mayday calls from within the structure and activated his PASS device. Confusion erupted exteriorly on scene when trying to verify who was calling the Mayday, what his exact location was, and how he got into the basement. The IC was aware that he had crews attempting to gain access into the basement from Side 2, but was unaware that there had been a floor collapse within the deli section of the structure. Simultaneously, Victim #2, a member of the firefighter assistance and search team (FAST), was standing by outside Victim #1's point of entry when the Mayday calls came out. It is believed that Victim #2 knew where Victim #1 was since he had gone in the structure with him earlier in the incident. Victim #2 grabbed a tool, went on air, and rushed into the structure. The FAST and additional personnel on scene concentrated on Side 2 initially while other firefighters followed an unmanned hose-line into the deli. Crews within the deli quickly discovered a floor collapse and reported hearing a PASS device alarming. Victim #1 was immediately identified as missing during the first accountability check, but Victim #2 was not accounted for as missing until the third accountability check, more than 50 minutes after Victim #1's Mayday. After the fire was controlled, both victims were discovered side by side in the basement where the first floor had partially collapsed. They were found without their face pieces on and with SCBA bottles empty. Victim #1's PASS device was still alarming. They were pronounced dead on scene. Four firefighters and one lieutenant suffered minor injuries during the incident. No civilians were discovered within the structure.

Key contributing factors identified in this investigation include working above an uncontrolled, free-burning basement fire; interior condition reports not communicated to Command; inadequate risk-versus-gain assessments; and crew integrity not maintained. **See Appendix B: NIOSH Reports in www.pearsonhighered.com/bradyresources to read the complete report and recommendations.**

FINAL REPORT

A written report of the critique's findings should be shared with the members who attended, other fire department members, and the mutual-aid departments that responded. The report can be divided into three sections. The first part is a narrative account describing conditions, problems encountered, life safety considerations, and fire department actions.

The middle section is a review of specific areas. It should start with the vital statistics of the incident: date, weather, times, dispatch numbers, specific location, who responded, injuries, and other pertinent information. This should be followed by addressing each area of the assignment. The content of this section can vary, depending on the assignment and the needs of each department. Some general areas are:

Communications

Did any problems arise? Did Dispatch receive proper and timely reports?

Size-Up

This should include dispatch information received and conditions observed during the course of the fire, as well as problems encountered.

Incident Management System

What staff positions, divisions, or groups were created? An addendum can be included showing an incident management organizational chart for the incident with the names of those who assumed the various positions.

Strategy and Tactics

Review the strategies developed and the tactics initiated. What changes, if any, had to be made to facilitate their accomplishment? What problems arose that required special attention?

Medical Assignments

What medical problems had to be handled? Was this a mass casualty incident? Were First-aid Stations set up? Was rehabbing of firefighters needed? If so, how was it accomplished? If the Medical Group consisted of units outside of the department, how did the fire department and these units interact?

Safety

What were the safety issues? Was a SO assigned? What problems confronted him or her? A list of all injuries and how they occurred can be included. Recommendations on the prevention of future injuries can be noted under lessons learned.

Apparatus and Equipment

Was apparatus properly placed and utilized? Could special equipment in either the department or a mutual-aid department have completed the assignment in a safer manner? Did the equipment on hand meet the needs of the incident?

Resources

Were requests for additional resources timely? Was there a time when certain functions could not be performed due to a lack of resources at the scene?

Outside Agencies

What agencies were requested and responded? Did they meet the needs of the incident? How can they better assist us in the future?

LESSONS LEARNED, LESSONS REINFORCED

The final component of the report should be a lessons learned section. Because the critique is a learning process, mistakes discovered can be beneficial to everyone. This section should be written in a positive way. For instance, if deluge guns were not properly secured, the report in the lessons learned area can state that the importance of securing all master streams was reinforced on this assignment. If improperly placed apparatus were relocated due to the instability of walls, it can be noted that the initial proper placement of apparatus negates the need to shut down hose-lines to move them later in an operation. The main purpose of this section is the sharing of information.

A committee should routinely compare the critique reports with previous ones to see if there are recurring problems or patterns developing that need to be addressed.

 NIOSH FIREFIGHTER FATALITY REPORT F2013-07

On April 6, 2013, a 53-year-old male career captain died from injuries suffered from a fall during roof operations at a commercial structure fire. The initial box alarm was for smoke in the basement of a fabric store. Approximately 30-45 minutes after smelling the odor of smoke, the store owner went to the basement to investigate and found a fire in the rear of the basement. He attempted to extinguish the fire with a portable fire extinguisher, but due to smoke and fire, he was forced to leave the basement. The first alarm for companies assigned to Box 1232 had a difficult time finding the seat of the fire due to the amount of heat and smoke. Crews also struggled to gain access to the basement due to fabric and other products limiting aisle space. After the second alarm was struck, the IC ordered Ladder 27 (L27) to the roof to make a trench cut between the fire building and the B exposure. The L27 crew accessed the roof from Side C and was preparing to walk to the B exposure. The captain of L27 took several steps towards Side A of the roof, which was obscured by smoke, and fell to the roof of a one-story storage building attached to the fire building. The captain died instantly. The other members of L27 heard the sound of the captain landing on the roof of a storage building. The driver/operator of L27 ordered the crew to their knees, conducted a personnel accountability report, and realized the captain was missing. The crew from L27 got off the roof and tried to locate the captain. They found him lying on the roof of the storage building and made several rescue attempts. Before the captain could be removed, the storage building roof collapsed into the basement of the fabric store. Rescue operations were started by breaching the wall of the storage building. Approximately two hours later, the captain was removed from the structure.

Critical Incident Stress

Firefighters are extremely dedicated individuals who hold themselves to a very high set of standards. They are action-oriented and seek immediate results. Firefighters have been looked upon as being tough as steel. No matter what they are confronted with, they accept the challenge and give their best effort. When dealing with serious injuries, they must block out the emotional aspects of mutilation and human suffering and continue to extricate and treat the injured or remove the dead. They are able to make difficult decisions under the pressure of time constraints and deteriorating incident scene conditions. They attempt to accomplish perfection at an emergency incident while operating under circumstances that would be overwhelming to others.

critical incident stress ■ Stress experienced by firefighters exposed to traumatic incidents.

Because firefighters are so highly motivated, they become frustrated when, despite their skills, rescues cannot be made and those they are trying to help are injured or killed. As a result, they can suffer from **critical incident stress**.

As much as firefighting is physical, it is also one of the most psychologically stressful jobs. This is reflected in the large number of injuries sustained by firefighters due to stress. A major source of stress comes from the many emotional occurrences confronting firefighters at incident scenes. Stress occurs when critical decisions have to be made in a short period of time. Firefighters rely on their experience and training to evaluate information and act upon it. If an operation is not completely successful, they will review every action and decision, seeking a reason for failure. This individual critique can mislead them into self-blame and acceptance of full responsibility. Firefighters will become frustrated and magnify what occurred. Stress will increase if they ponder only the outcome of the operation and not the many factors that affected their decisions and actions. In retrospect, many pieces of information known after an event were not available to the firefighters to assist them in their decision making during the operation.

Firefighters associate fireground events to their own situations. When a fellow firefighter is injured or killed at an incident, they question why the injury occurred to someone else and not them. They may falsely blame themselves for either causing the injury or not preventing it. This sense of guilt adds to their stress. Likewise, when confronted at an incident with the death of a child, firefighters may realize the vulnerability of their own children and the pain and suffering being felt by the child's parents. Their frustration in being unable to perform rescues will be replayed over and over in their minds.

Because firefighters have the same concerns as everyone, they must express their feelings. If kept bottled up, these feelings can become a destructive force. Traumatic incidents that affect firefighters psychologically have a direct impact on their performance.

INCIDENT COMMANDERS

ICs and SOs must be alert to the physical and emotional needs of firefighters. Being able to predict how a firefighter is affected by incident scene occurrences is difficult. Commanding officers must monitor their firefighters for symptoms of stress. Some immediate indicators are shaking or trembling, loss of coordination, blurred vision, respiratory problems, confusion, shock, disorientation, and frustration. If a firefighter's actions reflect signs of undue stress, the officer should act immediately to remove that individual to a rehab area for medical treatment and evaluation.

Some recommended incident scene procedures to alleviate stress are to:

■ Schedule five-minute breaks for units as often as possible
■ Rotate personnel (Though most firefighters do not want relief, it must be done.)
■ Provide a rehab area with hot and cold drinks (avoid caffeine and nicotine)
■ Have firefighters medically screened by EMS personnel or a physician

DEBRIEFING

Critical incident stress debriefing (CISD) will reduce incident stress. Mental health professionals and peer counselors are needed for a formal debriefing. This should occur as soon as possible after the event and must include all members who responded to the incident. The setting should allow each participant to talk (though no one should be forced). Discussing the incident brings into the open the fact that everyone has similar concerns. Members should be encouraged to vent their frustration, anger, fear, and other emotions. They should tell of any particularly painful occurrences and should be encouraged to talk of any feelings of pride that they felt due to their accomplishments.

The debriefing must be strictly confidential. No records should be kept and only those firefighters directly involved in the incident and the debriefing team should attend.

In the past, debriefing was often accomplished (though not realized by the firefighters) in an informal setting by firefighters gathering after an incident and talking about what happened. These discussions served a similar purpose as the CISD. A big difference, however, was the lack of trained professionals, and no follow-up was available to those firefighters needing it.

Ground rules should be set whereby CISD is mandatory for certain situations. Mandatory CISD should include the death or serious injury to a firefighter, death of a child, mass casualty incidents, incidents involving an extrication process exposing rescuers to trapped victims for a long period of time, or unusual occurrences causing stress to the firefighters.

DELAYED STRESS

Firefighters may have no immediate reaction to stress at the scene. Reactions can show up days or weeks after the incident because a firefighter may be unable to get over it and put it in the past. Firefighters who still revisit a tragic event in dreams or flashbacks after several weeks should seek professional help.

In severe cases, if stress persists, a firefighter who does not receive treatment can develop delayed symptoms or even a post-traumatic stress disorder. Potential problems that may require attention are upset stomach, sleep disturbances (nightmares, insomnia), displays of anger, irritability, rage, respiratory difficulties, headaches, rashes, loss of concentration, phobias, sexual disturbances, depression, increased use of alcohol and/or tobacco, changes in appearance and job performance, and other problems.

TAKING CARE OF OUR OWN

The aggressive attitude firefighters bring to the incident scene is to rescue everyone endangered while gaining control of the incident as quickly as possible. Real-life situations present barriers that must be overcome to achieve this. Delayed alarms, companies on other responses, severe weather conditions, heavy traffic, and other factors beyond the control of the fire department can complicate the department's efforts. Failure to accomplish tasks, along with deaths and injuries, can cause stress.

In the past, it was felt that anyone who did not bounce back immediately from these horrible incidents was not strong. This attitude is wrong and has no place in the fire service. No stigma should be attached to firefighters seeking assistance to help themselves in overcoming emotional feelings that develop from incident stress.

It is the fire department's responsibility to provide debriefing and ongoing medical treatment, if needed, to assist firefighters in coping with critical incident stress. For fire departments to provide psychological assistance to their firefighters, they must have a plan in place prior to an incident.

The plan should consist of two parts: education and treatment. A program should be initiated to educate firefighters on the effects of emergency scene stress. This will increase their awareness of the warning signs of stress and help them to cope when confronted with critical stress situations. An integral part of the plan should be the development of CISD teams or the ability to assemble them quickly from adjacent fire departments or a statewide system.

Summary

When the fire is out or the medical response has brought a patient to the hospital, there may be other factors that still need to be dealt with before the incident is complete. The type of incident may require some form of critique. The critique should be looked upon as a great learning tool that improves the way we operate.

In addition to a critique, the implementation of critical incident stress debriefings aids firefighters in putting a difficult incident in the past. The counseling that they receive allows them to function properly at future incidents and to maintain a healthy family life.

Review Questions

1. When should a formal critique be conducted?
2. What are the components of a self-critique?
3. List specific decisions that company officers should review in their self-critique.
4. List areas where improvement has occurred in your department due to problems found during critiques.
5. List incident scene occurrences that could trigger critical incident stress.
6. What are some immediate indicators of stress?
7. Does your department have a procedure in place to address critical incident stress? Are all members familiar with departmental guidelines?

Suggested Readings, References, or Standards for Additional Information

National Fallen Firefighters Foundation. "Training Opportunities." Web page lists training courses in firefighter behavioral health, LODD preparedness, LODD prevention, and community risk reduction/fire prevention. Available at: http://www.firehero.org/resources/department-resources/training/.

National Fallen Firefighters Foundation. "Taking Care of Our Own: A Fire Chief's Guide to Preparing for a Line-of-Duty Death." Outline of training program. Available at: http://www.firehero.org/resources/department-resources/programs/taking-care-of-our-own/.

Related Courses Presented by the National Fire Academy, Emmitsburg, Maryland

Fire Service Safety Culture: Who Protects Firefighters from Firefighters?
Shaping the Future
Politics and the White Helmet

In summarizing, I hope that the information contained within this book will be beneficial. A lifetime of experiences is condensed into the preceding pages.

My one wish for the fire service is to find ways to reduce injuries and eliminate firefighter deaths. I think if we follow the 16 Life Safety Initiatives, we can have a positive impact. If I could challenge all firefighters, it would be to never allow a serious firefighter injury or firefighter death to occur without learning something from it that allows them to perform their duties in a safer manner.

The strength of every fire department is its firefighters. They are, in my estimation, the finest group of individuals I have ever encountered. Regardless of the part of the country or the name of the individual fire departments, firefighters are a breed of their own. Dedication, commitment, perseverance, and loyalty are but a few of their universal traits. The names may change, but the basic mission remains the same.

I would like to leave you with a closing thought. In 1987, I was promoted to deputy chief and assigned to the First Division. My area of responsibility covered one half of the City of Philadelphia. It included hundreds of high-rise buildings, an international airport, miles of piers, crude oil refineries, a US naval base, and just about any problem associated with a large metropolitan area. On my first day, I was sitting at my desk checking everything out when, under the glass on the desktop, I spotted an index card yellowed with time. It put everything in perspective. It read:

The objective of all dedicated chief officers should be to thoroughly analyze all situations, anticipate all problems prior to their occurrence, and have answers for these problems when called upon. However, when you are up to your ass in alligators, it's difficult to remind yourself that your initial objective was to drain the swamp.

BE SAFE OUT THERE

ACRONYMS

BIA—Bureau of Indian Affairs
BLEVE—Boiling Liquid Expanding Vapor Explosion
BLM—Bureau of Land Management
B-NICE—Biological, Nuclear, Incendiary, Chemical, Explosive
BTU—British Thermal Unit
CCA—Casualty Collection Area
CDL—Clandestine Drug Labs & Commercial Driver's License
CHEMTREC—CHEMical TRansportation Emergency Center
CISD—Critical Incident Stress Debriefing
CFR—Code of Federal Regulations
COG—Continuity of Government
COOP—Continuity of Operations
CPSC—Consumer Product Safety Commission
CP—Command Post
CTA—Casualty Transportation Area
DHS—Department of Homeland Security
DLSC—Deputy Logistics Section Chief
DOJ—Department of Justice
DFSC—Deputy Finance Section Chief
DOSC—Deputy Operations Section Chief
DPSC—Deputy Planning Section Chief
DOT—Department of Transportation
EMI—Emergency Management Institute
EMCO—Emergency Medical Control Officer
EMS—Emergency Medical Services
EMSO—Emergency Medical Services Officer
EOC—Emergency Operating Center
EOP—Emergency Operating Plan
EPA—United States Environmental Protection Agency
FAST—Firefighter Assist and Search Team
FBI—Federal Bureau of Investigation
FEMA—Federal Emergency Management Agency
FSC—Finance Section Chief
FIRESCOPE—FIre RESources of California Organized for Potential Emergencies
GHS—Globally Harmonized System or Globally Harmonized System for the Classification and Labeling of Chemicals
GIS—Geographic Information System
GPS—Global Positioning System
HCS—Hazard Communication Standard (OSHA Standard)
HVAC—Heating, Ventilation, and Air-Conditioning
IAFF—International Association of Fire Fighters
IAP—Incident Action Plan
IC—Incident Commander
ICS—Incident Command System
IDLH—Immediately Dangerous to Life and Health
IEMS—Integrated Emergency Management System
IMS—Incident Management System
IMT—Incident Management Team
IST—Incident Support Team

JIC—Joint Information Center
JIS—Joint Information System
LOFR—Liaison Officer
LPG—Liquefied Petroleum Gas
LRC—Learning Resource Center at Emmitsburg, MD
LSC—Logistics Section Chief
MSDS—Material Safety Data Sheets
MACS—Multi-Agency Coordinating System
MCI—Mass Casualty Incident
NASF—National Association of State Foresters
NFA—National Fire Academy
NFF—Needed Fire Flow
NFIRS—National Fire Incident Reporting System
NFPA—National Fire Protection Association
NIMS—National Incident Management System
NIIMS—National Interagency Incident Management System
NIOSH—National Institute for Occupational Safety and Health
NPS—National Park Service
NIST—National Institute of Standards and Technology
NWCG—National Wildfire Coordinating Group
NWS—National Weather Service
OPS—Operations Section Chief
OSC—Operations Section Chief
OSHA—Occupational Safety and Health Administration
PAR—Personnel Accountability Report
PASS—Personal Alert Safety System
PIA—Postincident Analysis
PIO—Public Information Officer
PPE—Personal Protective Equipment
PPV—Positive Pressure Ventilation
PSC—Planning Section Chief
RECEO-VS—Rescue, Exposures, Confinement, Extinguishment, Overhaul, Ventilation, Salvage
RIC—Rapid Intervention Crew
RIT—Rapid Intervention Team
SCBA—Self-Contained Breathing Apparatus
SDS—Safety Data Sheet
SOG—Standard Operating Guideline
SOP—Standard Operating Procedure
UC—Unified Command
UL—Underwriters Laboratories
USCG—United States Coast Guard
USFA—United States Fire Administration
USFS—United States Forest Service
USFWS—United States Fish and Wildlife Service
WALLACE WAS HOT—Water, Area, Life, Location, Apparatus, Construction, Exposures, Weather, Auxiliary appliances, Special matters, Height, Occupancy, Time
WMD—Weapons of Mass Destruction

GLOSSARY

A

Active shooter—Violent acts perpetrated on the student body at schools, or occupants at other venues, with the intent of killing and maiming.

Agency Administrator or Executive—Chief executive officer (or designee) of the agency or jurisdiction that has responsibility for the incident.

Area Command (Unified Area Command)—An organization established to oversee the management of (1) multiple incidents that are each being handled by an ICS organization, or (2) large or multiple incidents to which several Incident Management Teams have been assigned. Area Command has the responsibility to set overall strategy and priorities, allocate critical resources according to priorities, and ensure that incidents are properly managed.

Assistant—Title for subordinates of the Command Staff positions. The title indicates a level of technical capability, qualifications, and responsibility subordinate to the primary positions.

Assisted living facilities—Designed to allow the occupants the privacy of their own living quarters and normally have a minimal staff.

Attic—In a structure with a peaked roof, the space between the topside of the ceiling on the top floor and the underside of the roof.

Auto-exposure—Extension of fire to the floor above by lapping out of windows on the floor below.

B

Backdraft—An atmosphere that is deficient in oxygen that creates a smoldering fire. If oxygen is then introduced, the carbon monoxide that has been generated by the smoldering fire reacts violently, causing a smoke explosion or backdraft.

Base—The location where primary logistics functions are coordinated and administered. Commonly found at wildland fires, hazardous materials incidents, active shooter incidents, and high-rise fires.

Basement and cellar fires—Fires occurring below ground level in a building.

Bearing wall—A wall that supports roofs, floors, or other walls.

Behavior of fire—Fire is a chemical process where fuel, oxygen, and heat come together in an uninhibited chain reaction. It involves the rapid oxidation of a combustible material producing heat and flame.

BLEVE—Boiling Liquid Expanding Vapor Explosion. A container failure with a release of energy, often rapid and violent, accompanied by a release of gas to the atmosphere, followed by ignition (fireball) and propulsion of the container or container pieces.

Blitz attack—Master streams utilized to knock down a large body of fire; can precede an offensive attack.

B-NICE—Pertaining to biological, nuclear, incendiary, chemical, and/or explosive weapons. The various forms of attack that could be used by terrorists.

Boilover—The violent expulsion of oil and froth from a tank containing heavy or unrefined oil. It results when a heat wave produced by the burning oil in the tank reaches water residue that is usually near the bottom of the tank. The water rapidly expands into steam, causing the boilover.

Branch—An organizational level between divisions/groups and the Incident Commander or Operations. In wildland firefighting, it is usually designated by a Roman numeral, for example, Branch I or Branch II. In the all-hazard application, it can have a geographical or functional designation, for example, Medical Branch, Hazmat Branch, Law Enforcement Branch, or West Branch. A branch can also be established under Logistics. Branches are utilized to ensure adequate span of control.

Bridging—Bracing placed between joists to stiffen them, hold them in place, and assist in distributing the load.

Building collapse—The partial or major collapse of a building.

C

CAN report—A method of giving a comprehensive report. The acronym stands for Conditions, Actions, and Needs.

Carcinogen—A substance that is capable of causing cancer in mammals. Examples include asbestos, benzene, and vinyl chloride.

CHEMTREC—CHEMical TRansportation Emergency Center. A public service provided by the Chemical Manufacturers Association. Provides immediate advice to those at an emergency scene for handling incidents involving shipments of hazardous materials and then contacts the shipper for more detailed assistance. Emergency phone: 1-800-424-9300.

Chord—An inclined or horizontal member that is the upper or lower member of a truss.

Clandestine drug labs (CDLs)—Illegal laboratories ranging from primitive to highly sophisticated facilities for the production of illegal drugs such as methamphetamines.

Clear text—The use of plain English in radio communications transmissions. No 10 codes or agency-specific codes are used when utilizing clear text.

Cockloft—In a structure with a flat roof, the space between the topside of the ceiling on the top floor and the underside of the roof.

Cold zone—At a hazardous materials incident, a support area outside the perimeter where Command and support functions take place. Special protective equipment is not required in this area.

Collapse search—Occurs when firefighters are called on to find and extricate those trapped. Requires skill and the implementation of a plan.

Collapse zone—A safety zone placed around a building to allow a minimum safe distance should a wall collapse. The minimum distance should be no less than 100 percent of the height of the wall.

Combination fire departments—Fire departments that are formed with both career and volunteer members.

Command Post (CP)—A location from which an incident is managed. At a fire scene it is typically in front of the fire building with a view of the incident. A location farther from the scene would be necessary at a high-rise fire, wildland fire, natural disaster, or a hazardous materials incident.

Command presence—The traits an Incident Commander exhibits, which will influence the conduct of everyone operating at an incident scene.

Commercial buildings and warehouses—Can range from one to eight stories and spread over hundreds of feet or a city block.

Consist—A railroad's shipping paper similar to a cargo manifest. It may contain a list of the cars in the train order, or a list of those cars carrying hazardous materials and their location in the train.

Corbelled brick—Brick built out upon itself to form an ornate cornice.

Core construction—The location in a high-rise building through which the utilities, shaftways, and elevators reach up the building. The core may be located in the center, front, rear, or side of the building.

Crew resource management (CRM)—A tool created to optimize human performance.

Critical incident stress—Stress experienced by firefighters exposed to traumatic incidents.

D

Dead load—The weight of the building and any part that is permanently attached or built in. This would include flooring, walls, roof, and paint.

Decontamination (decon)—The process of making any person, object, or area safe by absorbing, destroying, neutralizing, making harmless, or removing the hazardous material to which they have been exposed.

Defensive mode—An exterior fire attack. Usually initiated due to heavy fire involvement, lack of resources, insufficient water supply, or a high-risk situation that presents little or no gain to be achieved.

Defensive/offensive mode—An initial exterior attack that may involve a blitz attack to knock down a large body of fire, followed by an interior attack.

Deputy—In the incident command system, a deputy is a fully qualified individual who, in the absence of a superior, could be delegated the authority to manage a functional operation or perform a specific task. In some cases, a deputy could act as relief for a superior and therefore must be fully qualified in the position. Deputies can be assigned to the Incident Commander, General Staff, and Branch Directors.

Derecho—(deh REY cho). A complex of thunderstorms or a mesoscale convective system (MCS) that produces large swaths of severe, straight-line wind damage at Earth's surface.

Designed load—A building load that the building designer anticipated and planned for.

Dike—An embankment or ridge, natural or manmade, used to prevent the movement of liquids, solids, or other materials. Commonly used at refineries or tank farms.

Dirty bomb—A homemade bomb that uses conventional explosives and contains radioactive material that is intended to be dispersed as the bomb explodes.

Division—A designated geographic area in an incident management system, e.g., Bravo Division, Division 14.

Downwind—The area directly in the path of the wind from the incident site.

E

Eave—A building overhang that provides protection for the building walls and can provide ventilation openings for the roof space.

Eccentric load—A load perpendicular to the cross section of the supporting element that does not pass through the center. An example would be a wall sign or marquee hanging from the face of a wall exerting a downward pull on the wall that supports it.

Emergency Management Institute—Located in Emmitsburg, Maryland. It is a part of the Federal Emergency Management Agency.

Emergency medical services (EMS)—A group of personnel employed for the medical care and transportation of civilians and firefighters alike. It may be part of the fire department, a separate entity, or a combination of the two.

Emergency operating plan (EOP)—An operating plan to be used for emergency operations. An EOP should be maintained by each jurisdiction as well as facilities, such as hospitals, penal institutions, and schools.

Emergency operations center (EOC)— The physical location at which the coordination of information and resources to support domestic incident management activities normally takes place. An EOC may be a temporary facility or may be located in a more central or permanent facility, perhaps at a higher level of organization within a jurisdiction. EOCs may be organized by major functional disciplines (e.g., fire, law enforcement, and medical services), by jurisdiction (e.g., federal, state, regional, county, city), or some combination thereof.

Emergency Response Guidebook—A book produced by the US Department of Transportation to assist in identifying hazardous materials. It gives information on placards to assist in determining fire, explosion, or health hazards, and possible emergency actions. It should be carried on every emergency vehicle.

Emergency traffic/Mayday—A radio message used to clear a radio frequency to allow a critical or important message to take priority over routine messages. A Mayday indicates that a firefighter is in trouble.

Enclosed shopping mall—Found in urban and suburban areas. Merchandise may consist of flammable items.

Engine company operations—Provide the water to control and extinguish fires.

Exposures—Locations adjacent to a fire area or building that are threatened by the fire. They can consist of interior or exterior areas.

F

Federal Emergency Management Agency (FEMA)—A government agency responsible for preparing the nation for hazards; manages federal response and recovery efforts following any national incident and administers the National Flood Insurance Program.

Fire and Emergency Services Higher Education (FESHE)—A US Fire Administration initiative partnering with post-secondary institutions and fire-related book publishers to design model course outlines for fire-related and EMS Management degree programs. These national models allow for an integrated system of higher education from associate's to doctoral degrees.

Fire-cut beam—A 30-degree cut on each end of a wooden floor joist that allows the floor to collapse down into the building without pushing the masonry wall outward.

Fire load—All parts or contents of a building that are combustible; measured in British thermal units (BTUs).

Fire officer—Usually holds the rank of lieutenant or captain and has the responsibility for leading an engine, ladder truck company, specialized units, or administrative positions in a fire department.

Fire-resistive construction—Building construction in which the structural members are of noncombustible materials that meet or exceed requirements prescribed by the applicable code.

Fire under control—A relative term used to describe a fire situation in which the life and safety of civilians has been assured, no additional extension of fire will take place, and normally no additional firefighting equipment will be required to respond at emergency speed.

Firefighter Life Safety Initiatives—The First National Firefighter Life Safety Summit produced 16 major initiatives that will give the fire service a blueprint for making positive changes.

Flashover—The almost simultaneous ignition of all combustible materials in a room.

Flow path—The movement of heat and smoke from the higher pressure in the fire area towards the lower pressure areas accessible via doors, window openings, and roof structures. As the heated fire gases are moving towards the low pressure areas, the energy of the fire is pulling in additional oxygen from the low pressure areas.

Frame building—Wood frame building in which all members are wooden or of similar material.

Frothover—A steady, slow frothing over the rim of a tank without the sudden violent action that occurs in a boilover.

Function—Function refers to the five major activities in the incident command system: Command, Operations, Planning, Logistics, and Finance/Administration. The term *function* is also used when describing the activity involved, e.g., the planning function. A sixth function, Intelligence, may be established, if required, to meet incident management needs.

G

Garden apartments—Buildings that are set back from the roadway.

General Staff—A group of incident management personnel organized according to function and reporting to the Incident Commander. The General Staff normally consists of the Operations Section Chief, Planning Section Chief, Logistics Section Chief, and Finance/Administration Section.

Geographic information system (GIS)—A system that integrates hardware, software, and data for capturing, managing, analyzing, and displaying all forms of geographically referenced information.

Globally Harmonized System (GHS)—Officially known as the Globally Harmonized System of Classification and Labeling of Chemicals (or simply GHS). It is a worldwide standardized approach to hazard communication developed by the United Nations. GHS is designed to provide clear, consistent label messages to chemical handlers and users, emergency first responders, and the public. Signal words, pictograms, and hazard statements that are used with the system will have the same meaning in all settings, domestically and internationally.

Group—Groups are established to divide the incident into functional areas of operation. Groups are composed of resources assembled to perform a special function not necessarily within a single geographic division, e.g., ventilation group, salvage group.

Gusset plate—A flat wooden, iron, or steel plate of various sizes and thickness. It can be found in both heavy timber and lightweight construction. These plates are centered on the outside edges of the connection point of the truss members. The wooden gusset plates are usually held in place by nails or staples. Some plates of older design were glued in place. The steel plates are held in place by bolts in heavy timber truss. Lightweight truss plates are called *sheet metal surface fasteners*. These flat steel plates have triangle pieces punched into a 90-degree angle. They are then pressed into the truss at intersecting points. These plates are 16 to 22 gauge in thickness and are often referred to as *staple plates* or *gang nails*.

H

Hazard Communication Standard (HCS)—An OSHA standard that covers hazardous materials in the workplace to protect workers. It is currently aligned with the Globally Harmonized System.

HazMat 2012—Part of the Hazard Communication Standard that updated the HCS to include the Globally Harmonized System.

Hazardous materials—A substance or combination of substances that may cause or contribute to an increase in mortality or irreversible or incapacitating illness, or pose a present or potential hazard to health, safety, or the environment.

Hazardous materials incident—Can expose firefighters to uncontrolled situations that present a more complex set of occupational health and safety concerns than does traditional structural firefighting.

Hazardous materials unit—A team of trained technicians that responds to mitigate hazardous materials releases or spills. It can consist of firefighters or other trained professionals.

Heavy timber construction—Uses wooden timbers of large-dimension lumber; also called *mill buildings*.

High-rise building—This type of structure is more than 75 feet tall and presents numerous life safety problems to firefighters fighting a fire.

Hospitals—Range from full-service facilities handling hundreds of patients daily, to those with a few beds and limited care capabilities.

Hotels and motels—Classified as quarters that have more than 16 sleeping accommodations used by transients for less than 30 days. The quality of the hotel can vary from upscale to rundown.

Hotwash—A term similar to *critique* or *post-incident analysis* that is used to identify immediate after-action discussions and evaluations of performance at an incident scene.

Hot zone—An exclusion zone or area immediately around an incident where serious threat of harm exists. It should extend far enough to prevent adverse effects from hazardous materials to personnel outside the zone. Entry into the hot zone requires appropriately trained personnel using proper personal protective equipment.

Houses of worship—Churches. Gothic-style churches are one-story buildings that can be equal to four or more stories in height. They contain masonry walls and roofs constructed of large wooden beams formed into timber trusses.

Hydraulic ventilation—A form of ventilation performed by utilizing hose-lines with a nozzle set on a fog pattern to vent smoke through window or door openings.

I

Immediately dangerous to life and health (IDLH)—An atmosphere that poses an immediate hazard to life or produces immediate, irreversible, debilitating effects on health.

Impact load—A load that is in motion when applied.

Incident—An occurrence or event, natural or human-caused, that requires an emergency response to protect life or property. Incidents can include major disasters, emergencies, terrorist attacks, terrorist threats, wildland and urban fires, hazardous materials spills, nuclear accidents, aircraft accidents, earthquakes, hurricanes, tornadoes, tropical storms, war-related disasters, public health and medical emergencies, and other occurrences requiring an emergency response.

Incident Action Plan (IAP)—An oral or written plan containing general objectives reflecting the overall strategy for managing an incident. It may include the identification of operational resources and assignments. It may also include attachments that provide direction and important information for management of the incident during one or more operational periods.

Incident Commander (IC)—Has the responsibility for the overall management of an incident.

Incident command system (ICS)—The combination of facilities, equipment, personnel, procedures, and communications within a common organizational structure with responsibility for management of those resources at an incident scene.

Incident critique—A means to reconstruct events and assess how the fire department performed at an incident. The outcome should be the improvement of operations.

Incident Management Team (IMT)—The Incident Commander and appropriate Command and General Staff personnel assigned to an incident.

Incident scene control—A method used to ensure that all units are operating within a plan that provides a safe and successful conclusion to the emergency.

Incident scene decision making—The two basic methods are the classical method and the *naturalistic*, or *recognition prime*, or *cue-based method*.

Incident scene management—The utilization of an incident command system to achieve command, control, and coordination at an incident.

Incident Support Team (IST)—A team of personnel to support and mentor the local IMT and not to take over Command of an incident. These ISTs can support a local team and offer their expertise to provide various levels of documentation. A common arrangement is that the members of the IST report to the Incident Commander, and work in a support role for Planning, Logistics, Staging, Public Information Officer, Safety Officer, and other areas by serving at the Incident Commander's pleasure.

Initial responders—The first personnel to arrive at an incident scene. At a hazardous materials incident they will need a basic knowledge of what hazards may exist to ensure the safety of all personnel.

Intelligence/Investigations Officer—Responsible for managing internal information, intelligence, and operational security requirements supporting incident management activities. These may include information security and operational security activities, as well as the complex task of ensuring that sensitive information of all types (e.g., classified information, law enforcement-sensitive information, proprietary information, or export-controlled information) is handled in a way that not only safeguards the information, but also ensures that it gets to those who need access to it to perform their missions effectively and safely.

J

Joint information center (JIC)—A joint information center may be created at large or complex incidents. The PIO position may become part of the JIC. Through the use of a JIC, multiple organizations or agencies can work together to accomplish the goals of public information. The JIC is used to gather incident data, analyze public perception, and keep the public informed.

Joint operations center (JOC)—A JOC is usually pre-established, often operated 24/7, and allows multiple agencies to have a dedicated facility for assigning staff to interface and interact with their counterparts from other agencies. These centers provide a location where federal, state, and/or local agencies meet to exchange strategic information and develop and implement tactical plans.

K

Kerf cut—A carpentry term (kerf): the thickness of a saw blade. It can be used to observe whether fire has reached the point where the cut is located. It is used in conjunction with a trench cut in a roof.

L

Learning Resource Center (LRC)—Located at the National Emergency Training Center at Emmitsburg, Maryland, a library and research facility for the fire service.

Lightweight building components—Have replaced full-dimensional lumber in both frame and ordinary construction. Preferred components include triangular and parallel chord trusses, wooden I-beams, and steel bar joists.

Liquefied petroleum gas (LPG)—Propane, butane, etc.

Live load—Any load in a building that is not permanently attached or built-in. An example would be furniture or machinery.

Looped standpipe system—In a dry standpipe, all piping in the system is interconnected. This allows the system to be pressurized at any intake location and supply the entire system.

Lumberyard—May contain sawmills, millwork shops, woodworking factories, kilns, and a large quantity of highly combustible lumber in exposed exterior piles that create a severe fire hazard.

M

Managers—Individuals within ICS organizational units that are assigned specific managerial responsibilities, e.g., Staging Area Manager or Camp Manager.

Mass casualty incident (MCI)—Occurs when the number of patients exceeds the level of stabilization and care that can normally be provided in a fast and timely manner.

Master stream—A fixed device that is capable of water flows in excess of 300 gallons per minute, too powerful for handheld hose-lines.

Material safety data sheets (MSDS)—This is the former term for informational documents with 9 sections; now replaced with Safety data sheets (SDS). This form contains specific information on the properties, dangers, and correct handling procedures of hazardous materials. Copies must be kept at sites where hazardous materials are stored or processed.

Mayday/emergency traffic—A radio message used to clear a radio frequency to allow a critical or important message to take priority over routine messages. A Mayday indicates that a firefighter is in trouble.

Modes of fire attack—Encompasses the types of fire attack: offensive, defensive, or transitional attacks, which can be offensive/defensive or defensive/offensive. The use of nonintervention mode will also be a consideration when encountering hazardous materials incidents.

Motels—Classified as quarters that have more than 16 sleeping accommodations used by transients for less than 30 days. The quality of the motel can vary from upscale to rundown.

Multiple Agency Coordination Center (MACC)—The MACC is a central command and control facility responsible for the strategic "big picture" of a disaster. The common functions of all MACCs is to collect, gather, and analyze data; make decisions that protect life and property; maintain continuity of the government or corporation, within the scope of applicable laws; and disseminate those decisions to all concerned agencies and individuals. A MACC is often used when multiple incidents are occurring in one area, or incidents are particularly complex for various reasons, such as when scarce resources must be allocated across multiple requests. While often similar to an emergency operations center (EOC), the MACC is a separate entity with a defined area or mission and lifespan, whereas an EOC is a permanently established facility and operation for a political jurisdiction or agency.

N

National Emergency Training Center—Located in Emmitsburg, Maryland, it is the home of the US Fire Administration, the National Fire Academy, and the Emergency Management Institute.

National Fire Academy—Located in Emmitsburg, Maryland, it develops and delivers programs to firefighters and fire officers. It accomplishes this task through actual classroom delivery on its campus, through outreach programs delivered by adjunct instructors throughout the United States, and by handing off courses that it develops to state and local training groups.

National Incident Management System (NIMS)—A system mandated by Homeland Security Presidential Directive-5 (HSPD-5) that provides a consistent nationwide approach for federal, state, local, and tribal governments, the private sector, and nongovernmental organizations to work effectively and efficiently together to prepare for, respond to, and recover from domestic incidents, regardless of cause, size, or complexity. To provide for interoperability and compatibility among federal, state, local, and tribal capabilities, the NIMS includes a core set of concepts, principles, and terminology. HSPD-5 identifies these as the ICS; multiagency coordination systems; training; identification and management of resources (including systems for classifying types of resources); qualification and certification; and the collection, tracking, and reporting of incident information and incident resources.

National Institute for Occupational Safety and Health (NIOSH)—A federal agency that tests and certifies respiratory protection devices, recommends occupational exposure limits, and assists OSHA in safety and health research and investigations including firefighter deaths.

National Institute of Standards and Technology (NIST)—A non-regulatory federal agency within the US Department of Commerce. NIST's mission is to promote US innovation and industrial competitiveness by advancing measurement science, standards, and technology in ways that enhance economic security and improve our quality of life.

Natural disasters—Devastating natural events, such as earthquakes, tornadoes, and hurricanes, occurring with little or no warning that destroy properties and take many lives.

Nave—The seating or pew area in a church.

Needed fire flow (NFF)—The theoretical amount of water needed to control and extinguish a fire. It is derived by mathematical calculation.

Negative ventilation—Accomplished by placing exhaust fans into window or doorway openings and pulling the smoke or toxic fumes from a building.

National Integration Center (NIC)—Part of the US Department of Homeland Security, the NIC is responsible for managing the implementation and administration of the National Incident Management System.

Noncombustible/limited combustible construction—Construction in which structural members are of noncombustible materials and assemblies but which does not meet the requirements for fire-resistive construction in the applicable code.

Nonintervention mode—A passive mode of operation used when a situation presents itself in which there is nothing to be gained and there is a severe threat to the firefighters' safety. In this case, nothing should be risked, and firefighters should be withdrawn.

Nursing home—A facility in which the occupants are often bedridden, lack mobility, and may suffer from dementia or Alzheimer's disease.

O

Offensive mode—An interior attack on a fire. An offensive attack can include an offensive-interior attack, or an offensive-exterior attack where a brief application of water can be made through a window or door opening, followed by hose-lines being taken to the interior to complete extinguishment with an offensive-interior attack.

Offensive/defensive mode—A mode of attack whereby hose-lines enter a building and protect rescuers until rescues are completed. After the rescue efforts, all personnel are withdrawn and an exterior attack commences.

Off-gassing—The dispersion of a toxic product that was absorbed by a person's clothing. The product is gradually released from the clothing and can affect the health of those wearing or handling the clothing.

Officer—The ICS title for the personnel responsible for the Command Staff positions of Safety, Liaison, Public Information, and Intelligence.

Operational period— The period of time scheduled for execution of a given set of operation actions as specified in the Incident Action Plan. Operational periods can be of various lengths.

Ordinary construction—Constructed of wood, with exterior masonry walls.

Oriented strand board (OSB)—Widely used as construction sheathing, as the web materials for wooden I-beams, and for other applications. It is composed of compressed wood strands arranged in layers to form a mat. The individual strands are typically three to six inches long. These mats are then oriented at right angles to one another. The orientation of layers achieves the same advantages of cross-laminated veneers in plywood. Like the veneer in plywood, these mats are layered and oriented for maximum strength, stiffness, and stability.

P

Parapet—A wall that protrudes above a roofline. Its purpose may be decorative, or it may be functional as when utilized as a fire-stop.

Parging—A thin layer of mortar covering a wall.

Penal institutions—Prisons and detention centers.

Personal alert safety system (PASS)—The PASS device is a warning alarm that firefighters carry on their person or as part of an integrated system on their self-contained breathing apparatus. It will sound an alarm if the firefighter is motionlessness for a specified short period of time. The intent of the automatic alarm for motionlessness is to signify that the firefighter has been overcome or trapped. The warning system can also be activated by the firefighter to sound an alarm if he or she is in trouble.

Personnel accountability report (PAR)—A verbal report used to account for personnel assigned to a division, group or the entire incident.

Planning meeting—A meeting is held as needed throughout the duration of an incident to select specific strategies and tactics for incident control operations, and for service and support planning. On larger incidents, the planning meeting is a major element in the development of a written Incident Action Plan.

Planning "P"—A practical tool that is used in the planning process and for the development of the Incident Action Plan.

Plenum—A space created between a suspended ceiling and the original ceiling. The term is often associated with a high-rise building. This space can be used as part of an air return system.

Poke-through—A hole or opening in fire-stopping material, negating the compartmentation and creating a large non-fire-stopped area in the concealed space.

Positive pressure ventilation (PPV)—A method of ventilation utilizing powerful fans that pressurize a structure by driving fresh air into a building and forcing out the smoke via openings made by the fire department.

Pre-incident planning—A method of gathering facts about a building or a process prior to an emergency.

Public assembly buildings—Includes amusement park buildings, grandstand tents, auditoriums, dance halls, poolrooms, lecture halls, bowling alleys, gymnasiums, stadiums, skating rinks, houses of worship, restaurants, nightclubs, movie theaters, and museums.

Public Information Officer (PIO)—A member of the Command Staff responsible for interfacing with the public and media or with other agencies with incident-related information requirements.

Pyrolysis—A chemical change brought about by the action of heat. It can occur when wooden beams are in contact with a heater or heater flue, causing the wood to turn to charcoal. This action on the wood reduces the ignition temperature of the wood, and constant heating can ignite these beams.

R

Rapid Intervention Crews (RICs)—Also referred to as rapid intervention teams (RITs) or firefighter assist search teams (FAST). These are personnel or a unit that reports to the Incident Commander and stands by in the vicinity of the command post for immediate deployment should firefighters be in need of assistance due to being trapped, lost, or for any reason.

RECEO-VS—Rescue, exposures, confinement, extinguishment, overhaul, ventilation, salvage. The seven basic strategies for structural firefighting.

Rehabilitation (rehab)—A function and location that includes medical evaluation and treatment, food and fluid replenishment, and relief from extreme climatic conditions for emergency responders, depending on the circumstances of the incident.

Relay pumping—The movement of water a long distance using multiple apparatus to pump water from a source to the fire scene. The use of large-diameter hose-line can facilitate this operation. Relay pumping is usually performed when a nearby water supply is insufficient or nonexistent. The water source can be from a lake, pond, or municipal fire hydrants. This operation demands organization and management to ensure the proper size hose-lines, sufficient apparatus, and their correct placement.

Renovated buildings—Buildings in which changes have been made that can alter how a fire affects them structurally.

Rollover—The flame front burning that is often observed rolling along in front of burning material. This flame front consists of the gases being liberated from the burning material.

Row house—Vary in size and can be up to four stories high.

S

Safe operation of fire department apparatus—A shared responsibility by the driver and the officer to ensure safe driving procedures whenever a fire department apparatus is in motion.

Safety—The prime responsibility of the IC for the personnel operating at an incident scene.

Safety data sheets (SDSs)—Contain specific information on the properties, dangers, and correct handling procedures of hazardous materials. Copies must be kept at sites where hazardous materials are stored or processed. Formerly called material safety data sheets (MSDS), the SDS is a different format that has 16 sections.

Safety Officer (SO)—A member of the Command Staff responsible for monitoring and assessing safety hazards or unsafe situations, and for developing measures for ensuring personnel safety. The Safety Officer may have Assistants.

School and other active shooter incidents—Violent acts perpetrated on the student body at schools, or occupants at other venues, with the intent of killing and maiming.

School fire—A fire in an institution of learning.

Scuttles—Coverings over roof openings.

Secondary device—An explosive device placed by perpetrators at the scene of an incident specifically designed to harm responders.

Sector—A location or geographic area that is part of an incident management system. It is referred to as a division, e.g., Division 1 indicating the first floor of a building, or Division A indicating the exterior front of a building. A division can be established for any designated geographic area.

Sheet metal surface fastener—*See* gusset plate.

Shoring—Temporary support for framework. May be utilized in collapse rescue situations.

Size-up—Identifies problems at an incident scene.

Slopover—An expulsion of oil and froth from the surface of a tank, produced when water or foam is applied to a burning liquid surface.

Soffit—Interior soffit can be an enclosed area above kitchen cabinets. An exterior soffit can be found beneath the eaves of a roof.

Spalling—A condition in which a piece of brick or concrete separates from the wall, floor, or ceiling to which it was attached. It may fall or be propelled a distance from its attached point. When spalling occurs, the wall or ceiling from which the material separated is weakened.

Span of control—Refers to the number of personnel reporting to any given individual. Optimal span of control in the incident command system is five, with an acceptable spread of two to seven.

Spandrel space—The area between the top of the window on the lower floor to the bottom of the window on the floor above.

Staging—A location from which apparatus and personnel can be immediately deployed as directed by the Incident Commander. It consists of two levels. Level 1 Staging is accomplished by apparatus taking preassigned positions at an incident. Level 2 Staging is a specific location where apparatus go to await orders on deployment.

Status report—Report prepared by the IC to assess fire conditions for Dispatch; helps fulfill the initial and ongoing demands of constantly sizing up and reporting incident scene conditions.

Strategy—The overall goals that will solve the problems found.

Strike team—A specified combination of the same kind and type of resources (apparatus and staffing) with common communications and a Strike Team Leader. Commonly used for engine companies in wildland fires.

Strip mall—Row of stores with parking in front.

Supermarkets—Typically one-story self-service stores that stores that contain food supplies and household goods for customer purchase arranged in long aisles, and may include commercial kitchens and restaurant areas as well.

Suspended ceiling—A ceiling that is placed below the original ceiling. It can be found in any type of structure. It is supported by wires or steel rods connected to the original ceiling.

Synergy—The sum of the parts is greater when combined than as individual components. This is especially true with light-weight building construction components where small pieces of wood or steel when combined with other small pieces can support large loads.

T

Tactics—The way strategies or goals will be achieved.

Tank farm/refinery fires—Involve highly flammable products and there is the ever-present danger of rapid incident scene changes.

Target hazard—A facility that contains specific hazards that will severely tax responding firefighters. It will produce a significant negative impact on the community, involve atypical hazards, and initiate multi-agency involvement. Nursing homes, college dormitories, penal institutions, bulk storage facilities, public assemblies, and high-rise buildings are a few examples.

Task force—A combination of single resources with common communications and a Task Force Leader assembled for a specific assignment.

Tasks—Stipulate who will do which step and when.

Terrorism incidents—Violent acts, or acts dangerous to human life, in violation of criminal laws of the United States. The Federal Bureau of Investigation (FBI) recognizes two categories of terrorism: domestic and international.

Timber truss—Construction using wood trusses at least four inches by six inches in size.

Town house—Usually larger than row houses and not typically set in the same contiguous style.

Training—The backbone of every fire department; encompasses basic and advanced areas.

Transom—A small operable window above a doorway that can be opened to permit circulation of air between a room and the hallway. Previously found in hotel rooms and bedrooms in residential properties.

Truck company operations—Rescue, laddering, forcible entry, ventilation, overhaul, and salvage are the primary duties assigned to a truck company.

Truss void or truss loft—A concealed space created by the installation of open frame truss. This area lacks compartmentation due to the open web design of the truss.

Tuberculation—The reduction of the inside diameter of a pipe caused by deposits of chemicals or growth of organisms on the inside of iron pipe. These tubercules roughen the inside of the pipe, increasing its resistance to water flow.

Type—A classification of resources in the ICS that refers to capability. Type 1 is generally considered to be more capable than Types 2, 3, or 4, respectively, because of size, power, capacity, or, in the case of Incident Management Teams, experience and qualifications.

U

Undesigned load—A load placed on a building that it was not designed to support.

Unified Area Command—A Unified Area Command is established when incidents under an Area Command are multijurisdictional. (See Area Command and Unified Command.)

Unified Command—A method of coordinating command of an incident when multiple agencies have jurisdiction. It allows all agencies with responsibility for the incident, either geographic or functional, to manage an incident by establishing a common set of incident objectives and strategies. This is accomplished without abdicating agency authority, responsibility, or accountability.

Unity of Command—Dictates that no person reports to more than one person and that everyone has someone to answer to.

Upwind—The side from which the wind is blowing.

US Department of Homeland Security (DHS)—The mission statement of DHS is: "homeland security is a concerted national effort to prevent terrorist attacks within the United States, reduce America's vulnerability to terrorism, and minimize the damage and recover from attacks that do occur."

V

Vacant buildings—Can be categorized into two distinct types: vacated and awaiting resale, or vacant for a period of time and stripped of all contents.

Viscous—Refers to the resistance to flow of a liquid. It typically is used in reference to heavy liquids that do not flow easily.

W

Wall collapse—The failure of a wall. The basic types of wall collapse are 90-degree, inward-outward, and curtain collapse.

Wall spreader—A building feature in a masonry wall that can be utilized as a potential collapse indicator. It can be installed when the wall is initially constructed as additional support for the wall, or to stabilize a wall that has developed structural problems to stop the problem from becoming worse.

Warm zone—An area where decontamination is performed and entry is restricted. The hazardous materials safety officer will oversee safety in this area.

Water tender—An over-the-road tank truck pulled by a tractor. It contains a pump and carries water for firefighting. Typical sizes range from 1,000- to 4,000-gallon capacity. It is used to shuttle water to a fire scene.

Web—The intervening members that join the top and bottom chords to form the various patterns of a truss.

Wildland urban interface (WIU)—A mix of wildland with urban areas.

Wood lath—Thin layers of wood strips attached to wall studs and ceiling joists to allow plaster to adhere and form finished plaster walls and ceilings.

Wythe—The vertical section of a wall one masonry unit thick.

INDEX

A